PROJECT PHYSICS

HANDBOOK

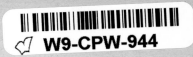

Directors of Harvard Project Physics

F. James Rutherford
Department of Science Education, New York University

Gerald Holton
Department of Physics, Harvard University

Fletcher G. Watson
Harvard Graduate School of Education

Editorial Development: William N. Moore, Roland Cormier, Lorraine Smith-Phelan
Editorial Processing: Margaret M. Byrne, Regina Chilcoat, Holly Massey
Art, Production, and Photo Resources: Vivian Fenster, Fred C. Pusterla, Robin M. Swenson, Annette Sessa, Beverly Silver, Anita Dickhuth, Dorina Virdo
Product Manager: Laura Zuckerman
Advisory Board: John Taggart, Maurice E. Fey, Norman Hughes, David J. Miller, John W. Griffiths, William L. Paul
Consultant: John Matejowsky
Researchers: Pamela Floch, Gerard LeVan

Acknowledgments appear on page 1.
Picture credits appear on page 1.

Printed in the United States of America

ISBN 0-03-055146-3
1234-059-98765432

Science is an adventure of the whole human race to learn to live in and perhaps to love the universe in which they are. To be a part of it is to understand, to understand oneself, to begin to feel that there is a capacity within man far beyond what he felt he had, of an infinite extension of human possibilities. . . .

I propose that science be taught at whatever level, from the lowest to the highest, in the humanistic way. It should be taught with a certain historical understanding, with a certain philosophical understanding, with a social understanding and a human understanding in the sense of the biography, the nature of the people who made this construction, the triumphs, the trials, the tribulations.

I. I. RABI
Nobel Laureate in Physics

PREFACE

The Project Physics Course is based on the ideas and research of a national curriculum development project that worked for eight years.

Preliminary results led to major grants from the U.S. Office of Education and the National Science Foundation. Invaluable additional financial support was also provided by the Ford Foundation, the Alfred P. Sloan Foundation, the Carnegie Corporation, and Harvard University. A large number of collaborators were brought together from all parts of the nation, and the group worked together intensively for over four years under the title Harvard Project Physics. The instructors serving as field consultants and the students in the trial classes were also of vital importance to the success of Harvard Project Physics. As each successive experimental version of the course was developed, it was tried out in schools throughout the United States and Canada. The instructors and students in those schools reported their criticisms and suggestions to the staff in Cambridge. These reports became the basis for the subsequent revisions of the course materials. In the Preface to the Text you will find a list of the major aims of the course.

Unhappily, it is not feasible to list in detail the contributions of each person who participated in some part of Harvard Project Physics. Previous editions of the Text have included a partial list of the contributors. We take particular pleasure in acknowledging the assistance of Dr. Andrew Ahlgren of the University of Minnesota. Dr. Ahlgren was invaluable because of his skill as a physics instructor, his editorial talent, his versatility and energy, and above all, his commitment to the goals of Harvard Project Physics.

We would also especially like to thank Ms. Joan Laws, whose administrative skills, dependability, and thoughtfulness contributed so much to our work. Holt, Rinehart and Winston, Publishers of New York, provided the coordination, editorial support, and general backing necessary to the large undertaking of preparing the final version of all components of the Project Physics Course. Damon-Educational Division, located in Westwood, Massachusetts, worked closely with us to improve the engineering design of the laboratory apparatus and to see that it was properly integrated into the program.

In the years ahead, the learning materials of the Project Physics Course will be revised as often as is necessary to remove remaining ambiguities, to clarify instructions, and to continue to make the materials more interesting and relevant to students.

F. James Rutherford
Gerald Holton
Fletcher G. Watson

CONTENTS

Introduction 1
Keeping Records 4
Using the Polaroid Land Camera 4

Unit 1/Concepts of Motion

EXPERIMENTS
1-1 Naked Eye Astronomy 7
1-2 Regularity and Time 13
1-3 Variations in Data 13
1-4 Measuring Uniform Motion 14
1-5 A Seventeenth-Century Experiment 19
1-6 Twentieth-Century Version of Galileo's Experiment 20
1-7 Measuring the Acceleration of Gravity a_g 21
1-8 Newton's Second Law 24
1-9 Mass and Weight 27
1-10 Curves of Trajectories 28
1-11 Prediction of Trajectories 30
1-12 Centripetal Force 32
1-13 Centripetal Force on a Turntable 33

ACTIVITIES
Checker Snapping 35
Beaker and Hammer 35
Pulls and Jerks 35
Experiencing Newton's Second Law 35
Make One of These Accelerometers 35
Projectile Motion Demonstration 38
Speed of a Stream of Water 38
Photographing a Waterdrop Parabola 39
Ballistic Cart Projectiles 39
Motion in a Rotating Reference Frame 40
Penny and Coat Hanger 41
Measuring Unknown Frequencies 41

FILM LOOP NOTES
L1 Acceleration Caused by Gravity. I 41
L2 Acceleration Caused by Gravity. II 42
L3 Vector Addition: Velocity of a Boat 42
L4 A Matter of Relative Motion 44
L5 Galilean Relativity: Ball Dropped from Mast of Ship 44
L6 Galilean Relativity: Object Dropped from Aircraft 45
L7 Galilean Relativity: Projectile Fired Vertically 46
L8 Analysis of a Hurdle Race. I 47
L9 Analysis of a Hurdle Race. II 48

Unit 2/Motion in the Heavens

EXPERIMENTS
2-1 Naked-Eye Astronomy 50
2-2 Size of the Earth 54
2-3 The Distance to the Moon 57
2-4 The Height of Piton, a Mountain on the Moon 57
2-5 Retrograde Motion 60
2-6 The Shape of the Earth's Orbit 61
2-7 Using Lenses to Make a Telescope 63
2-8 The Orbit of Mars 67
2-9 Inclination of Mars' Orbit 70
2-10 The Orbit of Mercury 72
2-11 Stepwise Approximation to an Orbit 75
2-12 Model of the Orbit of Halley's Comet 79

ACTIVITIES
Making Angular Measurements 83
Epicycles and Retrograde Motion 84
Celestial Sphere Model 86
How Long is a Sidereal Day? 87
Scale Model of the Solar System 88
Build a Sundial 88
Plot an Analemma 88
Stonehenge 88
Moon Crater Names 89
Literature 89
Frames of Reference 89
Demonstrating Satellite Orbits 90
Galileo 90
Conic-Section Models 90
Challenging Problem: Finding Earth–Sun Distance from Venus Photos 91

Measuring Irregular Areas 91
Other Comet Orbits 91
Drawing a Parabolic Orbit 91
Forces on a Pendulum 92
Trial of Copernicus 93

FILM LOOP NOTES

L10 Retrograde Motion: Geocentric Model 94
L11 Retrograde Motion: Heliocentric Model 94
L12 Jupiter Satellite Orbit 95
L13 Program Orbit. I 97
L14 Program Orbit. II 98
L15 Central Forces: Iterated Blows 98
L16 Kepler's Laws 99
L17 Unusual Orbits 100

Unit 3 / The Triumph of Mechanics

EXPERIMENTS

3-1 Collisions in One Dimension. I 102
3-2 Collisions in One Dimension. II 104
3-3 Collisions in Two Dimensions. I 110
3-4 Collisions in Two Dimensions. II 113
3-5 Conservation of Energy. I 118
3-6 Conservation of Energy. II 121
3-7 Measuring the Speed of a Bullet 122
3-8 Energy Analysis of a Pendulum Swing 124
3-9 Least Energy 124
3-10 Temperature and Thermometers 126
3-11 Calorimetry 128
3-12 Ice Calorimetry 131
3-13 Monte Carlo Experiment on Molecular Collisions 132
3-14 Behavior of Gases 137
3-15 Wave Properties 139
3-16 Waves in a Ripple Tank 140
3-17 Measuring Wavelength 141
3-18 Sound 142
3-19 Ultrasound 144

ACTIVITIES

Is Mass Conserved? 147
Exchange of Momentum Devices 147
Student Horsepower 148
Drinking Duck 148
Mechanical Equivalent of Heat 149
A Diver in a Bottle 149
How to Weigh a Car With a Tire Pressure Gauge 150
Perpetual Motion Machines? 150
Standing Waves on a Drum and a Violin 152
Reflection 152
Moiré Patterns 153
Music and Speech Activities 154
Measurement of the Speed of Sound 154
Mechanical Wave Machines 155

FILM LOOP NOTES

L18 One-Dimensional Collisions. I 156
L19 One-Dimensional Collisions. II 157
L20 Inelastic One-Dimensional Collisions 157
L21 Two-Dimensional Collisions. I 157
L22 Two-Dimensional Collisions. II 158
L23 Inelastic Two-Dimensional Collisions 158
L24 Scattering of a Cluster of Objects 158
L25 Explosion of a Cluster of Objects 159
L26 Finding the Speed of a Rifle Bullet. I 160
L27 Finding the Speed of a Rifle Bullet. II 161
L28 Recoil 162
L29 Colliding Freight Cars 162
L30 Dynamics of a Billiard Ball 163
L31 A Method of Measuring Energy: Nails Driven into Wood 164
L32 Gravitational Potential Energy 164
L33 Kinetic Energy 165
L34 Conservation of Energy: Pole Vault 166
L35 Conservation of Energy: Aircraft Takeoff 167
L36 Reversibility of Time 168
L37 Superposition 168
L38 Standing Waves on a String 170
L39 Standing Waves in a Gas 170
L40 Vibrations of a Wire 171
L41 Vibrations of a Rubber Hose 172
L42 Vibrations of a Drum 173
L43 Vibrations of a Metal Plate 173

Unit 4 / Light and Electromagnetism

EXPERIMENTS

4-1 Refraction of a Light Beam 174
4-2 Young's Experiment: The Wavelength of Light 177
4-3 Electric Forces. I 179
4-4 Electric Forces. II: Coulomb's Law 181
4-5 Forces on Currents 183
4-6 Currents, Magnets, and Forces 187
4-7 Electron Beam Tube. I 190
4-8 Electron Beam Tubes. II 192
4-9 Waves and Communication 195

ACTIVITIES

Thin Film Interference 200
Handkerchief Diffraction Grating 200
Photographing Diffraction Patterns 200
Poisson's Spot 201
Photographic Activities 201
Color 201
Polarized Light 202
Make an Ice Lens 203
Detecting Electric Fields 203
An 11¢ Battery 204
Voltaic Pile 204
Measuring Magnetic Field Intensity 204
More Perpetual Motion Machines 205
Transistor Amplifier 206
An Isolated North Magnetic Pole? 206
Faraday Disk Dynamo 206
Generator Jump Rope 207
Simple Meters and Motors 207
Simple Motor–Generator Demonstration 208
Physics Collage 209
Bicycle Generator 209
Lapis Polaris, Magnes 209
Microwave Transmission Systems 210
Good Reading 210

FILM LOOP NOTES

L44 Standing Electromagnetic Waves 211

Unit 5 / Models of the Atom

EXPERIMENTS

5-1 Electrolysis 212
5-2 The Charge-to-Mass Ratio for an Electron 214
5-3 The Measurement of Elementary Charge 217
5-4 The Photoelectric Effect 219
5-5 Spectroscopy 222

ACTIVITIES

Dalton's Puzzle 226
Electrolysis of Water 226
Single-Electrode Plating 226
Activities from *Scientific American* 227
Writing by or about Einstein 227
Measuring q/m for the Electron 227
Cathode Rays in a Crookes Tube 227
Lighting an Electric Lamp with a Match 227
X Rays from a Crookes Tube 228
Scientists on Stamps 228
Measuring Ionization: A Quantum Effect 229
Modeling Atoms with Magnets 230
"Black Box" Atoms 231
Standing Waves on a Band-Saw Blade 231
Turntable Oscillator Patterns Resembling de Broglie Waves 231
Standing Waves in a Wire Ring 232

FILM LOOP NOTES

L45 Production of Sodium by Electrolysis 233
L46 Thomson Model of the Atom 233
L47 Rutherford Scattering 234

Unit 6 / The Nucleus

EXPERIMENTS
6-1 Random Events **236**
6-2 Range of ∝ and β Particles **240**
6-3 Half-life. I **243**
6-4 Half-life. II **246**
6-5 Radioactive Tracers **248**
6-6 Measuring the Energy of β Radiation **250**

FILM LOOP NOTES
L48 Collisions with an Object of Unknown Mass **253**

ACKNOWLEDGMENTS

Page 52 Table 2-4 is reprinted from *Solar and Planetary Longitudes for Years −2500 to +2500* prepared by William D. Stalman and Owen Gingerich (University of Wisconsin Press, 1963).

Page 205 Smedlie, S. Raymond, *More Perpetual Motion Machines,* Science Publications of Boston, 1962.

Page 206 I.F. Stacy, *The Encyclopedia of Electronics* (Charles Susskind, Ed.), Reinhold Publishing Corp., New York, Fig. 1, p. 246.

PICTURE CREDITS

Unit 1, pp. 2, 15, 24, 25 HRW Photos by Russell Dian; pp. 10, 13, 34 (bottom), 35, 37 (cartoons) By permission of Johnny Hart and Field Enterprises, Inc.; p. 39 (left) Courtesy of Mr. Harold M. Waage, Palmer Physical Laboratory, Princeton University, (right and bottom) Courtesy of Educational Development Center, Newton, Mass. All photographs used with film loops courtesy of National Film Board of Canada. Photographs of laboratory equipment and of students using laboratory equipment were supplied with the cooperation of the Project Physics staff and Damon Corporation.

Unit 2, pp. 51, 65 (bottom), 66 (top), 67, 69, 75 Mount Wilson and Palomar Observatories; pp. 53, 55, 61, 67, 84, 87 (top) (cartoons) By permission of Johnny Hart and Field Enterprises, Inc.; p. 58 (top) Lick Observatory; p. 59 NASA; p. 61 sun film strip photograph courtesy of U.S. Naval Observatory; p. 65 (top) Yerkes Observatory; p. 73 Lowell Observatory; p. 85 (right) Photograph courtesy of Damon Corporation, Educational Division. All photographs used with film loops courtesy of National Film Board of Canada. Photographs of laboratory equipment and of students using laboratory equipment were supplied with the cooperation of the Project Physics staff and Damon Corporation.

Unit 3, p. 110 (bottom) J. Ph. Charbonnier/Photo Researchers; p. 112 Wide World Photo; p. 129 HRW Photo by Russell Dian; pp. 141, 150, 151 (cartoons) By permission of Charles Gary Solin; p. 153 (top) "Physics and Music," *Scientific American*, July 1948; p. 160 (cartoon) By permission of Johnny Hart and Field Enterprises, Inc.; p. 167 Cessna Aircraft. All Photographs used with Film Loops courtesy of National Film Board of Canada. Photographs of laboratory equipment and of students using laboratory equipment were supplied with the cooperation of the Project Physics staff and Damon corporation.

Unit 4, pp. 180 (bottom), 183 (bottom), 190, 210 (bottom) Cartoons by Charles Gary Solin and reproduced by his permission only; p. 197 HRW Photos by Russell Dian; p. 203 By permission of Johnny Hart and Field Enterprises, Inc.; p. 209 (bottom) "Physics" by Bob Lillich; p. 210 (top) Burndy Library. All photographs and notes with film loops courtesy of the National Film Board of Canada. Photographs of laboratory equipment and of students using laboratory equipment were supplied with the cooperation of the Project Physics staff and Damon Corporation.

Unit 5, p. 217 Courtesy L. J. Lippie, Dow Chemical Company, Midland, Michigan; p. 220 HRW Photo by Russell Dian; p. 229 From the cover of *The Science Teacher*, Vol. 31, No. 8, December 1964. All photographs used with film loops courtesy of National Film Board of Canada. Photographs of laboratory equipment and of students using laboratory equipment were supplied with the cooperation of the Project Physics staff and Damon Corporation.

Unit 6. All Photographs used with film loops courtesy of National Film Board of Canada. Photographs of laboratory equipment and of students using laboratory equipment were supplied with the cooperation of the Project Physics staff and Damon Corporation.

INTRODUCTION

This *Handbook* is your guide to observations, experiments, activities, and explorations, far and wide, in the realms of physics.

Prepare for challenging work, fun, and some surprises. One of the best ways to learn physics is by *doing* physics, in the laboratory and out. Do not rely on reading alone.

This *Handbook* is different from laboratory manuals you may have worked with before. Far more projects are described here than you alone can possibly do, so you will need to pick and choose.

Although only a few of the experiments and activities will be assigned, do any additional ones that interest you. Also, if an activity occurs to you that is not described here, discuss with your instructor the possibility of doing it. Some of the most interesting science you will experience in this course will be the result of the activities you choose to pursue beyond the regular laboratory assignments.

The many projects in this *Handbook* are divided into the following sections:

The **Experiments** contain full instructions for the investigations you can do alone or with others in the laboratory.

The **Activities** contain many suggestions for construction projects, demonstrations, and other activities you can do by yourself in the laboratory or at home.

The **Film Loop Notes** give instructions for the use of the variety of film loops that have been specially prepared for the course.

Do as many of these projects as you can. Each one will give you a better grasp of the physical principles involved.

EXPERIMENT NO. ______________ DATE ________________

This experiment is to see how a rubber band stretches under the influence of forces

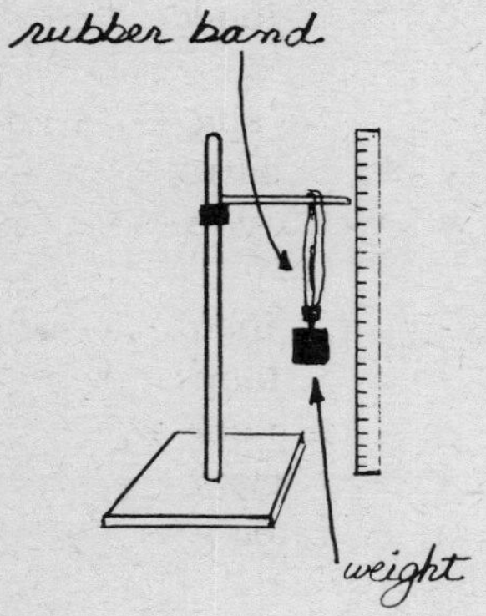

YOUR OWN SKETCH WILL ALWAYS HELP YOUR MEMORY

I put different masses on the end of the rubber band and recorded the position of the top of the hook that holds the weight

Room Temperature 26°C

Position of the top of rubber band 36.3 cm

Mass (g)	Force (n)	Pts of Bottom (cm)	Extension (cm)	INCLUDE ANY DATA YOU THINK MAY BE RELEVANT
(weights from set, no error) 0	0	44.0 ± .1	0	ALWAYS SHOW UNITS OF TABULATED QUANTITIES
10	.098	45.1 "	1.1 ± .2	ESTIMATE THE ERROR OF EVERY QUANTITY YOU MEASURE
20	.196	45.8	1.8	
30	.294	46.8	2.8	
50	.490	49.6	5.6	
60	.588	51.5	7.5	second 20g weight is missing from set
70	.686	53.7	9.7	INCLUDE COMMENTS ON YOUR DATA
80	.784	56.1	12.1	KEEP DATA IN NEAT TABLES
100	.980	60.6	16.6	
80	.784	56.2	12.2	recheck

On these two pages is shown an example of a student's lab notebook report. The table is used to record both observed quantities (mass, scale position) and calculated quantities (force, extension of rubber band). The graph shows at a glance how the extension of the rubber band changes as the force acting on it is increased. The notes in capital letters are comments.

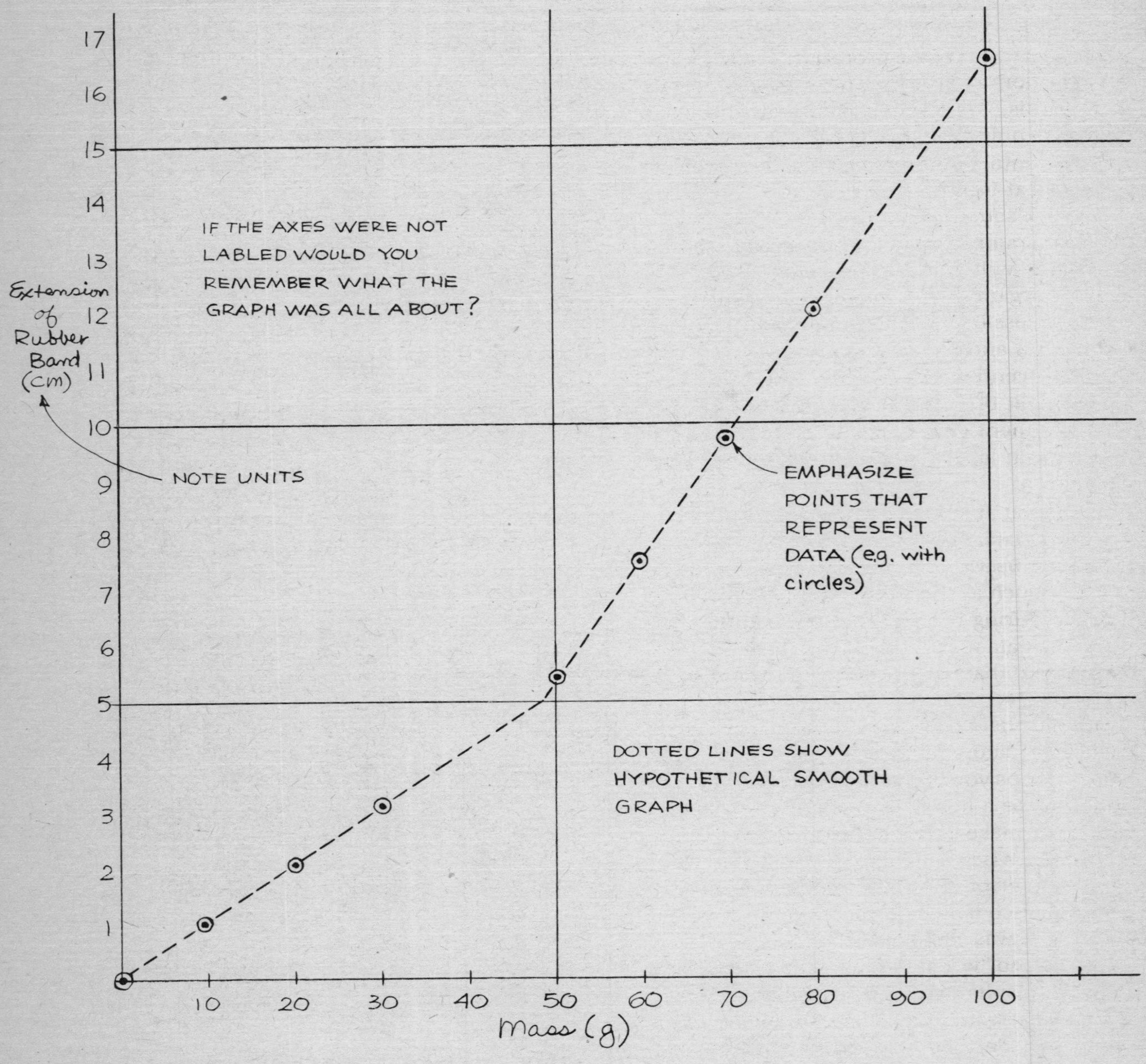

JOT DOWN YOUR THOUGHTS
INCLUDE POSSIBLE QUESTIONS → ESPECIALLY ONES YOU CAN'T ANSWER

There are obviously two different straight lines.
It would have been nice to see what it was at 40gm, since that's just where the two lines cross.

The slopes of the two lines are the force constants $F = -kx$. For the first line $k = .105$ n/cm and for the second $k = .0438$ n/cm

Keeping Records

Your records of observations made in the laboratory or at home can be kept in many ways. Regardless of the procedure followed, the key question for deciding what kind of record you need is: "Do I have a clear enough record so that I could pick up my lab notebook a few months from now and explain to myself or others what I did?"

Here are some general rules to be followed in every laboratory exercise. Your records should be neatly written without being fussy. You should organize all numerical readings in tables, if possible, as in the sample lab write up on pages 6 and 7. You should always identify the units (centimeters, kilograms, seconds, etc.) for each set of data you record. Also, identify the equipment you are using, so that you can find it again later if you need to recheck your work.

In general, it is better to record more rather than less data. Even details that may seem to have little bearing on the experiment you are doing — such as the temperature and whether it varied during the observations, and the time when the data were taken — may turn out to be information that has a bearing on your analysis of the results.

You may have some reason to suspect that a particular datum is less reliable than other data. Perhaps you had to make the reading very hurriedly, or a line on a photograph was very faint. If so, make a note of that fact. Never erase a reading. When you think an entry in your notes is in error, draw a single line through it; do not scratch it out completely or erase it. You may find it was significant after all.

There is no "wrong" result in an experiment, although results may be in considerable error. If your observations and measurements were carefully made, then your result will be reliable. Whatever happens in nature, including the laboratory, cannot be "wrong." It may have nothing to do with your investigation, or it may be mixed up with so many other events you did not expect, that your report is not useful. Therefore, you must think carefully about the *interpretation* of your results.

Finally, the cardinal rule in a laboratory is to choose in favor of "getting your hands dirty" instead of "dry-labbing." In 380 B.C., the Greek scientist, Archytas, summed it up this way:

In subjects of which one has no knowledge, one must obtain knowledge either by learning from someone else, or by discovering it for oneself. That which is learnt, therefore, comes from another and by outside help; that which is discovered comes by one's own efforts and independently. To discover without seeking is difficult and rare, but if one seeks, it is frequent and easy; if, however, one does not know how to seek, discovery is impossible.

Using the Polaroid Land Camera

You will find the Polaroid Land camera a very useful device for recording many of your laboratory observations. Your textbook shows how the camera is used to study moving objects. In the experiments and activities described in this *Handbook*, many suggestions are made for photographing moving objects, both with an electronic stroboscope (a rapidly flashing xenon light) and with a mechanical disk stroboscope (a slotted disk rotating in front of the camera lens). The setup of the rotating disk stroboscope with a Polaroid Land camera is shown below.

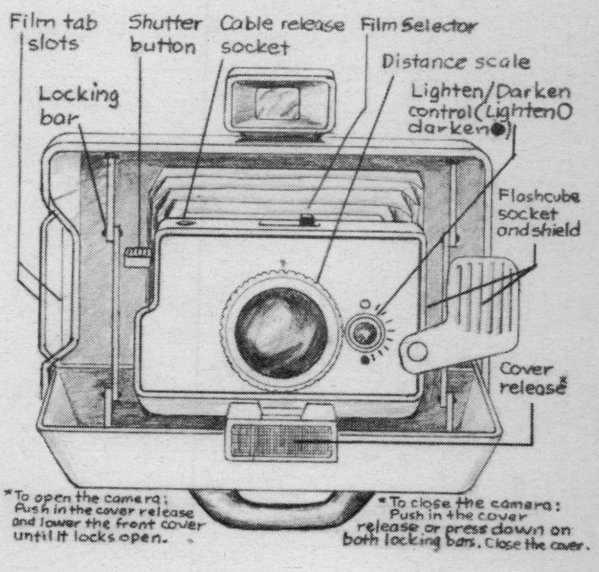

Below is a checklist of operations to help you use the modified Polaroid Land camera model 210. For other models, your instructor will provide instructions.

1. Make sure that there is film in the camera. If no white tab shows in the front of the door marked "4," you must put in new film.

2. Fasten camera to tripod or disk strobe base. If you are using the disk strobe technique, fix the clip-on slit in front of the lens.

3. Check film (speed) selector. Set to suggested position (75 for disk strobe or blinky; 3000 for xenon strobe).

4. If you are taking a "bulb" exposure, cover the electric eye.

5. Check distance from lens to plane of object to be photographed. Adjust focus if necessary. Work at the distance that gives an image just one-tenth the size of the object, if possible. This distance is about 120 cm.

6. Look through viewer to be sure that whatever part of the event you are interested in will be recorded. (At a distance of 120 cm, the field of view is just under 100 cm long.)

7. Make sure the shutter is cocked (by depressing the number 3 button).

8. Run through the experiment a couple of times without taking a photograph, to accustom yourself to the timing needed to photograph the event.

9. Take the picture; keep the cable release depressed only as long as necessary to record the event itself. Do not keep the shutter open longer than necessary.

10. Pull the white tab all the way out of the camera. Do not block the door (marked "4" on the camera).

11. Pull the large yellow tab straight out, all the way out of the camera. Begin timing development.

12. Wait 10 to 15 sec (for 3,000-speed black-and-white film).

13. Ten to 15 sec after removing the film from the camera, strip the white print from the negative.

14. Take measurements immediately. (The magnifier may be helpful.)

15. After initial measurements have been taken, coat your picture with the preservative supplied with each pack of film. Let the preservative dry thoroughly. Label the picture on the back for identification and mount the picture in your (or a partner's) lab report.

16. The negative can be used, too. Wash it carefully with a wet sponge, and coat with preservative.

17. Recock the shutter so it will be set for the next use.

18. Always be careful when moving around the camera that you do not inadvertently kick the tripod.

19. Always keep the electric eye covered when the camera is not in use. Otherwise the batteries inside the camera will run down.

Concepts of Motion

EXPERIMENTS

Experiment 1-1
NAKED-EYE ASTRONOMY

This first experiment will familiarize you with the continually changing appearance of the sky. By watching the heavenly bodies closely day and night over a period of time, you will begin to understand what is going on in the sky and gain the experience you will need in Unit 2, "Motion in the Heavens."

Do you know how the sun and the stars, the moon and the planets, appear to move through the sky? Do you know how to tell a planet from a star? Do you know when you can expect to see the moon during the day? Do you know how the sun and planets move in relation to the stars?

The Babylonians and Egyptians knew the answers to these questions over 5,000 years ago. They found the answers by watching the ever-changing sky. Thus, astronomy began with simple observations of the sort you can make with your unaided eye.

You know that the earth appears to be at rest while the sun, stars, moon, and planets are seen to move in various paths through the sky. The problem, as it was for the Babylonians, is to describe what these paths are and how they change from day to day, from week to week, and from season to season.

Some of these changes occur very slowly. In fact, that is why you may not have noticed them. You will need to watch the motions in the sky carefully, comparing them to fixed points of reference that you establish. You will need to keep a record of your observations for at least four to six weeks.

Choosing References

To locate objects in the sky accurately, you first need some fixed lines or planes to which your measurements can be referred, just as a map maker uses lines of latitude and longitude to locate places on the earth.

For example, you can establish a north–south line along the ground for your first reference. Then, with a protractor held horizontally, you can measure the direction of an object in the sky around the horizon from this north–south line. The angle of an object around the horizon from a north–south line is called the object's *azimuth*. Azimuths are measured from the north point (0°) through east (90°) to south

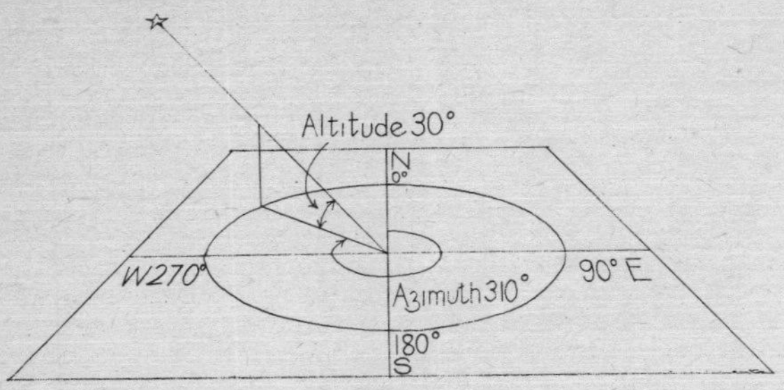

Fig. 1-1

(180°) and west (270°) and around to north again (360° or 0°). See Fig. 1-1.

To measure the height of an object in the sky, you can measure the angle between the object and the horizon. When your horizon is obscured by trees or buildings, you can measure from the zenith overhead (altitude 90°) down to the object; its altitude is then 90° minus its zenith distance, for your second

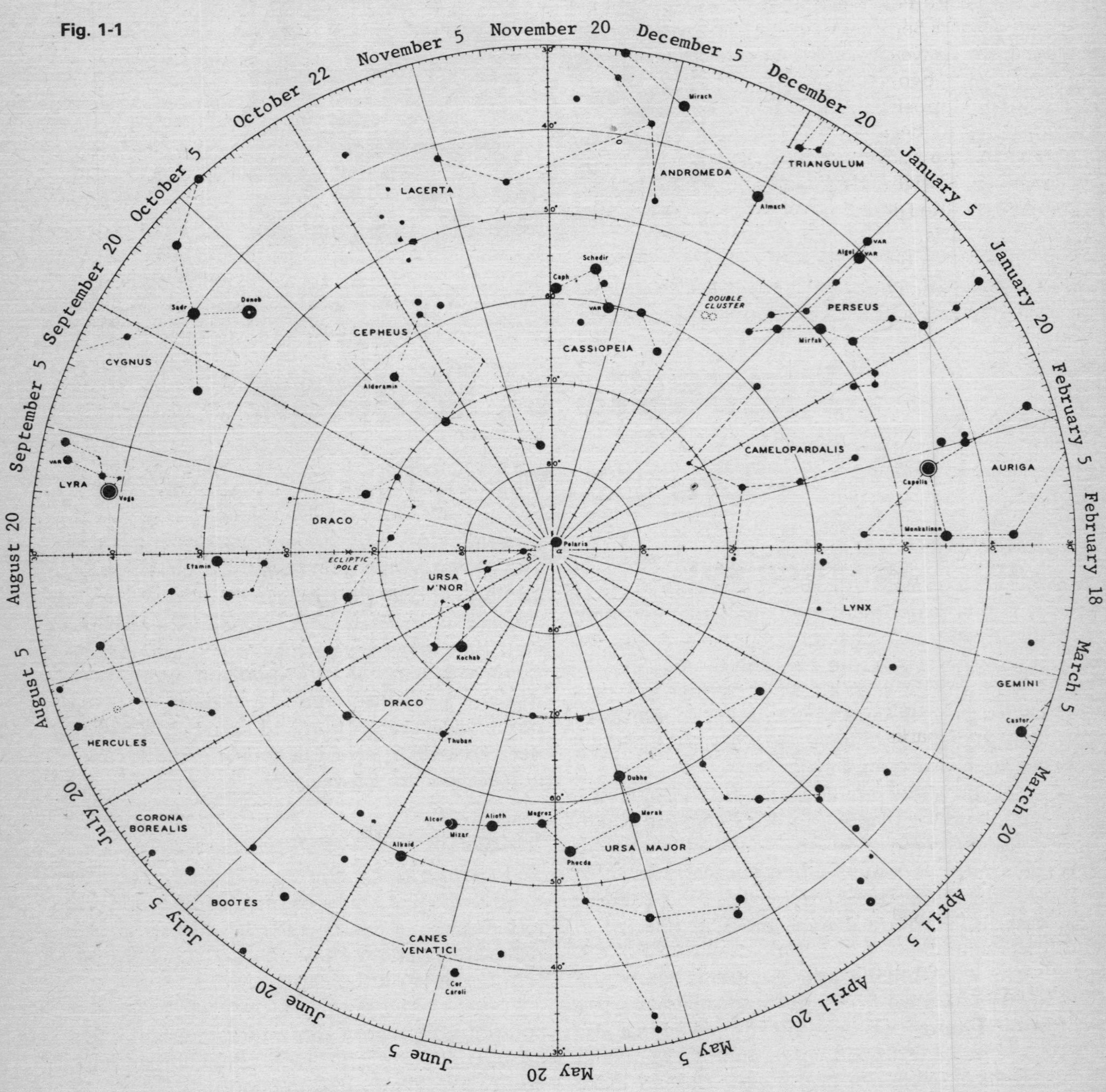

Fig. 1-2 This chart of the stars will help you locate some of the bright stars and the constellations. To use the map, face north and turn the chart until today's date is at the top. Then move the map up nearly over your head. The stars will be in these positions at 8 P.M. For each hour *earlier* than 8 P.M., rotate the chart 15 degrees (one sector) clockwise. For each hour *later* than 8 P.M., rotate the chart counterclockwise. If you are observing the sky outdoors with the map, cover the glass of a flashlight with fairly transparent red paper to look at the map. This will prevent your eyes from losing their adaptation to the dark when you look at the map.

coordinate. The angle between the horizontal plane and the line to an object in the sky is called the *altitude* of the object.

At night, you can use the North Star (Polaris) to establish the north–south line. Polaris is the one fairly bright star in the sky that moves least from hour to hour or with the seasons. It is almost due north of an observer anywhere in the northern hemisphere.

To locate Polaris, first find the "Big Dipper" which on a September evening is low in the sky and a little west of north. (See the star map, Fig. 1-2.) The two stars forming the end of the dipper opposite the handle are known as the "pointers," because they point to the North Star. A line passing through them passes very close to a bright star, the last star in the handle of the "Little Dipper." This bright star is the Pole Star, Polaris.

Imagine a line from Polaris straight down to the horizon. The point where this line meets the horizon is nearly due north of you. See Fig. 1-3.

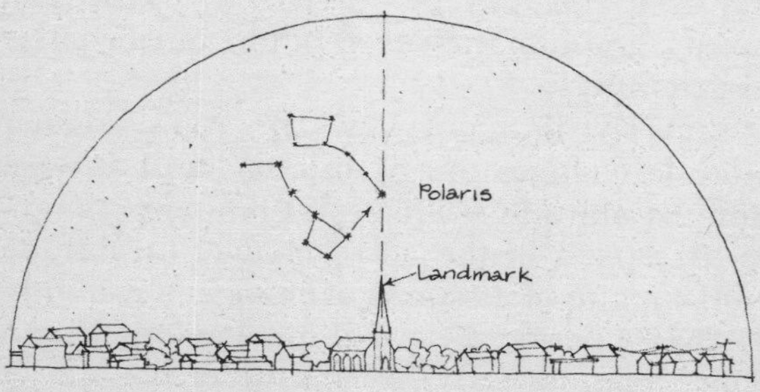

Fig. 1-3

Now that you have established a north–south line, note its position with respect to fixed landmarks, so that you can use it day or night.

You can establish the second reference, the plane of the horizon, and measure the altitude of objects in the sky from the horizon, with an *astrolabe*. An astrolabe is a simple instrument you can obtain easily or make yourself, very similar to those used by ancient viewers of the heavens. Use the astrolabe in your hand or on a flat table mounted on a tripod or on a permanent post. A simple hand astrolabe you can make is described in this *Handbook*, in the experiment dealing with the size of the earth.

Sight along the surface of the flat table to be sure it is horizontal, in line with the horizon in all directions. If there are obstructions on your horizon, a carpenter's level turned in all directions on the table will show when the table is level.

Fig. 1-4

Turn the base of the astrolabe on the table until the north–south line on the base points along your north–south line. You can also obtain the north–south line by sighting on Polaris through the astrolabe tube. Sight through the tube of the astrolabe at objects in the sky you wish to locate and obtain their altitude above the horizon in degrees from the protractor on the astrolabe. With some astrolabes, you can also obtain the azimuth of the objects from a scale around the base of the astrolabe.

To follow the position of the sun with the astrolabe, slip a large piece of cardboard with a hole in the middle over the sky-pointing end of the tube. (CAUTION: NEVER look directly at the sun; it can cause permanent eye damage.) Standing beside the astrolabe, hold a small piece of white paper in the shadow of the large cardboard, several centimeters from the sighting end of the tube. Move the tube about until the bright image of the sun appears through the tube on the paper. (See Fig. 1-5.) Then read

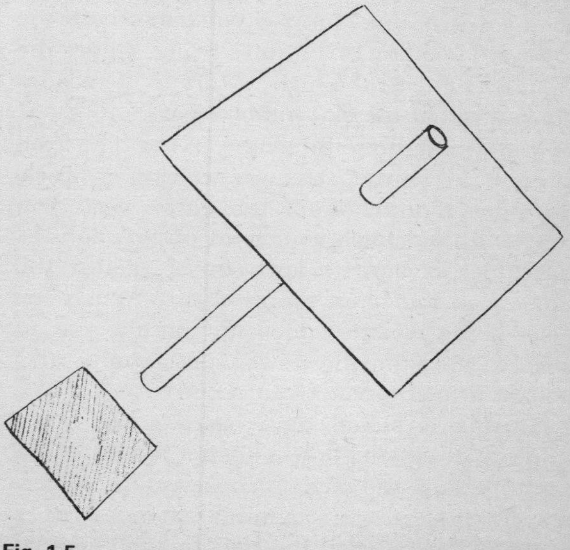

Fig. 1-5

the altitude of the sun from the astrolabe, and the sun's azimuth, if your instrument permits. Also record the date and time of your observation.

Observations

Now that you know how to establish your references for locating objects in the sky, here are suggestions for observations you can make on the sun, the moon, the stars, and the planets. Choose at lease one of these objects to observe. Record the date and time of all your observations. Later, compare notes with classmates who observed other objects.

A. Sun

CAUTION: NEVER look directly at the sun; it can cause permanent eye damage. Do not depend on sunglasses or fogged photographic film for protection. It is safest to make sun observations on projected images.

1. Observe the direction in which the sun sets. Always make your observation from the same observing position. If you do not have an unobstructed view of the horizon, note where the sun disappears behind the buildings or trees in the evening.

2. Observe the time the sun sets or disappears below your horizon.

3. Try to make these observations about once a week. The first time, draw a simple sketch of the horizon and the position of the setting sun.

4. Repeat the observations a week later. Note if the position or time of sunset has changed. Note if they change during a month. Try to continue these observations for at least two months.

5. If you are up at sunrise, you can record the time and position of the sun's rising. (Check the weather forecast the night before to be reasonably sure that the sky will be clear.)

6. Determine how the length of the day, from sunrise to sunset, changes during a week; during a month; or for the entire year. You might like to check your own observations of the times of sunrise and sunset against the times that are often reported in newspapers. Also, if the weather does not permit you to observe the sun, the newspaper reports may help you to complete your records.

7. During a single day, observe the sun's azimuth at various times. Keep a record of the azimuth and the time of observation. Determine whether the azimuth changes at a constant rate during the day, or whether the

B.C. By John Hart

sun's apparent motion is more rapid at some times than at others. Find how fast the sun moves in degrees per hour. See if you can make a graph of the speed of the sun's change in azimuth.

Similarly, find out how the sun's angular altitude changes during the day, and at what time its altitude is greatest. Compare a graph of the speed of the sun's change in altitude with a graph of its speed of change in azimuth.

8. Over a period of several months, or even an entire year, observe the altitude of the sun at noon or some other convenient hour. (Do not worry if you miss some observations.) Determine the date on which the altitude of the sun is at a minimum. On what date would the sun's altitude be at a maximum?

B. Moon

1. Observe and record the altitude and azimuth of the moon and draw its shape on successive evenings at the same hour. Carry your observations through at least one cycle of phases, or shapes, of the moon, recording in your data the dates of any nights that you missed.

For at least one week, make a daily sketch showing the appearance of the moon and another "overhead" sketch of the relative positions of the earth, moon, and sun. If the sun is below the horizon when you observe the moon, you will have to estimate the sun's position.

2. Locate the moon against the background of the stars, and plot its position and phase on a sky map supplied by your instructor.

3. Find the full moon's maximum altitude. Find how this compares with the sun's maximum altitude on the same day. Determine how the moon's maximum altitude varies from month to month.

4. There may be a total eclipse of the moon this year. Consult Table 1-1 on page 12, or the *Celestial Calendar and Handbook*, for the dates of lunar eclipses. Observe one if you possibly can.

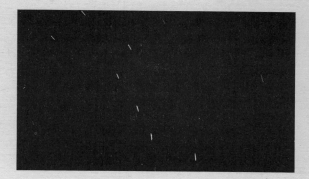

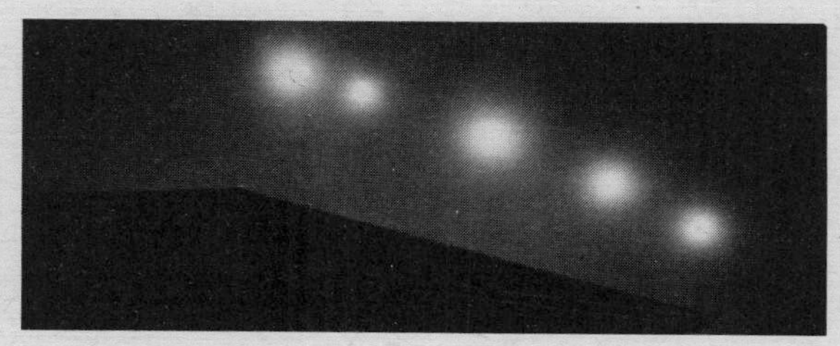

Fig. 1-6 This multiple-exposure picture of the moon was taken with a Polaroid Land camera. Each exposure was for 30 sec using 3,000-speed film. The time intervals between successive exposures were 15 min, 30 min, 30 min, and 30 min.

C. *Stars*

1. On the first evening of star observation, locate some bright stars that will be easy to find on successive nights. Later you will identify some of those groups with constellations that are named on the star map (Fig. 1-2), which shows the constellations around the North Star, or on another star map furnished by your instructor. Record how much the stars have changed their positions compared to your horizon after an hour; after 2 hours.

2. Take a time exposure photograph of several minutes of the night sky to show the motion of the stars. Try to work well away from bright street lights and on a moonless night. Include some of the horizon in the picture for reference. Prop up your camera so it will not move during the time exposures of an hour or more. Use a *small* camera lens opening (large *f*-number) to reduce fogging of your film by stray light.

Fig. 1-7 A time-exposure photograph of Ursa Major (The Big Dipper) taken with a Polaroid Land camera on an autumn evening.

3. Viewing at the same time each night, find whether the positions of the star groups are constant in the sky from month to month. Find if any new constellations appear in the eastern sky after one month; after 3 or 6 months. Over the same periods, find out if some constellations are no longer visible. Determine in what direction and by how much the positions of the stars shift per week and per month.

D. *Planets*

The planets are located within a rather narrow band across the sky (called the zodiac) within which the sun and the moon also move. For details on the location of planets, consult Table 1-1 on page 12, the *Celestial Calendar and Handbook*, or the magazine *Sky and Telescope*. Identify a planet and record its position in the sky relative to the stars at 2-week intervals for several months.

Additional sky observations you may wish to make are described in Unit 2 of this *Handbook*.

TABLE 1.1 A GUIDE FOR PLANET AND ECLIPSE OBSERVATIONS.*

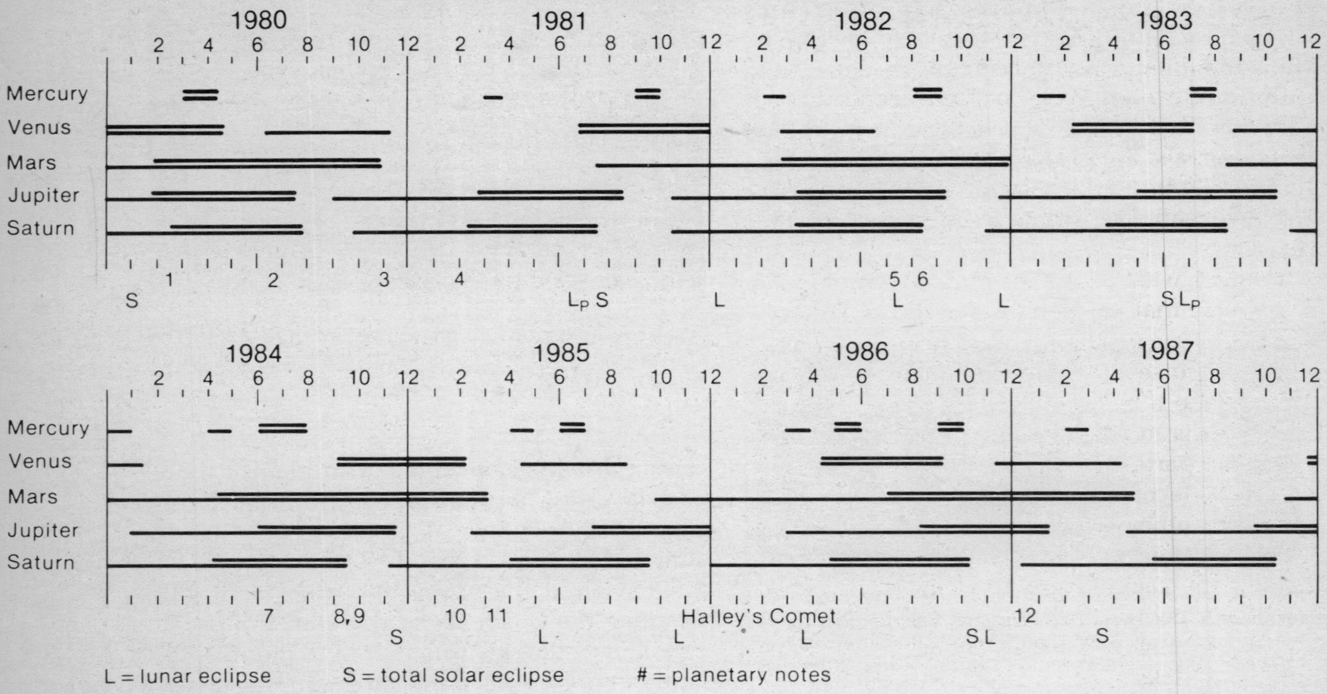

(a) Lunar Eclipses

Lunar Date	Type	Central Time (EST)	Visible from
1981			
July 17	partial	23:48	America (midnight)
1982			
Jan. 9	total	14:56	Asia
July 6	total	02:30	America (early morning)
Dec. 30	total	06:26	America (sunrise)
1983			
June 25	partial	02:25	America (early morning)
1985			
May 4	total	14:57	Asia
Oct. 28	total	12:43	Asia
1986			
April 24	total	07:44	Pacific Ocean
Oct. 17	total	14:19	Asia

(b) Solar Eclipses (Total)

Date	Visible from	Duration
1980		
Feb. 16	Central Africa, India	4.4 min
1981		
July 31	Siberia	2.2 min
1983		
June 11	Indonesia	5.4 min
1984		
Nov. 22	Indonesia, S. America	2.1 min
1986		
Oct. 3	near Greenland	very brief
1987		
March 29	Central Africa	0.3 min
1988		
March 18	Philippines, Indonesia	4.0 min
1990		
July 22	Finland, Arctic	2.6 min

(c) Planetary Notes

Date	Events
1980	1. February–March: Mars and Jupiter close, passing March 1 in retrograde near opposition.
	2. June 21: Mars passes Saturn.
	3. Nov. 1: Venus passes Jupiter and Saturn in early morning sky (spectacular).
1981	4. February: Jupiter and Saturn close, near opposition.
1982	5. July 9: Mars passes Saturn in evening sky.
	6. Aug. 12: Mars passes Jupiter in evening sky.
1984	7. June 18: Mars passes Saturn after opposition in evening sky.
	8. Oct. 8: Venus passes Saturn in evening sky.
	9. Oct. 10: Mars passes Jupiter in evening sky.
1985	10. Jan. 28: Venus passes Mars low in evening sky.
	11. March–April: Mercury and Venus close in early dawn.
1986	12. Dec. 20: Mars passes Jupiter in evening sky — nothing spectacular.

Experiment 1-2
REGULARITY AND TIME

You will often encounter regularity in your study of science. Many natural events occur regularly, that is, over and over again at equal time intervals. If you had no clock, how would you decide how regularly an event recurs? In fact, how can you decide how regular a clock is?

Working with a partner, find several recurring events that you can time in the laboratory. You might use such events as a dripping faucet, a human pulse, or the beat of recorded music. (Do *not* use a clock or watch.) Select pairs of these events to compare.

One lab partner marks each "tick" of Event A on one side of the strip chart recorder tape while the other partner marks each "tick" of Event B. After a long run has been taken, inspect the tape to see how the regularities compare. Record about 300 ticks of Event A. For each 50 ticks of that event, find on the tape the number of ticks of Event B; estimate to 1/10 of a tick. Record your results in a table something like this:

EVENT A	EVENT B
First 50 ticks	________ ticks
Second 50 ticks	________ ticks
Third 50 ticks	________ ticks
Fourth 50 ticks	________ ticks

Now repeat the procedure, comparing your Event A to at least one other periodic phenomenon, Event C, and prepare a similar table.

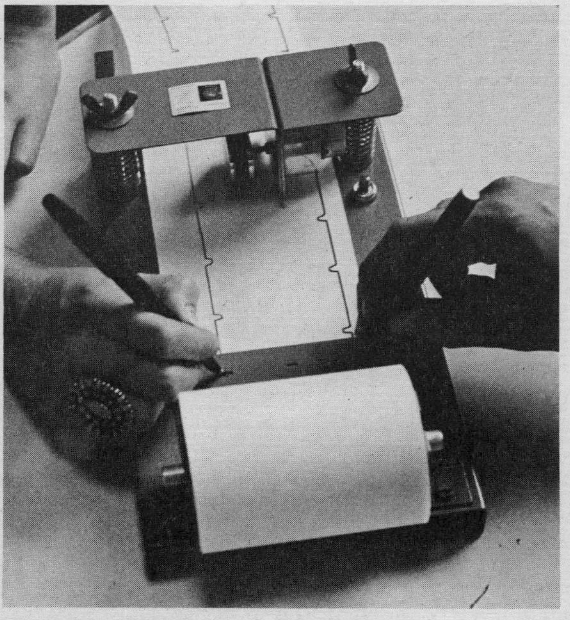

Fig. 1-8

?

1. What do you conclude about the regularity of Event B? If you think that the difference between A and B is larger than you would expect from measurement error, which of the two events is *not* regular? Explain.

2. Which is more regular, Event B or Event C? In answering, what assumptions were you making about Event A?

3. Now compare the regularity of one of your events to some device specifically designed to be regular, for example, an electric wall clock. What results do you get? How do you know the clock is regular? What standard could the clock be compared to? What about *that* standard?

Experiment 1-3
VARIATIONS IN DATA

If you count the number of chairs or people in an ordinary sized room, you will probably get exactly the right answer. But if you measure the length of this page with a ruler, your answer will have a small margin of uncertainty. That is, *numbers read from measuring instruments do not give the exact measurements.* Every measurement is to some extent uncertain.

B.C. By John Hart

Moreover, if your lab partner also measures the length of this page, the two answers will probably be different. Does this mean that the length of the page has changed? Hardly! Then can you possibly find the length of the page without *any* uncertainty in your measurement?

This lab exercise is intended to show you why the answer is "no."

Various stations have been set up around the room, and at each one you are to make some measurement. Record each measurement in a table like the one shown here. When you have completed the series, write your measurements on the board along with those of your classmates. Some interesting patterns should emerge if your measurements have not been influenced by anyone else. Therefore, do not talk about your results or how you got them until everyone has finished.

STATION	TO BE MEASURED	MEASUREMENT

?

1. How do you explain the variation in readings obtained by different students?
2. Is there any way of identifying which value is the "true" one?
3. Is the average value necessarily more correct than any of the actual measured values? If not, why is the average value often used in calculations?
4. How could you write the result of a series of differing measurements so as to indicate something about the range of values?

Experiment 1-4 MEASURING UNIFORM MOTION

If you roll a ball along a level floor or table, it eventually stops. Was the ball slowing down all the time, from the moment you gave it a push? Can you think of any things that have uniform motion in which their speed remains constant and unchanging? Could the dry-ice disk pictured in Sec. 1.3 of the text really be in uniform motion, even if the disk is called "frictionless"? Would the disk keep moving forever? Does everything eventually come to a stop?

In this experiment, you can check the answers to these questions for yourself. You will observe very simple motion and make a photo record of it, or work with similar photos. You will measure the speed of an object as precisely as you can, then record your data in tables and draw graphs from these data. From the graphs, you can decide whether the motion was uniform or not.

Your decision may be harder to make than you would expect, since your experimental measurements can never be exact. There are likely to be ups and downs in your final results. Your problem will be to decide whether the ups and downs are due partly to real changes in speed, to uncertainty in your measurements, or both.

If the speed of your object appears to be constant, does this mean that you have produced an example of uniform motion? Do you think it is possible to do so?

Doing the Experiment

Various setups for the experiment can be made. It takes two people working together to photograph a disk sliding on a smooth surface covered with fine plastic beads, or a glider on an air track, or a steadily flashing light (called a blinky) mounted on a small box pushed by a toy tractor. Your instructor will explain how to work with the setup you are using. Excellent photographs can be made of any of them.

If you do not use a camera at all, or if you work alone, then you may measure a transparency or a movie film projected on a large piece of paper. (See, for example, Film Loop 9 "Analysis of a Hurdle Race. II.") You may simply work from a previously prepared photograph such as Fig. 1-11. If there is time, you might try several of these methods.

Fig. 1-9

Fig. 1-10

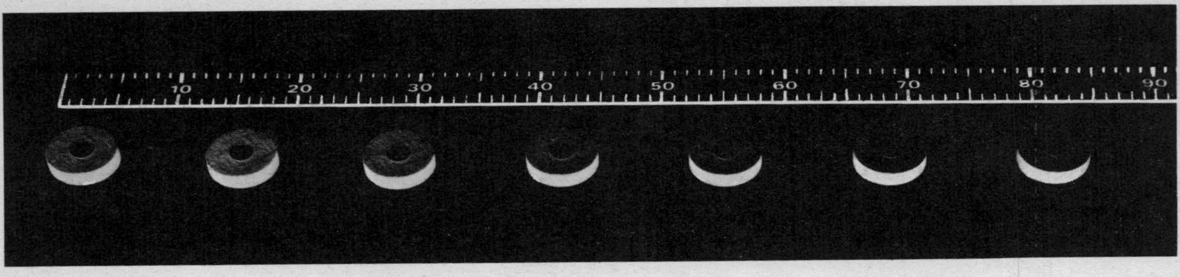

Fig. 1-11 Stroboscopic photograph of a moving CO_2 disk.

One setup uses for the moving object a disk made of metal or plastic. A few plastic beads sprinkled on a smooth, dust-free table top (or a sheet of glass) provide a surface on which the disk slides with almost no friction. Make sure the surface is quite level, so that the disk will not start to move once it is at rest.

Set up the Polaroid Land camera and the stroboscope equipment according to your instructor's directions. See the Introduction for instructions on operating the Polaroid Land model 210, and for a diagram for mounting this camera with a rotating disk stroboscope. A ruler need not be included in your photograph as in Fig. 1-11. Instead, you can use a magnifier with a scale that is more accurate than a ruler for measuring the photograph.

Either your instructor or a few trials will give you an idea of the camera settings and of the speed at which to launch the disk, so that the images of your disk are clear and well-spaced in the photograph. One student launches the disk while a second student operates the camera. A "dry run" or two without taking a picture will probably be needed for practice before you get a good picture. A good picture is one in which there are at least five sharp and clear images of your disk far enough apart for easy measuring on the photograph.

Making Measurements

Whatever method you have used, your next step is to measure the spaces between successive images of your moving object. For this, use a ruler with millimeter divisions and estimate the distances to the nearest tenth of a millimeter, as shown in Fig. 1-12. If you use a magnifier with a scale, rather than a ruler, you may be able to estimate these distances more precisely. List each measurement in a table like Table 1-2.

Since the intervals of time between one image and the next are equal, you can use that interval as a unit of time for analyzing the event. If the speed is constant, the distances of travel will be all the same, and the motion would be uniform.

How would you recognize motion that is not uniform?

Why is it unnecessary for you to know the time interval in seconds?

TABLE 1-2

Time Interval	Distance Traveled in Each Time Interval
1st	0.48 cm
2nd	0.48
3rd	0.48
4th	0.48
5th	0.48
6th	0.48

Table 1-2 has data that indicate uniform motion. Since the object traveled 0.48 cm during each time interval, the speed is 0.48 cm per unit time.

It is more likely that your measurements go up and down as in Table 1-3, particularly if you measure with a ruler.

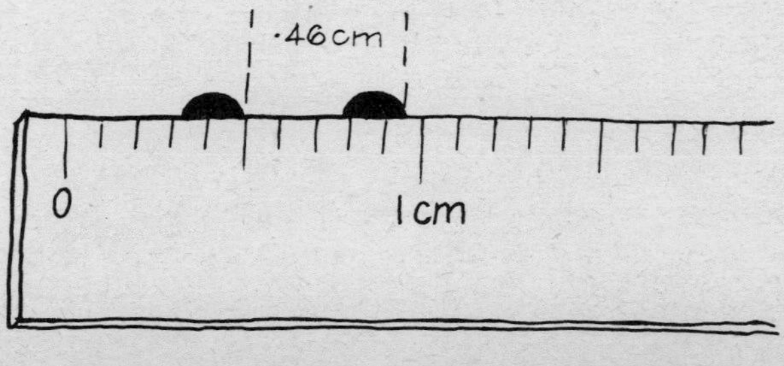

Fig. 1-12 Estimating to one-tenth of a scale division.

TABLE 1-3

Time Interval	Distance Traveled in Each Time Interval
1st	0.48 cm
2nd	0.46
3rd	0.49
4th	0.50
5th	0.47
6th	0.48

Is the speed constant in this case? Since the distances are *not* all the same, you might say "No, it is not constant." Perhaps you looked again at some of the more extreme data in Table 1-3, such as 0.46 cm and 0.50 cm, checked these measurements, and found them doubtful. Then you might say, "The ups and downs are because it is difficult to measure to 0.01 cm with the ruler. The speed really *is* constant as nearly as I can tell." Which statement is right?

Look carefully at the divisions or marks on your ruler. Can you read your ruler accurately to the nearest 0.01 cm? If you are like most people, you read it to the nearest mark of 0.01 cm (the nearest whole millimeter) and *estimate* the next digit between the marks for the nearest tenth of a millimeter (0.01 cm), as illustrated in Fig. 1-12.

In the same way, whenever you read the divisions of any measuring device, you should read accurately to the nearest division or mark and then estimate the next digit in the measurement. Then probably your measurement, including your estimate of a digit between divisions, is not more than half a division in error. It is not likely, for example, that in Fig. 1-12 you would read more than half a millimeter away from where the edge being measured comes between the divisions. In this case, in which the divisions on the ruler are millimeters, you are at most no more than 0.5 mm (0.05 cm) in error.

Suppose you assume that the motion really is uniform and that the slight differences between distance measurements are due only to the uncertainty in reading the ruler. What is then the best estimate of the constant distance the object traveled between flashes?

Usually, to find the "best" value of distance you must find the average of the values. The average for Table 1-3 is 0.48 cm, but the 8 is an uncertain measurement.

If the motion recorded in Table 1-3 really is uniform, the measurement of the distance traveled in each time interval is 0.48 cm plus or minus 0.05 cm, written as 0.48 ± 0.05 cm. The ± 0.05 is called the *uncertainty* of your measurement. The uncertainty for a single measurement is commonly taken to be half a scale division. With many measurements, this uncertainty may be less, but you can use it to be on the safe side.

Now you can return to the key question: Is the speed constant or not? Because the numbers go up and down you might suppose that the speed is constantly *changing*. Notice though that in Table 1-3 the changes of data above and below the average value of 0.48 cm are always smaller than the uncertainty, 0.05 cm. Therefore, the ups and downs *may* all be due to the difficulty in reading the ruler to better than 0.05 cm, and the speed may, in fact, be constant.

The conclusion from the data given here is that the *speed is constant to within the uncertainty of measurement, which is 0.05 cm per unit time.* If the speed goes up or down by less than this amount, you simply cannot reliably detect it with a ruler.

Study your own data in the same way.

Do they lead you to the same conclusion? If your data vary as in Table 1-3, can you think of anything in your setup that could have made the speed actually change? Even if you used a magnifier with a scale, do you still come to the same conclusion?

Measuring More Precisely

A more precise measuring instrument than a ruler or magnifier with a scale might show that the speed in this example was *not* constant. For example, if you used a measuring microscope whose divisions are 0.001 cm apart to measure the same picture again more precisely, you might arrive at the data in Table 1-4. Such precise measurement reduces the uncertainty from ± 0.05 cm to ± 0.0005 cm.

TABLE 1-4

Time Interval	Distance Traveled in Each Time Interval
1st	0.4826 cm
2nd	0.4593
3rd	0.4911
4th	0.5032
5th	0.4684
6th	0.4779

Is the speed constant when you measure to such high precision as this?

The average of these numbers is 0.4804. All the numbers are presumably correct within half a division, which is 0.0005 cm. Thus, the best estimate of the true value is 0.4804 ± 0.0005 cm.

Drawing a Graph

If you have read Sec. 1.5 in the text, you have seen how speed data can be graphed. Your data provide an easy example to use in drawing a graph.

Just as in the example on text page 19, mark off time intervals along the horizontal axis of the graph. Your units are probably not seconds; they are "blinks" if you used a stroboscope or simply "arbitrary time units" which mean here the equal time intervals between positions of the moving object.

Next, mark off the *total* distances traveled along the vertical axis. The beginning of each scale is in the lower left-hand corner of the graph.

Choose the spacing of your scale division so that your data will, if possible, spread across most of the graph paper.

?

1. Does your graph show uniform motion? Explain.
2. If the motion in your experiment was not uniform, review Sec. 1.5 of the Text. Then from your graph find the average speed of your object over the whole trip. Is the average speed for the whole trip the same as the average of the speeds between successive measurements?
3. Could you use the same methods you used in this experiment to measure the speed of a bicycle? a car? a person running? (Assume they are moving uniformly.)
4. The divisions on the speedometer scale of many cars are 5 km/hr in size. You can estimate the reading to the nearest 1 km/hr. (a) What is the uncertainty in a speed measurement by such a speedometer? (b) Could you reliably measure speed changes as small as 2 km/hr? 1 km/hr? 0.5 km/hr? 0.3 km/hr?

Experiment 1-5
A SEVENTEENTH-CENTURY EXPERIMENT

This experiment is similar to the one discussed by Galileo in the *Two New Sciences*. It will give you firsthand experience in working with tools similar to those of a seventeenth-century scientist. You will make quantitative measurements of the motion of a ball rolling down an incline, as described by Galileo.

From these measurements, you should be able to decide for yourself whether Galileo's definition of acceleration was appropriate or not. You should then be able to tell whether Aristotle or Galileo was correct in his conclusion about the acceleration of objects of different sizes.

Reasoning Behind the Experiment

You have read in Sec. 2.6 of the Text how Galileo expressed his belief that the speed of free-falling objects increases in proportion to the time of fall, in other words, that they accelerate uniformly. But since free fall was much too rapid to measure, Galileo assumed that the speed of a ball rolling down an incline increased in the same way as an object in free fall did, only more slowly.

However, even a ball rolling down a low incline moved too fast to measure the speed along different parts of the descent accurately. So Galileo used the relationship $d \propto t^2$ (*or* d/t^2 = constant), an expression in which speed differences have been replaced by the *total time* t and *total distance* d rolled by the ball. Both these quantities can be measured.

Be sure to study Text Sec. 2.7 in which the derivation of this relationship is described. If Galileo's original assumptions were true, this relationship would hold for both freely falling objects and rolling balls. Since total distance and total time are not difficult to measure, seventeenth-century scientists had a secondary hypothesis they could test by experiment; and so have you. Sec. 2.8 of the text discusses much of this material.

Apparatus

The apparatus that you will use is shown in Fig. 1-13. It is similar to that described by Galileo.

You will let a ball roll various distances down a channel about 2 m long and time the motion with a water clock.

You will use a water clock to time this experiment because that was the best timing device available in Galileo's time. The way your water clock works is very simple. Since the volume of water is proportional to the time of flow, you can measure daytime in milliliters of water. Start and stop the flow with your fingers over the upper end of the tube inside the funnel. Whenever you refill the clock, let a little water run through the tube to clear out the bubbles.

Fig. 1-13

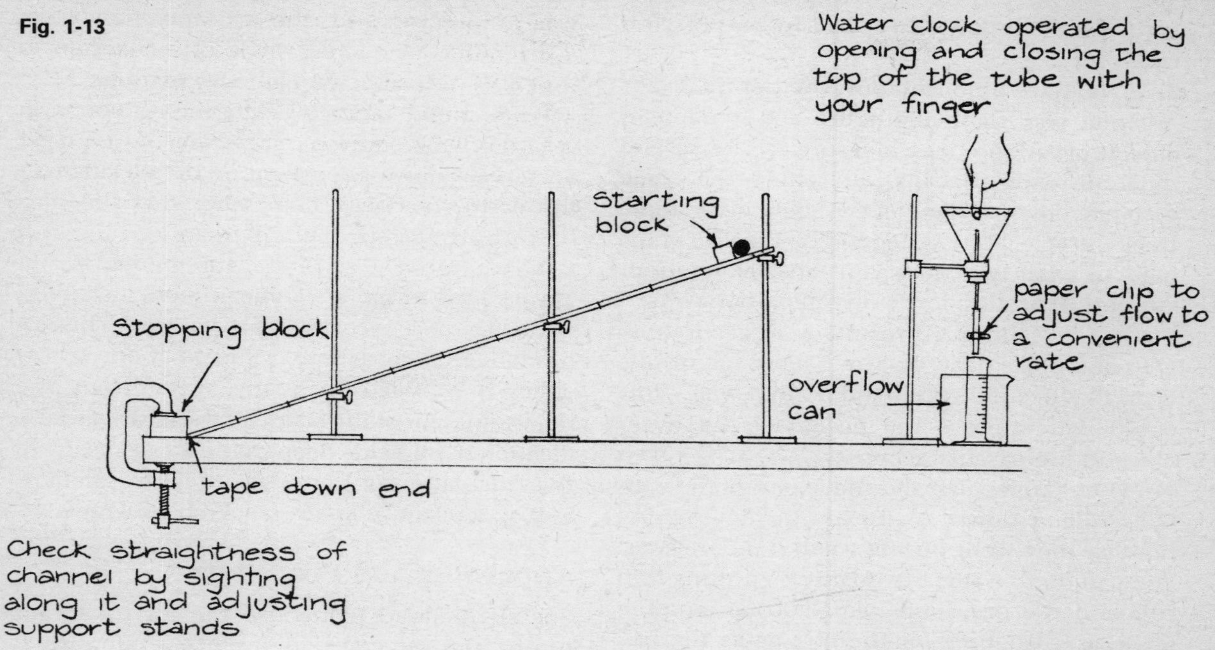

Compare your water clock with a stopwatch when the clock is full and when it is nearly empty to determine how accurate it is. Does the clock's timing change? If so, by how much? Record this information in your notebook.

It is almost impossible to release the ball with your fingers without giving it a slight push or pull. Therefore, restrain the ball with a ruler or pencil, and release it by quickly moving this barrier down the inclined plane. The end of the run is best marked by the sound of the ball hitting the stopping block.

Brief Comment on Recording Data

You should always keep neat, orderly records. Orderly work looks better and is more pleasing to you and everyone else. It may also save you from extra work and confusion. If you have an organized table of data, you can easily record and find your data. This will leave you free to think about your experiment or calculations rather than having to worry about which of two numbers on a scrap of paper is the one you want, or whether you made a certain measurement or not. A few minutes' preparation before you start work will often save you an hour or two of checking in books and with friends.

Operating Suggestions

You should measure times of descent for several different distances, keeping the inclination of the plane constant and using the same ball. Repeat each descent about four times, and average your results. Best results are found for very small angles of inclination (the top of the channel raised less than 30 cm). At steeper inclinations, the ball tends to slide as well as to roll.

From Data to Calculations

Galileo's definition of uniform acceleration (text, page 53) was "equal increases in speed in equal times." Galileo showed that if an object actually moved in this way, the total distance of travel should be directly proportional to the square of the total time of fall, or $d \propto t^2$.

If two quantities are proportional, a graph of one plotted against the other will be a straight line. Thus, making a graph is a good way to check whether two quantities are proportional. Make a graph of d plotted against t^2 in your notebook using your data.

?

1. Does your graph support the hypothesis? Explain.

2. How accurate is the water clock you have been using to time this experiment? If you have not already done so, check your water clock against a stopwatch for timing. In your judgment, how does the inaccuracy of your water clock affect your conclusion to Question 1 above?

Going Further

1. In Sec. 2.7 of the text, you learned that $a = 2d/t^2$. Use this relation to calculate the actual acceleration of the ball in one of your runs.

2. If you have time, find out whether Galileo or Aristotle was right about the acceleration of objects of various sizes. Measure d/t^2 for several different sizes of balls, all rolling the same distance down a plane of the same inclination. Does the acceleration depend on the size of the ball? In what way does your answer refute or support Aristotle's ideas on falling bodies?

3. Galileo claimed his results were accurate to 1/10 of a pulse beat. Do you believe his results were that accurate? Did you do that well? How could you improve the design of the water clock to increase its accuracy?

4. When Galileo first did the experiments with balls rolling down an incline he determined how far they went during equal time intervals. You can do this at least roughly by putting thin rubber bands or a similar small obstacle on the track, and listening for the bumps as the ball rolls down the track. Adjust the position of the rubber bands until the bumps all come at equal times. You might try to keep a regular rhythm by tapping on the table, or by listening to the drip of a faucet or other spigot (as in Method D of Experiment 1-7). Adjust the rubber bands until the bumps are at the same time intervals as the taps or drips. First try a few runs without the rubber bands, making chalk marks at the position of the ball, so that you have a good idea of where to put the rubber bands.

When the rubber bands are properly spaced, measure the distance between them. Are these distances from the start in the ratio of 1:3:5:7...? (There are several ways you can check. You could, for example, divide all the distances by the shortest one and see how close the ratios are to the odd integers. You could also divide each interval by the appropriate odd integer and see how similar the ratios are.)

This version of Galileo's experiments is described by Stillman Drake in his article "The Role of Music in Galileo's Experiments," *Scientific American*, Vol. 232, No. 6 (June, 1975), pp. 98–104.

Experiment 1-6 TWENTIETH-CENTURY VERSION OF GALILEO'S EXPERIMENT

Galileo's seventeenth-century experiment had its limitations, as you read in the text, Sec. 2.9. The measurement of time with a water clock was imprecise and the extrapolation from acceleration at a small angle of inclination to that at a vertical angle (90°) was extreme.

With more modern equipment, you can verify Galileo's conclusions; further, you can get an actual value for acceleration in free fall (near the earth's surface). Remember that the idea behind the improved experiment is still Galileo's. More precise measurements do not always lead to more significant conclusions.

Determine a_g as carefully as you can. This is a fundamental measured value in modern science. It is used in many ways, from the determination of the shape of the earth and the location of oil fields deep in the earth's crust to the calculation of the orbits of earth satellites and spacecraft in space research programs.

Apparatus and Procedure

For an inclined plane use the air track. For timing the air track glider use a stopwatch instead of the water clock; otherwise, the procedure is the same as that used in Experiment 1-5. As you go to higher inclinations, you should stop the glider by hand before it is damaged by hitting the stopping block.

Instead of a stopwatch, you may wish to use the Polaroid Land camera to make a strobe photo of the glider as it descends. A piece of white tape on the glider will show up well in the photograph; or you can attach a small light source to the glider. You can use a magnifier with a scale attached to it to measure the glider's motion as recorded on the photograph. Here the values of d will be millimeters on the photograph and t will be measured in an arbitrary unit, the "blink" of the stroboscope, or the "slot" of the strobe disk.

?

1. Plot your data as before on a graph of d versus t^2. Compare your plotted lines with graphs of the preceding cruder seventeenth-century experiment, if they are available. Are there differences between them? Explain.

2. Is d/t^2 constant for your air track glider? What is the significance of your answer?

3. As a further challenge, if time permits, try to predict the value of a_g, which the glider approaches as the air track becomes vertical. What values do you get? The accepted value of a_g is 9.8 m/sec^2 near the earth's surface.

4. What is the percentage error in your calculated value? That is, what percent of the accepted value is your error?

percentage error =

$$\frac{\text{accepted value} - \text{calculated value}}{\text{accepted value}} \times 100$$

Therefore, if your value of a_g is 9.5m/sec², your percentage error is

$$\frac{9.8\ \text{m/sec}^2 - 9.5\ \text{m/sec}^2}{9.8\ \text{m/sec}^2} \times 100\%$$

$$(3/98) \times 100\% = 3\%$$

Notice that you *cannot* carry this 3% out to 3.06% because you only know the 3 in the fraction 3/98 to one digit. Therefore, you can only know one digit in the answer, 3%. A calculated value like this is said to have one significant digit. You cannot know the second digit in the answer until you know the digit following the 3. To be significant, this digit would require a third digit in the calculated values of 9.5 and 9.8.

5. What are some of the sources of your error?

Experiment 1-7 MEASURING THE ACCELERATION OF GRAVITY$_g$*

Aristotle's idea that bodies falling to the earth are seeking out their natural places sounds strange today. After all, you know that gravity makes things fall.

But just what is gravity? Newton tried to give operational meaning to the idea of gravity by seeking out the laws according to which it acts. Bodies near the earth fall toward it with a certain acceleration due to the gravitational "attraction" of the earth. How can the earth make a body at a distance fall toward it? How is the gravitational force transmitted? Has the acceleration due to gravity always remained the same? These and many other questions about gravity have yet to be answered satisfactorily.

Whether you do one or several parts of this experiment, you will become more familiar with the effects of gravity by finding the acceleration of bodies in free fall yourself. You will learn more about gravity in later chapters.

METHOD A: a_g by Direct Fall*

In this experiment, you will measure the acceleration of a falling object. Since the distance and therefore the speed of fall is too small for air resistance to become important, and since other sources of friction are very small, the acceleration of the falling weight is very nearly a_g.

*Adapted from R. F. Brinckerhoff and D. S. Taft, *Modern Laboratory Experiments in Physics*, by permission of Science Electronics, Inc., Nashua, New Hampshire.

Doing the Experiment

The falling object is an ordinary laboratory hooked weight of at least 200 g mass. (The drag on the paper tape has too great an effect on the fall of lighter weights.) The weight is suspended from about 1 m of paper tape. Reinforce the tape by doubling a strip of masking tape over one end and punch a hole in the reinforcement 1 cm from the end. With careful handling, this tape can support at least 1 kg.

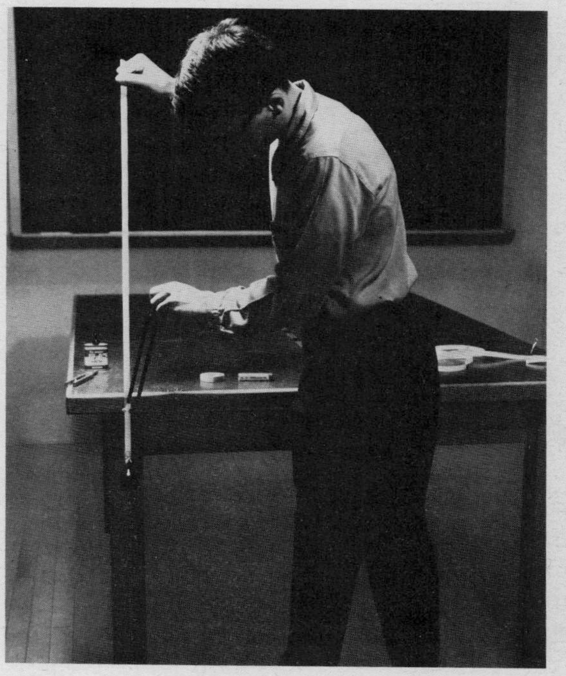

Fig. 1-14

When the suspended weight is allowed to fall, a vibrating tuning fork will mark equal time intervals on the tape pulled down after the weight.

The tuning fork must have a frequency between about 100 Hz (vibrations/second) and about 400 Hz. In order to mark the tape, the fork must have a tiny felt cone (cut from a marking pen tip) glued to the side of one of its prongs close to the end. Such a small mass affects the fork frequency by much less than 1 Hz. Saturate this felt tip with a drop or two of marking pen ink, set the fork in vibration, and hold the tip very gently against the tape. The falling tape is conveniently guided in its fall by two thumbtacks in the edge of the table. The easiest procedure is to have an assistant hold the weighted tape straight up until you have touched the vibrating tip against it and said

"Go." After a few practice runs, you will become expert enough to mark many centimeters of tape with a wavy line as the tape is accelerated past the stationary vibrating fork.

Instead of using the inked cone, you may press a corner of the vibrating tuning fork gently against a 2-cm square of carbon paper which the thumbtacks hold ink-surface-inwards over the falling tape. With some practice, this method can be made to yield a series of dots on the tape without seriously retarding its fall.

Analyzing Your Tapes

Label with an A one of the first wave crests (or dots) that is clearly formed near the beginning of the pattern. Count 10 intervals between wave crests (or dots), and mark the end of the tenth space with a B. Continue marking every tenth crest with a letter throughout the length of the record, which ought to be at least 40 waves long.

At A, the tape already had a speed of v_0. From this point to B, the tape moved a distance d_1 in a time t. The distance d_1 is described by the equation for free fall:

$$d_1 = v_0 t + \frac{a_g t^2}{2}$$

In covering the distance from A to C, the tape took a time exactly twice as long, $2t$, and fell a distance d_2 described (by substituting $2t$ for t and simplifying) by the equation:

$$d_2 = 2v_0 t + \frac{4a_g t^2}{2}$$

In the same way, the distances AD, AE, etc., are described by the equations:

$$d_3 = 3v_0 t + \frac{9a_g t^2}{2}$$

$$d_4 = 4v_0 t + \frac{16a_g t^2}{2}$$

and so on.

All of these distances are measured from A, the arbitrary starting point. To find the distances fallen in each 10-crest interval, you must subtract each equation from the one before it, getting:

$$AB = v_0 t + \frac{a_g t^2}{2}$$

$$BC = v_0 t + \frac{3a_g t^2}{2}$$

$$CD = v_0 t + \frac{5a_g t^2}{2}$$

and $$DE = v_0 t + \frac{7a_g t^2}{2}$$

From these equations, you can see that the weight falls farther during each later time interval. Moreover, when you subtract each of these distances, AB, BC, CD, ... from the subsequent distance, you find that the *increase* in distance fallen is a constant. That is, each difference BC − AB = CD − BC = DE − CD = $a_g t^2$. This quantity is the increase in the distance fallen in each successive 10-wave interval and thus is an acceleration. The formula describes a body falling with a constant acceleration.

From your measurements of AB, AC, AD, etc., make a column of AB, BC, CD, DE, etc., and in the next column record the resulting values of $a_g t^2$. The values of $a_g t^2$ should all be equal (within the accuracy of your measurements). Why? Make all your measurements as precisely as you can with the equipment you are using.

Find the average of all your values of $a_g t^2$, the acceleration in centimeters/(10-wave interval)^2. You want to find the acceleration in cm/sec^2. If you call the frequency of the tuning fork n per second, then the length of the time interval t is $10/n$ sec. Replacing t of 10 waves by $10/n$ sec gives you the acceleration a_g in cm/sec^2.

?

1. What value do you get for a_g? What is the percentage error? (The ideal value of a_g is close to 9.8 m/sec^2.)

METHOD B: a_g from a Pendulum

You can easily measure the acceleration due to gravity by timing the swinging of a pendulum. Of course, the pendulum is not falling straight down, but the time it takes for a round-trip swing still depends on a_g. The time T it takes for a round-trip swing is

$$T = 2\pi\sqrt{\frac{l}{a_g}}$$

In this formula, l is the length of the pendulum. If you measure l with a ruler and T with a clock, you should be able to solve for a_g.

You may learn in a later physics course how to derive the formula. Scientists often use formulas they have not derived themselves, as long as they are confident of their validity.

Making the Measurements

The formula is derived for a pendulum with all the mass concentrated in the weight at the bottom, called the bob. Therefore, the best pendulum to use is one whose bob is a metal sphere hung by a fine thread. In this case, you can be sure that almost all the mass is in the bob. The pendulum's length, l, is the distance from the point of suspension to the *center* of the bob.

Your suspension thread can have any convenient length. Measure l as accurately as possible in meters.

Set the pendulum swinging with *small* swings. The formula does not work well for large swings, as you can test for yourself later.

Time at least 20 complete round trips, preferably more. By timing many round trips instead of just one, you make the error in starting and stopping the clock a smaller fraction of the total time being measured. (When you divide by 20 to get the time for a single round trip, the error in the calculated value for one trip will be only 1/20 as large as if you had measured only one trip.)

Divide the total time by the number of swings to find the time T of one swing.

Repeat the measurement at least once as a check.

Finally, substitute your measured quantities into the formula and solve for a_g.

If you measured l in meters, the accepted value of a_g is 9.80 m/sec^2.

?

1. What value did you get for a_g?
2. What was your percentage error? You find *percentage error* by dividing your error by the accepted value and multiplying by 100:

$$\frac{\text{accepted value} - \text{your value}}{\text{accepted value}} \times 100$$
$$= \frac{\text{your error}}{\text{accepted value}} \times 100$$

With care, your value of a_g should agree within about 1%.

3. Which of your measurements do you think was the least accurate?

If you believe the answer to question 3 was your measurement of length and you think you might be off by as much as 0.5 cm, change your value of l by 0.5 cm and calculate once more the value of a_g. Has a_g changed enough to account for your error? (If a_g went up and your value of a_g was already too high, then you should have altered your measured l in the opposite direction. Try again!)

If your possible error in measuring is not enough to explain your difference in a_g, try changing your *total* time by a few tenths of a second; there may be a possible error in timing. Then you must recalculate T and therefore a_g.

If neither of these attempts works (nor both taken together in the appropriate direction), then you almost certainly have made an error in arithmetic or in reading your measuring instruments. It is most unlikely that a_g in your school differs from 9.80 m/sec^2 by more than $\pm$0.01.

METHOD C: a_g with Slow-Motion Photography (Film Loop)

With a high-speed movie camera you could photograph an object falling along the edge of a vertical measuring stick. Then you could determine a_g by projecting the film at standard speed and measuring the time for the object to fall specified distance intervals.

A somewhat similar method is used in Film Loops 4 and 5. Detailed directions are given for their use in the *Film Loop Notes* on pages 59 and 60.

METHOD D: a_g from Falling Water Drops

You can measure the acceleration due to gravity a_g simply with drops of water falling on a pie plate.

Put the pie plate or a metal dish or tray on the floor. Set up a glass tube with a stopcock, valve, or spigot so that drops of water from the valve will fall at least 1 m to the plate. Support the plate on three or four pencils so that each drop sounds distinctly, like a drum beat.

Adjust the valve carefully until one drop strikes the plate at the same instant the next drop from the valve begins to fall. You can do this most easily by *watching* the drops on the valve while *listening* for the drops hitting the plate. When you have exactly set the valve, the time it takes a drop to fall to the plate is equal to the time interval between one drop and the next.

With the drip rate adjusted, find the time interval t between drops. For greater accuracy, you may want to count the number of drops that fall in 30 sec or 60 sec, or to time the number of seconds for 50 to 100 drops to fall.

Your results are likely to be more accurate if you run a number of trials, adjusting drip rate each time, and average your counts of drops or seconds. The average of several trials should be closer to actual drip rate, drop count, and time intervals than one trial would be.

Now you have all the data you need. You know the time t it takes a drop to fall a distance d from rest. From these data you can calculate a_g, since you know that $d = \frac{1}{2}a_g t^2$ for objects falling from rest.

?

1. What value did you get for a_g?
2. What is your percentage error? How does this compare with your percentage error by any other methods you have used?
3. What do you think led to your error? Could it be leaking connections, allowing more water to escape sometimes? How does this affect your answer?

Suppose the distance of fall was lessened by a puddle forming in the plate; how would this change your results?

There is less water pressure in the tube after a period of dripping; would this increase or decrease the rate of dripping? Do you get the same counts when you refill the tube after each trial?

Would the starting and stopping of your counting against the watch or clock affect your answer? What else may have added to your error?

4. Can you adapt this method of measuring the acceleration of gravity so that you can do it at home? Would it work in the kitchen sink? Would your results be more accurate if the water fell a greater distance, such as down a stairwell?

METHOD E: a_g with Falling Ball and Turntable

You can measure a_g with a record-player turntable, a ring stand and clamp, carbon paper, two balls with holes in them, and thin thread.

Ball X and ball Y are draped across the prongs of the clamp. Line up the balls along a radius of the turntable, and make the lower ball hang just above the paper, as shown in Fig. 1-15.

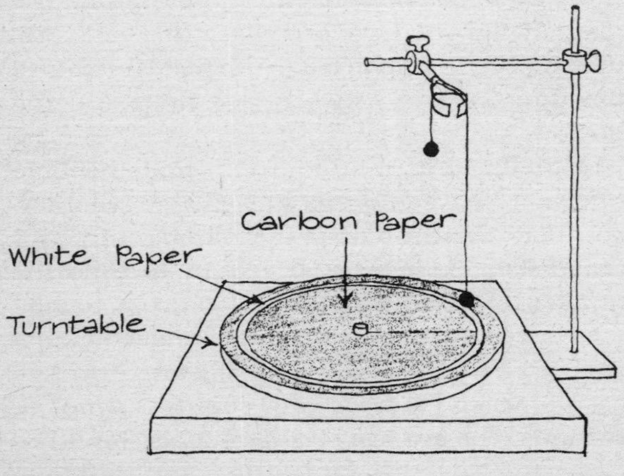

Fig. 1-15

With the table turning, the thread is burned and each ball, as it hits the carbon paper, will leave a mark on the paper under it.

Measure the vertical distance between the balls and the angular distance between the marks. With these measurements and the speed of the turntable, determine the free-fall time.

?

1. What value do you get for a_g?
2. What was your percentage error?
3. What is the most probable source of error? Explain.

METHOD F: a_g with Strobe Photography

Photographing a falling light source with the Polaroid Land camera provides a record that can be graphed and analyzed to give an average value of a_g. The 12-slot strobe disk gives a very accurate 60 slots per second. (A neon bulb can also be connected to the ac line outlet in such a way that it will flash a precise 60 times per second, as determined by the line frequency. Your instructor has a description of the approximate circuit for doing this.)

?

1. What value do you get for a_g?
2. What was your percentage error?
3. What is the most probable source of error? Explain.

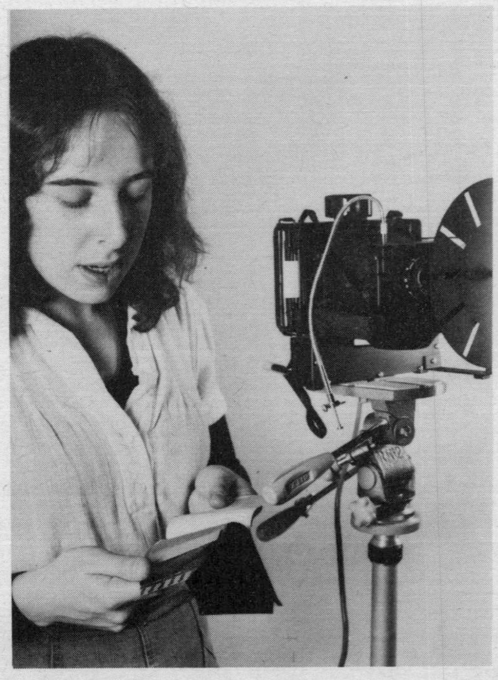

Fig. 1-16

Experiment 1-8
NEWTON'S SECOND LAW

Newton's second law of motion is one of the most important and useful laws of physics. Review text Sec. 3.7 on Newton's second law to make sure you are familiar with it.

Newton's second law is part of a much larger body of theory than can be studied with a simple set of laboratory experiments. Our experiment on the second law has two purposes.

First, because the law *is* so important, it is useful to get a feeling for the behavior of objects in terms of force (F), mass (m), and acceleration (a). You will do this in the first part of the experiment.

Second, the experiment permits you to consider the uncertainties of your measurements. This is the purpose of the latter part of the experiment.

You will apply different forces to carts of different masses and measure the acceleration.

How the Apparatus Works

You are about to find the mass of a loaded cart on which you then exert a measurable force. From Newton's second law you can predict the resulting acceleration of the loaded cart.

Arrange the apparatus as shown in Fig. 1-17. A spring scale is firmly taped to a dynamics cart. The cart, carrying a blinky, is pulled along by a cord attached to the hook of the spring scale. The scale therefore measures the force exerted on the cart.

Fig. 1-17

The cord runs over a pulley at the edge of the lab table, and from its end hangs a weight. The hanging weight can be changed so as to produce various tensions in the cord and thus various accelerating forces on the cart.

Now You Are Ready to Go

Measure the total mass of the cart, the blinky, the spring scale, and any other weights you

Fig. 1-18

may want to include with it to vary the mass m of the cart being accelerated.

Release the cart and allow it to accelerate. Repeat the motion several times while watching the spring-scale pointer. You may notice that the pointer has a range of positions. The midpoint of this range is a fairly good measurement of the average force F_{av} producing the acceleration. Record F_{av} in newtons (N).

Faith in Newton's law is such that you can assume the acceleration is the same and is constant every time this particular F_{av} acts on the mass m.

Substituting your known values of F and m, use Newton's law to predict what the average acceleration a_{av} was during the run.

Then, from your record of the cart's motion, find a directly to see how accurate your prediction was.

To measure the average acceleration a_{av}, take a Polaroid photograph through a rotating disk stroboscope of a light source mounted on the cart. As alternatives, you might use a liquid surface accelerometer, described in detail on page 42, or a blinky. Analyze your results just as in the experiments on uniform and accelerated motion (1-4 and 1-5) to find a_{av}.

This time, however, you must know the distance traveled in meters and the time interval in seconds, not just in blinks, flashes, or other arbitrary time units.

You may wish to observe the following effects without actually making numerical measurements: (a) Keep the mass of the cart constant and observe how various forces affect the acceleration. (b) Keep the force constant and observe how various masses of the cart affect the acceleration.

?

1. Does F_{av} (as measured) equal ma_{av} (as computed from measured values)?
2. Do your other observations support Newton's second law? Explain.

Experimental Errors

It is unlikely that your values of F_{av} and ma_{av} were equal.

Does this mean that you have done a poor job of taking data? Not necessarily. As you think about it, you will see that there are at least two other possible reasons for the inequality. One may be that you have not yet measured everything necessary in order to get an accurate value for each of your three quantities.

In particular, the force used in the calculation ought to be the net, or resultant, force on the cart, not just the towing force that you measured. Friction force also acts on your cart, opposing the accelerating force. You can measure it by reading the spring scale as you tow the cart by hand at *constant speed*. Do it several times and take an average, F_f. Since F_f acts in a direction opposite to the towing force F_T,

$$F_{net} = F_T - F_f$$

If F_f is too small to measure, then $F_{net} = F_T$, and no correction for friction is needed.

Another reason for the inequality of F_{av} and ma_{av} may be that your value for each of these quantities is based on *measurements* and every measurement is uncertain to some extent.

You need to estimate the uncertainty of each of your measurements.

Uncertainty in Average Force F_{av}

Your uncertainty in the measurement of F_{av} is the amount by which your reading of your spring scale varied above and below the average force, F_{av}. Thus, if your scale reading ranged from 1.0 to 1.4 N, the average is 1.2 N, and the range of uncertainty is 0.2 N. The value of F_{av} would be reported as 1.2 ± 0.2 N. Record *your* value of F_{av} and its uncertainty.

Uncertainty in Mass m

Your uncertainty in m is roughly half the smallest scale reading of the balance with which you measured it. The mass consisted of a cart, a blinky, and a spring scale (and possibly an additional mass). If the smallest scale reading is 0.1 kg, your record of the mass of each of these in kilograms might be as follows:

$$m_{cart} = 0.90 \pm 0.05 \text{ kg}$$
$$m_{blinky} = 0.30 \pm 0.05 \text{ kg}$$
$$m_{scale} = 0.10 \pm 0.05 \text{ kg}$$

The total mass being accelerated is the sum of these masses. The uncertainty in the total mass is the sum of the three uncertainties. Thus, in this example, $m = 1.30 \pm 0.15$ kg. Record *your* value of m and its uncertainty.

Uncertainty in Average Acceleration a_{av}

Finally, consider a_{av}. You found this by measuring $\Delta d/\Delta t$ for each of the intervals between the points on your blinky photograph.

Suppose the points in Fig. 1-19 represent images of a light source photographed through a single slot, giving 5 images per second. Calculate $\Delta d/\Delta t$ for several intervals.

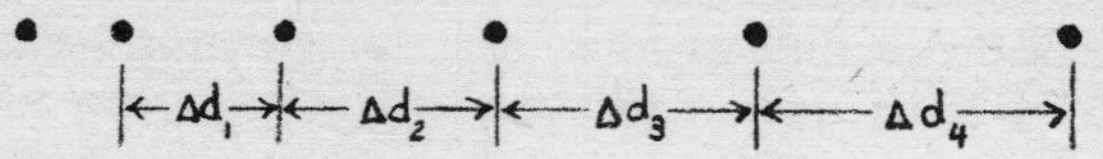

Fig. 1-19

If you assume the time between blinks to have been equal, the uncertainty in each value of $\Delta d/\Delta t$ is due primarily to the fact that the photographic images are a bit fuzzy. Suppose that the uncertainty in locating the distance between the centers of the dots is 0.1 cm as shown in the first column of Table 1-5.

TABLE 1-5

Average Speeds	Average Accelerations
$\Delta d_1/\Delta t = 2.5 \pm 0.1$ cm/sec	
$\Delta d_2/\Delta t = 3.4 \pm 0.1$ cm/sec	$\Delta v_1/\Delta t = 0.9 \pm 0.2$ cm/sec²
$\Delta d_3/\Delta t = 4.0 \pm 0.1$ cm/sec	$\Delta v_2/\Delta t = 0.6 \pm 0.2$ cm/sec²
$\Delta d_4/\Delta t = 4.8 \pm 0.1$ cm/sec	$\Delta v_3/\Delta t = 0.8 \pm 0.2$ cm/sec²
	Average $= 0.8 \pm 0.2$ cm/sec²

When you take the differences between successive values of the speeds, $\Delta d/\Delta t$, you get the accelerations, $\Delta v/\Delta t$, which are recorded in the second column. When a difference in two measurements is involved, you find the uncertainty of the differences (in this case, $\Delta v/\Delta t$) by *adding* the uncertainties of the two measurements. This results in an uncertainty in acceleration of $(\pm 0.1) + (\pm 0.1)$ or $\pm$ 0.2 cm/sec² as recorded in the table. Determine and record *your* value of a_{av} and its uncertainty.

Comparing Your Results

You now have the values of F_{av}, m, and a_{av}, their uncertainties, and you considered the uncertainty of ma_{av}. When you have a value for the uncertainty of this *product* of two quantities, you will then compare the value of ma_{av} with the value of F_{av} and draw your final conclusions. For convenience, the "av" has been dropped from the symbols in the equations in the following discussion. When two quantities are multiplied, the *percentage* uncertainty in the product never exceeds the sum of the percentage uncertainties in each of the factors. In the example, $m \times a = 1.30$ kg $\times$ 0.8 cm/sec² $= 1.04$ N. The uncertainty in a (0.8 ± 0.2 cm/sec²) is 25% (since 0.2 is 25% of 0.8). The uncertainty in m is 11%. Thus, the uncertainty in ma is 25% + 11% = 36%. The product can be written as $ma = 1.04$ N $\pm$ 36% which is, to two significant figures,

$$ma = 1.04 \pm 0.36 \text{ N}$$

(The error is so large here that it really is not appropriate to use the two decimal places; round off to 1.0 ± 0.4 N.) In the example, from direct measurement, $F_{net} = 1.2 \pm 0.2$ N. Are these two results equal within their uncertainties?

Although 1.0 does not equal 1.2, the range of 1.0 ± 0.4 overlaps the range of 1.2 ± 0.2. Therefore, the two numbers agree within the range of uncertainty of measurement.

An example of the lack of agreement would be 1.0 ± 0.2 and 1.4 ± 0.1. These are presumably not the same quantity since there is no overlap of expected uncertainties.

In a similar way, work out your own values of F_{net} and ma_{av}.

?

3. Do your own values agree within the range of uncertainty of your measurement?
4. Is the relationship $F_{net} = ma_{av}$ consistent with your observations?

Experiment 1-9
MASS AND WEIGHT

You know from your own experience that an object that is pulled strongly toward the earth (for example, an automobile) is difficult to accelerate by pushing. In other words, objects with great weight also have great inertia. Is there some simple, exact relationship between the masses of objects and the gravitational forces acting on them? For example, if one object has twice the mass of another, does it also weigh twice as much?

Measuring Mass

The masses of two objects can be compared by observing the accelerations each experiences when acted on by the same force. Accelerating an object in one direction with a constant force for long enough to take measurements is often not practical in the laboratory. Fortunately there is an easier way. If you rig up a puck and springs between two rigid supports as shown in Fig. 1-20, you can attach objects to the puck

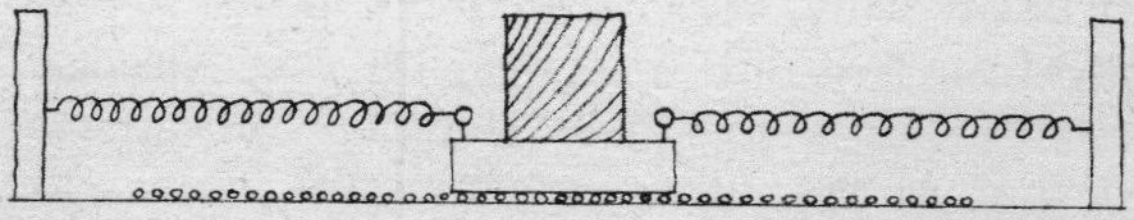

Fig. 1-20

and have the springs accelerate the object back and forth. The greater the inertial mass of the object, the less the magnitude of acceleration will be, and the longer it will take to oscillate back and forth.

To "calibrate" your oscillator, first time the oscillations. The time required for five complete round trips is a convenient measure. Next tape pucks on top of the first one, and time the period for each new mass. (The units of mass are not essential here; you will be interested only in the ratio of masses.) Then plot a graph of mass against the oscillation period, drawing a smooth curve through your experimental plot points. Do not leave the pucks stuck together.

From your results, try to determine the relationship between inertial mass and the oscillation period. If possible, write an algebraic expression for the relationship.

Weight

To compare the gravitational forces on two objects, they can be hung on a spring scale. In this investigation, the units on the scale are not important because you are interested only in the ratio of the weights.

Comparing Mass and Weight

Use the puck and spring oscillator, and the calibration graph to find the masses of two objects (say, a dry cell and a stapler). Find the gravitational pulls on these two objects by hanging each from a spring scale.

?

1. How does the *ratio* of the gravitational forces compare to the *ratio* of the masses?
2. How would you conduct a similar experiment to compare the masses of two iron objects to the magnetic forces exerted on them by a large magnet?

Comment

You probably will not be surprised to find that, to within your uncertainty of measurement, the ratio of gravitational forces is the same as the ratio of masses. Is this really worth doing an experiment to find out, or is the answer obvious to begin with? Newton did not think it was obvious. He did a series of very precise experiments using many different substances to find out whether gravitational force was always proportional to inertial mass. To the limits of his precision, Newton found the proportionality to hold exactly. (Newton's results have been confirmed to a precision of ±0.000000001%.)

Newton could offer no explanation from his physics as to why the attraction of the earth for an object should increase in exact proportion to the object's inertial mass. No other forces bear such a simple relation to inertia, and this remained a complete puzzle for two centuries until Einstein related inertia and gravitation theoretically. Even before Einstein, Ernst Mach made the ingenious suggestion that inertia is not the property of an object by itself, but is the result of the gravitational forces exerted on an object by everything else in the universe.

Experiment 1-10
CURVES OF TRAJECTORIES

Picture a ski jumper. He leans forward at the top of the slide, grasps the railing on each side, and propels himself out onto the track. Streaking down the trestle, he crouches and gives a mighty leap at the takeoff lip, soaring up and out, over the snow-covered fields far below. The hill flashes into view and he lands on its steep incline, bobbing to absorb the impact.

Like so many interesting events, this one involves a more complex set of forces and motions than you can conveniently deal with in the laboratory at one time. Therefore, concentrate on just one aspect: the flight through the air. What kind of a path, or trajectory, would a ski-jumping flight follow?

At the moment of projection into the air a skier has a certain velocity (that is, a certain

speed in a given direction), and throughout the flight must experience the downward acceleration due to gravity. These are circumstances that can be duplicated in the laboratory. To be sure, the flight path of an actual ski jumper is probably affected by other factors, such as air velocity and friction; but you now know that it usually pays to begin experiments with a simplified approximation that allows you to study the effects of a few factors at a time. Thus, in this experiment you will launch a steel ball from a ramp into the air and try to determine the path it follows.

How to Use the Equipment

If you are assembling the equipment for this experiment for the first time, follow the manufacturer's instructions.

The apparatus consists primarily of two ramps down which you can roll a steel ball. Adjust one of the ramps (perhaps with the help of a level) so that the ball leaves it horizontally.

Tape a piece of squared graph paper to the plotting board with its left-hand edge behind the end of the launching ramp.

To find a path that extends fully across the graph paper, release the ball from various points up the ramp until you find one from which the ball falls close to the bottom right-hand corner of the plotting board. Mark the point of release on the ramp and release the ball each time from this point.

Attach a piece of carbon paper to the impact board, with the carbon side facing the ramp. Then tape a piece of thin onionskin paper over the carbon paper.

Now when you put the impact board in its way, the ball hits it and leaves a mark that you can see through the onionskin paper, and automatically records the point of impact between ball and board. (Make sure that the impact board does not move when the ball hits it; steady the board with your hand if necessary.) Transfer the point to the plotting board by making a mark on it just next to the point on the impact board.

Do not hold the ball in your fingers to release it; it is impossible to let go of the ball in the same way every time. Instead, restrain it with a ruler held at a mark on the ramp and release the ball by moving the ruler quickly away from it down the ramp.

Try releasing the ball several times (always from the same point) for the same setting of the impact board. Do all the impact points exactly coincide?

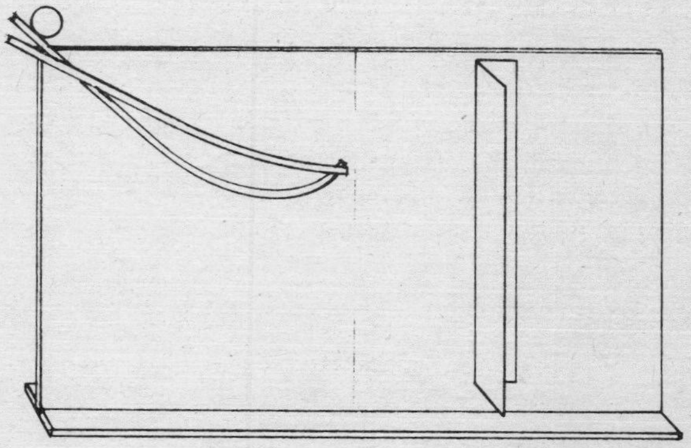

Fig. 1-21

Repeat this for several positions of the impact board to record a number of points on the ball's path. Move the board *equal distances* every time and always release the ball from the same spot on the ramp. Continue until the ball does not hit the impact board any longer.

Now remove the impact board, release the ball once more, and watch carefully to see that the ball moves along the points marked on the plotting board.

The curve traced out by your plotted points represents the *trajectory* of the ball. By observing the path the ball follows, you have completed the first phase of the experiment.

If you have time, you will find it worthwhile to go further and explore some of the properties of your trajectory.

Analyzing Your Data

To help you analyze the trajectory, draw a horizontal line on the paper at the level of the end of the launching ramp. Then remove the paper from the plotting board and draw a smooth continuous curve through the points as shown in Fig. 1-22.

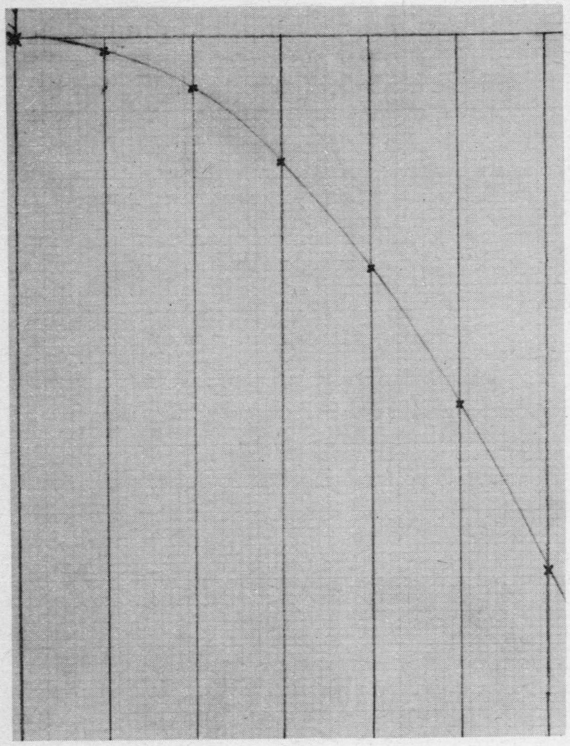

Fig. 1-22

You already know that a moving object on which no net force is acting will move at constant speed. There is no appreciable horizontal force acting on the ball during its fall, so you can make an *assumption* that its horizontal motion is at a constant speed. Then, equally spaced vertical lines will indicate equal time intervals.

Draw vertical lines through the points on your graph. Make the first line coincide with the end of the launching ramp. *Because of your plotting procedure*, these lines should be equally spaced. If the horizontal speed of the ball is uniform, these vertical lines are drawn through positions of the ball separated by equal time intervals.

Now consider the vertical distances fallen in each time interval. Measure down from your horizontal line the vertical fall to each of your plotted points. Record your measurements in a column. Alongside them record the corresponding horizontal distances measured from the first vertical line.

?

1. What would a graph look like on which you plot horizontal distance against time?
2. Earlier in your work with accelerated motion, you learned how to recognize uniform acceleration (see Secs. 2.5–2.8 in the Text and Experiment 1-4). Use the data you have just collected to decide whether the vertical motion of the ball was uniformly accelerated motion. What do you conclude?
3. Do the horizontal and the vertical motions affect each other in any way?
4. What equation describes the horizontal motion in terms of horizontal speed, v, the horizontal distance, Δx, and the time of travel, Δt?
5. What equation describes the vertical motion in terms of the distance fallen vertically, Δy, the vertical acceleration, a_g, and the time of travel, Δt?

Try These Yourself

There are many other things you can do with this apparatus. Some of them are suggested by the following questions:

1. What do you expect would happen if you repeated the experiment with a glass marble of the same size instead of a steel ball?
2. What will happen if you next try to repeat the experiment starting the ball from a different point on the ramp?
3. What do you expect if you use a smaller or larger ball, starting always from the same reference point on the ramp?
4. Plot the trajectory that results when you use a ramp that launches the ball at an angle to the horizontal. In what way is this curve similar to your first trajectory?

Experiment 1-11
PREDICTION OF TRAJECTORIES

You can predict the landing point of a ball launched horizontally from a tabletop at any speed. If you know the speed v of the ball as it leaves the table, the height of the table above the floor, Δy, and a_g, you can then use the equation for projectile motion to predict where on the floor the ball will land.

You know an equation for horizontal motion:

$$\Delta x = v\ \Delta t$$

and you know an equation for free fall from rest:

$$\Delta y = \tfrac{1}{2}a_g(\Delta t)^2$$

The time interval is difficult to measure. Besides, in talking about the *shape* of the path, all you really need to know is how Δy relates to Δx. Since, as you found in the previous

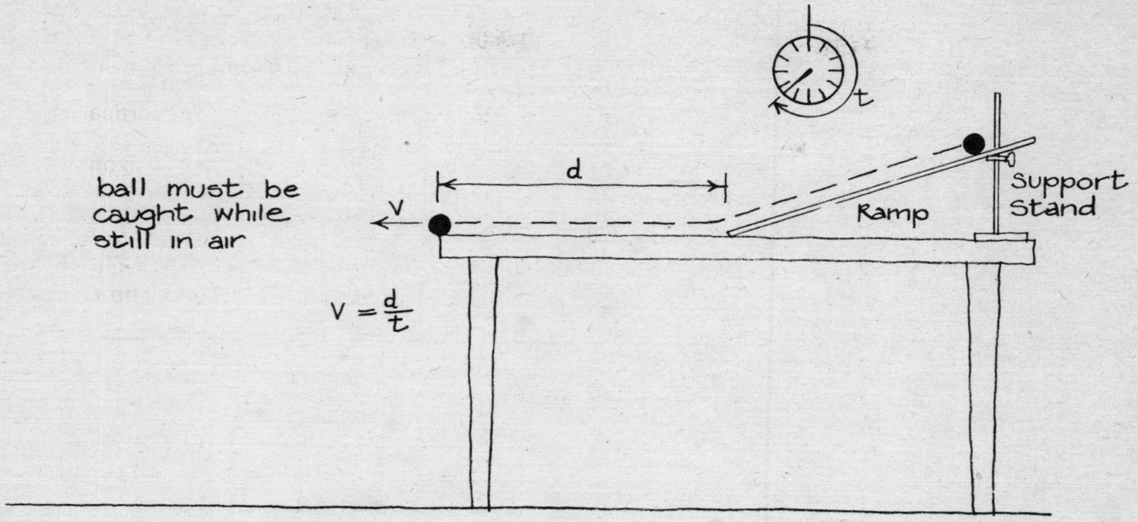

Fig. 1-23

experiment, these two equations still apply when an object is moving horizontally and falling at the same time, you can combine them to get an equation relating Δy and Δx, without Δt appearing at all. You can rewrite the equation for horizontal motion as:

$$\Delta t = \frac{\Delta x}{v}$$

Then you can substitute this expression for Δt into the equation for free fall, obtaining:

$$\Delta y = \tfrac{1}{2} a_g \frac{(\Delta x)^2}{v^2}$$

Thus, the derived equation should describe how Δy changes with Δx: that is, it should give the shape of the trajectory. If you want to know how far out from the edge of the table the ball will land (Δx), you can calculate it from the height of the table (Δy), a_g, and the ball's speed v along the table.

Doing the Experiment

Find v by measuring with a stopwatch the time t that the ball takes to roll a distance d along the tabletop. (See Fig. 1-23.) Be sure to have the ball caught as it comes off the end of the table. Repeat the measurement a few times, always releasing the ball from the same place on the ramp, and take the average value of v.

Measure Δy and then use the equation for Δy to calculate Δx. Place a target, a paper cup, for example, on the floor at your predicted landing spot as shown in Fig. 1-24. How confident are you of your prediction? Since it is based on *measurement*, some uncertainty is involved. Mark an area around the spot to indicate your uncertainty.

Now release the ball once more. This time, let it roll off the table; if your measurements were accurate, it should land on the target as shown in Fig. 1-24.

If the ball actually does fall within the range of values you have estimated for x, then you have supported the assumption on which your calculation was based, that vertical and horizontal motion are not affected by each other.

Measuring Δ x

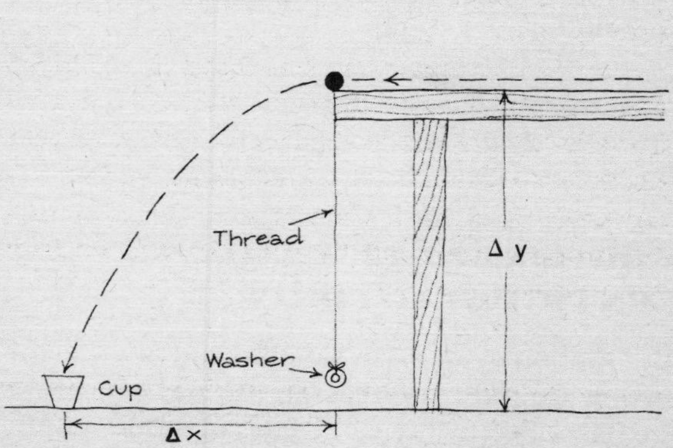

Fig. 1-24

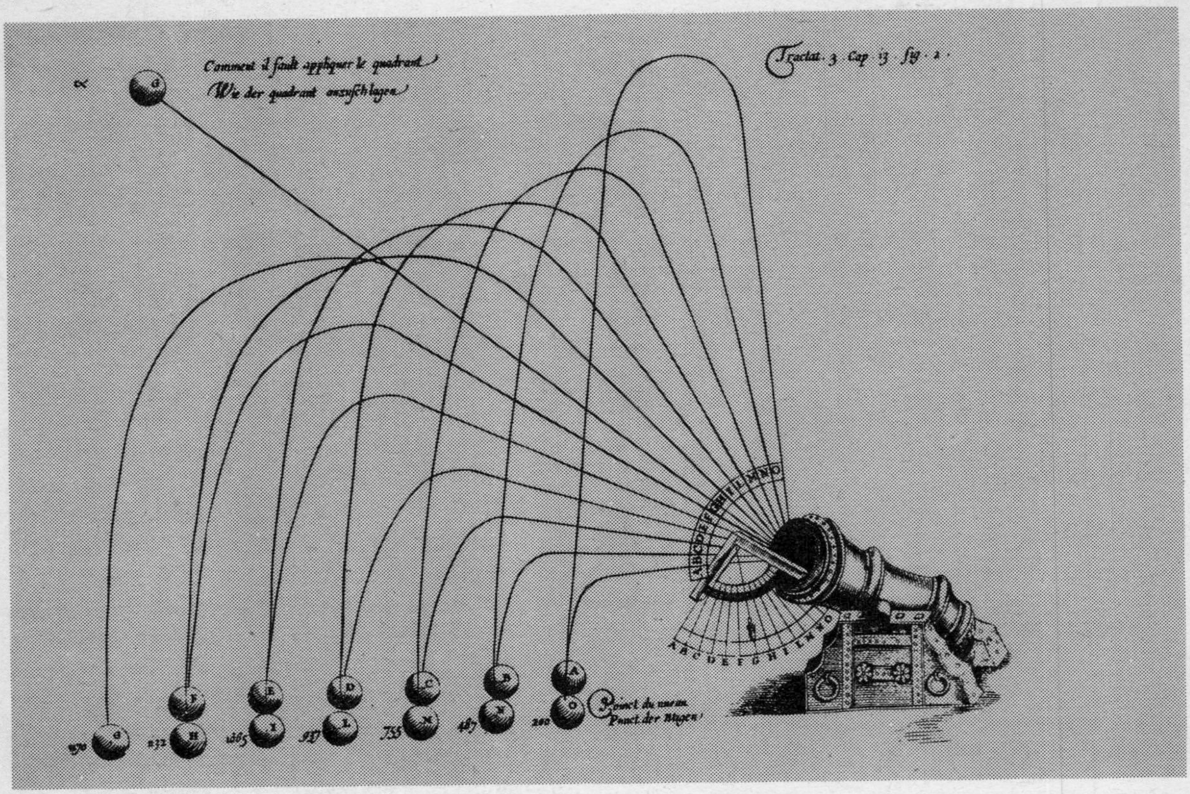

Fig. 1-25 The path taken by a cannon ball according to a drawing by Ufano (1621). He shows that the same horizontal distance can be obtained by two different firing angles. Gunners had previously found this by experience. What angles give the maximum range? What is wrong with the way Ufano drew the trajectories?

?

1. How could you determine the range of a ball launched horizontally by a slingshot?
2. Assume you can throw a baseball 40 m on the earth's surface. How far could you throw the same ball on the surface of the moon, where the acceleration of gravity is one-sixth what it is on the surface of the earth?
3. Will the assumptions made in the equations $\Delta x = v\,\Delta t$ and $\Delta y = \frac{1}{2}a_g(\Delta t)^2$ hold for a ping-pong ball? If the table were 1,000 m above the floor, could you still use these equations? Why or why not?

Experiment 1-12 CENTRIPETAL FORCE

The motion of an earth satellite and of a weight swung around your head on the end of a string are described by the same laws of motion. Both are accelerating toward the center of their orbit due to the action of an unbalanced force.

In the following experiment, you can discover for yourself how this centripetal force depends on the mass of the satellite and on its speed and distance from the center.

How the Apparatus Works

Your "satellite" is one or more rubber stoppers. When you hold the apparatus in both hands and swing the stopper around your head, you can measure the centripetal force on it with a spring scale at the base of the stick. The scale should read in newtons; otherwise its readings should be converted to newtons.

You can change the length of the string to vary the radius R of the circular orbit. Tie on more stoppers to vary the satellite mass m.

The best way to set the frequency f is to swing the apparatus in time with some periodic sound from a metronome or an earphone attached to a blinky. Keep the rate constant by adjusting the swinging until you see the stopper cross the same point in the room every tick.

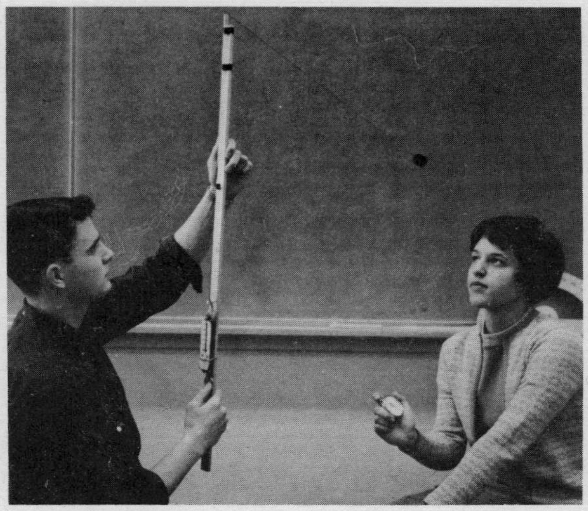

Fig. 1-26

Hold the stick vertically and have as little motion at the top as possible, since this would change the radius. Because the stretch of the spring scale also alters the radius, it is helpful to have a marker (knot or piece of tape) on the string. You can move the spring scale up or down slightly to keep the marker in the same place.

Doing the Experiment

The object of the experiment is to find out how the force F, read on the spring scale, varies with m, with f, and with R.

You should only change *one* of these three quantities at a time so that you can investigate the effect of each quantity independently of the others. Either double or triple m, f, and R (or halve them, and so on, if you started with large values).

Two or three different values should be enough in each case. Make a table and clearly record your numbers in it.

?

1. How do changes in m affect F when R and f are kept constant? Write a formula that states this relationship.
2. How do changes in f affect F when m and R are kept constant? Write a formula to express this relationship.
3. What is the effect of R on F?
4. Now, put m, f, and R all together in a single formula for centripetal force, F. How does your formula compare with the expression derived in Sec. 4-6 of the Text?

Experiment 1-13
CENTRIPETAL FORCE ON A TURNTABLE

You may have had the experience of spinning around on an amusement park contraption known as the Whirling Platter. The riders seat themselves at various places on a large, flat, polished wooden turntable about 12 m in diameter. The turntable gradually rotates faster and faster until everyone (except for the person at the center of the table) has slid off. The people at the edge are the first to go. Why do the people slide off?

Unfortunately, you probably do not have a Whirling Platter in your classroom, but you do have a Masonite disk that fits on a turntable. The object of this experiment is to predict the maximum radius at which an object can be placed on the rotating turntable without sliding off.

If you do this under a variety of conditions, you will see for yourself how forces act in circular motion.

Before you begin, be sure you have studied Sec. 4-6 in your text where you learned that the centripetal force needed to hold a rider in a circular path is given by $F = mv^2/R$.

Studying Centripetal Force

For these experiments, it is more convenient to rewrite the formula $F = mv^2/R$ in terms of the frequency f. This is because f can be measured more easily than v. You can rewrite the formula as follows:

$$v = \frac{\text{distance traveled}}{\text{in one revolution}} \times \frac{\text{number of revolutions}}{\text{per second}}$$

$$= 2\pi R \times f$$

Substituting this expression for v in the formula for centripetal force gives:

$$F = \frac{m \times (2\pi Rf)^2}{R}$$

$$= \frac{4\pi^2 mR^2 f^2}{R}$$

$$= 4\pi^2 mRf^2$$

You can measure all the quantities in this equation.

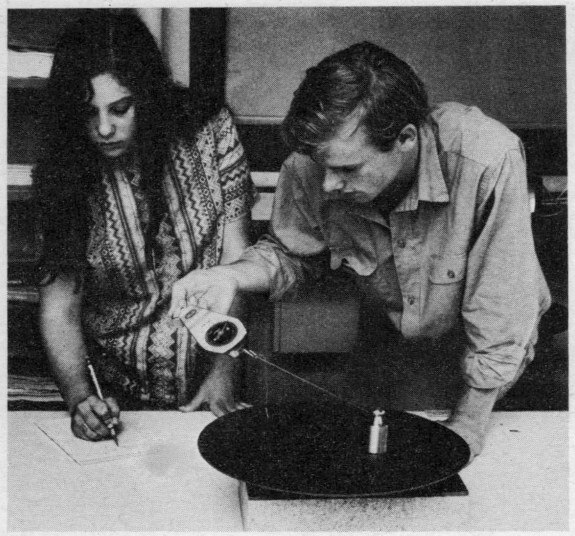

Fig. 1-27

Friction on a Rotating Disk

For objects on a rotating disk, the centripetal force is provided by friction. On a frictionless disk, there could be no such centripetal force. As you can see from the equation just derived, the centripetal acceleration is proportional to R and to f^2. Since the frequency of f is the same for any object moving around on a turntable, then the centripetal acceleration is directly proportional to R, the distance from the center. The further an object is from the center of the turntable, therefore, the greater the centripetal force must be to keep it in a circular path.

You can measure the maximum force F_{max} that friction can provide on the object, measure the mass of the object, and then calculate the maximum distance from the center R_{max} that the object can be without sliding off. Solving the centripetal force equation for R gives:

$$R_{max} = \frac{F_{max}}{4\pi^2 m f^2}$$

Use a spring scale to measure the force needed to make some object (of mass m from 0.2 kg to 1.0 kg) start to slide across the motionless disk. This will be a measure of the maximum friction force that the disk can exert on the object.

Make a chalk mark on the turntable and time it (say, for 100 sec) or accept the marked value of revolutions per minute and calculate the frequency in hertz (Hz).

Make your predictions of R_{max} for turntable frequencies of 33 revolutions per minute (rpm), 45 rpm, and 78 rpm.

Then try the experiment!

?

1. How great is the percentage difference between prediction and experiment for each turntable frequency? Is this reasonable agreement?
2. What effect would decreasing the mass have on the predicted value of R? Careful! Decreasing the mass has an effect on F also. Check your answer by doing an experiment.
3. What is the smallest radius in which you can turn a car if you are moving 100 km/hr and the friction force between tires and road is one-third the weight of the car? (Remember that weight is equal to $a_g \times m$.)

B.C. by John Hart

ACTIVITIES

CHECKER SNAPPING

Stack several checkers. Put another checker on the table and snap it into the stack. On the basis of Newton's first law, can you explain what happened?

BEAKER AND HAMMER

Place a glass beaker half full of water on top of a pile of three wooden blocks. Three quick back-and-forth swipes (NOT FOUR!) of a hammer on the blocks leave the beaker sitting on the table.

PULLS AND JERKS

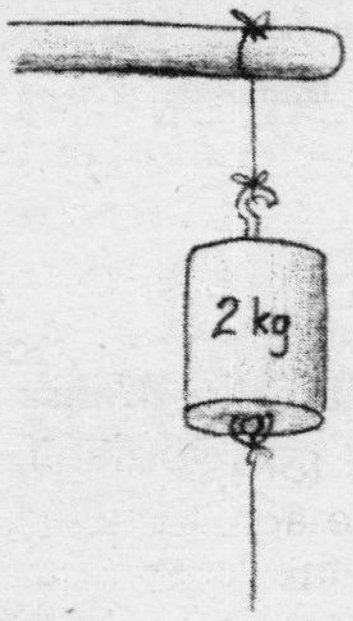

Fig. 1-28

Hang a weight (such as a heavy wooden block) by a string that just barely supports it, and tie another identical string below the weight. A slow, steady pull on the string below the weight breaks the string *above* the weight. A quick jerk breaks it *below* the weight. *Why?*

EXPERIENCING NEWTON'S SECOND LAW

One way for you to get the feel of Newton's second law is actually to pull an object with a constant force. Load a cart with a mass of several kilograms. Attach one end of a long rubber band to the cart and, pulling on the other end, move at such a speed that the rubber band is maintained at a constant length, for example, 70 cm. Holding a meter stick above the band with its 0-cm end in your hand will help you to keep the length constant.

The acceleration will be very apparent to the person applying the force. Vary the mass on the cart and the number of rubber bands (in parallel) to investigate the relationship between F, m, and a.

MAKE ONE OF THESE ACCELEROMETERS

An accelerometer is a device that measures acceleration. Actually, anything that has mass could be used for an accelerometer. Because you have mass, you were acting as an accelerometer the last time you lurched forward in the seat of a car as the brakes were applied. With a knowledge of Newton's laws and certain information about yourself, anybody who measured how far you leaned forward and how tense your muscles were would get a good idea of the magnitude and direction of the acceleration that you were undergoing; but it would be complicated.

Here are four accelerometers of a much simpler kind. With a little practice, you can learn to read accelerations from them directly, without making any difficult calculations.

A. The Liquid-Surface Accelerometer

This device is a hollow, flat plastic container partly filled with a colored liquid. When it is not being accelerated, the liquid surface is horizontal, as shown by the dotted line in Fig. 1-29. But when it is accelerated toward the left (as

B.C. by John Hart

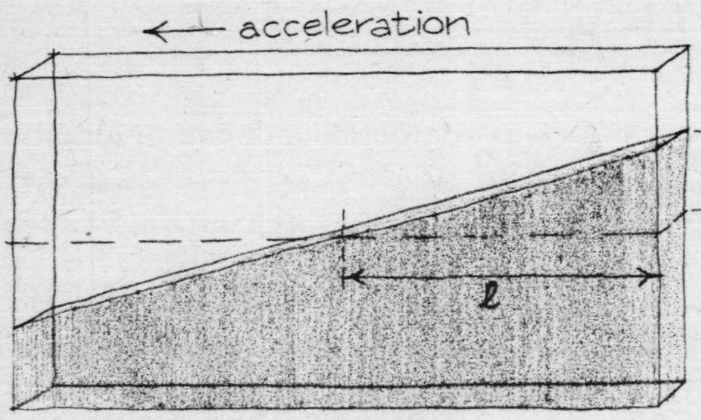

Fig. 1-29

shown) with a uniform acceleration a, the surface becomes tilted. The level of the liquid rises a distance h above its normal position at one end of the accelerometer and falls the same distance at the other end. The greater the acceleration, the more steeply the surface of the liquid is slanted. This means that the slope of the surface is a measure of the magnitude of the acceleration a.

The length of the accelerometer is $2l$, as shown in Fig 1-29. So the slope of the surface may be found by:

$$\text{slope} = \frac{\text{vertical distance}}{\text{horizontal distance}} = \frac{2h}{2l} = \frac{h}{l}$$

Theory gives you a very simple relationship between this slope and the acceleration a:

$$\text{slope} = \frac{h}{l} = \frac{a}{a_g}$$

Notice what this equation tells you. It says that if the instrument is accelerating in the direction shown with just a_g (one common way to say this is that it has a "one-G acceleration" — the acceleration of gravity), then the slope of the surface is just 1; that is, $h = l$ and the surface makes a 45° angle with its normal, horizontal direction. If it is accelerating with $\frac{1}{2}a_g$, then the slope will be $\frac{1}{2}$; that is $h = \frac{1}{2}l$. In the same way, if $h = \frac{1}{4}l$, then $a = \frac{1}{4}a_g$, and so on with any acceleration you care to measure.

To measure h, stick a piece of centimeter tape on the front surface of the accelerometer as shown in Fig. 1-30. Then stick a piece of white paper or tape to the back of the instrument to make it easier to read the level of the liquid. Solving the equation above for a gives:

$$a = a_g \times \frac{h}{l}$$

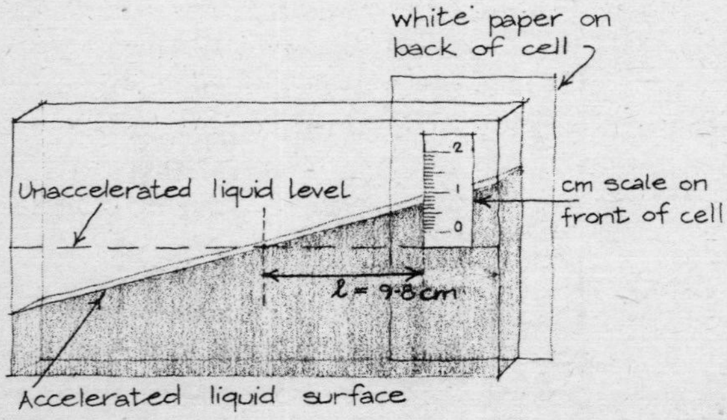

Fig. 1-30

Since a_g is very close to 9.8 m/sec^2 at the earth's surface, if you place the scale 9.8 scale units from the center, you can read accelerations directly in meters per second^2. For example, if you stick a centimeter tape just 9.8 cm from the center of the liquid surface, 1 cm on the scale is equivalent to an acceleration of 1 m/sec^2.

Calibration of the Accelerometer

You do not have to trust blindly the theory just mentioned. You can test it for yourself. Does the accelerometer really measure accelerations directly in meters per second^2? Stroboscopic methods give you an independent check on the correctness of the prediction.

Set the accelerometer on a dynamics cart and arrange strings, pulleys, and masses as you did in Experiment 1-9 to give the cart a uniform acceleration on a long tabletop. Put a block of wood at the end of the cart's path to stop it. Make sure that the accelerometer is fastened firmly enough so that it will not fly off the cart when it stops suddenly. Make the string as long as you can, so that you use the entire length of the table.

Give the cart a wide range of accelerations by hanging different weights from the string. Use a stroboscope to record each motion. To measure the accelerations from your strobe records, plot t^2 against d, as you did in Experiment 1-5. (What relationship did Galileo discover between d/t^2 and the acceleration?) Or use the method of analysis you used in Experiment 1-9.

Compare your stroboscopic measurements with the readings on the accelerometer during each motion. It takes some cleverness to read the accelerometer accurately, particularly near the end of a high-acceleration run. One way is to have several students along the table observe the reading as the cart goes by; use the average

By permission of John Hart and Field Enterprises, Inc.

of their reports. If you are using a xenon strobe, of course, the readings on the accelerometer will be visible in the photograph; this is probably the most accurate method.

Plot the accelerometer readings against the stroboscopically measured accelerations. This graph is called a *calibration curve.* If the two methods agree perfectly, the graph will be a straight line through the origin at a 45° angle to each axis. If your curve has some other shape, you can use it to convert "accelerometer readings" to "accelerations" if you are willing to assume that your strobe measurements are more accurate than the accelerometer. (If you are not willing, what can you do?)

B. Automobile Accelerometer. I

With a liquid-surface accelerometer mounted on the front-to-back line of a car, you can measure the magnitude of acceleration along its path. Here is a modification of the liquid-surface design that you can build for yourself. Bend a small glass tube (about 30 cm long) into a U-shape, as shown in Fig. 1-31.

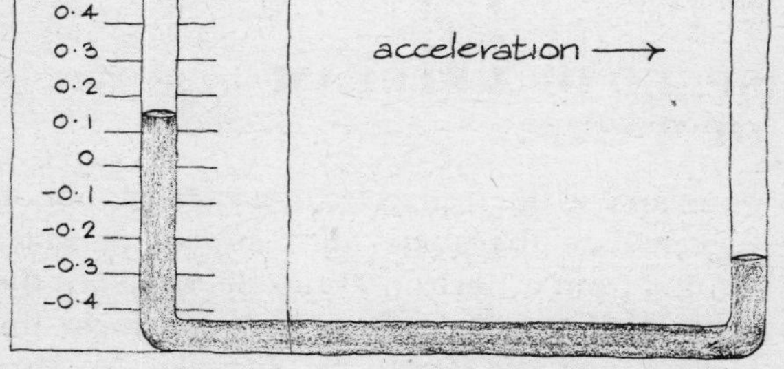

Fig. 1-31

Calibration is easiest if you make the long horizontal section of the tube just 10 cm long; then each 5 mm on a vertical arm represents an acceleration of 0.1g = (about) 1 m/sec^2, by the same reasoning as before. The two vertical arms should be at least three-fourths as long as the horizontal arm (to avoid splashing out the liquid during a quick stop). Attach a scale to one of the vertical arms, as shown. Holding the long arm horizontally, pour colored water into the tube until the water level in the arm comes up to the zero mark. How can you be sure the long arm is horizontal?

To mount your accelerometer in a car, fasten the tube with staples (carefully) to a piece of plywood or cardboard a little bigger than the U-tube. To reduce the hazard from broken glass while you do this, cover all but the scale (and the arm by it) with cloth or cardboard, but leave both ends open. It is essential that the accelerometer be horizontal if its readings are to be accurate. When you are measuring acceleration in a car, be sure the road is level. Otherwise, you will be reading the tilt of the car as well as its acceleration. When a car accelerates, in any direction, it tends to tilt on the suspension. This will introduce error in the accelerometer readings. Can you think of a way to avoid this kind of error?

C. Automobile Accelerometer. II

An accelerometer that is more directly related to $F = ma$ can be made from a 1-kg cart and a spring scale marked in newtons. The spring scale is attached between a wood frame and the cart as shown in Fig. 1-32. If the frame is kept level, the acceleration of the system can be read directly from the spring scale, since 1 N

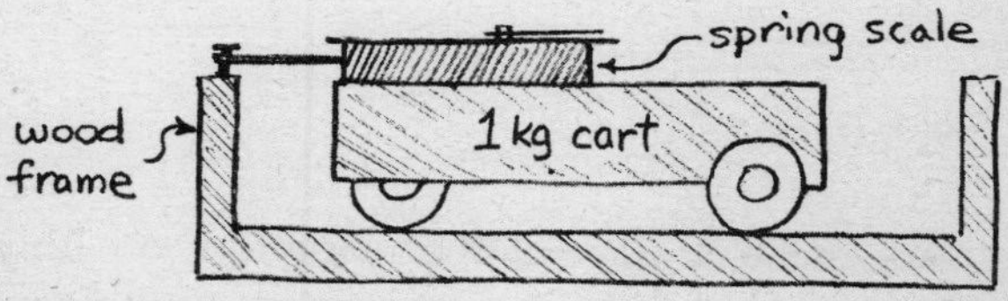

Fig. 1-32

of force on the 1-kg mass indicates an acceleration of 1 m/sec². (Instead of a cart, any 1-kg object can be used on a layer of low-friction plastic beads.)

D. Damped-Pendulum Accelerometer

One advantage of liquid-surface accelerometers is that it is easy to put a scale on them and read accelerations directly from the instrument. They have a drawback, though; they give only the component of acceleration that is parallel to their horizontal side. If you accelerate one at right angles to its axis, it does not register any acceleration at all. Also, if you do not know the direction of the acceleration, you have to use trial-and-error methods to find it with the accelerometers discussed up to this point.

A damped-pendulum accelerometer, on the other hand, indicates the direction of any horizontal acceleration; it also gives the magnitude, although less directly than the previous instruments do.

Hang a small metal pendulum bob by a short string fastened to the middle of the lid of a 1-L wide-mouthed jar as shown on the left-hand side of the sketch in Fig. 1-33. Fill the jar with water and screw the lid on tight. For any position of the pendulum, the angle that it makes with the vertical depends upon your position. What would you see, for example, if the bottle were accelerating straight toward you? Away from you? Along a table with you standing at the side? (Careful: This last question is trickier than it looks.)

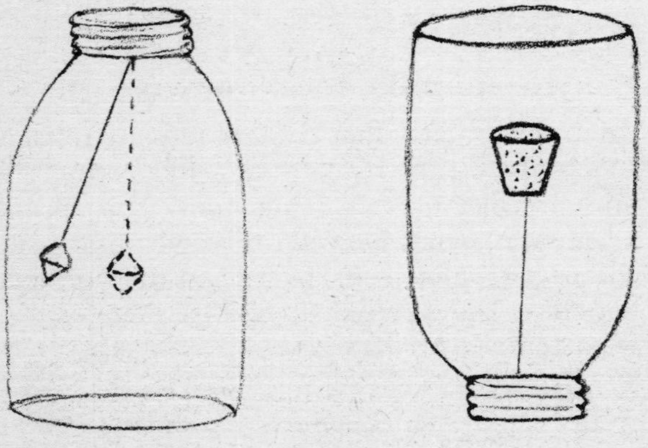

Fig. 1-33

To make a fascinating variation on the damped-pendulum accelerometer, simply replace the pendulum bob with a cork and turn the bottle upside down as shown on the right-hand side of the sketch. If you have punched a hole in the bottle lid to fasten the string, you can prevent leakage with the use of sealing wax, paraffin, or tape.

This accelerometer will do just the opposite from what you would expect. The explanation of this odd behavior is a little beyond the scope of this course; it is thoroughly explained in *The Physics Teacher*, Vol. 2, No. 4 (April, 1964), p. 176.

PROJECTILE MOTION DEMONSTRATION

Here is a simple way to demonstrate projectile motion. Place one coin near the edge of a table. Place an identical coin on the table and snap it with your finger so that it flies off the table, just ticking the first coin enough that it falls almost straight down from the edge of the table. The fact that you hear only a single ring as both coins hit shows that both coins took the same time to fall to the floor from the table. Incidentally, do the coins *have* to be identical? Try different ones.

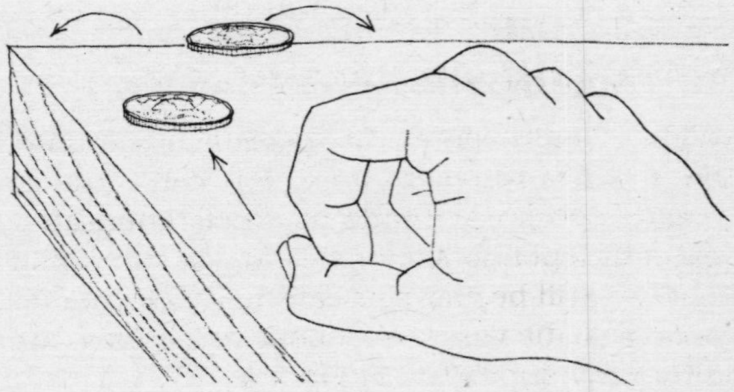

Fig. 1-34

SPEED OF A STREAM OF WATER

You can use the principles of projectile motion to calculate the speed of a stream of water issuing from a horizontal nozzle. Measure the vertical distance Δy from the nozzle to the ground, and the horizontal distance Δx from the nozzle to where the water hits the ground.

Use the equation relating Δx and Δy that was derived in Experiment 1-11, solving it for v:

$$y = \tfrac{1}{2} a_g \frac{(\Delta x)^2}{v^2}$$

so

$$v^2 = \tfrac{1}{2}a_g \frac{(\Delta x)^2}{y}$$

and

$$v = \Delta x \sqrt{\frac{a_g}{2\,\Delta y}}$$

The quantities on the right can all be measured and used to compute v.

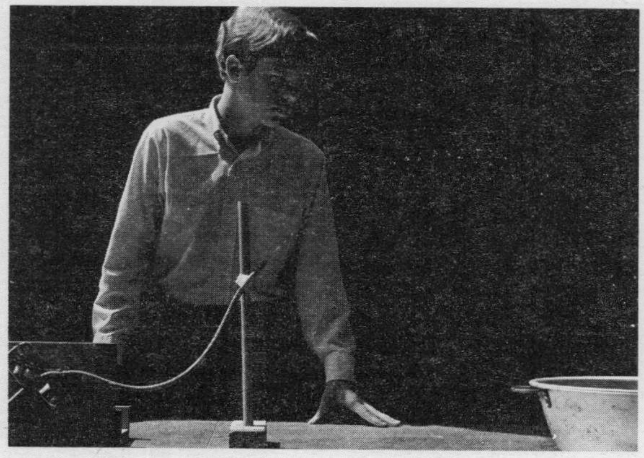

Fig. 1-35

PHOTOGRAPHING A WATERDROP PARABOLA

Using an electronic strobe light, a doorbell timer, and water from a faucet, you can photograph a waterdrop parabola. The principle of independence of vertical and horizontal motions will be clearly evident in your picture.

Remove the wooden block from the timer. Fit an "eye dropper" barrel in one end of some tubing and fit the other end of the tubing onto a water faucet. (Instead of the timer you can use a doorbell without the bell.) Place the tube through which the water runs under the clapper so that the tube is given a steady series of sharp taps. This has the effect of breaking the stream of water into separate, equally spaced drops (see Fig. 1-36).

To get more striking power, run the vibrator from a variable transformer (Variac) connected to the 110-volt ac, gradually increasing the Variac from zero just to the place where the striker vibrates against the tubing. Adjust the water flow through the tube and eye dropper nozzle. By viewing the drops with the xenon strobe set at the same frequency as the timer, a parabola of motionless drops is seen. A spotlight and disk strobe can be used instead of the electronic strobe light, but it is more difficult to match the frequencies of vibrator and strobe. The best photos are made by lighting the parabola from the side (that is, putting the light source in the plane of the parabola). Figure 1-36 was made in that way. With front lighting, the shadow of the parabola can be projected onto graph paper for more precise measurement.

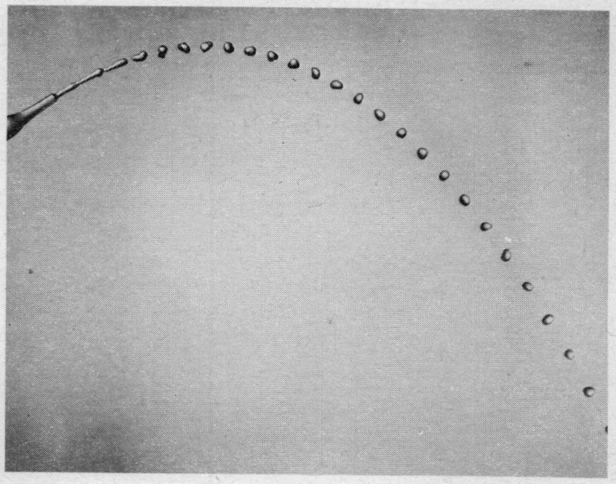

Fig. 1-36

Some heating of the doorbell coil results, so the striker should not be run continuously for long periods of time.

BALLISTIC CART PROJECTILES

Fire a projectile straight up from a cart or toy locomotive (as shown in Fig. 1-37) that is rolling across the floor with nearly uniform velocity. You can use a commercial device called a ballistic cart or make one yourself. A spring-loaded piston fires a steel ball when you pull a string attached to a trigger pin. Use the electronic strobe to photograph the path of the ball.

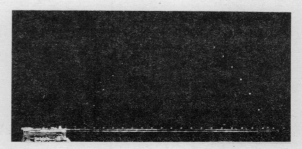

Fig. 1-37

Fig. 1-38

Projectile trajectories of any object thrown into the air can be photographed using the electronic strobe and Polaroid Land camera. By fastening the camera (securely!) to a pair of carts, you can photograph the action from a moving frame of reference.

MOTION IN A ROTATING REFERENCE FRAME

Here are three ways you can show how a moving object would appear in a rotating reference frame.

METHOD I

Attach a piece of paper to a phonograph turntable. Draw a line across the paper as the turntable is turning (see Fig. 1-39), using as a guide a meter stick supported on books at either side of the turntable. The line should be drawn at a constant speed.

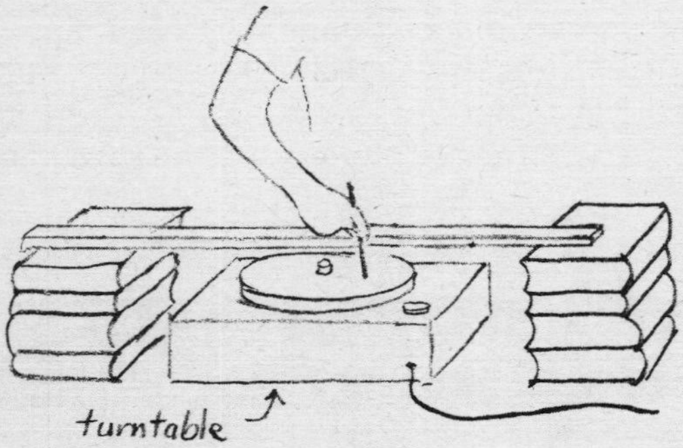

Fig. 1-39

METHOD II

Place a Polaroid Land camera on the turntable on the floor and let a Uniform Motor Device (UMD) run along the edge of a table, with a flashlight bulb on a pencil taped to the UMD so that it sticks out over the edge of the table. (See Fig. 1-40.)

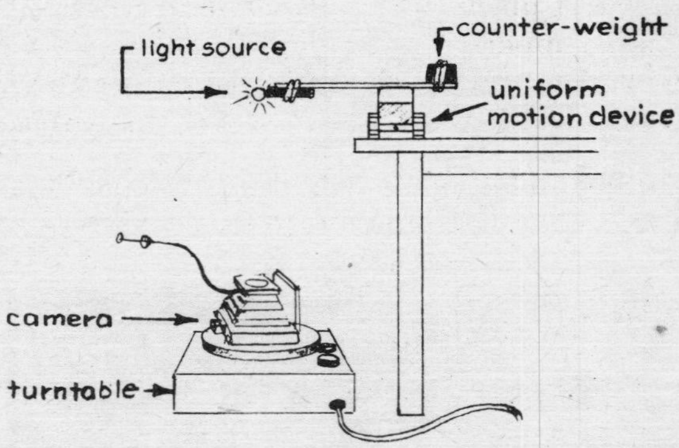

Fig. 1-40

METHOD III

How would an elliptical path appear if you were to view it from a rotating reference system? You can find out by placing a Polaroid Land camera on a turntable on the floor, with the camera aimed upwards. (See Fig. 1-41.) For a pendulum, hang a flashlight bulb and an AA dry cell. Make the pendulum long enough so that the light is about 120 cm from the camera lens.

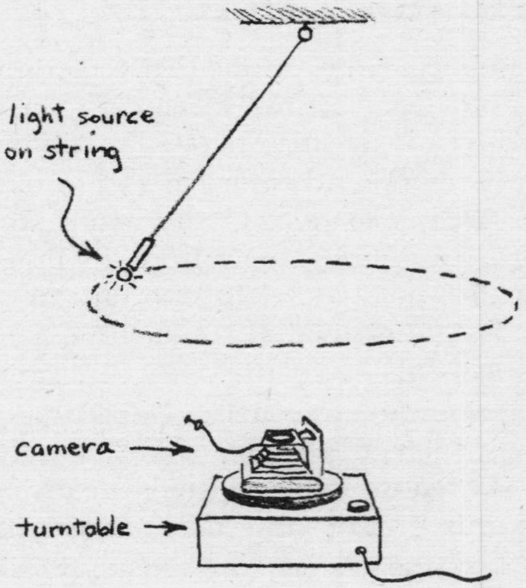

Fig. 1-41

With the lights out, give the pendulum a swing so that it swings in an elliptical path. Hold the shutter open while the turntable makes one revolution. You can get an indication of how fast the pendulum moves at different points in its swing by using a motor strobe in front of the camera, or by hanging a blinky.

PENNY AND COAT HANGER

Bend a coat hanger into the shape shown in Fig. 1-42. Bend the end of the hook slightly with a pair of pliers so that it points to where the finger supports the hanger. File the end of the hook flat. Balance a penny on the hook. Move your finger back and forth so that the hanger (and balanced penny) starts swinging like a pendulum. Some practice will enable you to swing the hanger in a vertical circle, or around your head, and still keep the penny on the hook. The centripetal force provided by the hanger keeps the penny from flying off on a straight-line path. Some people have done this demonstration successfully with a pile of as many as five pennies at once.

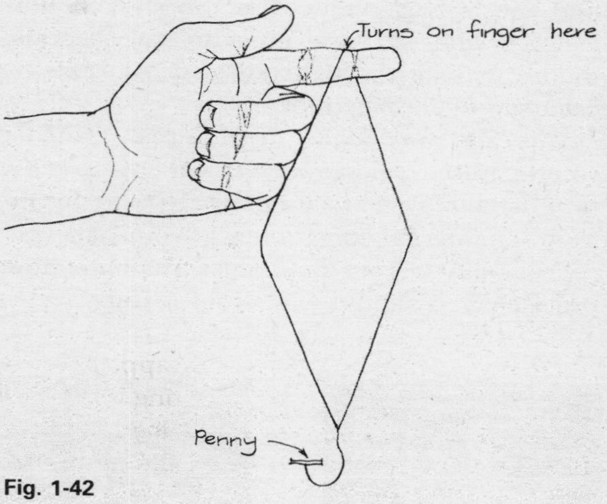

Fig. 1-42

MEASURING UNKNOWN FREQUENCIES

Use a calibrated electronic stroboscope or a hand stroboscope and stopwatch to measure the frequencies of various motions. Look for such examples as an electric fan, a doorbell clapper, and a banjo string.

On page 111 of the text you will find tables of frequencies of rotating objects. Notice the enormous range of frequencies listed, from the electron in the hydrogen atom to the rotation of the Milky Way galaxy.

FILM LOOP NOTES

Film Loop 1
ACCELERATION CAUSED BY GRAVITY. I

A bowling ball in free fall was filmed in real time and in slow motion. Using the slow-motion sequence, you can measure the acceleration of the ball as caused by gravity. This film was exposed at 3,900 frames/sec and is projected at about 18 frames/sec; therefore, the slow-motion factor is 3,900/18, or about 217. However, your projector may not run at exactly 18 frames/sec. To calibrate your projector, time the length of the entire film containing 3,331 frames. (Use the yellow circle as the zero frame.)

To find the acceleration of the falling body using the definition

$$\text{acceleration} = \frac{\text{change in speed}}{\text{time interval}}$$

you need to know the instantaneous speed at two different times. You cannot directly measure instantaneous speed from the film, but you can determine the average speed during small intervals. Suppose the speed increases steadily, as it does for freely falling bodies. During the first half of any time interval, the instantaneous speed is less than the average speed; during the second half of the interval, the speed is greater than average. Therefore, for uniformly accelerated motion, the average speed v_{av} for the interval is the same as the instantaneous speed at the midtime of the interval.

If you find the instantaneous speed at the midtimes of each of two intervals, you can calculate the acceleration a from

$$a = \frac{v_2 - v_1}{t_2 - t_1}$$

where v_1 and v_2 are the average speeds during the two intervals, and where t_1 and t_2 are the midtimes of these intervals.

Two intervals 0.5 m in length are shown in the film. The ball falls 1 m before reaching the first marked interval, so it has some initial speed when it crosses the first line. Using a watch or clock with a sweep second hand, time the ball's motion and record the times at which the ball crosses each of the four lines. You can make measurements using either the bottom edge of the ball or the top edge. With this information, you can determine the time (in apparent seconds) between the midtimes of the two intervals and the time required for the ball to move through each 0.5-m interval. Repeat these measurements at least once and then find the average times. Use the slow-motion factor to convert these times to real seconds; then, calculate the two values of v_{av}. Finally, calculate the acceleration a_g.

This film was made in Montreal, Canada, where the acceleration caused by gravity, rounded off to ±1%, is 9.8 m/sec^2. Try to decide from the internal consistency of your data (the repeatability of your time measurements) how precisely you should write your result.

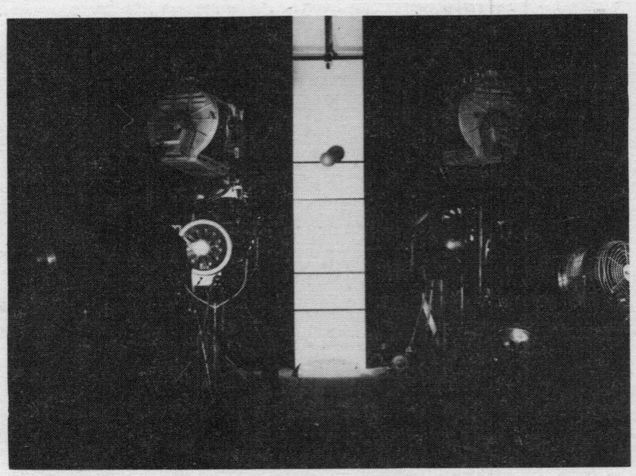

Fig. 1-43

Film Loop 2
ACCELERATION CAUSED BY GRAVITY. II

A bowling ball in free fall was filmed in slow motion. The film was exposed at 3,415 frames/sec and it is projected at about 18 frames/sec. You can calibrate your projector by timing the length of the entire film, 3,753 frames. (Use the yellow circle as a reference mark.)

If the ball starts from rest and steadily acquires a speed v after falling through a distance d, the change in speed Δv is $v - 0$, or v, and the average speed is:

$$v_{av} = \frac{0 + v}{2} = \tfrac{1}{2}v$$

The time required to fall this distance is:

$$\Delta t = \frac{d}{v_{av}} = \frac{d}{\frac{1}{2}v} = \frac{2d}{v}$$

The acceleration a is given by:

$$a = \frac{\text{change of speed}}{\text{time interval}} = \frac{\Delta v}{\Delta t} = \frac{v}{2d/v} = \frac{v^2}{2d}$$

Thus, if you know the instantaneous speed v of the falling body at a distance d below the starting point, you can find the acceleration. Of course, you cannot directly measure the instantaneous speed but only average speed over the interval. For a small interval, however, you can make the approximation that the average speed is the instantaneous speed at the midpoint of the interval. (The average speed is the instantaneous speed at the mid*time*, not the mid*point*; but the error is small if you use a short enough interval.)

In the film, small intervals of 20 cm are centered on positions 1 m, 2 m, 3 m, and 4 m below the starting point. Determine four average speeds by timing the ball's motion across the 20-cm intervals. Repeat the measurements several times and average out errors of measurement. Convert your measured times into real times using the slow-motion factor. Compute the speeds, in meters per second, and compute the value of $v^2/2d$ for each value of d.

Make a table of calculated values of a, in order of increasing values of d. Is there any evidence for a systematic trend in the values? Would you expect any such trend? State the results by giving an average value of the acceleration and an estimate of the possible error. This error estimate is a matter of judgment based on the consistency of your four measured values of the acceleration.

Film Loop 3
VECTOR ADDITION: VELOCITY OF A BOAT

A motorboat was photographed from a bridge in this film. The boat heads upstream, then downstream, then directly across stream, and at an angle across the stream. The operator of the boat tried to keep the throttle at a constant setting to maintain a steady speed relative to the water. Your task is to find out if he succeeded.

First project the film on graph paper and mark the lines along which the boat's image

Fig. 1-44 This photograph was taken from one bank of the stream. It shows the motorboat heading across the stream and the camera filming this loop fixed on the scaffolding on the bridge.

moves. You may need to use the reference crosses on the markers. Then measure speeds by timing the motion through some predetermined number of squares. Repeat each measurement several times, and use the average times to calculate speeds. Express all speeds in the same unit, such as "squares per second" (or "squares per centimeter", where centimeter refers to measured separations between marks on the moving paper of a dragstrip recorder). Why is there no need to convert the speeds to meters per second? Why is it a good idea to use a large distance between the timing marks on the graph paper?

The head-to-tail method of adding vectors is illustrated in Fig. 1-45. Since velocity is a vector with both magnitude and direction, you can study vector addition by using velocity vectors. An easy way of keeping track of the velocity vectors is by using subscripts:

$\vec{v}_{BE}$ velocity of boat relative to earth

$\vec{v}_{BW}$ velocity of boat relative to water

$\vec{v}_{WE}$ velocity of water relative to earth

Then,

$$\vec{v}_{BE} = \vec{v}_{BW} + \vec{v}_{WE}$$

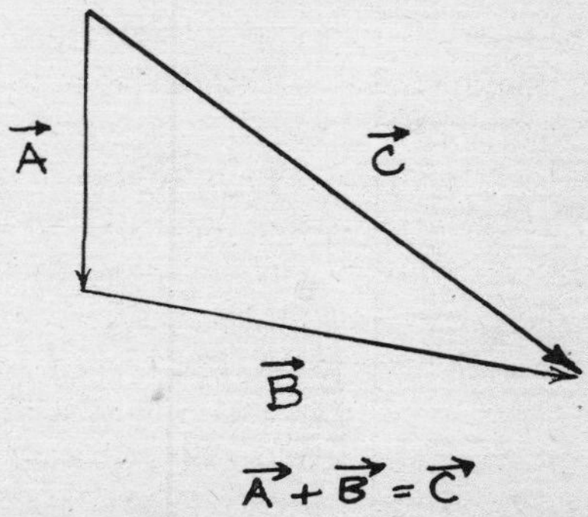

Fig. 1-45 The head-to-tail method of adding vectors.

For each heading of the boat, a vector diagram can be drawn by drawing the velocities to scale. A suggested procedure is to record data (direction and speed) for each of the five scenes in the film, and then draw the vector diagram for each.

Scene 1

Two blocks of wood are dropped overboard. Time the blocks. Find the speed of the river, the magnitude of $\vec{v}_{WE}$.

Scene 2

The boat heads upstream. Measure $\vec{v}_{BE}$, then find $\vec{v}_{BW}$ using a vector diagram similar to Fig. 1-46.

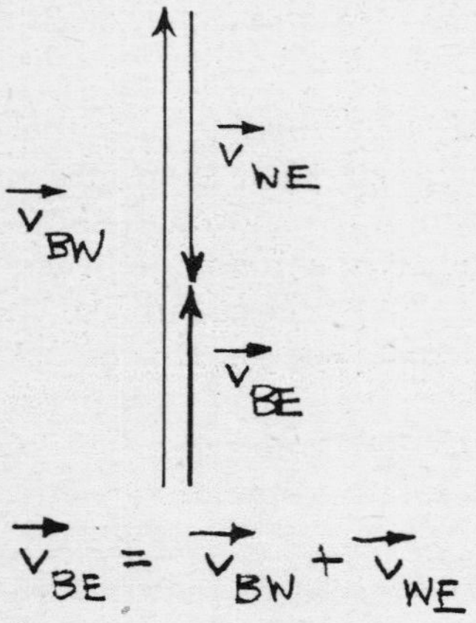

Fig. 1-46

Scene 3

The boat heads downstream. Measure $\vec{v}_{BE}$, then find $\vec{v}_{BW}$ using a vector diagram.

Scene 4

The boat heads across stream and drifts downstream. Measure the speed of the boat and the direction of its path to find $\vec{v}_{BE}$. Also measure the *direction* of $\vec{v}_{BW}$, the direction the boat points. One way to record data is to use a set of axes with the 0°–180° axis passing through the markers anchored in the river. A diagram, such as Fig. 1-47, will help you record and analyze your measurements. (Note that the numbers in the diagram are deliberately not correct.) Your vector diagram should be something like Fig. 1-48.

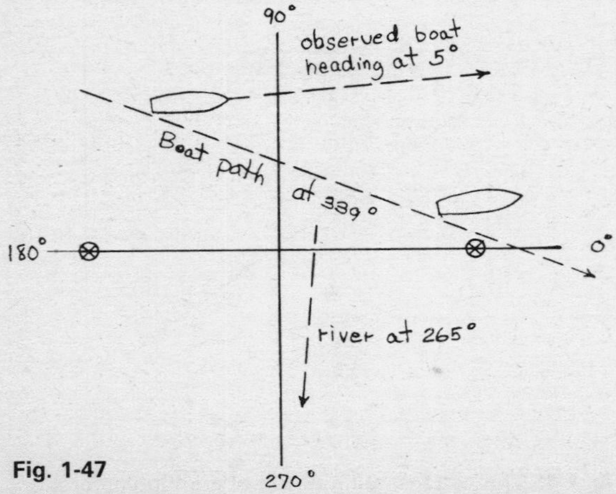

Fig. 1-47

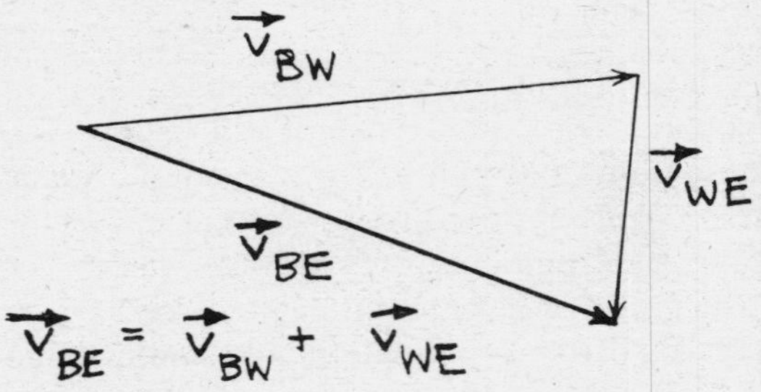

Fig. 1-48

Scene 5

The boat heads upstream at an angle, but moves directly across stream. Again find a value for $\vec{v}_{BW}$.

Checking Your Work

(a) How well do the four values of the magnitude of $\vec{v}_{BW}$ agree with each other? Can you suggest reasons for any discrepancies? **(b)** From Scene 4, you can *calculate* the heading of the boat. How well does this angle agree with the *observed* boat heading? **(c)** In Scene 5, you determined a direction for $\vec{v}_{BW}$. Does this angle agree with the observed boat heading?

Film Loop 4
A MATTER OF RELATIVE MOTION

Two carts of equal mass collide in this film. Three sequences labeled Event A, Event B, and Event C are shown. Stop the projector after *each* event and describe these events in words, as they appear to you. View the loop now, before reading further.

Even though Events A, B, and C are visibly different to the observer, in each the carts interact similarly. The laws of motion apply for each case. Thus, these events could be the *same* event observed from different reference frames. They *are* closely similar events photographed from different frames of reference, as you see after the initial sequence of the film.

The three events are photographed by a camera on a cart, which is on a second ramp parallel to the one on which the colliding carts move. The camera is your frame of reference, your coordinate system. This frame of reference may or may not be in motion with respect to the ramp. As photographed, the three events appear to be quite different. Do such concepts as position and velocity have a meaning independently of a frame of reference, or do they take on a precise meaning only when a frame of reference is specified? Are these three events really similar events, viewed from different frames of reference?

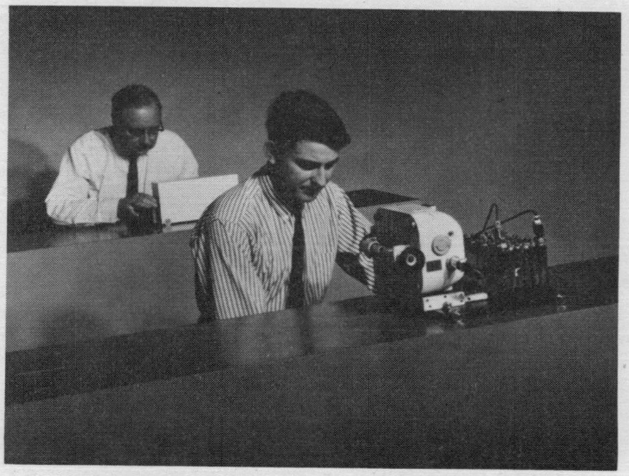

You might think that the question of which cart is in motion is resolved by sequences at the end of the film in which an experimenter, Franklin Miller of Kenyon College, stands near the ramp to provide a reference object. Other visual clues may already have provided this information. The events may appear different when this reference object is present. But is this *fixed* frame of reference any more fundamental than one of the moving frames of reference? Fixed relative to what? Is there a "completely" fixed frame of reference?

If you have studied the concept of momentum, you can also consider each of these three events from the standpoint of momentum conservation. Does the total momentum depend on the frame of reference? Does it seem reasonable to assume that the carts would have the same *mass* in all the frames of reference used in the film?

Film Loop 5
GALILEAN RELATIVITY: BALL DROPPED FROM MAST OF SHIP

This film is a partial actualization of an experiment described by Sagredo in Galileo's *Two New Sciences:*

If it be true that the impetus with which the ship moves remains indelibly impressed in the stone after it is let fall from the mast; and if it be further true that this motion brings no impediment or retardment to the motion directly downwards natural to the stone, then there ought to ensue an effect of a very wondrous nature. Suppose a ship stands still, and the time of the falling of a stone from the mast's round top to the deck is two beats of the pulse. Then afterwards have the ship under sail and let the same stone depart from the same place. According to what has been premised, it shall take up the time of two pulses in its fall, in which time the ship will have gone, say, twenty yards. The true motion of the stone will then be a transverse line (i.e., a curved line in the vertical plane), considerably longer than the first straight and perpendicular line, the height of the mast, and yet nevertheless the stone will have passed it in the same time. Increase the ship's velocity as much as you will, the falling stone shall describe its transverse lines still longer and longer and yet shall pass them all in those selfsame two pulses.

In the film a ball is dropped three times:

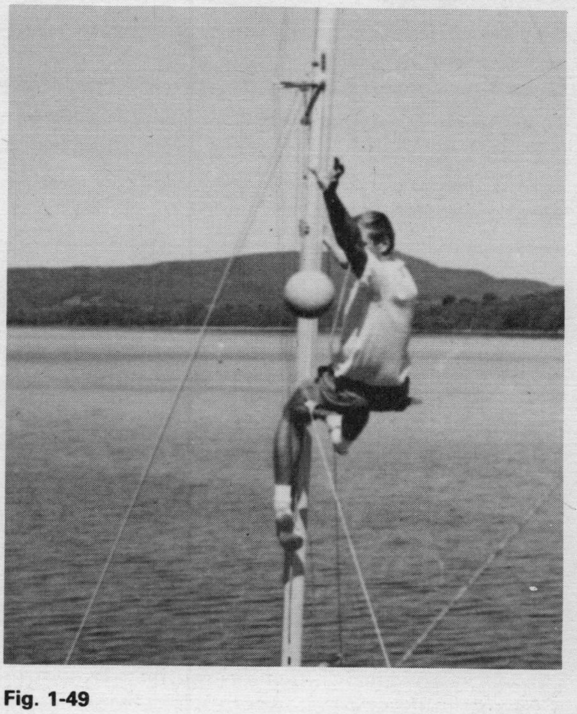

Fig. 1-49

Scene 1

The ball is dropped from the mast. As in Galileo's discussion, the ball continues to move horizontally with the boat's velocity, and also it falls vertically relative to the mast.

Scene 2

The ball is tipped off a stationary support as the boat goes by. It has no forward velocity, and it falls vertically relative to the water surface.

Scene 3

The ball is picked up and held briefly before being released.

The ship and earth are frames of reference in constant relative motion. Each of the three events can be described as viewed in either frame of reference. The laws of motion apply for all six descriptions. The fact that the laws of motion work for *both* frames of reference, one moving at constant velocity with respect to the other, is what is meant by *Galilean relativity.* (The positions and velocities are relative to the frame of reference, but the *laws of motion* are not. A "relativity" principle also states what is *not* relative.)

Scene 1 can be described from the boat frame as follows: "A ball, initially at rest, is released. It accelerates downward at 9.8 m/sec^2 and strikes a point directly beneath the starting point." Scene 1 described differently from the earth frame is: "A ball is projected horizontally toward the left; its path is a parabola and it strikes a point below and to the left of the starting point."

To test your understanding of Galilean relativity, you should describe the following: Scene 2 from the boat frame; Scene 2 from the earth frame; Scene 3 from the boat frame; Scene 3 from the earth frame.

Film Loop 6
GALILEAN RELATIVITY: OBJECT DROPPED FROM AIRCRAFT

A Cessna 150 aircraft 7 m long is moving about 30 m/sec at an altitude of about 60 m. The action is filmed from the ground as a flare is dropped from the aircraft. Scene 1 shows part of the flare's motion; Scene 2, shot from a greater distance, shows several flares dropping into a lake; Scene 3 shows the vertical motion viewed head-on. Certain frames of the film are "frozen" to allow measurements. The time interval between freeze frames is always the same.

Seen from the earth's frame of reference, the motion is that of a projectile whose original velocity is the plane's velocity. If gravity is the only force acting on the flare, its motion should be a parabola. (Can you check this?) Relative to the airplane, the motion is that of a body falling freely from rest. In the frame of reference of the plane, the motion is vertically downward.

Fig. 1-50

The plane is flying approximately at uniform speed in a straight line, but its path is not necessarily a horizontal line. The flare starts with the plane's velocity, in both magnitude and in direction. Since it also falls freely under the action of gravity, you expect the flare's downward displacement below the plane to be $d = \frac{1}{2}at^2$. The trouble is that you cannot be sure that the first freeze frame occurs at the very instant the flare is dropped. However, there is a way of getting around this difficulty. Suppose a time B has elapsed between the release of the flare and the first freeze frame. This time must be added to each of the freeze frame times (conveniently measured from the first freeze frame) and so you would have

$$d = \tfrac{1}{2}a(t + B)^2$$

To see if the flare follows an equation such as this, take the square root of each side:

$$\sqrt{d} = (\text{constant})\,(t + B)$$

If you plot $\sqrt{d}$ against t, you expect a straight line. Moreover, if $B = 0$, this straight line will also pass through the 0-0 point.

Suggested Measurements

(a) Vertical Motion

Project Scene 1 on paper. At each freeze frame, when the motion on the screen is stopped briefly, mark the positions of the flare and of the aircraft cockpit. Measure the displacement d of the flare below the plane. Use any convenient units. The times can be taken as integers, $t = 0, 1, 2, \ldots$, designating successive freeze frames. Plot $\sqrt{d}$ versus t. Is the graph a straight line? What would be the effect of air resistance, and how would this show up in your graph? Can you detect any signs of this? Does the graph pass through the 0-0 point? Analyze Scene 2 in the same way.

(b) *Horizontal Motion*

Use another piece of graph paper with time (in intervals) plotted horizontally and displacement (in squares) plotted vertically. Using measurements from your record of the flare's path, make a graph of the two motions in Scene 2. What are the effects of air resistance on the horizontal motion? the vertical motion? Explain your findings between the effect of air friction on the horizontal and vertical motions.

(c) *Acceleration Caused by Gravity*

The "constant" in your equation, $d = (\text{constant})(t + B)^2$, is $\frac{1}{2}a$; this is the slope of the straight-line graph obtained in part (**a**). The square of the slope gives $\frac{1}{2}a$. Therefore, the acceleration is twice the square of the slope. In this way you can obtain the acceleration in squares per $(\text{interval})^2$. To convert your acceleration into meters per second^2, you can estimate the size of a "square" from the fact that the length of the plane is 7 m. The time interval in seconds between freeze frames can be found from the slow-motion factor.

Film Loop 7
GALILEAN RELATIVITY: PROJECTILE FIRED VERTICALLY

A rocket tube is mounted on bearings that leave the tube free to turn in any direction.

When the tube is hauled along the snow-covered surface of a frozen lake by a "ski-doo," the bearings allow the tube to remain pointing vertically upward in spite of some roughness of path. Equally spaced lamps along the path allow you to judge whether the ski-doo has constant velocity or whether it is accelerating. A preliminary run shows the entire scene; the setting is in the Laurentian Mountains in the Province of Quebec at dusk.

Four scenes are photographed. In each case a rocket flare is fired vertically upward. With care you can trace a record of the trajectories.

Scene 1

The ski-doo is stationary relative to the earth. How does the flare move?

Scene 2

The ski-doo moves at uniform velocity relative to the earth. Describe the motion of the flare to the earth; describe the motion of the flare relative to the ski-doo.

Scenes 3 and 4

The ski-doo's speed changes after the shot is fired. In each case describe the motion of the ski-doo and describe the flare's motion relative to the earth and relative to the ski-doo. In which cases are the motions a parabola?

Fig. 1-51

How do the events shown in this film illustrate the principle of Galilean relativity? In which frames of reference does the rocket flare behave the way you would expect it to behave in all four scenes, knowing that the force is constant, and assuming Newton's laws of motion? In which systems do Newton's laws fail to predict the correct motion in some of the scenes?

Film Loop 8
ANALYSIS OF A HURDLE RACE. I

The initial scenes in this film show a regulation hurdle race, with 1-m high hurdles spaced 9 m apart. (Judging from the number of hurdles knocked over, the competitors were of something less than Olympic caliber!) Next, a runner, Frank White, a 75-kg student at McGill University, is shown in medium slow motion (slow-motion factor 3) during a 50-m run. His time was 8.1 sec. Finally, the beginning of the run is shown in extreme slow motion (slow-motion factor 80). "Analysis of a Hurdle Race. II" has two more extreme slow-motion sequences.

Fig. 1-52

To study the runner's motion, measure the average speed for each of the 1-m intervals in the slow-motion scene. A "drag-strip" chart recorder is particularly convenient for recording the data on a single viewing of the loop. Whatever method you use for measuring time, the small but significant variations in speed will be lost in experimental uncertainty unless you work very carefully. Repeat each measurement several times.

The extreme slow-motion sequence shows the runner from 0 m to 6 m. The seat of the runner's white shorts might serve as a reference mark. (What are other reference points on the runner that could be used? Are all the reference points equally useful?) Measure the time to cover each of the distances: 0 – 1, 1 – 2, 2 – 3, 3 – 4, 4 – 5, and 5 – 6 m. Repeat the measurements several times, view the film over again, and average your results for each interval. Your accuracy might be improved by forming a grand average that combines your average with others in the class. (Should you use *all* the measurements in the class?) Calculate the average speed for each interval, and plot a graph of speed versus displacement. Draw a smooth graph through the points. Discuss any interesting features of the graph.

You might assume that the runner's legs push between the time when a foot is directly beneath his hip and the time when that foot leaves the ground. Is there any relationship between your graph of speed and the way the runner's feet push on the track?

The initial acceleration of the runner can be estimated from the time to move from the starting point to the 1-m mark. You can use a watch with a sweep second hand. Calculate the average acceleration, in meters per second^2, during this initial interval. How does this forward acceleration compare with the magnitude of the acceleration of a falling body? How much force was required to give the runner this acceleration? What was the origin of this force?

Film Loop 9
ANALYSIS OF A HURDLE RACE. II

This film loop, which is a continuation of "Analysis of a Hurdle Race. I," shows two scenes of a hurdle race photographed at a slow-motion factor of 80.

In Scene 1, the hurdler moves from 20 m to 26 m, clearing a hurdle at 23 m. In Scene 2, the runner moves from 40 m to 50 m, clearing a hurdle at 41 m and sprinting to the finish line at 50 m. Plot graphs of these motions, and discuss any interesting features. The seat of the runner's pants furnishes a convenient reference point for measurements. (See the film notes about "Analysis of a Hurdle Race. I" for further details.)

No measurement is entirely precise; measurement error is always present, and it cannot be ignored. Thus, it may be difficult to tell if the small changes in the runner's speed are significant, or are only the result of measurement uncertainties. You are in the best tradition of experimental science when you pay close attention to errors.

It is often useful to display the experimental uncertainty graphically, along with the measured or computed values.

For example, say that the dragstrip timer was used to make three different measurements of the time required for the first meter of the run: 13.7 units, 12.9 units, and 13.5 units, which gives an average time of 13.28 units. (If you wish to convert these dragstrip units to seconds, it will be easier to wait until the graph has been plotted using your units, and then add a seconds scale to the graph.) The lowest and highest values are about 0.4 units on either side of the average, so you could report the time as 13.3 ± 0.4 units. The uncertainty 0.4 is about 3% of 13.3; therefore, the percentage uncertainty in the time is 3%. If you assume that the distance was exactly 1 m, so that all the uncertainty is in the time, then the percentage uncertainty in the speed will be the same as for the time, 3%. The slow-motion speed is 100 cm/13.3 time units, which equals 7.53 cm/unit. Since 3% of 7.53 is 0.23, the speed can be reported as 7.53 ± 0.23 cm/unit. In graphing this speed value, plot a point at 7.53 and draw an *error bar*

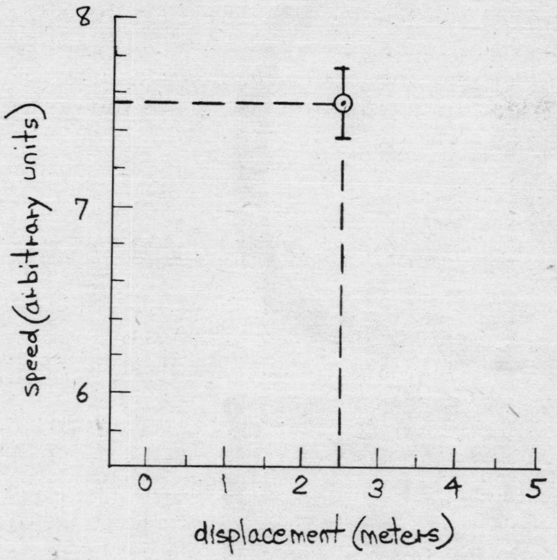

Fig. 1-53

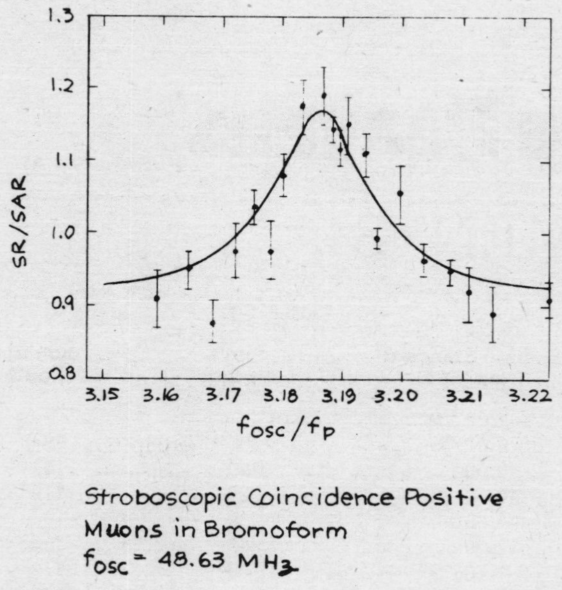

Fig. 1-54

extending 0.23 above and below the point, as shown in Fig. 1-54. Now, estimate the limit of error for a typical point on your graph and add error bars showing the range to each plotted point.

Your graph for this experiment may well look like some graphs commonly obtained in scientific research. For example, in Fig. 1-54, a research team has plotted its experimental data; they published their results in spite of the considerable scattering of plotted points and even though some of the plotted points have errors as large as 5%.

How would you represent the uncertainty in measuring *distance*, if there were significant errors here also?

Motion in the Heavens

EXPERIMENTS

Experiment 2-1
NAKED-EYE ASTRONOMY

Weather permitting, you have been watching events in the day and night sky since this course started. Perhaps you have followed the sun's path, or viewed the moon, planets, or stars.

From observations much like your own, scientists in the past developed a remarkable sequence of theories. The more aware you are of the motions in the sky and the more you interpret them yourself, the more easily you can follow the development of these theories. If you do not have your own data, you can use the results provided in the following sections.

A. One Day of Sun Observations

A student made the observations of the sun's position on September 23 as shown in Table 2-1.

If you plot altitude (vertically) against azimuth (horizontally) on a graph and mark the hours for each point, it will help you to answer the questions that follow.

TABLE 2-1

Eastern Standard Time (EST)	Sun's Altitude	Sun's Azimuth
7:00 A.M.	—	—
8:00	08°	097°
9:00	19	107
10:00	29	119
11:00	38	133
12:00	45	150
1:00 P.M.	49	172
2:00	48	197
3:00	42	217
4:00	35	232
5:00	25	246
6:00	14	257
7:00	03	267

?

1. What was the sun's greatest altitude during the day?
2. What was the latitude of the observer?
3. At what time (EST) was the sun highest?
4. When during the day was the sun's direction (azimuth) changing the fastest?
5. When during the day was the sun's altitude changing the fastest?
6. At what time of day did the sun reach its greatest altitude? How do you explain the fact that it is not exactly at 12:00?

B. A Year of Sun Observations

A student made the following monthly observations of the sun through a full year.

TABLE 2-2

Dates	Sun's Noon Altitude	Sunset Azimuth	Time Between Noon and Sunset
Jan 1	20°	238°	4h25m*
Feb 1	26	245	4 50
Mar 1	35	259	5 27
Apr 1	47	276	6 15
May 1	58	291	6 55
Jun 1	65	300	7 30
Jul 1	66	303	7 40
Aug 1	61	295	7 13
Sept 1	52	282	6 35
Oct 1	40	267	5 50
Nov 1	31	250	5 00
Dec 1	21	239	4 30

*h = hours, m = minutes.

Use these data to make three plots (different colors or marks on the same sheet of graph paper) of the sun's noon altitude, direction at sunset, and time of sunset after noon. Place these data on the vertical axis and the dates on the horizontal axis.

?

7. What was the sun's noon altitude at the equinoxes (March 21 and September 23)?
8. What was the observer's latitude?
9. If the longitude was 71°W, what city was the observer near?
10. Through what range (in degrees) did the sunset point change during the year?
11. By how much did the observer's time of sunset change during the year?
12. If the time from sunrise to noon was always the same as the time between noon and sunset, how long was the sun above the horizon on the shortest day? on the longest day?

C. Moon Observations

During October, a student in Las Vegas made the following observations of the moon at sunset when the sun had an azimuth of about 255°.

TABLE 2-3

Date	Angle from Sun to Moon	Moon Altitude	Moon Azimuth
Oct 16	032°	17°	230°
18	057	25	205
20	081	28	180
22	104	30	157
24	126	25	130
26	147	16	106
28	169	05	083

?

13. Plot these positions of the moon on a chart similar to Fig. 2-1.
14. From the data and your plot, estimate the dates of new moon, first quarter moon, and full moon.
15. For each of the points you plotted, sketch the shape of the lighted area of the moon.

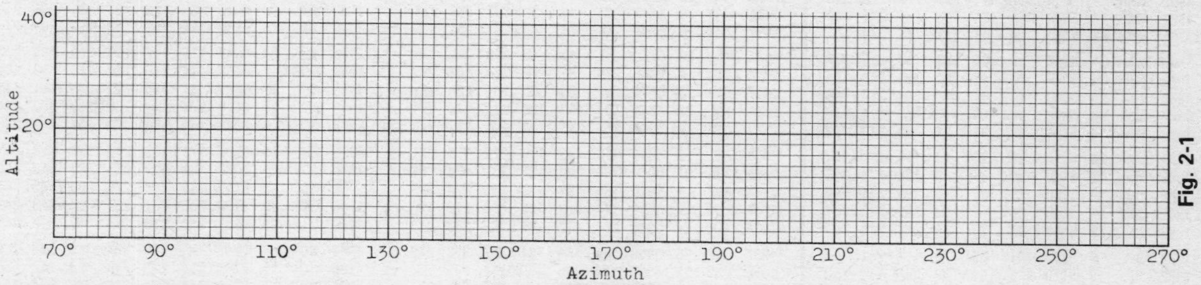

Fig. 2-1

Fig. 2-2 Phases of the moon: (1) 26 days, (2) 23 days, (3) 17 days, (4) 5 days, (5) 3 days after new moon.

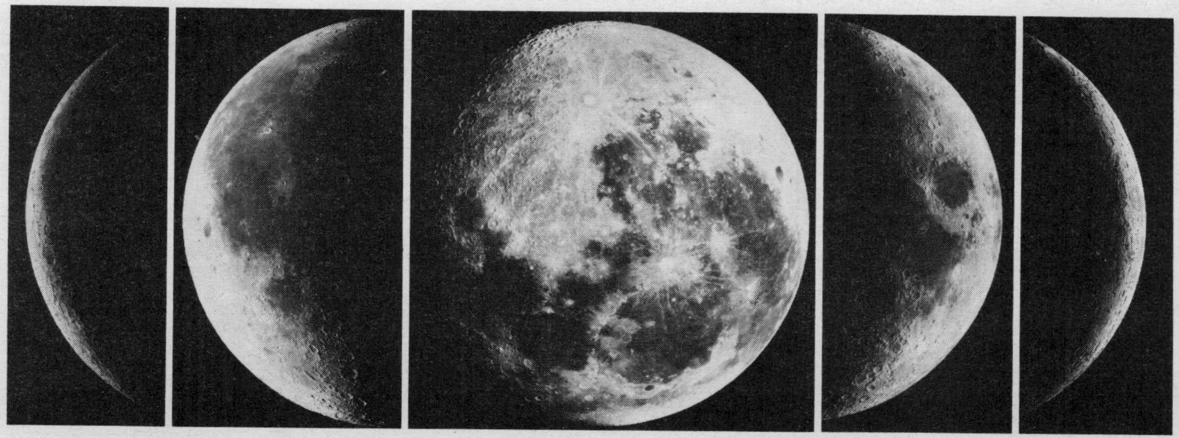

1 2 3 4 5

TABLE 2-4. PLANETARY LONGITUDES AT TEN-DAY INTERVALS

Year	Date	J.D.	Sun	Mer	Ven	Mar	Jup	Sat
1981	JAN 5	4610	285	288	262	304	189	190
1981	JAN 15	4620	295	305	275	313	190	190
1981	JAN 25	4630	305	321	288	320	190	190
1981	FEB 4	4640	316	334	300	328	190	190
1981	FEB 14	4650	326	333	313	336	190	190
1981	FEB 24	4660	336	323	325	344	189	190
1981	MAR 6	4670	346	321	338	352	188	189
1981	MAR 16	4680	356	328	350	0	187	188
1981	MAR 26	4690	6	340	3	7	185	187
1981	APR 5	4700	16	355	15	15	184	187
1981	APR 15	4710	25	12	27	23	183	186
1981	APR 25	4720	35	32	40	30	182	185
1981	MAY 5	4730	45	54	52	38	181	185
1981	MAY 15	4740	54	73	64	45	181	184
1981	MAY 25	4750	64	87	77	52	181	184
1981	JUN 4	4760	74	94	89	59	181	184
1981	JUN 14	4770	83	94	102	66	181	185
1981	JUN 24	4780	93	88	114	73	182	185
1981	JUL 4	4790	102	85	126	81	182	185
1981	JUL 14	4800	112	91	139	88	183	186
1981	JUL 24	4810	121	104	151	94	185	187
1981	AUG 3	4820	131	123	162	101	186	187
1981	AUG 13	4830	140	144	174	107	188	188
1981	AUG 23	4840	150	163	186	114	190	189
1981	SEP 2	4850	160	179	198	120	192	190
1981	SEP 12	4860	169	194	210	127	194	191
1981	SEP 22	4870	179	206	221	133	196	192
1981	OCT 2	4880	189	213	233	139	198	193
1981	OCT 12	4890	199	212	244	145	200	194
1981	OCT 22	4900	209	200	255	151	202	195
1981	NOV 1	4910	219	200	266	156	204	197
1981	NOV 11	4920	229	212	276	162	207	197
1981	NOV 21	4930	239	228	286	168	209	198
1981	DEC 1	4940	249	244	295	172	211	199
1981	DEC 11	4950	259	260	302	177	212	200
1981	DEC 21	4960	270	275	307	182	214	201
1981	DEC 31	4970	280	292	310	187	216	201
1982	JAN 10	4980	290	307	308	190	217	202
1982	JAN 20	4990	300	318	303	193	218	202
1982	JAN 30	5000	310	315	297	196	219	203
1982	FEB 9	5010	320	304	293	199	219	203
1982	FEB 19	5020	330	305	295	200	220	203
1982	MAR 1	5030	341	314	299	198	220	203
1982	MAR 11	5040	351	326	306	197	220	202
1982	MAR 21	5050	0	341	314	194	220	201
1982	MAR 31	5060	10	359	323	190	219	200
1982	APR 10	5070	20	18	324	186	218	199
1982	APR 20	5080	30	40	344	184	217	199
1982	APR 30	5090	40	59	355	182	214	198
1982	MAY 10	5100	49	71	6	180	213	197
1982	MAY 20	5110	59	75	17	180	212	197
1982	MAY 30	5120	69	72	29	182	211	197
1982	JUN 9	5130	78	66	40	185	211	197
1982	JUN 19	5140	88	67	52	189	210	197
1982	JUN 29	5150	97	75	64	193	211	197
1982	JUL 9	5160	107	89	76	197	211	197
1982	JUL 19	5170	116	109	88	202	211	198
1982	JUL 29	5180	126	131	100	208	212	198
1982	AUG 8	5190	135	150	112	214	212	199
1982	AUG 18	5200	145	166	124	219	214	199
1982	AUG 28	5210	155	181	137	225	215	200
1982	SEP 7	5220	164	191	149	232	217	201
1982	SEP 17	5230	174	198	162	239	218	202
1982	SEP 27	5240	184	194	174	245	220	203
1982	OCT 7	5250	194	184	187	252	222	204
1982	OCT 17	5260	204	185	199	259	224	205
1982	OCT 27	5270	214	199	212	267	227	207
1982	NOV 6	5280	224	215	224	274	229	208
1982	NOV 16	5290	234	232	237	282	231	209
1982	NOV 26	5300	244	248	249	289	233	210
1982	DEC 6	5310	254	263	262	297	235	211
1982	DEC 16	5320	264	279	274	305	238	212
1982	DEC 26	5330	274	294	287	313	240	213
1983	JAN 5	5340	285	303	299	321	242	213
1983	JAN 15	5350	295	297	312	329	243	214
1983	JAN 25	5360	305	287	325	336	245	214
1983	FEB 4	5370	315	290	337	344	247	215
1983	FEB 14	5380	325	300	350	352	248	215
1983	FEB 24	5390	335	313	2	359	249	215
1983	MAR 6	5400	345	328	15	7	250	215
1983	MAR 16	5410	355	346	27	15	251	215
1983	MAR 26	5420	5	5	39	23	251	214
1983	APR 5	5430	15	26	51	30	251	213
1983	APR 15	5440	25	44	63	37	251	213
1983	APR 25	5450	35	54	75	45	250	212
1983	MAY 5	5460	44	55	86	52	249	211
1983	MAY 15	5470	54	49	97	59	248	210
1983	MAY 25	5480	64	46	108	66	247	210
1983	JUN 4	5490	73	49	119	73	245	209
1983	JUN 14	5500	83	60	129	80	244	209
1983	JUN 24	5510	92	75	138	87	243	209
1983	JUL 4	5520	102	95	146	94	242	209
1983	JUL 14	5530	111	117	153	101	242	209
1983	JUL 24	5540	121	136	158	107	241	210
1983	AUG 3	5550	130	153	160	113	241	210
1983	AUG 13	5560	140	167	157	120	242	211
1983	AUG 23	5570	150	177	152	127	242	211
1983	SEP 2	5580	159	181	146	133	243	212
1983	SEP 12	5590	169	176	142	139	244	212
1983	SEP 22	5600	179	167	144	145	245	213
1983	OCT 2	5610	189	170	148	151	246	214
1983	OCT 12	5620	198	185	154	158	248	215
1983	OCT 22	5630	208	203	163	164	250	216
1983	NOV 1	5640	218	219	172	170	252	218
1983	NOV 11	5650	228	235	182	176	254	219
1983	NOV 21	5660	239	251	193	182	256	220
1983	DEC 1	5670	249	266	204	188	258	221
1983	DEC 11	5680	259	280	215	193	261	222
1983	DEC 21	5690	269	287	227	199	263	223
1983	DEC 31	5700	279	280	239	204	266	224
1984	JAN 10	5710	289	270	251	209	268	225
1984	JAN 20	5720	300	275	263	214	270	225
1984	JAN 30	5730	310	286	275	219	272	226
1984	FEB 9	5740	320	300	288	224	274	226
1984	FEB 19	5750	330	316	300	228	276	227
1984	FEB 29	5760	340	333	312	232	278	227
1984	MAR 10	5770	350	351	324	235	279	227
1984	MAR 20	5780	360	12	337	238	281	227
1984	MAR 30	5790	10	29	349	239	282	226
1984	APR 9	5800	20	37	2	239	282	226
1984	APR 19	5810	29	34	14	238	283	225
1984	APR 29	5820	39	28	26	236	283	224
1984	MAY 9	5830	49	26	39	233	283	223
1984	MAY 19	5840	58	32	51	229	283	222
1984	MAY 29	5850	68	45	63	226	282	222
1984	JUN 8	5860	78	61	76	224	281	221
1984	JUN 18	5870	87	81	88	222	280	221
1984	JUN 28	5880	97	103	100	222	279	221
1984	JUL 8	5890	106	123	113	223	277	220
1984	JUL 18	5900	116	140	125	226	276	221
1984	JUL 28	5910	125	152	137	231	274	221
1984	AUG 7	5920	135	161	150	235	274	221
1984	AUG 17	5930	144	163	162	240	273	222
1984	AUG 27	5940	154	156	174	246	273	223
1984	SEP 6	5950	164	149	187	252	273	223
1984	SEP 16	5960	174	155	199	258	273	224
1984	SEP 26	5970	183	171	211	265	274	225
1984	OCT 6	5980	193	190	223	271	275	225
1984	OCT 16	5990	203	207	236	278	276	226
1984	OCT 26	6000	213	223	248	285	277	227
1984	NOV 5	6010	223	239	260	292	279	228
1984	NOV 15	6020	233	253	272	300	281	230
1984	NOV 25	6030	243	265	284	307	283	231
1984	DEC 5	6040	253	271	296	315	285	232
1984	DEC 15	6050	264	262	307	323	287	233
1984	DEC 25	6060	274	254	319	330	289	234
1985	JAN 4	6070	284	261	330	338	292	239
1985	JAN 14	6080	294	273	341	345	294	236
1985	JAN 24	6090	304	287	352	353	297	237
1985	FEB 3	6100	315	303	1	0	299	237
1985	FEB 13	6110	325	320	10	8	303	238
1985	FEB 23	6120	335	338	17	15	304	239
1985	MAR 5	6130	345	357	21	23	304	239
1985	MAR 15	6140	355	13	23	31	308	239
1985	MAR 25	6150	5	19	20	38	310	239
1985	APR 4	6160	15	13	14	45	312	239
1985	APR 14	6170	24	7	8	52	313	238
1985	APR 24	6180	34	8	5	59	314	237
1985	MAY 4	6190	44	17	7	66	316	236
1985	MAY 14	6200	53	30	11	73	317	235
1985	MAY 24	6210	63	47	18	80	317	234
1985	JUN 3	6220	73	67	26	86	317	234
1985	JUN 13	6230	82	90	36	92	317	233
1985	JUN 23	6240	92	109	46	99	317	232
1985	JUL 3	6250	101	125	56	106	316	232
1985	JUL 13	6260	111	137	67	112	315	232
1985	JUL 23	6270	120	145	78	119	314	232
1985	AUG 2	6280	130	144	89	126	312	232
1985	AUG 12	6290	139	137	101	132	311	233
1985	AUG 22	6300	149	132	113	139	310	233
1985	SEP 1	6310	159	141	124	145	309	234
1985	SEP 11	6320	168	158	136	151	308	234
1985	SEP 21	6330	178	177	148	157	307	234
1985	OCT 1	6340	188	195	161	164	307	236
1985	OCT 11	6350	198	211	173	170	307	237
1985	OCT 21	6360	208	226	186	176	307	237
1985	OCT 31	6370	218	240	198	182	308	238
1985	NOV 10	6380	229	251	211	188	309	239
1985	NOV 20	6390	238	256	223	195	311	240
1985	NOV 30	6400	248	245	236	201	312	242
1985	DEC 10	6410	258	239	249	207	314	243
1985	DEC 20	6420	269	247	261	213	316	244
1985	DEC 30	6430	279	260	274	220	318	245
1986	JAN 9	6440	289	275	286	226	320	246
1986	JAN 19	6450	299	291	299	232	322	247
1986	JAN 29	6460	309	307	311	238	324	247
1986	FEB 8	6470	319	325	324	243	327	248
1986	FEB 18	6480	329	343	336	249	329	249
1986	FEB 28	6490	339	358	349	255	332	250
1986	MAR 10	6500	349	1	1	260	335	250
1986	MAR 20	6510	359	353	14	266	337	250
1986	MAR 30	6520	9	348	26	271	339	250
1986	APR 9	6530	19	352	39	276	341	250
1986	APR 19	6540	29	2	51	281	343	250
1986	APR 29	6550	39	16	64	286	345	250
1986	MAY 9	6560	49	33	76	290	347	249
1986	MAY 19	6570	58	53	88	293	349	248
1986	MAY 29	6580	68	76	100	294	350	247
1986	JUN 8	6590	77	95	112	295	352	246
1986	JUN 18	6600	87	110	124	295	353	245
1986	JUN 28	6610	96	122	135	294	353	244
1986	JUL 8	6620	106	126	147	291	353	244
1986	JUL 18	6630	115	123	158	285	353	244
1986	JUL 28	6640	125	117	169	293	353	243
1986	AUG 7	6650	134	116	180	282	352	243
1986	AUG 17	6660	144	126	190	282	351	244
1986	AUG 27	6670	154	144	200	283	350	244
1986	SEP 6	6680	163	164	209	285	348	244
1986	SEP 16	6690	173	182	217	288	348	245
1986	SEP 26	6700	183	199	225	293	346	246
1986	OCT 6	6710	193	214	229	298	345	246
1986	OCT 16	6720	203	227	231	304	346	247
1986	OCT 26	6730	213	236	229	310	343	248
1986	NOV 5	6740	223	239	224	316	343	249
1986	NOV 15	6750	233	228	219	323	343	250
1986	NOV 25	6760	243	224	215	329	343	251
1986	DEC 5	6770	253	233	217	336	344	253
1986	DEC 15	6780	263	247	222	343	345	253
1986	DEC 25	6790	273	263	229	350	347	255
1987	JAN 4	6800	284	279	237	357	348	256
1987	JAN 14	6810	294	295	247	4	350	257
1987	JAN 24	6820	304	312	257	11	352	258
1987	FEB 3	6830	314	329	268	18	354	258
1987	FEB 13	6840	324	342	279	25	354	259
1987	FEB 23	6850	334	343	291	32	358	260
1987	MAR 5	6860	344	333	302	39	0	261
1987	MAR 15	6870	354	330	314	46	3	261
1987	MAR 25	6880	4	336	326	52	5	262
1987	APR 4	6890	14	347	338	59	8	262
1987	APR 14	6900	24	2	350	66	11	262
1987	APR 24	6910	34	19	2	73	13	262
1987	MAY 4	6920	43	39	14	79	15	261
1987	MAY 14	6930	53	62	25	86	17	261
1987	MAY 24	6940	63	81	38	92	19	260
1987	JUN 3	6950	72	95	50	90	21	259
1987	JUN 13	6960	82	105	62	105	23	258
1987	JUN 23	6970	91	106	74	112	25	257

B.C. By John Hart

D. Locating the Planets

Table 2-4, "Planetary Longitudes," lists the position of each major planet along the ecliptic. The positions are given, accurate to the nearest degree, for every 10-day interval. By interpolation, you can find a planet's position on any given day.

The column headed "J.D." shows the corresponding Julian Day calendar date for each entry. This calendar is simply a consecutive numbering of days that have passed since an arbitrary "Julian Day 1" in 4713 B.C.: September 22, 1983, for example, is the same as J.D. 2,445,600.

Julian dates are used by astronomers for convenience. For example, the number of days between March 8 and September 26 of this year is troublesome to figure out, but it is easy to find by simple subtraction if the Julian Days are used instead.

Look up the sun's present longitude in the table. Locate the sun on your SC-1 Constellation Chart. The sun's path, the *ecliptic*, is the curved line marked off in 360 degrees of longitude.

A planet that is just to the west of the sun's position (to the right on the chart) is "ahead of the sun," that is, it rises and sets just before the sun does. One that is 180° from the sun rises near sundown and is in the sky all night.

When you have decided which planets may be visible, locate them along the ecliptic shown on your sky map SC-1. Unlike the sun, they are not exactly ecliptic, but they are never more than 8° from it. Once you have located the ecliptic on the Constellation Chart, you know where to look for a planet among the fixed stars.

E. Graphing the Position of the Planets

Here is a useful way to display the information in Table 2-4, "Planetary Longitudes." On ordinary graph paper, plot the sun's longitude versus time. Use Julian Day numbers along the horizontal axis, beginning close to the present date. The plotted points should fall on a nearly straight line, sloping up toward the right until they reach 360° and then starting again at zero.

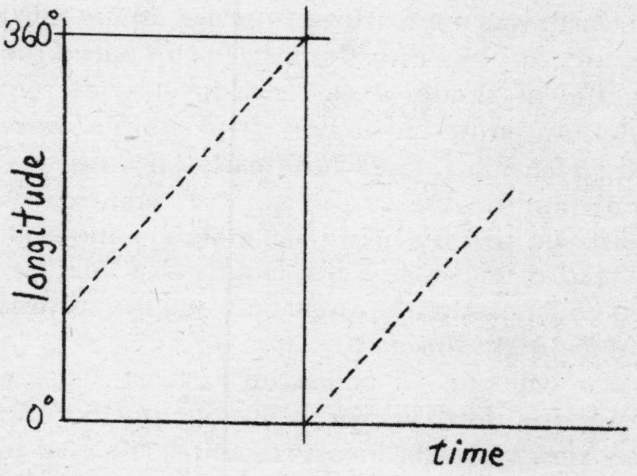

Fig. 2-3

How long will it be before the sun again has the same longitude it has today? Would the answer to that question be the same if it were asked three months from now? What is the sun's average angular speed (in degrees per day) over a whole year? When is its angular speed greatest?

Plot Mercury's longitudes on the same graph (use a different color or shape for the points). According to your plot, how far (in longitude) does Mercury get from the sun? (This is Mercury's *maximum elongation*.) At what time interval does Mercury pass between the earth and the sun?

Plot the positions of the other planets using a different color for each one. The data on the resulting chart are much like the data that puzzled the ancients. In fact, the table of longitudes is just an updated version of the tables made by Ptolemy, Copernicus, and Tycho.

The graph contains a good deal of useful information. For example, when will Mercury and Venus next be close enough to each other so that you can use bright Venus to help you find Mercury? Where are the planets, relative to the sun, when they go through their retrograde motions?

Experiment 2-2
SIZE OF THE EARTH

You probably know that the earth has a diameter of about 12,800 km and a circumference of about 40,000 km. Suppose someone challenged you to prove it? How would you go about it?

The first recorded calculation of the size of the earth was made a long time ago, in the third century B.C., by Eratosthenes. He compared the lengths of shadows cast by the sun at two different points in Egypt. The points were rather far apart, but were nearly on a north–south line on the earth's surface. The experiment you will do here uses a similar method. Instead of measuring the length of a shadow, you will measure the angle between the vertical and the sight line to a star.

You will need a colleague at least 300 km away, due north or south of your position, to take simultaneous measurements. The two of you will need to agree in advance on the star, the date, and the time for your observations. See how close you can come to calculating the actual size of the earth.

Assumptions and Theory of the Experiment

The experiment is based on the following assumptions:

1. The earth is a perfect sphere.

2. A plumb line points toward the center of the earth.

3. The distance from the earth to the stars and sun is very great compared with the earth's diameter.

The two observers must be located at points nearly north and south of each other. Suppose they are at points A and B, separated by a distance s, as shown in Fig. 2-4. The observer at A and the observer at B both sight on the same star at the prearranged time, when the star is on or near their meridian, and measure the angle between the vertical of the plumb line and the sight line to the star.

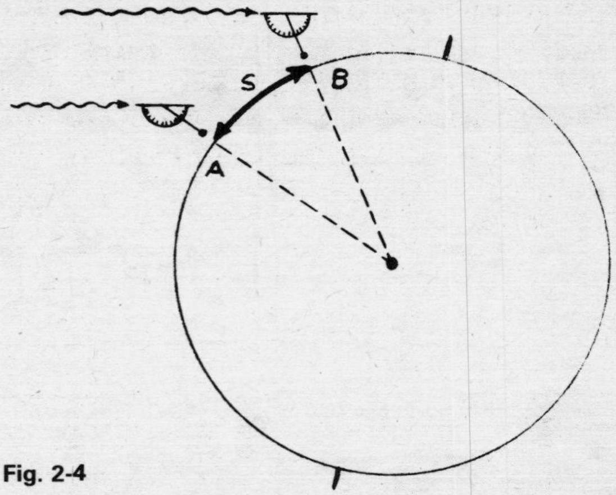

Fig. 2-4

Light rays from the star reaching locations A and B are parallel (this is implied by assumption 3).

The difference between the angle θ_A at A and the angle θ_B at B, is the angle ϕ between the two radii, as shown in Fig. 2-5.

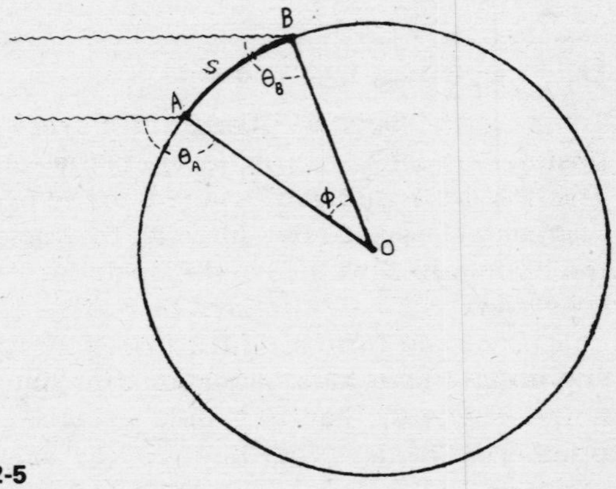

Fig. 2-5

In the triangle ABO

$$\phi = (\theta_A - \theta_B) \qquad (1)$$

If C is the circumference of the earth, and s is an arc of the meridian, you can make the proportion

$$\frac{s}{C} = \frac{\phi}{360^\circ} \qquad (2)$$

Combining equations (1) and (2), you have

$$C = \frac{360^\circ}{(\theta_A - \theta_B)} s,$$

where θ_A and θ_B are measured in degrees.

Doing the Experiment

For best results, the two locations A and B should be directly north and south of each other, and the observations should be made when the star is near its highest point in the sky.

You will need some kind of instrument to measure the angle θ. Such an instrument is called an astrolabe. If your instructor does not have an astrolabe, you can make one fairly easily from a protractor, a small sighting tube, and a weighted string assembled according to the design in Fig. 2-6.

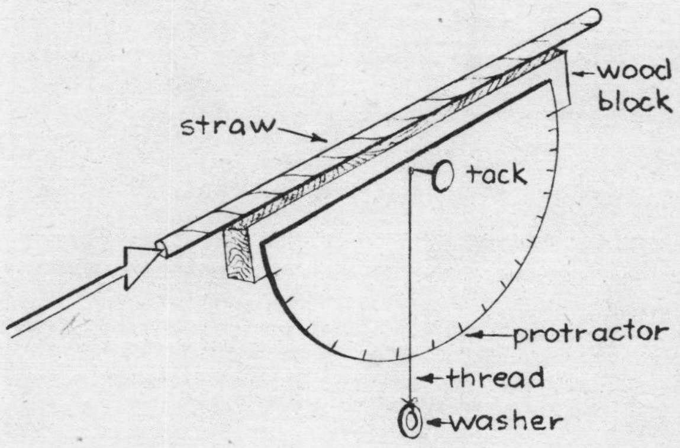

Fig. 2-6

Aim your astrolabe along the north–south line and measure the angle from the vertical to the star as it crosses the north–south line.

If the astrolabe is not aimed along the north–south line or meridian, the star will be observed before or after it is highest in the sky. An error of a few minutes from the time of crossing the meridian will make little difference in the angle measured.

B.C. By John Hart

An estimate of the uncertainty in your measurement of θ is important. Take several measurements on the same star and take the average value of θ. Use the spread in values of θ to estimate the uncertainty of your observations and of your result.

Your value for the earth's circumference depends on the over-the-earth distance between the two points of observation. You should get this distance from a map, using its scale.

?

1. How does the uncertainty of the over-the-earth distances compare with the uncertainty in your value for θ?
2. What is your calculated value for the circumference of the earth and what is the uncertainty of your value?
3. Astronomers have found that the average circumference of the earth is about 40,000 km. What is the percentage error of your result?
4. Is this acceptable, in terms of the uncertainty of your measurement?

The Size of the Earth; Simplified Version

Perhaps, for lack of a distant colleague, you were unable to determine the size of the earth as described above. You may still do so if you measure the maximum altitude of one of the objects on the following list and then use the attached data as described below.

In Santiago, Chile, Maritza Campusano Reyes made the following observations of the maximum altitude of stars and of the sun (all were observed *north* of her zenith):

Antares (Alpha Scorpio)		83.0°
Vega (Alpha Lyra)		17.5
Deneb (Alpha Cygnus)		11.5
Altair (Alpha Aquila)		47.5
Fomalhaut (Alpha Pisces Austr.)		86.5
Sun: October 1	59.4°	
15	64.8°	
November 1	70.7°	
15	74.8°	

Since Ms. Reyes made her observations when the objects were highest in the sky, the values depend only upon her latitude and not upon her longitude or the time at which the observations were made.

From a world atlas, find how far north you are from Santiago. Next, measure the maximum altitude of one or more of these objects at your location. Then calculate a value for the circumference of the earth.

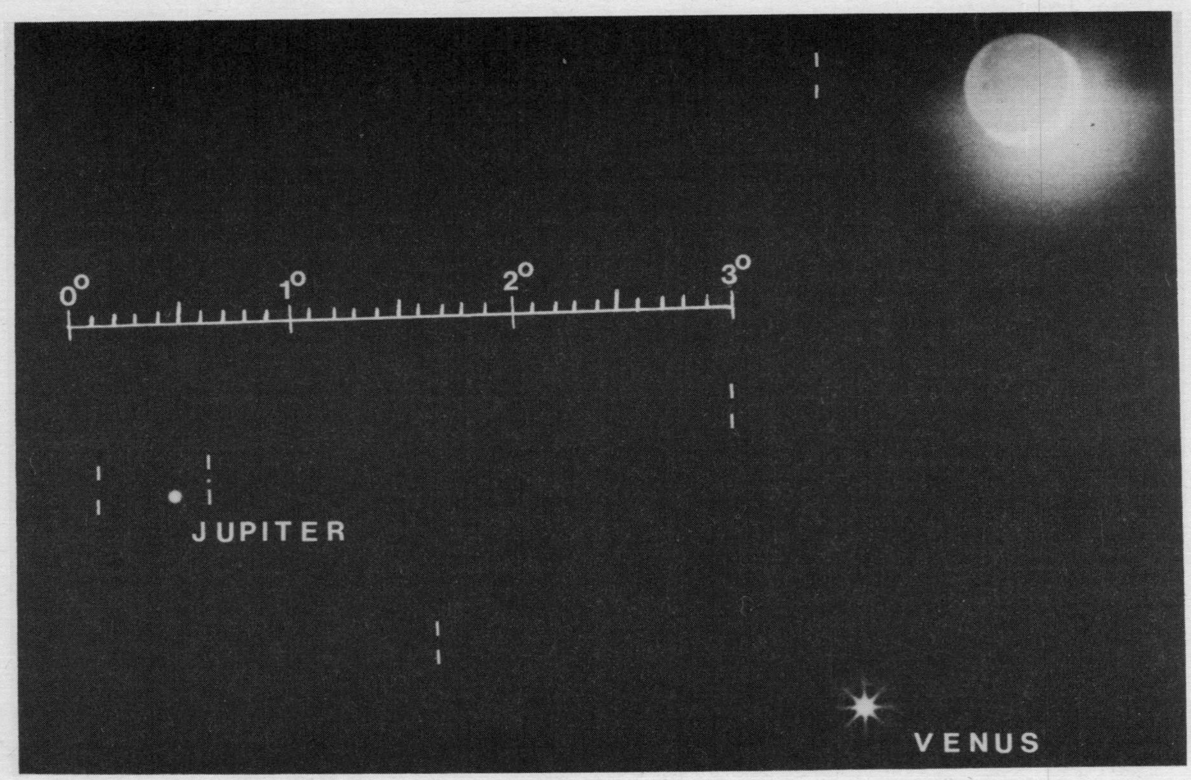

Fig. 2-7 Photographed by Kenneth R. Polley, 6:00 P.M. CST on December 27, 1973, at Finley Air Force Station, North Dakota (Lat. 47.5°N, Long. 97.9°W). Exposure: 4 sec at f5.6, 135 mm.

Fig. 2-8 Photographed by David Farley, 6:00 P.M. CST on December 27, 1973, at Starkville, Mississippi (Lat. 33.5°N, Long. 88.7°W). Exposure: 4 sec at f5.6, 105 mm.

Experiment 2-3
THE DISTANCE TO THE MOON

The moon is so near the earth that two widely separated observers see the moon in different positions against the background of fixed stars. (If you hold your thumb at arm's length and view it first with one eye and then the other, the apparent position of your thumb will also shift against the distant background.) This shift in apparent position from the two ends of the baseline is an angle called *parallax*.

?

1. If the object is moved farther away, will the parallax angle become larger or smaller?
2. How will the parallax angle change if the baseline is made longer? shorter?

On December 27, 1973, the moon and the planets Venus and Jupiter were close together in the sky. Simultaneous photographs of the objects were taken by two amateur astronomers in North Dakota and Mississippi (Figs. 2-7 and 2-8). The photographers were 1,857 km apart.

The moon was a thin crescent facing the sun; the remainder of the moon's disk was illuminated by earthshine. The "star points" of Venus were caused by internal reflections in the camera.

At first glance the two pictures appear identical, but notice the difference in the moon's position relative to the nearby star located between two short vertical lines. The apparent shift of the moon's position for these two observers can be found by tracing on thin paper or plastic the image of the moon, the centers of the images of Venus and Jupiter, and a star or two. Then match the tracing over the second picture.

?

3. Why should you measure the displacement of several matching points on the two pictures of the moon?
4. How do you know that the apparent position of the moon has changed and not the positions of Venus and Jupiter?

Record the parallax angle in decimal degrees from the scale. Now you can use your own measurements of the parallax angle to obtain your distance to the moon in kilometers. How do your results compare with those of others?

The linear diameter of the moon can also be found from measurements of its angular diameter on the photographs. In this geometrical relation, the moon's diameter is now the baseline.

TABLE 2-5. TABLE FOR CONVERTING SMALL ANGLES TO LENGTHS

$\theta°$		$\sin\theta$
0.00		0.00000
0.05		.00087
0.10		.00175
0.15		.00262
0.20		.00349
0.25		.00436
0.30		.00524
0.35		.00611
0.40		.00698
0.45		.00785
0.50		.00873
$\Delta 0.01°$	=	0.000175

?

5. What value do you compute for the diameter of the moon?
6. How many significant figures should you report?
7. If another object, such as the sun, a comet, or a star shows only a very small or no parallax, what may you conclude about its distance?
8. Several approximations have been made in this analysis. How does each of the following affect your estimate of the distance to the moon?
   a. The baseline is not quite perpendicular to the direction of the moon.
   b. The over-land distance between observers is not the shortest distance between the observers.
   c. The moon's lighted crescent was overexposed.
   d. The observer in Mississippi saw the moon several degrees higher in the sky than did the observer in North Dakota.

Experiment 2-4
THE HEIGHT OF PITON, A MOUNTAIN ON THE MOON

Closeup photographs of the moon's surface have been brought back to earth by the Apollo astronauts (Fig. 2-9). Scientists are discovering a great deal about the moon from such photographs, as well as from the landings made by astronauts in Apollo spacecraft.

But long before the Space Age, indeed since Galileo's time, astronomers have been learning about the moon's surface. In this experiment, you will use a photograph (Fig. 2-10) taken with a large telescope in California to estimate the height of a mountain on the moon. You will use a method similar to that used by Galileo, although you should be able to get a more accurate value than he could working with his small telescope (and without photographs!).

The photograph of the moon in Fig. 2-9 was taken at the Lick Observatory very near the time of the third quarter. The photograph does not show the moon as you see it in the sky at third quarter because an astronomical telescope gives an inverted image, reversing top-and-bottom and left-and-right. (Thus, north is at the bottom.) Figure 2-18 is a 10× enlargement of the area within the white rectangle in Fig. 2-9.

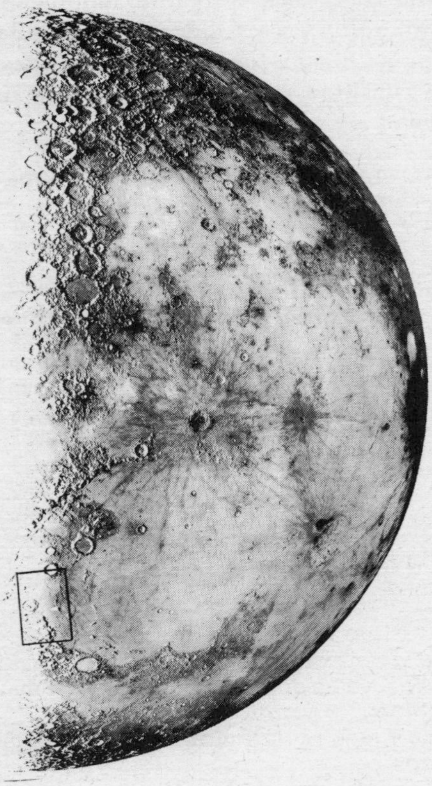

Fig. 2-9

Fig. 2-10

Why Choose Piton?

Piton, a mountain in the moon's northern hemisphere, is a slab-like pinnacle in a fairly flat area. When the photograph was made, with the moon near third-quarter phase, Piton was quite close to the line separating the lighted portion from the darkened portion of the moon. (This line is called the *terminator.*)

Assumptions and Relations

Figure 2-11 represents the third-quarter moon of radius r, with Piton P, its shadow of length l, at a distance d from the terminator.

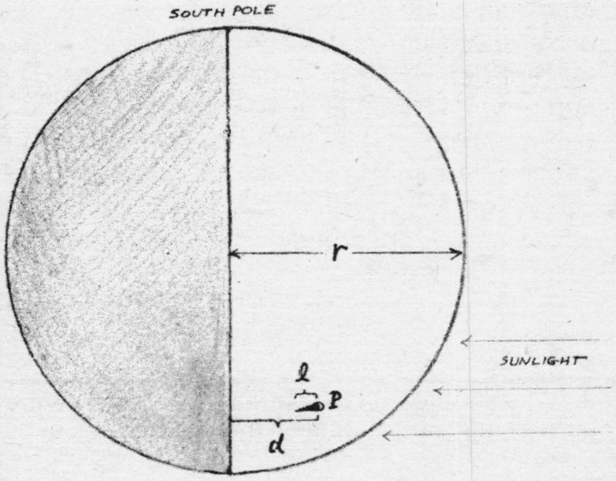

Fig. 2-11

The rays of light from the sun can be considered to be parallel because the moon is a great distance from the sun. Therefore, the angle at which the sun's rays strike Piton will not change if, in imagination, you rotate the moon on an axis that points toward the sun. In

Fig. 2-12, the moon has been rotated to put Piton on the lower edge. In this position, it is easier to work out the geometry of the shadow.

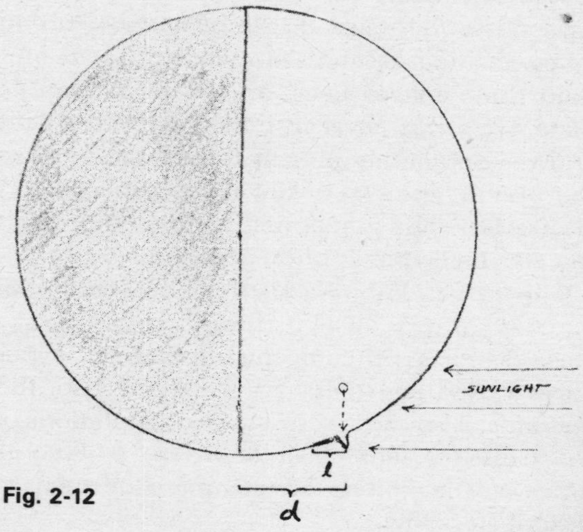

Fig. 2-12

Figure 2-13 shows how the height of Piton can be found from similar triangles; h represents the height of the mountain, l is the apparent length of its shadow, d is the distance of the mountain from the terminator, r is a radius of the moon (drawn from Piton at P to the center of the moon's outline at O).

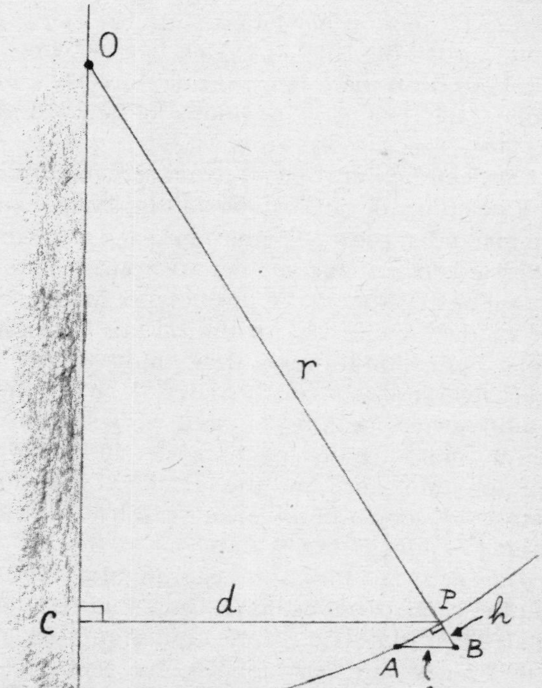

Fig. 2-13

It can be proven geometrically (and you can see from the drawing) that the small triangle BPA is similar to the large triangle PCO. The corresponding sides of similar triangles are proportional, so you can write

$$\frac{h}{l} = \frac{d}{r} \text{ then, } h = \frac{l \times d}{r}$$

All of the quantities on the right can be measured from the photograph.

The curvature of the moon's surface introduces some error into the calculations, but as long as the height and shadow are small compared to the size of the moon, the error is not great.

Fig. 2-14 A 130-km² area of the moon's surface near the large crater, Goclenius. An unusual feature of this crater is the prominent rille that crosses the crater rim.

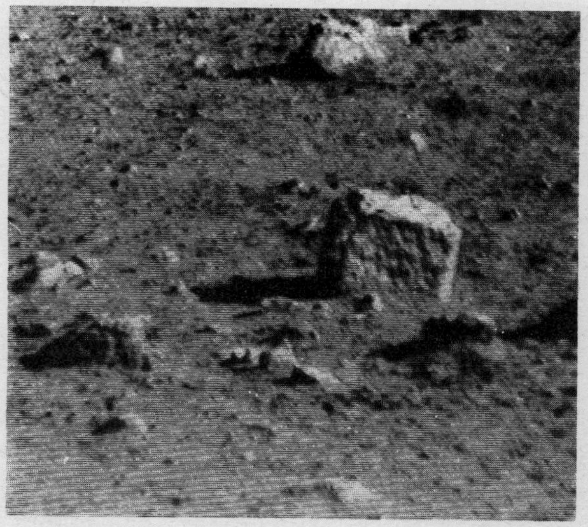

Fig. 2-15 A 10-cm rock photographed on the lunar surface.

Measurements and Calculations

Unless you are instructed otherwise, you should work on a tracing of the moon picture rather than in the book itself. Trace the outline of the moon and the location of Piton. If the photograph was made when the moon was exactly at third-quarter phase, then the moon was divided exactly in half by the terminator. The terminator appears ragged because highlands cast shadows across the lighted side and peaks stick up out of the shadow side. Estimate the best overall straight line for the terminator and draw it on your tracing. Use a millimeter scale to measure the length of Piton's shadow and the distance from the terminator to Piton's peak.

It probably will be easiest for you to do all the calculations in the scale of the photograph. Find the height of Piton in centimeters, and then change to the real scale of the moon.

?

1. How high is Piton in centimeters on the photograph scale?
2. The diameter of the moon is 3,476 km. What is the scale of the photograph?
3. What value do you get for the actual height of Piton?
4. Which of your measurements is the least certain? What is your estimate of the uncertainty of your height for Piton?
5. Astronomers, using more complicated methods than you used, find Piton to be about 2.3 km high (and about 22 km across at its base). Does your value differ from the accepted value by more than your experimental uncertainty? If so, can you suggest why?

Experiment 2-5
RETROGRADE MOTION

The filmstrip you will use in this experiment presents photographs of the positions of Mars, from the files of the Harvard College Observatory, for three oppositions of Mars, in 1941, 1943, and 1946. The first series of 12 frames shows the positions of Mars before and after the opposition of October 10, 1941. The series begins with a photograph on August 3, 1941, and ends with one on December 6, 1941. The second series shows positions of Mars before and after the opposition of December 5, 1943. This second series of seven photographs begins on October 28, 1943, and ends on February 19, 1944. The third set of 11 pictures, which shows Mars during 1945–46, around the opposition of January 14, 1946, begins with October 16, 1945, and ends with February 23, 1946. Jupiter also shows in the second and third series.

The photographs were taken by the routine Harvard Sky Patrol with a camera of 15-cm focal length and a field of 55°. During each exposure, the camera was driven by a clockwork to follow the daily western motion of the stars and hold their images fixed on the photographic plate. Mars was never in the center of the field and was sometimes almost at the edge because the photographs were not made especially to show Mars. The planet just happened to be in the star fields being photographed.

The images of the stars and planets are not of equal brightness on all pictures because the sky was less clear on some nights and the exposures varied somewhat in duration. Also, the star images show distortions from limitations of the camera's lens. Despite these limitations, however, the pictures are adequate for the uses described below.

Some of the frames show beautiful pictures of the Milky Way in Taurus (1943) and Gemini (1945).

Using the Filmstrip

1. The star fields for each series of frames have been carefully positioned so that the star positions are nearly identical. If the frames of each series are shown in rapid succession, the stars will be seen as stationary on the screen, while the motion of Mars among the stars is quite apparent. This would be like viewing a flip-book. Run the frames through quickly and notice the changing positions of Mars and Jupiter.

2. Project the frames on a paper screen where the positions of various stars and of Mars can be marked. If the star pattern for each frame is adjusted to match that plotted from the first frame of that series, the positions of Mars can be marked accurately for the various dates. A continuous line through these points will be a track for Mars. Estimate the dates of the turning points, when Mars begins and ends its retrograde motion. By using the scale (10°) shown on one frame, also find the angular size of the retrograde loop. Compare your results with the average values in the Text, page 140, Unit 2.

During 1943–1944 and again in 1945–1946, Jupiter came to opposition several months later than Mars did. As a result, Jupiter appears in the frames and also shows its retrograde motion. Jupiter's oppositions were: January 11, 1943; February 11, 1944; March 13, 1945; and April 13, 1946.

Track Jupiter's position and find the duration and size of its retrograde loop. Compare your results with the average values listed in the Text. This is the type of observational information that Ptolemy, Copernicus, and Kepler attempted to explain by their theories.

Experiment 2-6
THE SHAPE OF THE EARTH'S ORBIT

Ptolemy and most of the Greeks thought that the sun revolved around the earth. But after the time of Copernicus, the idea gradually became accepted that the earth and other planets revolve around the sun. Although you probably believe the Copernican model, the evidence of your senses gives you no reason to prefer one model over the other.

With your unaided eyes you see the sun going around the sky each day in what appears to be a circle. This apparent motion of the sun is easily accounted for by imagining that it is the *earth* that rotates once a day. But the sun also has a *yearly* motion with respect to the stars. Even if you argue that the daily motion of objects in the sky is due to the turning of the earth, it is still possible to think of the earth as being at the center of the universe, and to imagine the sun moving in a year-long orbit around the earth. Simple measurements show that the sun's angular size increases and decreases slightly during the year as if it were alternately changing its distance from the earth. An interpretation that fits these observations is that the sun travels around the earth in a slightly off-center circle.

During this laboratory exercise you will plot the sun's apparent orbit with as much accuracy as possible.

Plotting the Orbit

You know the sun's direction among the stars on each date that the sun is observed. From its observed diameter on that date, you can find its relative distance from the earth. So, date by date, you can plot the sun's direction and relative distance. When you connect your plotted points by a smooth curve, you will have drawn the sun's apparent orbit.

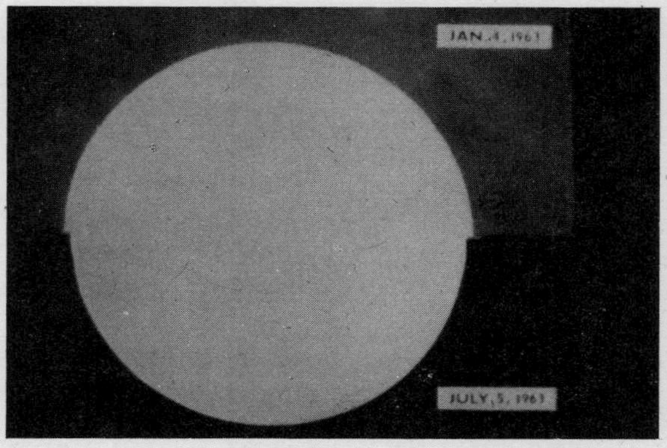

Fig. 2-16 Frame 4 of the Sun Filmstrip.

For observations you will use a series of sun photographs taken by the U.S. Naval Observatory at approximately one-month intervals and printed on a filmstrip. Frame 4, in which the images of the sun in January and in July are placed adjacent to each other, has been reproduced in Fig 2-16 so you can see how much the apparent size of the sun changes during the year. Note also how the apparent size of an object is related to its distance from you.

B.C. **By John Hart**

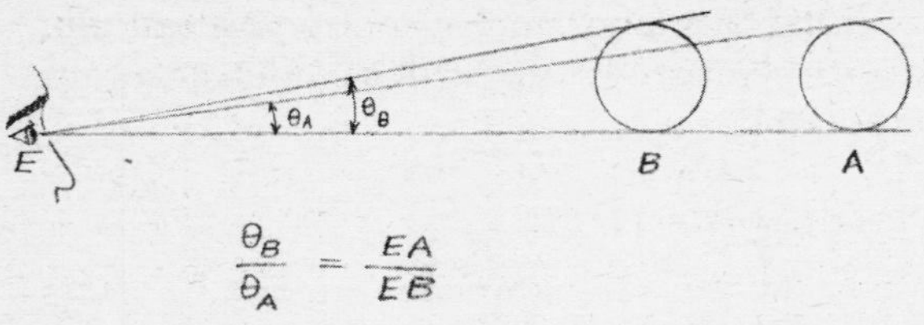

Fig. 2-17 As shown in the diagram, the angular size is inversely proportional to the distance — the farther away, the smaller the image.

Procedure

On a large sheet of graph paper (40 cm × 50 cm) make a dot at the center to represent the earth. It is particularly important that the graph paper be very large if you later plot the orbit of Mars in Experiments 2-8 and 2-9, which use the results of the present experiment.

Take the 0° direction (toward a reference point among the stars) to be along the graph-paper lines toward the right. This will be the direction of the sun as seen from the earth on March 21 (Fig. 2-18). The dates of all the photographs and the directions to the sun, measured counterclockwise from this 0° direction, are given in Table 2-6. Use a protractor to draw accurately a fan of lines radiating from the earth in these different directions.

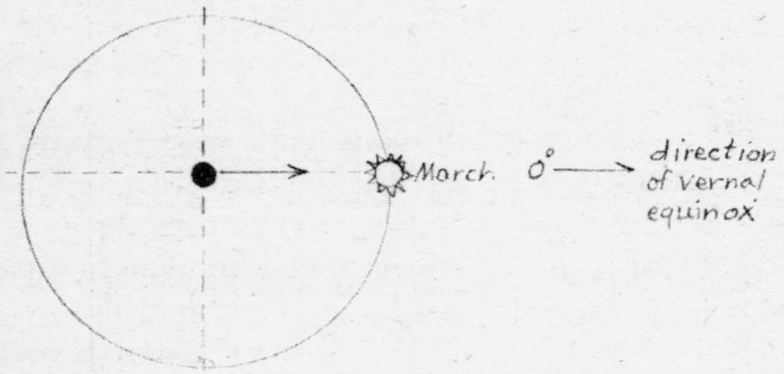

Fig. 2-18

TABLE 2-6

Date	Direction from Earth to Sun	Date	Direction from Earth to Sun
March 21	000°	Oct. 4	191°
April 6	015	Nov. 3	220
May 6	045	Dec. 4	250
June 5	074	Jan. 4	283
July 5	102	Feb. 4	315
Aug. 5	132	March 7	346
Sept. 4	162		

Measure carefully the diameter of the projected image of each frame of the filmstrip. You can get a set of relative distances to the sun by choosing a constant and then dividing it by the apparent diameters. An orbit with a radius of about 10 cm will be a particularly convenient size for later use. If you measure the sun's diameter to be about 50 cm, a convenient constant to choose would be 500, since $\frac{500}{50} = 10$. A larger image 51.0 cm in diameter leads to a smaller earth – sun distance:

$$\frac{500}{51.0} = 9.8 \text{ cm}$$

Make a table of the relative distances for each of the 13 dates.

Along each of the direction lines you have drawn, measure the relative distance to the sun for that date. Through the points located in this way draw a smooth curve. This is the apparent orbit of the sun relative to the earth. (Since the distances are only relative, you cannot find the actual distance in kilometers from the earth to the sun from this plot.)

Fig. 2-19

?

1. Is the orbit a circle? If so, where is the center of the circle? If the orbit is not a circle, what shape is it?
2. Locate the major axis of the orbit through the points where the sun passes closest to and farthest from the earth. What are the approximate dates of closest approach and greatest distance? What is the ratio of the largest distance to the smallest distance?

A Heliocentric System

Copernicus and his followers adopted the sun-centered model because they believed that the solar system could be described more simply that way. They had no new data that could not be accounted for by the old model.

Therefore, you should be able to use the same data to turn things around and plot the earth's orbit around the sun. Clearly, the two plots will be similar.

You already have a table of the relative distances between the sun and the earth. The dates of largest and smallest distances from the earth will not change, and your table of relative distances is still valid because it was not based on which body was moving, only on the distance between them. Only the directions used in your plotting will change.

To determine how the angles will change, remember that when the earth was at the center of the plot, the sun was in the direction 0° (to the right) on March 21.

?

3. This being so, what is the direction of the earth as seen from the sun on that date? (Figure 2-20 will help explain how to change directions for a sun-centered diagram.)

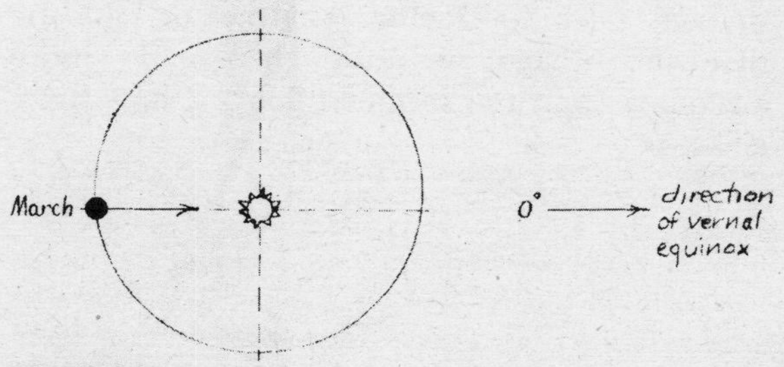

Fig. 2-20

If the sun is in the 0° direction from the earth, then from the sun the earth will appear to be in just the opposite direction, 180° away from 0°. You could make a new table of data giving the earth's apparent direction from the sun on the 13 dates, just by changing all the directions 180° and then making a new sun-centered plot. Another way is to rotate your plot until top and bottom are reversed; this will change all of the directions by 180°. Relabel the 0° direction; since it is toward a reference point among the distant stars, it will still be toward the right. You can now label the center as the sun, and the orbit as the earth's.

Experiment 2-7
USING LENSES TO MAKE A TELESCOPE

In this experiment, you will first examine some of the properties of single lenses. Then, you will combine these lenses to form a telescope, which you can use to observe the moon, the planets, and other heavenly (as well as earth-bound) objects.

The Simple Magnifier

You know something about lenses already, for instance, that the best way to use a magnifier is to hold it immediately in front of the eye and then move the object you want to examine until its image appears in sharp focus.

Examine some objects through several different lenses. Try lenses of various shapes and sizes. Separate the lenses that magnify from those that do not. What is the difference between lenses that magnify and those that do not?

?

1. Arrange the lenses in order of their magnifying powers. Which lens has the highest magnifying power?
2. What physical feature of a lens seems to determine its power or ability to magnify; is it diameter, thickness, shape, or the curvature of its surface? To vary the diameter, simply put pieces of paper or cardboard with various sizes of holes in them over the lens.

Sketch side views of a high-power lens, of a low-power lens, and of the highest-power and lowest-power lenses you can imagine.

Real Images

With one of the lenses you have used, project an image of a ceiling light or an outdoor scene on a sheet of paper. Describe all the properties of the image that you can observe. An image that can be projected is called a *real image*.

?

3. Do all your lenses from real images?
4. How does the size of the image depend on the lens?
5. If you want to look at a real image without using the paper, where do you have to put your eye?
6. The image (or an interesting part of it) may be quite small. How can you use a second lens to inspect it more closely? Try it.
7. Try using other combinations of lenses. Which combination gives the greatest magnification?

Making a Telescope

With two lenses properly arranged, you can magnify distant objects. Figure 2-21 shows a simple assembly of two lenses to form a telescope. It consists of a large lens (called the *objective*) through which light enters and either of two interchangeable lenses for eyepieces.

The following notes will help you assemble your telescope.

1. If you lay the objective down on a flat clean surface, you will see that one surface is more curved than the other. The more curved surface should face the front of the telescope.

2. Clean and dust off the lenses (using lens tissue or a clean handkerchief) before assembling and try to keep fingerprints off of it during assembly.

3. Wrap rubber bands around the slotted end of the main tube to give a convenient amount of friction with the draw tube, tight enough so as not to move once adjusted, but loose enough to adjust without sticking. Focus by sliding the draw tube with a rotating motion, not by moving the eyepiece in the tube.

4. To use high power satisfactorily, a steady support (a tripod) is essential.

5. Be sure that the lens lies flat in the high-power eyepiece.

Use your telescope to observe objects inside and outside the lab. Low power gives about 12× magnification. High power gives about 30× magnification, like Galileo's best telescope.

Mounting the Telescope

If no tripod mount is available, the telescope can be held in your hands for low-power observations. Grasp the telescope as far forward and as far back as possible and brace both arms firmly against a car roof, telephone pole, or other rigid support.

With the higher power you must use a mounting. If a swivel-head camera tripod is available, the telescope can be held in a wooden saddle by rubber bands, and the saddle attached to the tripod head by the head's standard mounting screw. Because camera tripods are usually too short for comfortable viewing while you are standing, you should be seated in a reasonably comfortable chair.

Aiming and Focusing

You may have trouble finding objects, especially with the high-power eyepiece. One technique is to sight over the tube, aiming slightly below the object, and then to tilt the tube up slowly while looking through it and sweeping left and right. To do this well, you will need some practice.

Focusing by pulling or pushing the sliding tube tends to move the whole telescope. To avoid this, rotate the sliding tube as if it were a screw.

Eyeglasses will keep your eye farther from the eyepiece than the best distance. Far-sighted or near-sighted observers are generally able to view more satisfactorily by removing their glasses and refocusing. Observers with astigmatism have to decide whether or not the distorted image (without glasses) is more annoying than the reduced field of view (with glasses).

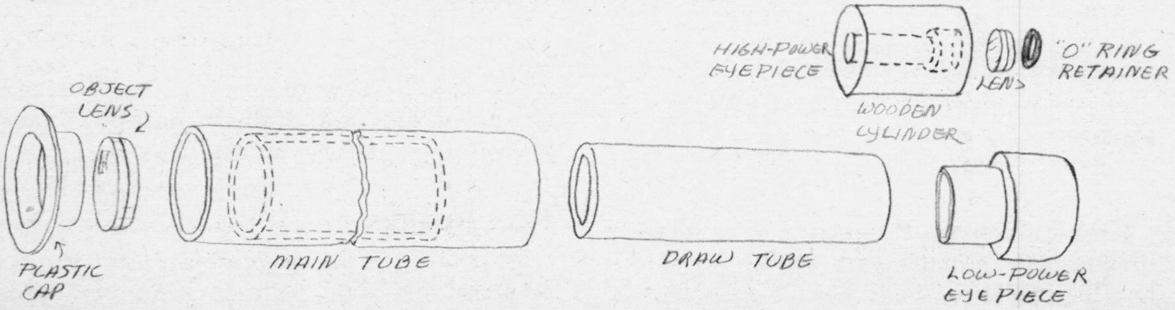

Fig. 2-21

Many observers find that they can keep their eye in line with the telescope while aiming and focusing if the brow and cheek rest lightly against the forefinger and thumb.

To minimize shaking the instrument when using a tripod mounting, remove your hands from the telescope while actually viewing.

Limitations of Your Telescope

You can get some idea of how much fine detail to expect when observing the planets by comparing the angular sizes of the planets with the resolving power of the telescope. For a telescope with a 2.5-cm diameter object lens, to distinguish between two details, they must be at least 0.001° apart as seen from the location of the telescope. The low-power *Project Physics* eyepiece may not quite show this much detail, but the high power will be more than sufficient.

The angular sizes of the planets as viewed from the earth are:

Venus	0.003°	(minimum)
	0.016	(maximum)
Mars	0.002	(minimum)
	0.005	(maximum)
Jupiter	0.012	(average)
Saturn	0.005	(average)
Uranus	0.001	(average)

Galileo's first telescope gave 3× magnification, and his "best" gave about 30× magnification. (But he used a different kind of eyepiece that gave a much smaller field of view.) You should find it challenging to see whether you can observe all the phenomena he saw, which are mentioned in Sec. 7.7 of the Text.

Observations You Can Make

The objects suggested for observation have been chosen because they are (1) fairly easy to find, (2) representative of what is to be seen in the sky, and (3) very interesting. You should observe all objects with the low power first and then the high power. For additional information on current objects to observe, see the paperback *New Handbook of the Heavens*, or the last few pages of each monthly issue of the magazines *Sky and Telescope*, *Natural History*, or *Science News*.

Venus

No features will be visible on this planet, but you can observe its phases, as shown in Fig. 2-22 (enlarged to equal sizes) and on page 196 of the Text. When Venus is very bright you may need to reduce the amount of light coming through the telescope in order to see the shape of the image. A paper lens cap with a round hole in the center will reduce the amount of light (and the resolution of detail!). You might also try using sunglasses as a filter.

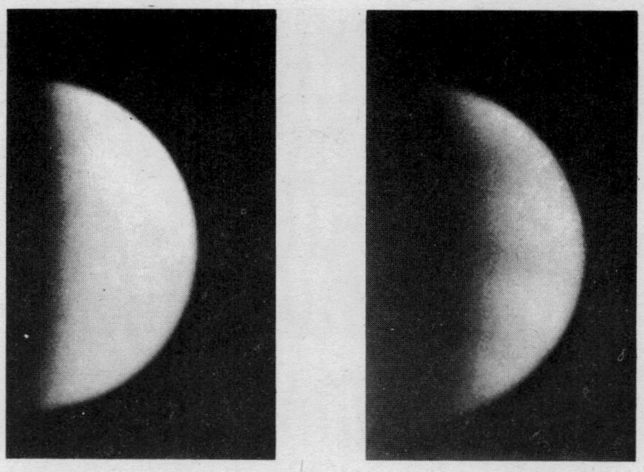

Fig. 2-22 Venus, photographed at Yerkes Observatory with the 205-cm telescope.

Saturn

This planet is so large that you can resolve the projection of the rings beyond the disk, but you probably cannot see the gap between the rings and the disk with your 30× telescope (Fig. 2-23). Compare your observations to the sketches on page 73 of the Text.

Fig. 2-23 Saturn photographed with the 250-cm telescope at Mount Wilson.

Jupiter

Observe the four satellites that Galileo discovered. Observe them several times, a few hours or a day apart, to see changes in their positions. By keeping detailed data over several months time, you can determine the period for each of the moons, the radii of their orbits, and then the mass of Jupiter. (See the notes for the Film Loop, "Jupiter Satellite Orbit," in this *Handbook* for directions on how to analyze your data.)

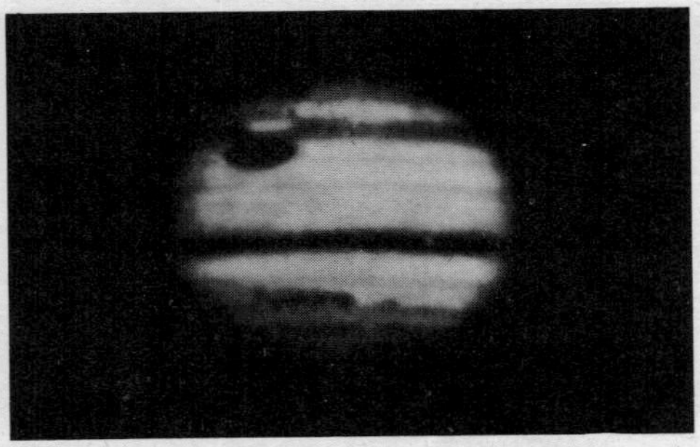

Fig. 2-24 Jupiter photographed with the 500-cm telescope at Mount Palomar.

Jupiter is so large that some of the detail on its disk, like a broad, dark, equatorial cloud belt (Fig. 2-24), can be detected (especially if you know it should be there!).

Moon

Moon features stand out mostly because of shadows. The best observations are made around the first and third quarters. Make sketches of your observations, and compare them to Galileo's sketch on page 194 of your Text. Look carefully for walls, mountains in the centers of craters, bright peaks on the dark side beyond the terminator, and craters within other craters.

The Pleiades

The Pleiades, a beautiful little star cluster, is located on the right shoulder of the bull in the constellation Taurus. These stars are almost directly overhead in the evening sky in December. The Pleiades were among the objects Galileo studied with his first telescope. He counted 36 stars, which the poet Tennyson described as "a swarm of fireflies tangled in a silver braid."

The Hyades

This cluster of stars is also in Taurus, near the star Aldebaran, which forms the bull's eye. The Hyades look like a *v*. The high power may show that several stars are double.

The Great Nebula in Orion

Look about halfway down the row of stars that form the sword of Orion. It is in the southeastern sky during December and January. Use low power.

Algol

This famous variable star is in the constellation Perseus, south of Cassiopeia. Algol is high in the eastern sky in December, and nearly overhead during January. Generally, it is a second-magnitude star, like the Pole Star. After remaining bright for more than 2½ days, Algol fades for 5 hours and becomes a fourth-magnitude star, like the faint stars of the Little Dipper. Then, the variable star brightens during the next 5 hours to its normal brightness. From one minimum to the next, the period is 2 days, 20 hours, 49 minutes.

Great Nebula in Andromeda

Look high in the western sky in the early evening in December for this nebula, for by January it is low on the horizon. It will appear as a fuzzy patch of light, and is best viewed with low power. The light you see from this galaxy has been on its way for nearly 2 million years.

The Milky Way

This is particularly rich in Cassiopeia and Cygnus (if air pollution in your area allows it to be seen at all).

Observing Sunspots

CAUTION: Do not look at the sun through the telescope. The sunlight will injure your eyes. Figure 2-25 shows an arrangement of a tripod, the low-power telescope, and a sheet of paper for projecting sunspots. Cut a hole in a piece of cardboard so it fits snugly over the object end of the telescope. This acts as a shield so there is a shadow area where you can view the sunspots. First focus the telescope, using the high-power eyepiece, on some distant object. Then, project the image of the sun on a piece of

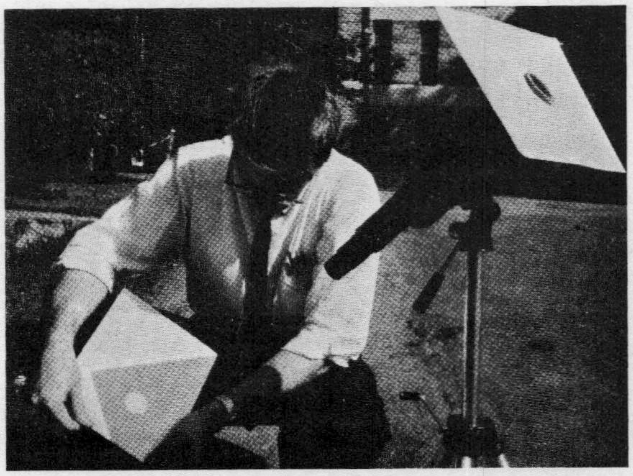

Fig. 2-25

B.C. By John Hart

white paper about 60 cm behind the eyepiece. Focus the image by moving the draw tube slightly further out. When the image is in focus, you may see some small dark spots on the paper. To distinguish marks on the paper from sunspots, jiggle the paper back and forth. How can you tell that the spots are not on the lenses?

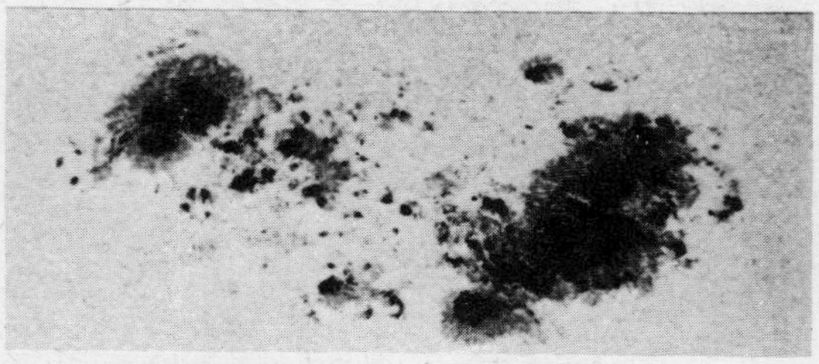

Fig. 2-26 The sunspots of April 7, 1947.

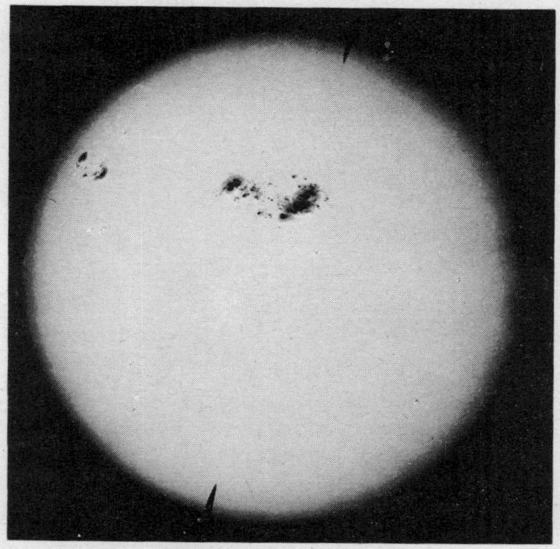

By focusing the image farther from the telescope, you can make the image larger and not so bright. It may be easier to get the best focus by moving the paper rather than the eyepiece tube.

Experiment 2-8
THE ORBIT OF MARS

In this laboratory activity you will derive an orbit for Mars around the sun by the same method that Kepler used in discovering that planetary orbits are elliptical. Since the observations are made from the earth, you will need the orbit of the earth that you developed in Experiment 2-6, "The Shape of the Earth's Orbit." Make sure that the plot you use for this experiment represents the orbit of the earth around the sun, not the sun around the earth.

If you did not do the earth-orbit experiment, you may use, for an approximate orbit, a circle of 10-cm radius drawn in the center of a large sheet of graph paper. Because the eccentricity of the earth's orbit is very small (0.017), you can place the sun at the center of a circular orbit without introducing a significant error in this experiment.

From the sun (at the center), draw a line to the right, parallel to the grid of the graph paper (Fig. 2-27). Label the line 0°. This line is directed toward a point on the celestial sphere called the *vernal equinox* and is the reference direction from which angles in the plane of the earth's orbit (the ecliptic plane) are measured. The earth crosses this line on September 23. When the earth is on the other side of its orbit on March 21, the sun is between the earth and the vernal equinox.

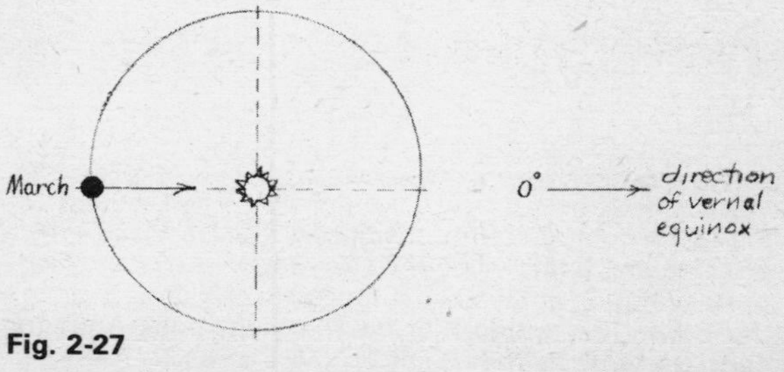

Fig. 2-27

Photographic Observations of Mars

You will use a booklet containing 16 enlarged sections of photographs of the sky showing Mars among the stars at various dates between 1931 and 1950. All were made with a small camera used for the Harvard Observatory Sky Patrol. In some of the photographs, Mars was near the center of the field. In many other photographs Mars was near the edge of the field where the star images are distorted by the camera lens. Despite these distortions the photographs can be used to provide positions of Mars that are satisfactory for this study. Photograph P is a double exposure, but it is still quite satisfactory.

Changes in the positions of the stars relative to each other are extremely slow. Only a few stars near the sun have motions large enough to be detected after many years observations with the largest telescopes. Thus, you can consider the pattern of stars as fixed.

Finding Mars' Location

Mars is continually moving among the stars but is always near the ecliptic. From several hundred thousand photographs at the Harvard Observatory, 16 were selected, with the aid of a computer, to provide pairs of photographs separated by 687 days, the period of Mars' journey around the sun as determined by Copernicus. Thus, each pair of photographs shows Mars at one place in its orbit.

During these 687 days, the earth makes nearly two full cycles of its orbit, but the interval is short of two full years by 43 days. Therefore, the position of the earth, from which you can observe Mars, will not be the same for the two observations of each pair. If you can determine the direction from the earth towards Mars for each of the pairs of observations, the two sight lines must cross at a point on the orbit of Mars. (See Fig. 2-28.)

Coordinate System Used

When you look into the sky you see no coordinate system. Coordinate systems are created for various purposes. The one used here centers on the ecliptic. Remember that the ecliptic is the imaginary line on the celestial sphere along which the sun appears to move.

Along the ecliptic, *longitudes* are always measured eastward from the 0° point (the vernal equinox). This is toward the left on star maps. *Latitudes* are measured perpendicular to the ecliptic north or south to 90°. The small movement of Mars above and below the ecliptic is considered in Experiment 2-9, "Inclination of Mars' Orbit."

To find the coordinates of a star or of Mars, you must project the coordinate system upon the sky. To do this you are provided with transparent overlays that show the coordinate system of the ecliptic for each frame, A to P. The positions of various stars are circled. Adjust the overlay until it fits the star positions. Then you can read off the longitude and latitude of the position of Mars. Figure 2-29 shows how you can interpolate between marked coordinate lines. Because you are interested in only a small section of the sky on each photograph, you can draw each small section of the ecliptic as a straight line. For plotting, an accuracy of 0.5° is satisfactory.

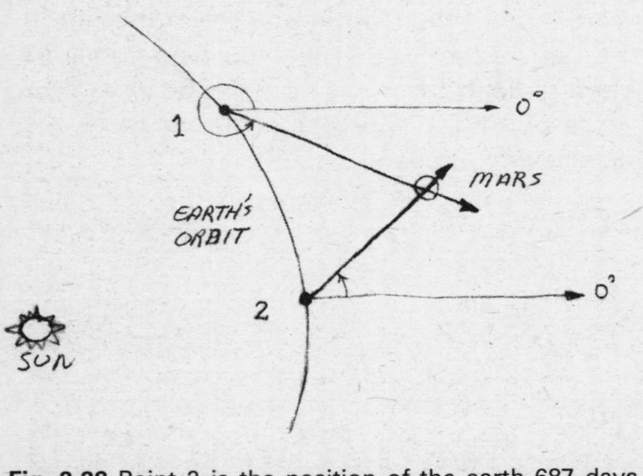

Fig. 2-28 Point 2 is the position of the earth 687 days after leaving point 1. In 687 days, Mars has made exactly one revolution and so has returned to the same point on the orbit. The intersection of the sight lines from the earth determines that point on Mars' orbit.

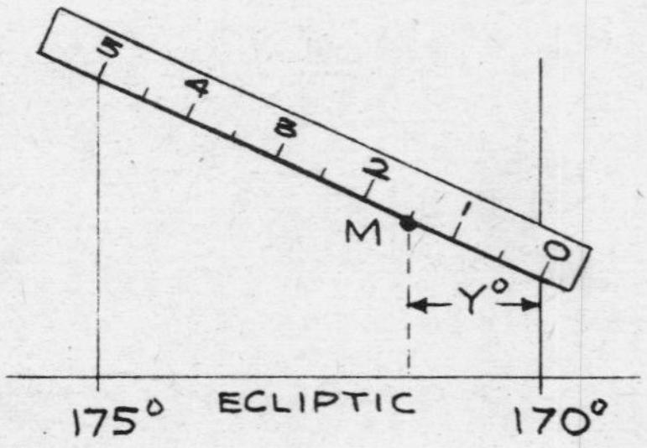

Fig. 2-29 Interpolation between coordinate lines. In the sketch, Mars (M), is at a distance *y*° from the 170° line. Take a piece of paper or card at least 10 cm long. Make a scale divided into 10 equal parts and label alternate marks 0, 1, 2, 3, 4, 5. This gives a scale in 0.5° steps. Notice that the numbering goes from right to left on this scale. Place the scale so that the edge passes through the position of Mars. Now tilt the scale so that the 0 and 5 marks each fall on a grid line. Read off the value of *y* from the scale. In the sketch *y* = 1.5°, so that the longitude of M is 171.5°.

TABLE 2-7. OBSERVED POSITIONS OF MARS

Frame	Date	Geocentric Long.	Geocentric Lat.	Mars-to-Earth Distance	Mars-to-Sun Distance	Heliocentric Long.	Heliocentric Lat.
A	Mar. 21, 1931						
B	Feb. 5, 1933						
C	Apr. 20, 1933						
D	Mar. 8, 1935						
E	May 26, 1935						
F	Apr. 12, 1937						
G	Sept. 16, 1939						
H	Aug. 4, 1941						
I	Nov. 22, 1941						
J	Oct. 11, 1943						
K	Jan. 21, 1944						
L	Dec. 9, 1945						
M	Mar. 19, 1946						
N	Feb. 3, 1948						
O	Apr. 4, 1948						
P	Feb. 21, 1950						

In a table similar to Table 2-7, record the longitude and latitude of Mars for each photograph. For a simple plot of Mars' orbit around the sun, you will use only the first column, the longitude of Mars. You will use the columns for latitude, Mars' distance from the sun, and the sun-centered coordinates if you also investigate the inclination, or tilt, of Mars' orbit.

Finding Mars' Orbit

When your table is completed for all eight pairs of observations, you are ready to locate points on the orbit of Mars.

1. On the plot of the earth's orbit, locate the position of the earth for each date given in the 16 photographs. You may do this by interpolating between the dates given for the earth's orbit experiment. Since the earth moves through 360° in about 365 days, you may use ±1° for each day ahead or behind the date given in the previous experiment. For example, frame A is dated March 21. The earth was at 166° on March 7; 14 days later on March 21, the earth will have moved 14° from 166° to 180°. Always work from the earth-position date nearest the date of the Mars photograph.

2. Through each earth-position point, draw a "0° line" parallel to the line you drew from the sun toward the vernal equinox (the grid on the graph paper is helpful). Use a protractor and a sharp pencil to mark the angle between the 0°-line and the direction to Mars on that date as seen from the earth (longitude of Mars). The two lines drawn from the earth's positions for each pair of dates will intersect at a point. This is a point on Mars' orbit. Figure 2-31 shows one

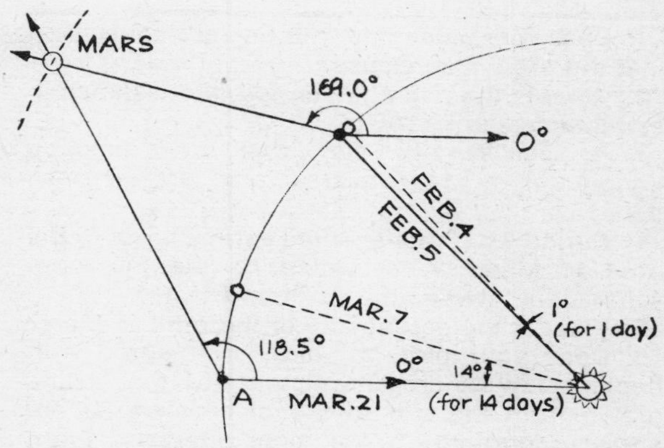

Fig. 2-31

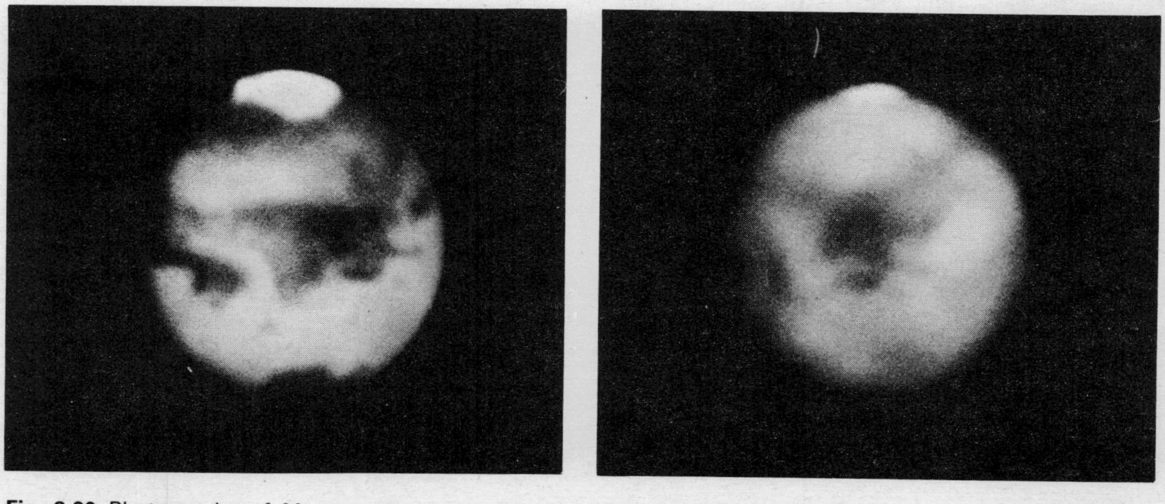

Fig. 2-30 Photographs of Mars made with a 150-cm reflecting telescope (Mount Wilson and Palomar Observatories) during closest approach to the earth in 1956. *Left*: August 10; *right*: Sept. 11. Note the shrinking of the polar cap.

point on Mars' orbit obtained from the data of the first pair of photographs. By drawing the intersecting lines from the eight pairs of positions, you establish eight points on Mars' orbit.

3. You will notice that there are no points in one section of the orbit. You can fill in the missing part because the orbit is symmetrical about its major axis. Use a compass and, by trial and error, find a circle that best fits the plotted points.

Now that you have plotted the orbit, you have achieved what you set out to do: You have used Kepler's method to determine the path of Mars around the sun.

Kepler's Law from Your Plot

If you have time to go on, it is worthwhile to see how well your plot agrees with Kepler's generalization about planetary orbits.

?

1. Does your plot agree with Kepler's conclusion that the orbit is an ellipse?
2. What is the average sun-to-Mars distance in astronomical units (AU)?
3. As seen from the sun, what is the direction (longitude) of Mars' nearest and farthest positions?
4. During what month is the earth closest to the orbit of Mars? What would be the minimum separation between the earth and Mars?
5. What is the eccentricity of the orbit of Mars?
6. Does your plot of Mars' orbit agree with Kepler's law of areas, which states that a line drawn from the sun to the planet sweeps out areas proportional to the time intervals? From your orbit, you see that Mars was at point B′ on February 5, 1933, and at point C′ on April 20, 1933, as shown in Fig. 2-32. There are eight such pairs of dates in your data. The time intervals are different for each pair.

Connect these pairs of positions with a line to the sun, Fig. 2-32. Find the areas of squares on the graph paper (count a square when more than half of it lies within the area). Divide the area (in squares) by the number of days in the interval to find an "area per day" value. Are these values nearly the same?

?

7. How much (by what percentage) do they vary?
8. What is the uncertainty in your area measurements?
9. Is the uncertainty the same for large areas as for small?
10. Do your results bear out Kepler's law of areas?

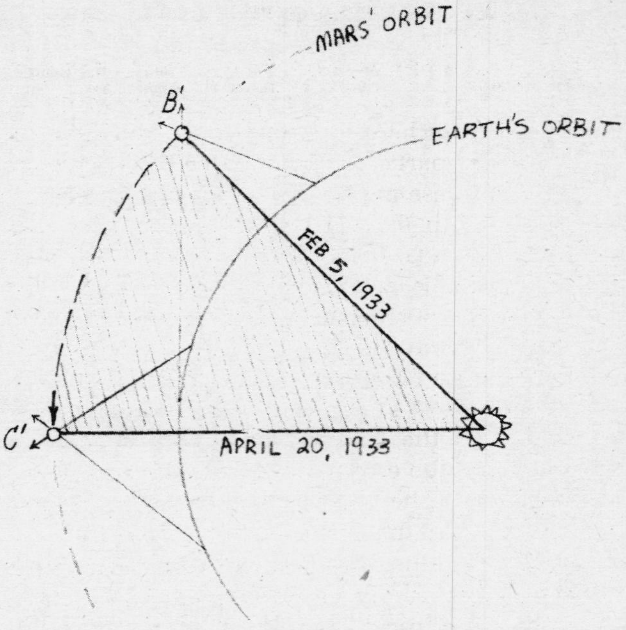

Fig. 2-32 In this example, the time interval is 74 days.

Experiment 2-9
INCLINATION OF MARS' ORBIT

When you plotted the orbit of Mars in Experiment 2-8, you ignored the slight movement of the planet above and below the ecliptic. This movement of Mars north and south of the ecliptic shows that the plane of its orbit is slightly inclined to the plane of the earth's orbit. Now you may use the table of values for Mars' latitude (which you made in Experiment 2-8) to determine the inclination of Mars' orbit.

First make a three-dimensional model of two orbits to see what is meant by the inclination of orbits. You can do this quickly with two small pieces of cardboard (or index cards). On each card draw a circle or ellipse, but have one larger than the other. Mark clearly the position of the focus (sun) on each card. Make a straight cut *to the sun*, on one card from the left, on the other from the right. Slip the cards together until the sun-points coincide. Tilt the two cards (orbit planes) at various angles (Fig. 2-33).

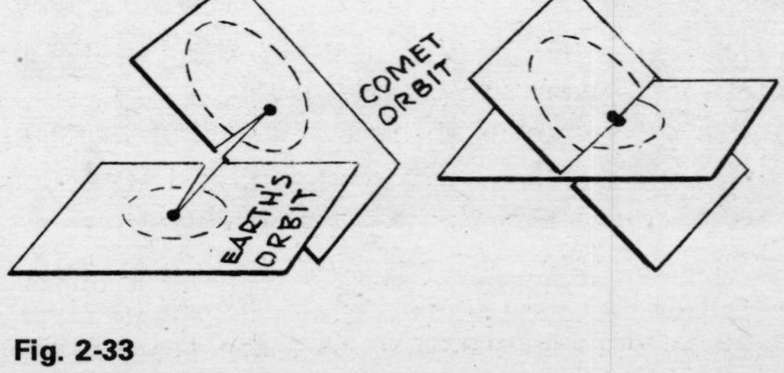

Fig. 2-33

Theory

From each of the photographs in the set of 16 that you used in Experiment 2-8, you can find the observed latitude (angle from the ecliptic) of Mars at a particular point in its orbital plane. Each of these angles is measured on a photograph taken *from the earth*. As you can see from Fig. 2-33, however, it is the *sun*, not the earth, that is at the center of the orbit. The inclination of Mars' orbit must, therefore, be an angle measured *at the sun*. It is this angle (the heliocentric latitude) that you wish to find.

Figure 2-34 shows that Mars can be represented by the head of a pin whose point is stuck into the ecliptic plane. Mars is seen from the earth to be north or south of the ecliptic, but you want the north–south angle of Mars as seen from the sun. The following example shows how you can derive the angles as if you were seeing them from the sun.

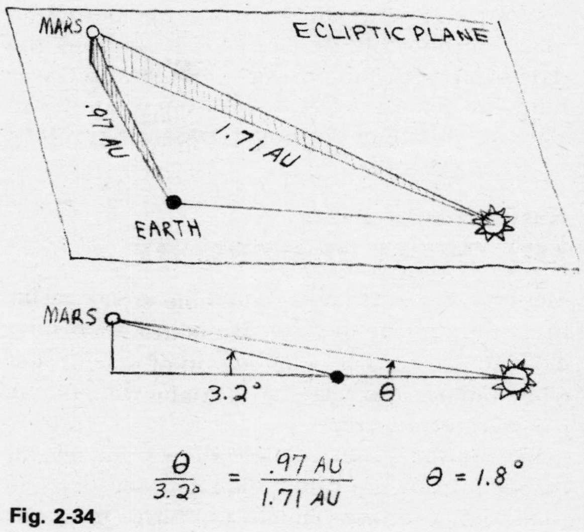

Fig. 2-34

In plate A (March 21, 1933), Mars was about 3.2° north of the ecliptic *as seen from the earth*. But the earth was considerably closer to Mars on this date than the sun was. The angular elevation of Mars above the ecliptic plane as seen from the sun will therefore be considerably less than 3.2°.

For very small angles, the apparent angular sizes are inversely proportional to the distances. For example, if the sun were twice as far from Mars as the earth, the angle at the sun would be one-half the angle at the earth.

Measurement on the plot of Mars' orbit (Experiment 2-8) gives the earth–Mars distance as 9.7 cm (0.97 AU) and the sun–Mars distance as 17.1 cm (1.71 AU) on the date of the photograph. The heliocentric latitude of Mars is, therefore,

$$\frac{9.7}{17.1} \times 3.2^\circ\text{N} = 1.8^\circ\text{N}$$

You can check this value by finding the heliocentric latitude of this same point in Mars' orbit on photograph B (February 5, 1933). The earth was in a different place on this date so the geocentric latitude and the earth–Mars distance will both be different, but the heliocentric latitude should be the same to within your experimental uncertainty (Fig. 2-35).

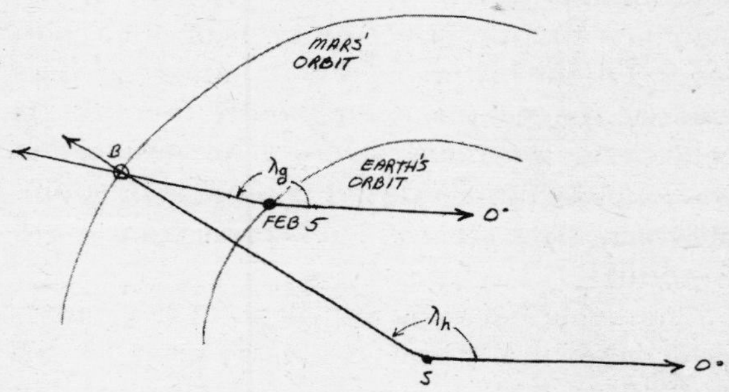

Fig. 2-35 On February 5, the heliocentric longitude (λ_h) of point B on Mars' orbit is 150°; the geocentric longitude (λ_g) measured from the earth's position is 169°.

Making the Measurements

Turn to the table of data you made for Experiment 2-8, on which you recorded the geocentric latitudes λ_g of Mars. On your Mars' orbit plot from Experiment 2-8, measure the corresponding earth–Mars and sun–Mars distances and note them in the same table.

From these two sets of values, calculate the heliocentric latitudes as explained above. The values of heliocentric latitude calculated from the two plates in each pair (A and B, C and D, etc.) should agree within the limits of your experimental procedure.

On the plot of Mars' orbit, measure the *heliocentric longitude* λ_h for each of the eight Mars positions. Heliocentric longitude is measured from the sun, counterclockwise from the 0° direction (direction toward vernal equinox).

Complete the table given in Experiment 2-8 by entering the earth–Mars and sun–Mars distances, the geocentric and heliocentric latitudes, and the geocentric and heliocentric longitudes for all 16 plates.

Draw a graph that shows how the heliocentric latitude of Mars changes with its heliocentric longitude (see Fig. 2-36).

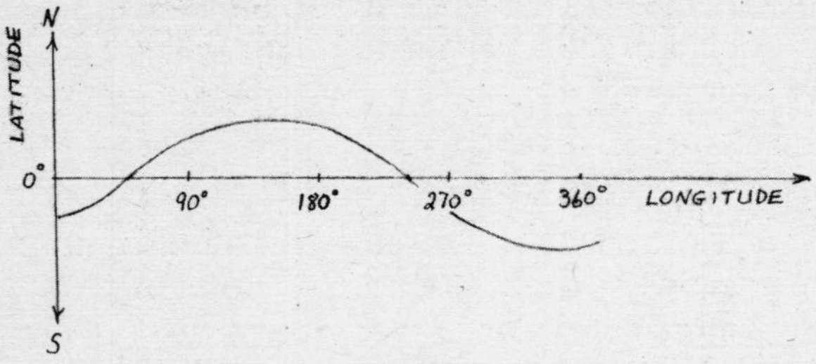

Fig. 2-36 Change of Mars' heliocentric latitude with heliocentric longitude. Label the ecliptic, latitude, ascending node, descending node, and inclination of the orbit in this drawing.

From this graph, you can find two of the elements that describe the orbit of Mars with respect to the ecliptic. The point at which Mars crosses the ecliptic from south to north is called the *ascending node*. (The *descending node*, on the other side of the orbit, is the point at which Mars crosses the ecliptic from north to south.)

The angle between the plane of the earth's orbit and the plane of Mars' orbit is the *inclination* of Mars' orbit i. When Mars reaches its maximum latitude above the ecliptic, which occurs at 90° beyond the *ascending node*, the planet's maximum latitude equals the *inclination* of the orbit i.

Elements of an Orbit

Two angles, the *longitude of the ascending node*, Ω, and the *inclination* i, locate the plane of Mars' orbit with respect to the plane of the ecliptic. One more angle is needed to orient the orbit of Mars in its orbital plane. This is the *argument of perihelion* ω, shown in Fig. 2-37, which is the angle in the *orbit plane* between the ascending node and perihelion point. On your plot of Mars' orbit, measure the angle from the ascending node Ω to the direction of the perihelion to obtain the argument of the perihelion ω.

If you have worked along this far, you have determined five of the six elements that define *any* orbit:

a: semimajor axis, or average distance (which determines the period)
e: eccentricity (shape of orbit as given by c/a in Fig. 2-37)
i: inclination (tilt of orbital plane)
Ω: longitude of ascending node (where orbital plane crosses ecliptic)
ω: argument of perihelion (which orients the orbit in its plane)

These five elements (shown in Fig. 2-37) fix the orbital plane of any planet or comet in space, tell the size and shape of the orbit, and also give its orientation within the orbital plane. To compute a complete timetable, or *ephemeris*, for the body, you need only to know T, a zero date when the body was at a particular place in the orbit. Generally, T is given as the date of a perihelion passage. Photograph G was made on September 16, 1933. From this you can estimate a date of perihelion passage for Mars.

Experiment 2-10
THE ORBIT OF MERCURY

Mercury, the innermost planet, is never very far from the sun in the sky. It can be seen only close to the horizon, just before sunrise or just after sunset, and viewing is made difficult by the glare of the sun.

Except for Pluto, which differs in several respects from the other planets, Mercury has the most eccentric planetary orbit in the solar

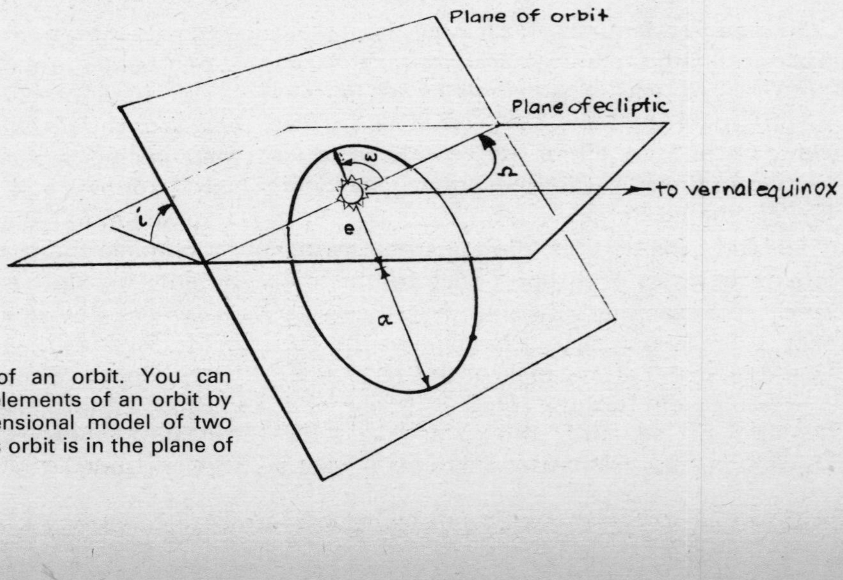

Fig. 2-37 The five elements of an orbit. You can familiarize yourself with these elements of an orbit by adding them to the three-dimensional model of two orbits, assuming that the earth's orbit is in the plane of the ecliptic.

system ($e = 0.206$). The large eccentricity of Mercury's orbit has been of particular importance, since it has led to one of the tests for Einstein's general theory of relativity. For a planet with an orbit inside the earth's, there is a simpler way to plot the orbit than by the pair observations you used for Mars. In this experiment, you will use this simpler method to get the approximate shape of Mercury's orbit.

Mercury's Elongations

Assume a heliocentric model for the solar system. Mercury's orbit can be found from Mercury's maximum angles of elongation east and west from the sun as seen from the earth on various known dates.

The angle (Fig. 2-38) between the sun and Mercury, as seen from the earth, is called the *elongation*. Note that when the elongation reaches its maximum value, the sight lines from the earth are tangent to Mercury's orbit.

Since the orbits of Mercury and the earth are both elliptical, the greatest value of the elongation varies from revolution to revolution. The 28° elongation given for Mercury in the Text refers to the maximum value. Table 2-8 gives the angles of a number of these greatest elongations.

TABLE 2-8. SOME DATES AND ANGLES OF GREATEST ELONGATION FOR MERCURY*

Date	Elongation
January 4, 1963	19° E
February 14	26 W
April 26	20 E
June 13	23 W
August 24	27 E
October 6	18 W
December 18	20 E
January 27, 1964	25 W
April 8	19 E
May 25	25 W

*from the American Ephemeris and Nautical Almanac

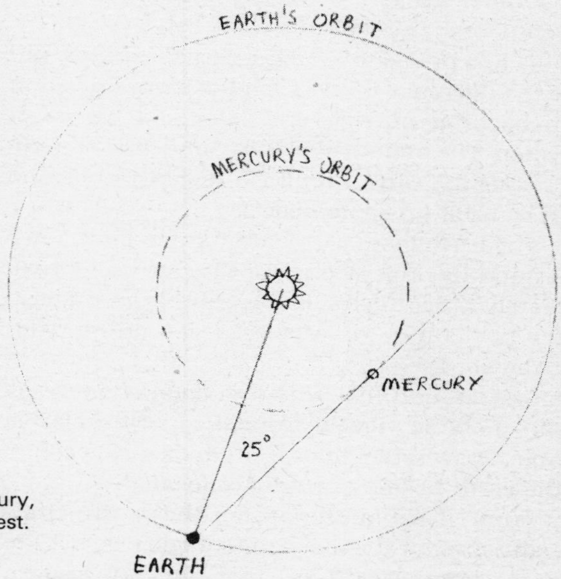

Fig. 2-38 The greatest western elongation of Mercury, May 25, 1964. The elongation had a value of 25° West.

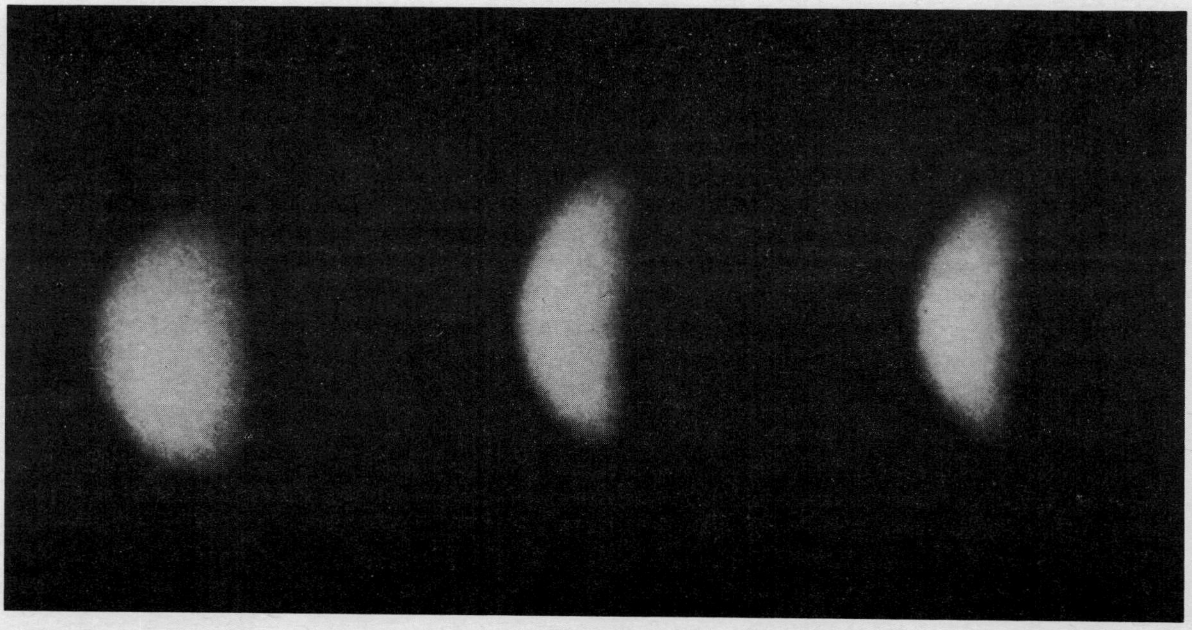

Fig. 2-39 Mercury, first quarter phase, taken June 7, 1934, at the Lowell Observatory, Flagstaff, Arizona.

Plotting the Orbit

You can work from the plot of the earth's orbit that you established in Experiment 2-6. Make sure that the plot you use for this experiment represents the orbit of the earth around the sun, not of the sun around the earth.

If you did not do the earth's orbit experiment, you may use, for an approximate earth orbit, a circle of 10-cm radius drawn in the center of a sheet of graph paper. Because the eccentricity of the earth's orbit is very small (0.017), you can place the sun at the center of the orbit without introducing a significant error in the experiment.

Draw a reference line horizontally from the center of the circle to the right. Label the line 0°. This line points toward the vernal equinox and is the reference from which the earth's position in its orbit on different dates can be established. The point where the 0° line from the sun crosses the earth's orbit is the earth's position in its orbit on September 23.

The earth takes about 365 days to move once around its orbit (360°). Use the rate of approximately 1° per day, or 30° per month, to establish the position of the earth on each of the dates given in Table 2-8. Remember that the earth moves around this orbit in a *counterclockwise* direction, as viewed from the north celestial pole. Draw radial lines from the sun to each of the earth positions you have located.

From these positions for the earth, draw sight lines for the elongation angles. Be sure to note, from Fig. 2-38, that for an *eastern* elongation, Mercury is to the *left* of the sun as seen from the earth. For a *western* elongation, Mercury is to the right of the sun.

You know that on a date of greatest elongation Mercury is somewhere along the sight line, but you do not know exactly where on the line to place the planet. You also know that the sight line is tangent to the orbit. A reasonable assumption is to put Mercury at the point along the sight line closest to the sun.

You can now find the orbit of Mercury by drawing a smooth curve through, or close to, these points. Remember that the orbit must *touch* each sight line without crossing any of them.

Finding R_{av}

The average distance of a planet in an elliptical orbit is equal to one-half the long diameter of the ellipse, the *semimajor axis*.

To find the size of the semimajor axis a of Mercury's orbit (Fig. 2-40), relative to the earth's semimajor axis, you must first find the aphelion and perihelion points of the orbit. You can use a drawing compass to find these points on the orbit farthest from and closest to the sun.

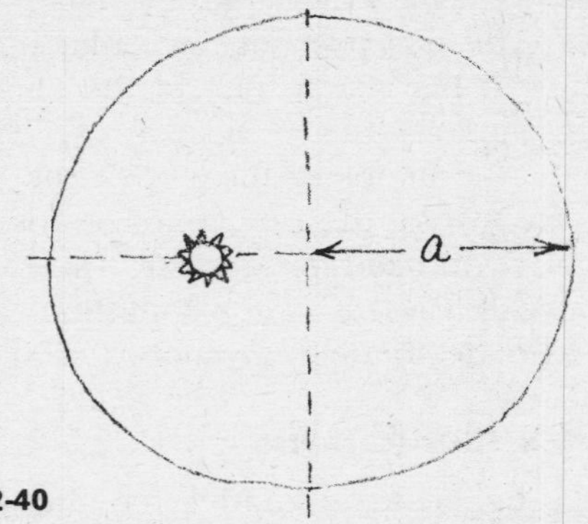

Fig. 2-40

Measure the greatest diameter of the orbit along the line perihelion – sun – aphelion. Since 10 cm corresponds to 1 AU (the semimajor axis of the earth's orbit), you can now obtain the semimajor axis of Mercury's orbit in astronomical units.

Calculating Orbital Eccentricity

Eccentricity is defined as $e = c/a$ (Fig. 2-41). Since c, the distance from the center of Mercury's ellipse to the sun, is small on your plot, you lose accuracy if you try to measure it directly.

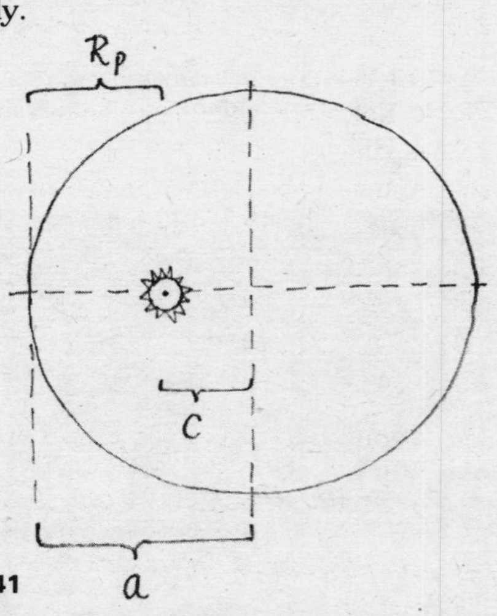

Fig. 2-41

From Fig. 2-41, you can see that c is the difference between Mercury's perihelion distance R_p and the semimajor axis a. That is:

$$c = a - R_p$$

Therefore,

$$e = \frac{c}{a}$$

$$e = \frac{a - R_p}{a}$$

$$e = 1 - \frac{R_p}{a}$$

You can measure R_p and a with reasonable accuracy from your plotted orbit. Compute e, and compare your value with the accepted value, $e \approx 0.206$.

Kepler's Second Law

You can test Kepler's equal-area law on your Mercury orbit in the same way as that described in Experiment 2-8, "The Orbit of Mars." By counting squares, you can find the area swept out by the radial line from the sun to Mercury between successive dates of observation, such as January 4 to February 14, and June 13 to August 24. Divide the area by the number of days in the interval to get the "area per day." This should be constant, if Kepler's law holds for your plot. Is it constant?

Experiment 2-11
STEPWISE APPROXIMATION TO AN ORBIT*

You have seen in the Text how Newton analyzed the motions of bodies in orbit, using the concept of a centrally directed force. On the basis of the discussion in the Text Sec. 8.4, you are now ready to apply Newton's method to develop an approximate orbit of a satellite or a comet around the sun. You can also, from your orbit, check Kepler's law of areas and other relationships discussed in the Text.

Imagine a ball rolling over a smooth level surface such as a piece of plate glass.

?

1. What would you predict for the path of the ball, based on your knowledge of Newton's laws of motion?
2. Suppose you were to strike the ball from one side. Would the path direction change?
3. Would the speed change? Suppose you gave the ball a series of "sideways" blows of equal force as it moves along. What do you predict its path might be?

Reread Sec. 8.4 of the Text if you have difficulties answering these questions.

*This experiment is based on a similar one developed by Dr. Leo Lavatelli, University of Illinois, *American Journal of Physics*, Vol. 33, p. 605.

Fig. 2-42 Photograph of the comet Cunningham made at Mount Wilson and Palomar Observatories December 21, 1940. Why do the stars leave trails and the comet does not?

Your Assumptions

A planet or satellite in orbit has a continuous force acting on it. As the body moves, the magnitude and direction of this force change. To predict exactly the orbit under the application of this continually changing force requires advanced mathematics. However, you can get a reasonable approximation of the orbit by breaking the continuous attraction into many small steps, in which the force acts as a sharp "blow" toward the sun once every 60 days. (See Fig. 2-43.)

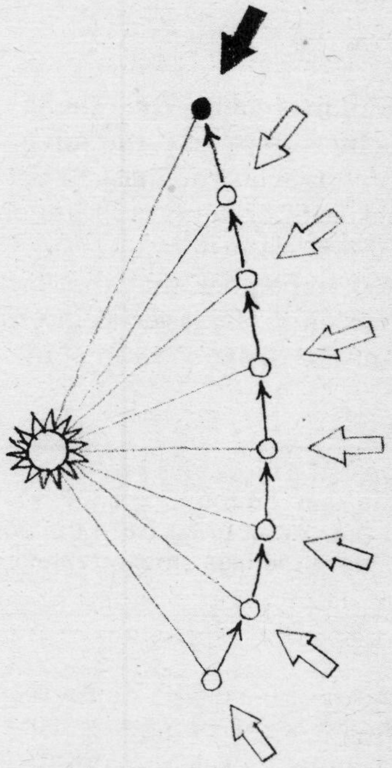

Fig. 2-43 A body, such as a comet, moving in the vicinity of the sun will be deflected from its straight-line path by a gravitational force. The force acts continuously but Newton has shown that we can think about the orbit as though it were produced by a series of sharp blows.

The application of repeated steps is known as *iteration*; it is a powerful technique for solving problems. Modern high-speed digital computers use repeated steps to solve complex problems, such as the best path (or paths) for a space probe to follow between earth and another planet.

You can now proceed to plot an approximate comet orbit if you make these additional assumptions:

1. The force on the comet is an attraction toward the sun.
2. The force of the blow varies inversely with the square of the comet's distance from the sun.
3. The blows occur regularly at equal time intervals — in this case, 60 days. The magnitude of each brief blow is assumed to equal the total effect of the continuous attraction of the sun throughout a 60-day interval.

Effect of the Central Force

From Newton's second law you know that the gravitational force will cause the comet to accelerate toward the sun. If a force $\vec{F}$ acts for a time interval Δt on a body of mass m, you know that

$$F = m\vec{a} = m\frac{\Delta \vec{v}}{\Delta t}$$

and, therefore, $$\Delta \vec{v} = \frac{\vec{F}}{m}\Delta t$$

This equation relates the change in the body's velocity to its mass, the force, and the time for which it acts. The mass m is constant; so is Δt (assumption 3 above). The change in velocity is therefore proportional to the force, $\Delta \vec{v} \propto \vec{F}$. But remember that the force is *not* constant in magnitude; it varies inversely with the square of the distance from comet to sun.

?

4. Is the force of a blow given to the comet when it is near the sun grerater or smaller than one given when the comet is far from the sun?
5. Which blow causes the greatest velocity change?

In Fig. 2-44, the vector $\vec{v}_0$ represents the comet's velocity at the point A. During the first 60 days, the comet moves from A to B (Fig. 2-45). At B a blow causes a velocity change $\Delta\vec{v}_1$ (Fig. 2-46). The new velocity after the blow is $\vec{v}_1 = \vec{v}_0 + \Delta\vec{v}_1$, and is found by completing the vector triangle (Fig. 2-47).

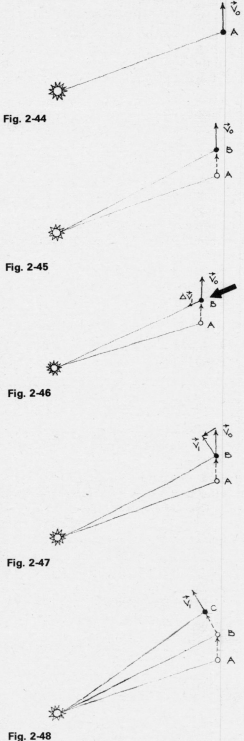

Fig. 2-44

Fig. 2-45

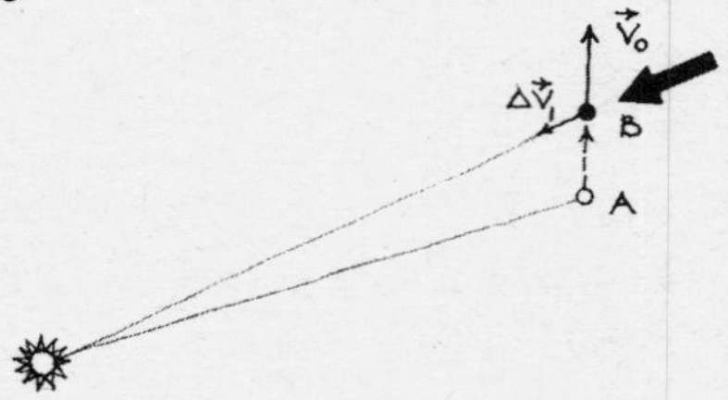

Fig. 2-46

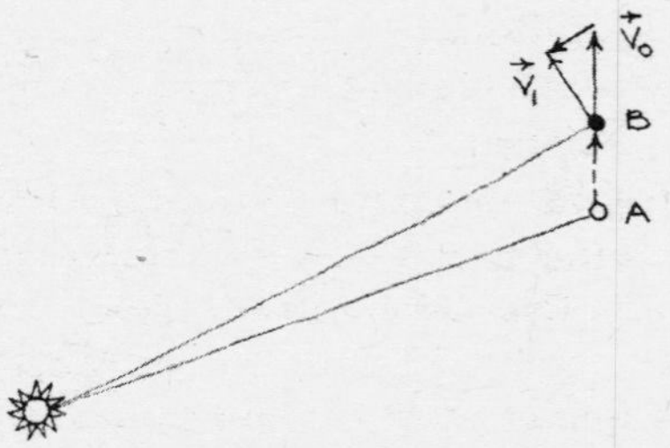

Fig. 2-47

Fig. 2-48

The comet therefore leaves point B with velocity $\vec{v}_1$ and continues to move with this velocity for another 60-day interval. Because the time interv•ls between blows are always the same (60 days), the displacement along the path is proportional to the velocity, $\vec{v}$. You therefore use a length proportional to the comet's velocity to represent its displacement during each time interval (Fig. 2-48).

Each new velocity is found, as above, by adding to the previous velocity the $\Delta\vec{v}$ given by the blow. In this way, step by step, the comet's orbit is built up.

Scale of the Plot

The shape of the orbit depends on the initial position and velocity, and on the acting force. Assume that the comet is first spotted at a distance of 4 AU from the sun. Also assume that the comet's velocity at this point is $\vec{v} = 2$ AU/yr at right angles to the sun–comet distance R.

The following scale factors will reduce the orbit to a scale that fits conveniently on a 40 × 50 cm piece of graph paper.

1. Let 1 AU be scaled to 6.3 cm so that 4 AU becomes about 25 cm.

2. Since the comet is hit every 60 days, it is convenient to express the velocity in astronomical units per 60 days. Adopt a scale factor in which a velocity vector of 1 AU/60 days is represented by an arrow 6.3 cm long.

The comet's initial velocity of 2 AU/yr can be given as 2/365 AU per day, or 2/365 × 60 = 0.33 AU/60 days. This scales to an arrow 2.14 cm long. This is the *displacement* A to B of the comet in the first 60 days.

Computing Δv

On the scale, and with the 60-day iteration interval that has been chosen, the force field of the sun is such that the $\Delta\vec{v}$ given by a blow when the comet is 1 AU from the sun is 1 AU/60 days.

To avoid computing Δv for each value of R, you can plot Δv against R on a graph. Then for any value of R, you can immediately find the value of Δv.

Table 2-9 gives values of R in astronomical units and in centimeters to fit the scale of your orbit plot. The table also gives for each value of R the corresponding value of Δv in AU/60 days and in centimeters to fit the scale of your orbit plot.

TABLE 2-9. SCALES FOR R AND Δv

Distance from the Sun, R		Change in Speed, Δv	
AU	cm	AU/60 days	cm
0.75	4.75	1.76	11.3
0.8	5.08	1.57	9.97
0.9	5.72	1.23	7.80
1.0	6.35	1.00	6.35
1.2	7.62	0.69	4.42
1.5	9.52	0.44	2.82
2.0	12.7	0.25	1.57
2.5	15.9	0.16	1.02
3.0	19.1	0.11	0.71
3.5	22.2	0.08	0.51
4.0	25.4	0.06	0.41
4.5	28.6	0.05	0.38

Graph these values on a separate sheet of paper at least 25 cm long, as illustrated in Fig. 2-49, and carefully connect the points with a smooth curve.

You can use this curve as a simple graphical computer. Cut off the bottom margin of the graph paper, or fold it under along the R axis. Lay this edge on the orbit plot and measure the distance from the sun to a blow point (such as

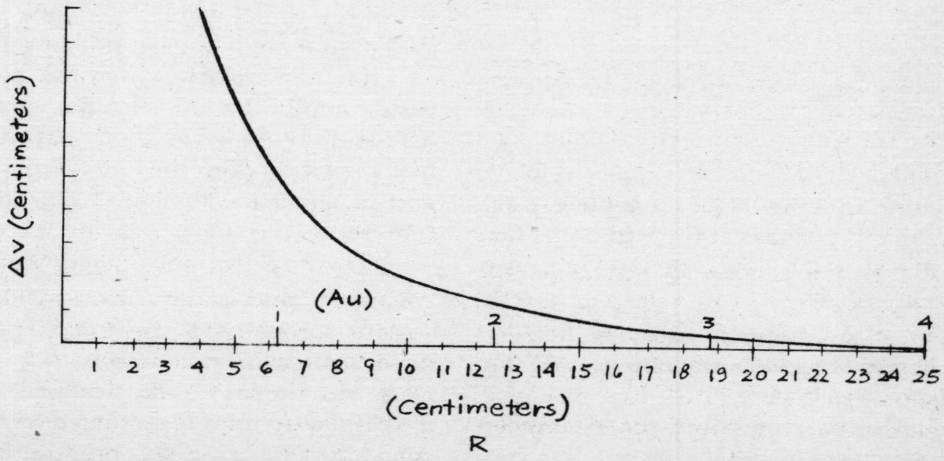

Fig. 2-49

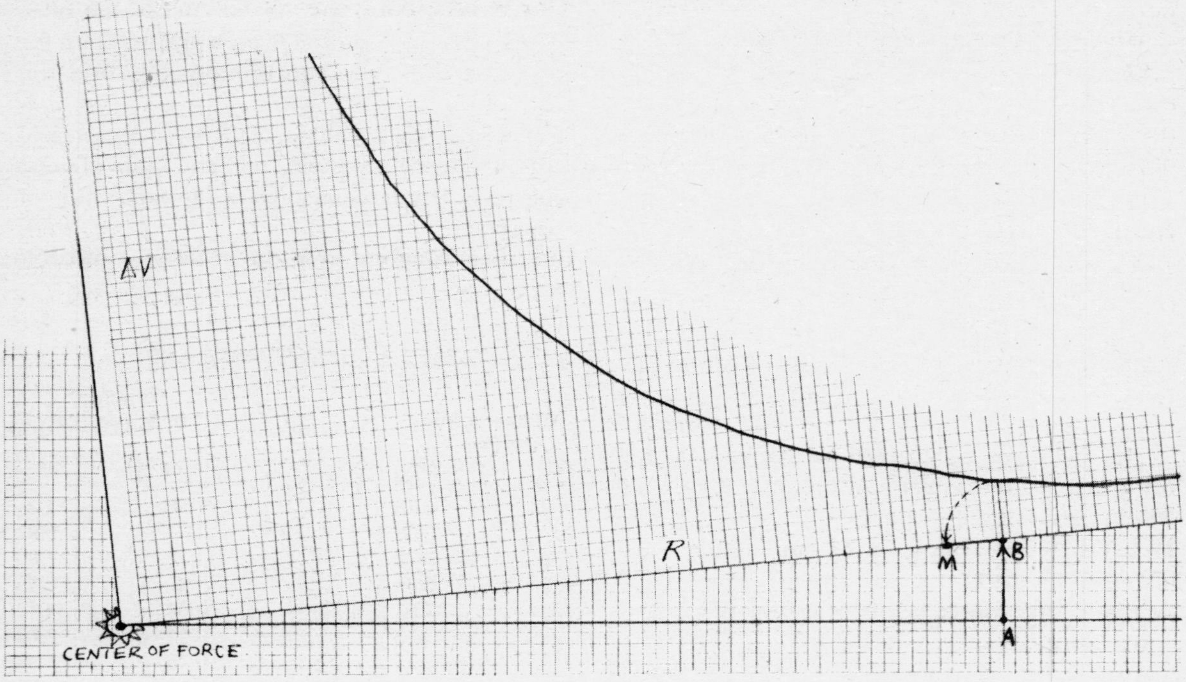

Fig. 2-50

B in Fig. 2-50). With dividers or a drawing compass, measure the value of Δv corresponding to this R and plot this distance along the radius line toward the sun (see Fig. 2-50).

Making the Plot

1. Mark the position of the sun S halfway up the large graph paper (held horizontally) and 30 cm from the right edge.

2. Locate a point 25 cm (4 AU) to the right from the sun S. This is point A where you first find the comet (Fig. 2-51).

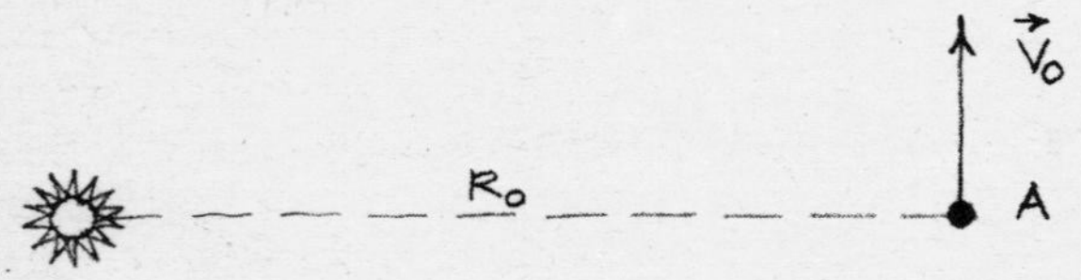

Fig. 2-51

3. To represent the comet's initial velocity, draw vector AB perpendicular to SA (Fig. 2-52). B is the comet's position at the end of the first 60-day interval. At B a blow is struck that causes a change in velocity Δv_1.

4. Use your Δv graph to measure the distance of B from the sun at S, and to find Δv_1 for this distance (Fig. 2-50).

5. The force, and therefore the change in velocity, is always directed toward the sun.

Fig. 2-52

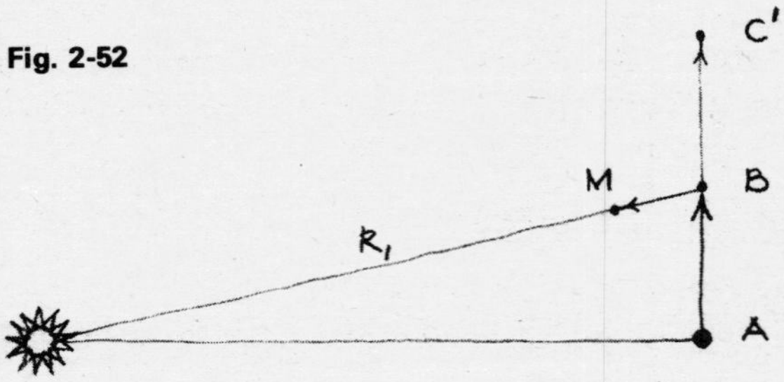

From B lay off Δv_1 toward S. Call the end of this short line M (Fig. 2-52).

6. Draw the line BC′, which is a continuation of AB and has the same length as AB. That is where the comet would have gone in the next 60 days if there had been no blow at B.

7. The new velocity after the blow is the vector sum of the old velocity (represented by BC′) and Δv (represented by BM). To find the new velocity v_1 draw the line C′C parallel to BM and of equal length (Fig. 2-53). The line BC represents the new velocity vector v_1, the velocity with which the comet leaves point B.

8. Again the comet moves with uniform velocity for 60 days, arriving at point C. Its displacement in that time is $\Delta d_1 = v_1 \times 60$ days, and because of the scale factor chosen, the displacement is represented by the line BC (Fig. 2-54).

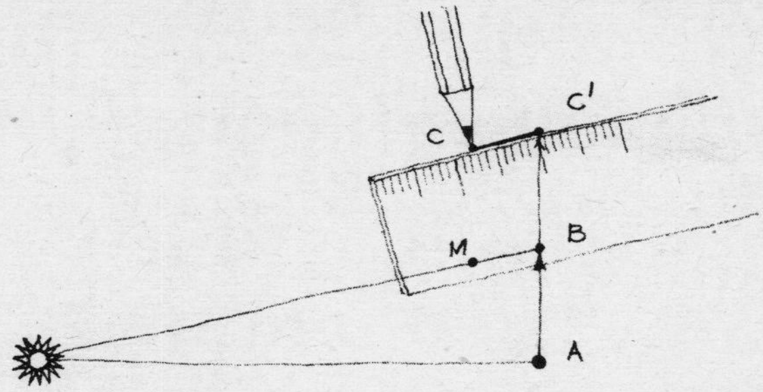

Fig. 2-53

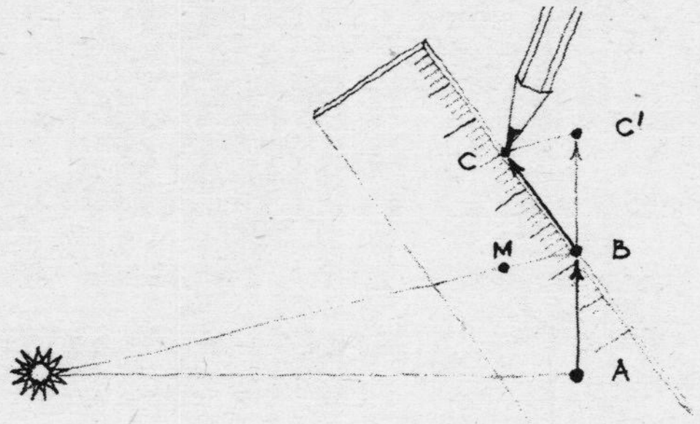

Fig. 2-54

9. Repeat steps 1 through 8 to establish point D, and so forth, for at least 14 or 15 steps (25 steps gives the complete orbit).

10. Connect points A, B, C . . . with a smooth curve. Your plot is finished.

Prepare for Discussion

?

6. From your plot, find the perihelion distance.
7. Find the center of the orbit and calculate the eccentricity of the orbit.
8. What is the period of revolution of your comet? (Refer to Text, Sec. 7.3.)
9. How does the comet's speed change with its distance from the sun?

It is interesting to go on to see how well the orbit obtained by iteration obeys Kepler's laws.

10. Is Kepler's law of ellipses confirmed? (Can you think of a way to test your curve to see how nearly it is an ellipse?)
11. Is Kepler's law of equal areas confirmed?

To answer question 11, remember that the time interval between blows is 60 days, so the comet is at positions B, C, D . . . , etc., after equal time intervals. Draw a line from the sun to each of these points (include A), and you have a set of triangles.

Find the area of each triangle. The area A of a triangle is given by $A = \frac{1}{2}ab$, where a and b are altitude and base, respectively. You can also count squares to find the areas.

More Things to Do

1. The graphical technique you have practiced can be used for many problems. You can use it to find out what happens if different initial speeds and/or directions are used. You may wish to use the $1/R^2$ graph, or you may construct a new graph. To do this, use a different law (for example, force proportional to $1/R^3$, to $1/R$, or to R) to produce different paths; actual gravitational forces are *not* represented by such force laws.

2. If you use the same force graph but reverse the direction of the force to make it a *repulsion*, you can examine how bodies move under such a force. Do you know of the existence of any such repulsive force?

Experiment 2-12
MODEL OF THE ORBIT OF HALLEY'S COMET

Halley's comet is referred to several times in your Text. If you construct a model of it, you will find that its orbit has a number of interesting features.

Since the orbit of the earth around the sun lies in one plane and the orbit of Halley's comet lies in another plane intersecting it, you will need two large pieces of stiff cardboard for planes on which to plot these orbits.

The Earth's Orbit

Draw the earth's orbit first. In the center of one piece of cardboard, draw a circle with a radius of 5 cm (1 AU) for the orbit of the earth. On the same piece of cardboard, also draw approximate (circular) orbits for Mercury (radius 0.4 AU) and Venus (radius 0.7 AU). For this plot, you can consider that all of the planet orbits lie in the same plane. Draw a line from the sun at the center and mark this line as 0° longitude.

The table on page 4 of this *Handbook* lists the apparent position of the sun in the sky on 12 dates. By adding 180° to each of the tabled values, you can get the positions of the earth in its orbit as seen from the sun. Mark these positions on your drawing of the earth's orbit. (If you wish to mark more than those 12 positions, you can do so by using the technique described on page 4.)

The Comet's Orbit

Figure 2-55 shows the positions of Halley's comet near the sun in its orbit, which is very

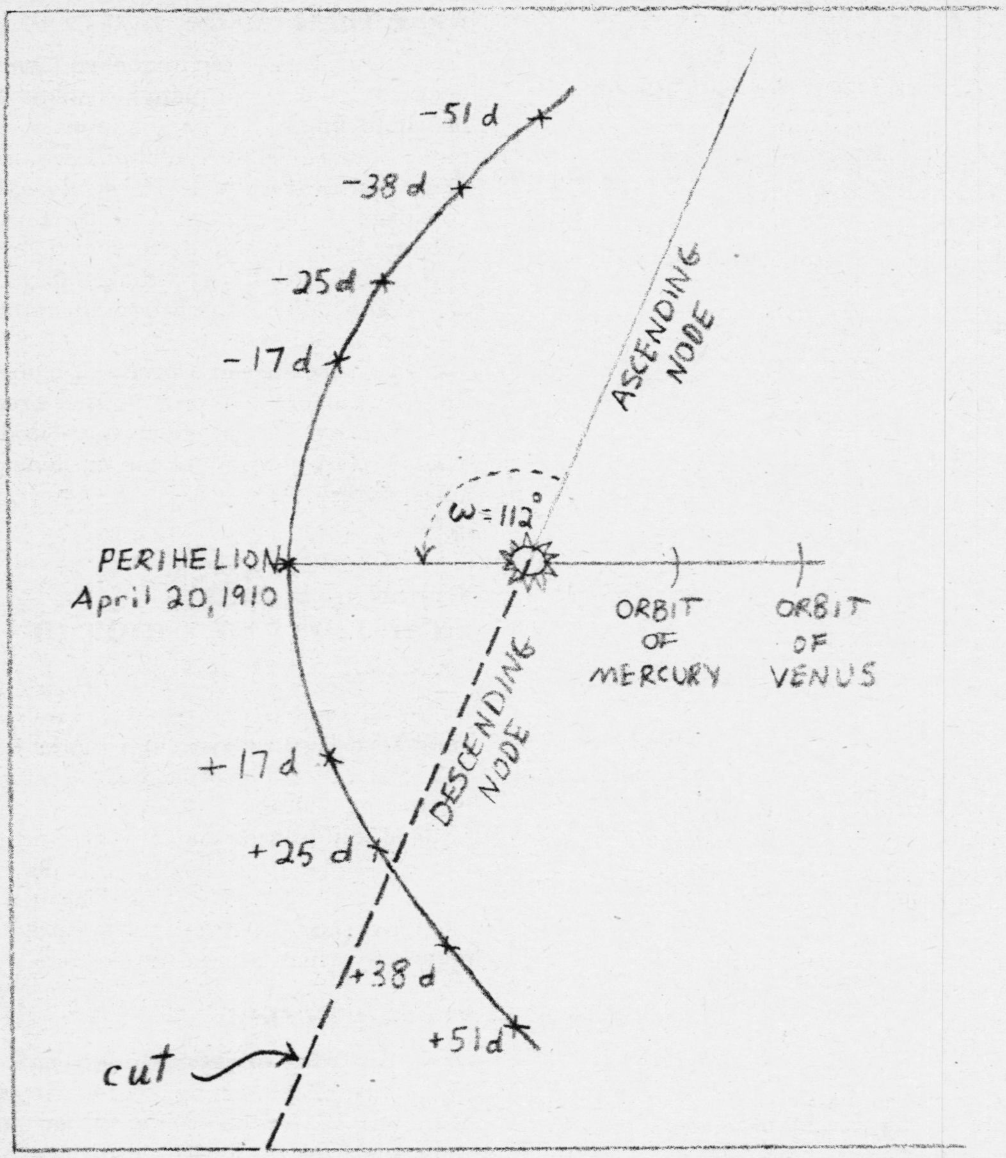

Fig. 2-55

nearly a parabola. You will construct your own orbit of Halley's comet by tracing Fig. 2-55 and mounting the tracing on stiff cardboard.

Combining the Two Orbits

Now you have the two orbits, the comet's and the earth's, in their planes, each of which contains the sun. You need only to fit the two together in accordance with the elements of orbits that you may have used in the experiment on the "Inclination of Mars' Orbit."

The line along which the comet's orbital plane cuts the ecliptic plane is called the *line of nodes*. Since you have the major axis drawn, you can locate the ascending node, in the orbital plane, by measuring ω, the angle from perihelion in a direction *opposite* to the comet's motion (see Fig. 2-55).

To fit the two orbits together, cut a narrow slit in the ecliptic plane (earth's orbit) along the line of the *ascending* node in as far as the sun. The longitude of the comet's ascending node Ω was at 57° as shown in Fig. 2-56. Then slit the comet's orbital plane on the side of the *descending* node in as far as the sun (see Fig. 2-55). Slip one plane into the other along the cuts until the sun-points on the two planes come together.

To establish the model in three dimensions you must now fit the two planes together at the

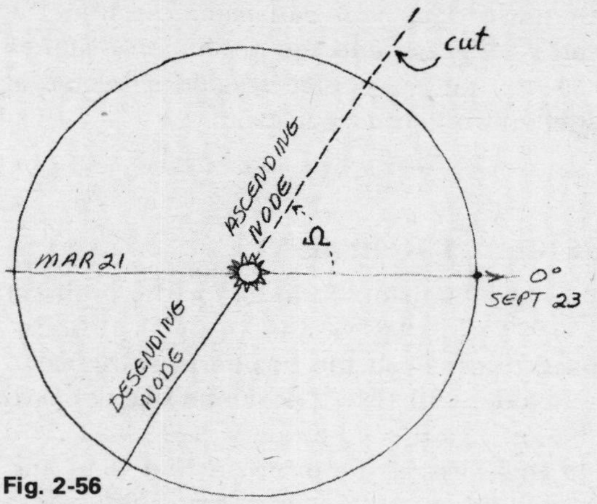

Fig. 2-56

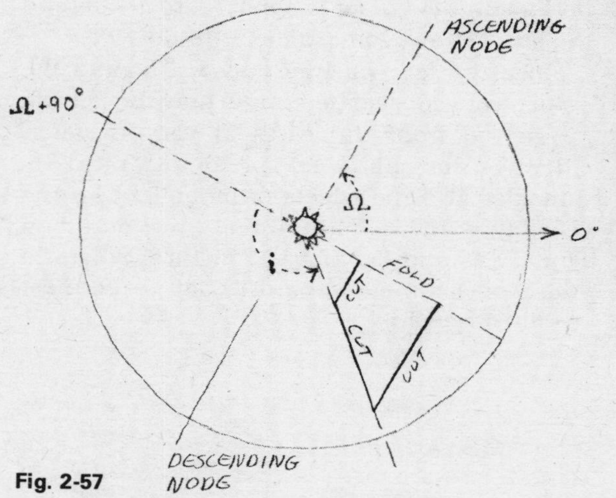

Fig. 2-57

correct angle. Remember that the inclination i, 162°, is measured upward (northward) from the ecliptic in the direction of $\Omega + 90°$ (see Fig. 2-57). When you fit the two planes together, you will find that the comet's orbit is on the underside of the cardboard. The simplest way to transfer the orbit to the top of the cardboard is to prick through with a pin at enough points so that you can draw a smooth curve through them. Also, you can construct a small tab to support the orbital plane at the correct angle of 18° (180° − 162°) as shown in Fig. 2-57.

Halley's comet moves in the opposite sense to the earth and other planets. Whereas the earth and planets move counterclockwise when viewed from above (north of) the ecliptic, Halley's comet moves clockwise.

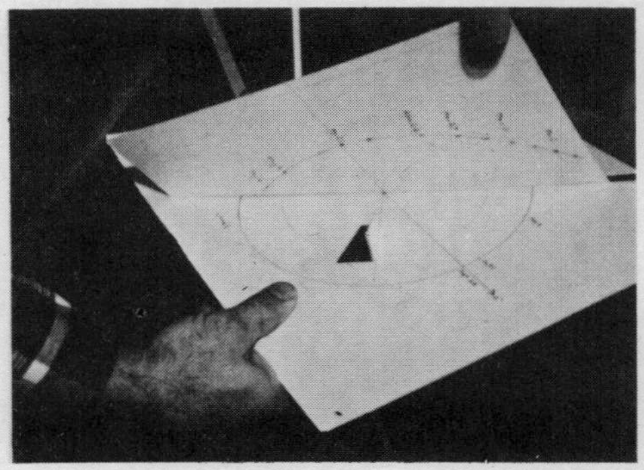

Fig. 2-58

If you have persevered this far, and your model is a fairly accurate one, it should be easy to explain the comet's motion through the sky shown in Fig. 2-59. The dotted line in the figure is the ecliptic.

With your model of the comet orbit you can now answer some very puzzling questions about the behavior of Halley's comet in 1910.

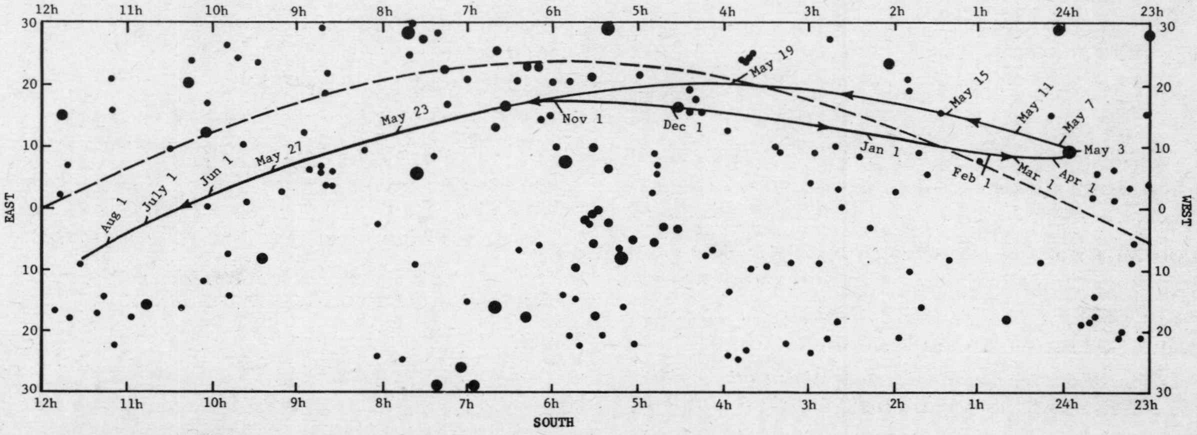

Fig. 2-59 Motion of Halley's comet in 1909-1910.

?

1. Why did the comet appear to move westward for many months?
2. How could the comet hold nearly a stationary place in the sky during the month of April 1910?
3. After remaining nearly stationary for a month, why did the comet move nearly halfway across the sky during the month of May 1910?
4. What was the position of the comet in space relative to the earth on May 19th?
5. If the comet's tail was many millions of miles long on May 19th, is it likely that the earth passed through part of the tail?
6. Were people worried about the effect a comet's tail might have on life on the earth? (See newspapers and magazines of 1910!)
7. Did anything unusual happen? How dense is the material in a comet's tail? Would you expect anything to have happened?

The elements of Halley's comet are, approximately:

a (semi-major axis)	17.9 AU
e (eccentricity)	0.967
i (inclination of orbit plane)	162°
Ω (longitude of ascending node)	057°
ω (angle to perihelion)	112°
Most recent perihelion date	April 20, 1910

From these data, you can calculate that the period is 76 years, and the perihelion distance is 0.59 AU. Halley's comet is again expected at perihelion on February 9, 1986.

1986 Comet Return

The three-dimensional model for the orbits of the earth and Halley's comet in 1909 – 1910 can be used to establish the positions of the earth and the comet in 1986. The comet will come to perihelion again on February 9, 1986. If you locate the earth in its orbit on that date and change the dates of the comet's position before and after perihelion, you will find that at perihelion the comet is almost directly behind the sun, far away and difficult to observe.

Since the tail will trail outward behind the comet, you can also determine how the tail will be viewed from the earth at various dates: sideways, foreshortened, or almost head-on. Consider also the effects of moonlight on the visibility of the tail; a full moon will occur on January 25 and February 22 in 1986. When do you expect the comet and its tail to be most readily observed?

ACTIVITIES

MAKING ANGULAR MEASUREMENTS

For astronomical observations, and often for other purposes, you need to estimate the angle between two objects. You have several instant measuring devices handy once you calibrate them. Held out at arm's length in front of you, they include:

1. your thumb;
2. your fist, not including thumb knuckle;
3. two of your knuckles; and
4. the span of your hand from thumb-tip to tip of little finger when your hand is opened wide.

For a first approximation, your fist is about 8° and thumb-tip to little finger is between 15° and 20°.

However, since the lengths of people's arms and the sizes of their hands vary, you can calibrate yours using the following method.

To find the angular size of your thumb, fist, and hand span at your arm's length, you can make use of one simple relationship. An object viewed from a distance that is 57.4 times its diameter covers an angle of 1°. For example, a 1-cm circle viewed from 57.4 cm away has an angular size of 1°.

Set a 10-cm scale on the blackboard chalk tray. Stand with your eye at a distance of 5.74 m from the scale. From there, observe how many centimeters of the scale are covered by your thumb, etc. Make sure that you have your arm straight out in front of your nose. Each 10 cm covered corresponds to 1°. Find some convenient measuring dimensions on your hand.

A Mechanical Aid

You can use an index card and a meter stick to make a simple instrument for measuring angles. Remember that when an object with a given diameter is placed at a distance from your eye equal to 57.4 times its diameter, it forms an angle of 1°. This means that a 1-cm object placed at a distance of 57.4 cm from your eye would form an angle of 1°. If a 1-cm diameter object placed at a distance of 57.4 cm from your eye covers an angle of 1°, at this same distance, a 2-cm diameter object would cover 2°, and a 5-cm object 5°.

Now you can make a simple device that you can use to estimate angles of a few degrees.

Cut a series of stepwise slots (as shown in Fig. 2-60) in the index card. Mount the card vertically at the 57-cm mark on a meter stick. Cut flaps in the bottom of the card, fold them to fit along the stick, and tape the card to the stick (Fig. 2-61). Hold the zero end of the stick against your upper lip and observe. (Keep a stiff upper lip!)

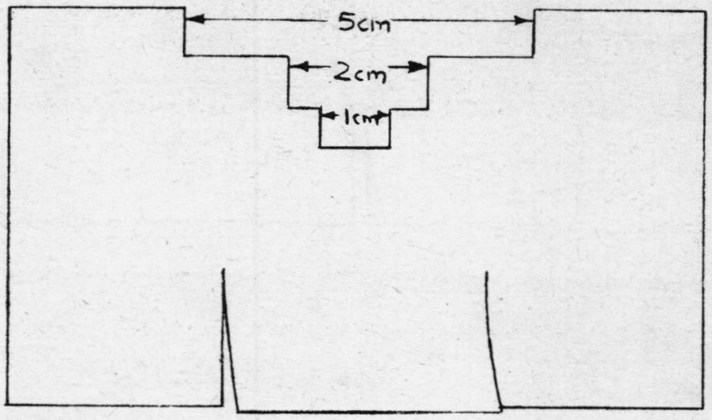

Fig. 2-60

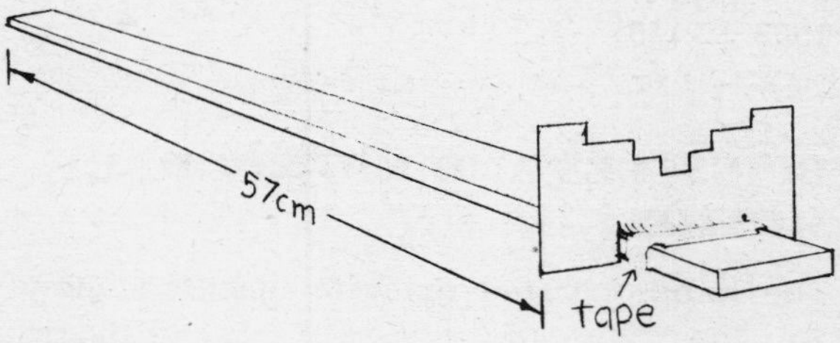

Fig. 2-61

Things to Observe

1. What is the visual angle between the pointers of the Big Dipper (see Fig. 2-62)?

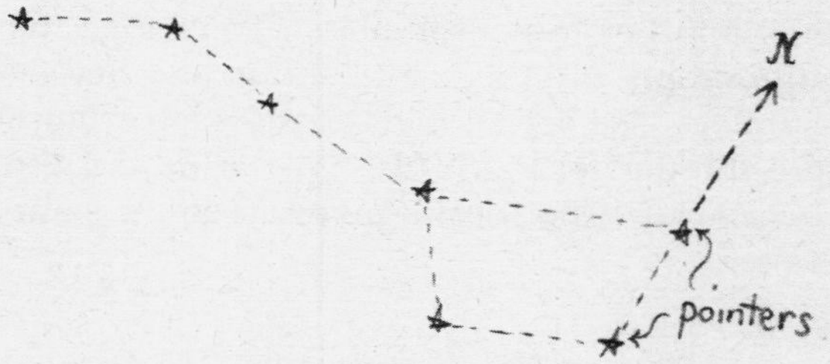

Fig. 2-62

2. What is the angular length of Orion's belt?
3. How many degrees away from the sun is the moon? Observe on several nights at sunset.
4. What is the angular diameter of the moon? Does it change between the time the moon

B.C. By John Hart

By permission of John Hart and Field Enterprises, Inc.

rises and the time when it is highest in the sky on a given night? To most people, the moon seems larger when near the horizon. Is it? See "The Moon Illusion," *Scientific American*, July, 1962, p. 120.

EPICYCLES AND RETROGRADE MOTION

The hand-operated epicycle machine allows you to explore the motion produced by two circular motions. You can vary both the ratio of the turning rates and the ratio of the radii to find the forms of the different curves that may be traced out.

The epicycle machine has three possible gear ratios: 2 to 1 (producing two loops per revolution), 1 to 1 (one loop per revolution), and 1 to 2 (one loop per two revolutions). To change the ratio, simply slip the drive band to another set of pulleys. The belt should be twisted in a figure 8 so the deferent arm (the long arm) and the epicycle arm (the short arm) rotate in the same direction.

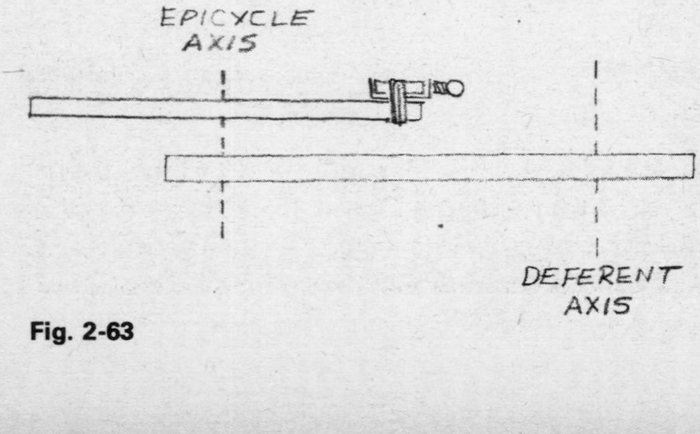

Fig. 2-63

Tape a light source (penlight cell, holder, and bulb) securely to one end of the short, epicycle arm and counterweight the other end of the arm with another (unlit) light source (Fig. 2-63). If you use a fairly high rate of rotation in a darkened room, you and other observers should be able to see the light source move in an epicycle.

The form of the curve traced depends not only on the gear ratio but also on the relative lengths of the arms. As the light is moved closer to the center of the epicycle arm, the epicycle loop decreases in size until it becomes a cusp (Fig. 2-64). When the light is very close to the center of the epicycle arm, as it would be for

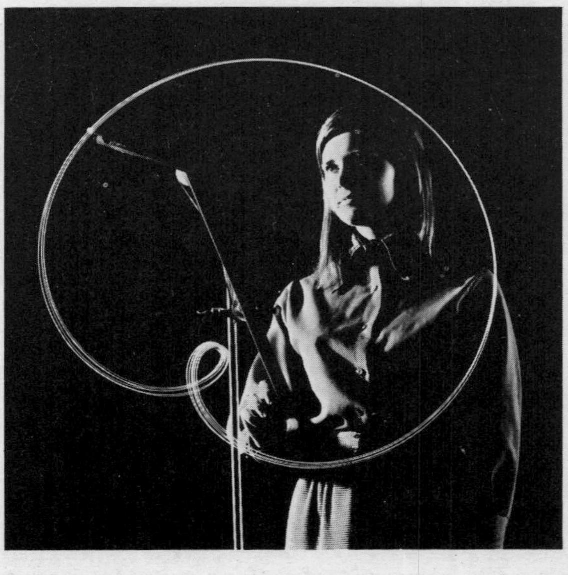

Fig. 2-64

the motion of the moon around the earth, the curve will be a slightly distorted circle, as shown in Fig. 2-65.

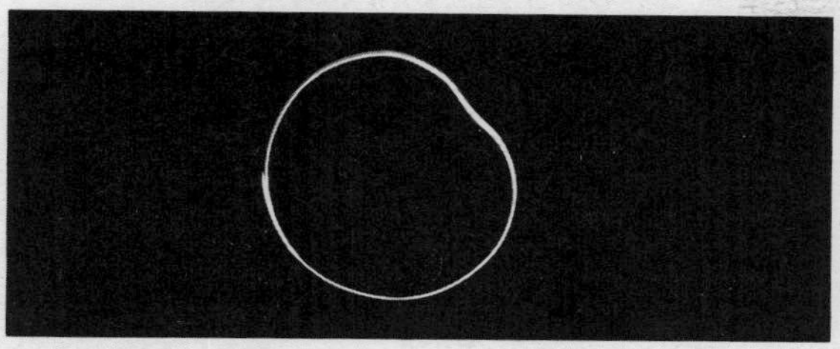

Fig. 2-65

To relate this machine to the Ptolemaic model, in which planets move in epicycles around the earth as a center, you should really stand at the center of the deferent arm (earth) and view the lamp against a distant fixed background. The size of the machine, however, does not allow you to do this, so you must view the motion from the side. (Or, you can glue a spherical glass Christmas-tree ornament at the center of the machine; the reflection you see in the bulb is just what you would see if you were at the center.) The lamp then goes into retrograde motion each time an observer in front of the machine sees a loop. The retrograde motion is most pronounced when the light source is far from the center of the epicycle axis.

Photographing Epicycles

The motion of the light source can be photographed by mounting the epicycle machine on a turntable and holding the center pulley stationary with a clamp (Fig. 2-66). Alternatively, the machine can be held in a burette clamp on a ringstand and turned by hand.

Fig. 2-66 An epicycle demonstrator connected to a turntable.

Fig. 2-67

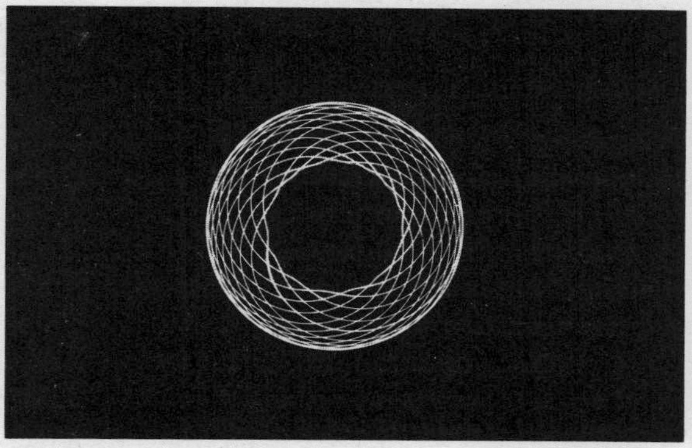

Fig. 2-68

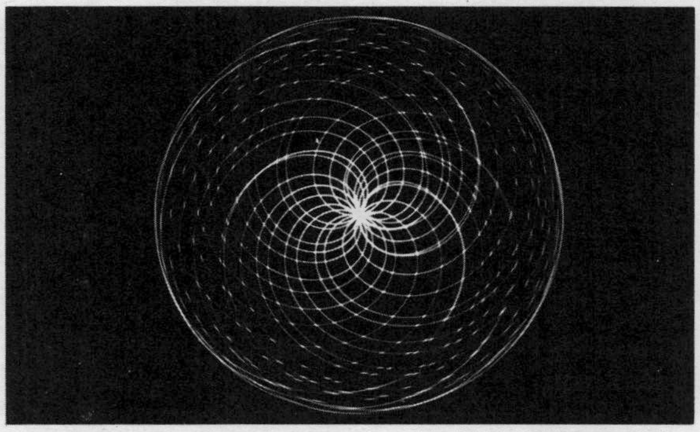

Fig. 2-69

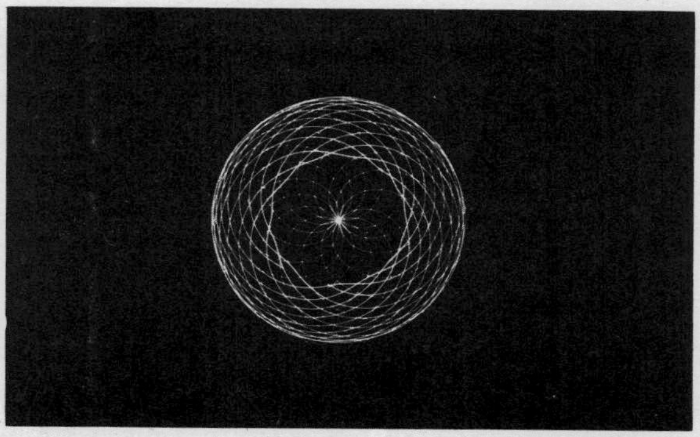

Fig. 2-70

Running the turntable for more than one revolution may show that the traces do not exactly overlap. (This probably occurs because the drive band is not of uniform thickness, particularly at its joint, or because the pulley diameters are not in exact proportion.) As the joining seam in the band runs over either pulley, the ratio of speeds changes momentarily and a slight displacement of the axes takes place. By letting the turntable rotate for some time, the pattern will eventually begin to overlap.

A time photograph of this motion can reveal very interesting geometric patterns. You might enjoy taking such pictures as an after-class activity. Figures 2-67 to 2-70 show four examples of the many different patterns that can be produced.

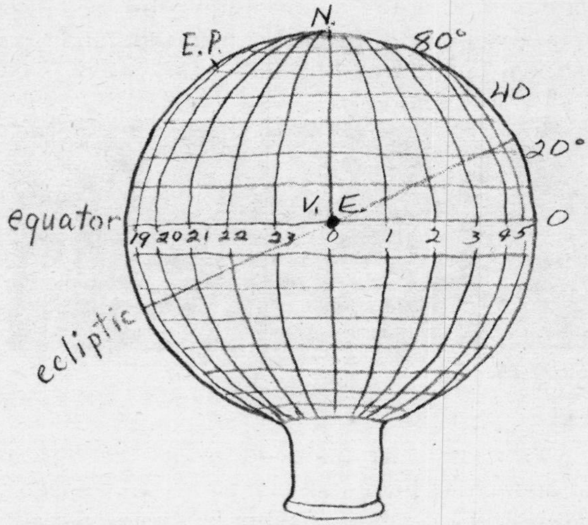

Fig. 2-71

CELESTIAL SPHERE MODEL*

You can make a model of the celestial sphere with a round-bottom flask. With it, you can see how the appearance of the sky changes as you go northward or southward and how the stars appear to rise and set.

To make this model, you will need, in addition to the round-bottom flask, a one-hole rubber stopper to fit its neck, a piece of glass tubing, paint, a fine brush (or grease pencil), a star map or a table of star positions, and considerable patience.

On the bottom of the flask, locate the point opposite the center of the neck. Mark this point and label it N for north celestial pole (Fig. 2-71). With a string or tape, determine the circumference of the flask, the greatest distance around it. This will be 360° in your model. Then, starting at the north celestial pole, mark points that are one-quarter of the circumference, or 90°, from the North Pole point. These points lie around the flask on a line that is the celestial equator. You can mark the equator with a grease pencil (china marking pencil), or with paint.

To locate the stars accurately on your "globe of the sky," you will need a coordinate system. If you do not wish to have the coordinate system marked permanently on your model, put on the lines with a grease pencil.

Mark a point 23.5° from the North Pole (about one-quarter of 90°). This will be the *pole of the ecliptic,* marked E.P. in Fig. 2-71. The ecliptic (path of the sun) will be a great circle 90° from the ecliptic pole. The point where the ecliptic crosses the equator from south to north is called the *vernal equinox*, the position of the sun on March 21. All positions in the sky are located eastward from this point, and north or south from the equator.

To set up the north–south scale, measure off eight circles, 10° apart, that run east and west in the northern hemisphere parallel to the equator. These lines are like latitude on the earth but are called *declination* in the sky. Repeat the construction of these lines of declination for the southern hemisphere.

A star's east–west position, called its *right ascension*, is recorded in hours eastward from the vernal equinox. To set up the east–west scale, mark intervals of 1/24th of the total circumference starting at the vernal equinox. These marks are 15° apart (rather than 10°) since the sky turns through 15° each hour.

From a table of star positions or a star map, you can locate a star's coordinates, then mark the star on your globe. All east–west positions are expressed eastward, or to the right of the vernal equinox as you face your globe.

To finish the model, put the glass tube into the stopper so that it almost reaches across the flask and points to your North Pole point. Then put enough inky water in the flask so that, when you hold the neck straight down, the water just comes up to the line of the equator. For safety,

*Adapted from *You and Science*, by Paul F. Brandwein, *et al.*, Harcourt, Brace, Jovanovich.

B.C.

By John Hart

YOU CERTAINLY PICKED A LOUSY NIGHT TO NAME THE CONSTELLATIONS.

By permission of John Hart and Field Enterprises, Inc.

wrap wire around the neck of the flask and over the stopper so it will not fall out (Fig. 2-72).

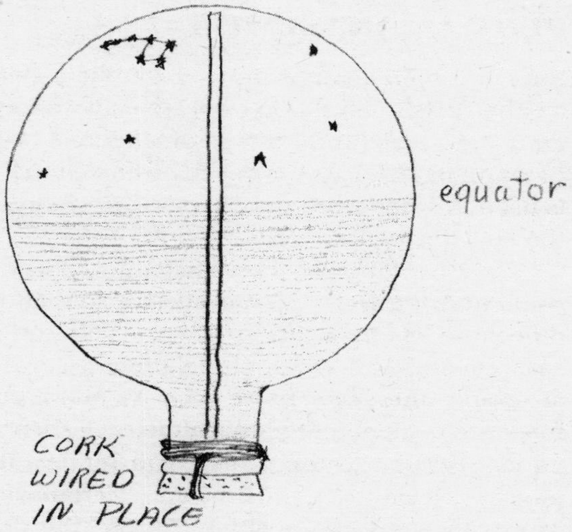
equator
CORK WIRED IN PLACE

Fig. 2-72

Now, as you tip the flask, you have a model of the sky as you would see it from different latitudes in the Northern Hemisphere. If you were at the earth's North Pole, the north celestial pole would be directly overhead and you would see only the stars in the northern half of the sky. If you were at latitude 45° N, the north celestial pole would be halfway between the horizon and the point directly overhead. You can simulate the appearance of the sky at 45° N by tipping the axis of your globe to 45° and rotating it. If you hold your globe with the axis horizontal, you would be able to see how the sky would appear if you were at the equator.

HOW LONG IS A SIDEREAL DAY?

A *sidereal day* is the time interval in which a star travels completely around the sky. To measure a sidereal day you need an electric clock and a screw eye.

Choose a neighboring roof or fence towards the west. Then fix a screw eye as an eyepiece in some rigid support such as a post or a tree so that a bright star, when viewed through the screw eye, will be a little above the roof (Fig. 2-73).

Fig. 2-73

Record the time when the star viewed through the screw eye just disappears behind the roof, then record the time again on the next night. How long did it take to go around? What is the uncertainty in your measurement? If you doubt your result, you can record times for several nights in a row and average the time intervals; this should give you a very accurate measure of a sidereal day. (If your result is not exactly 24 hours, calculate how many days would be needed for the difference to add up to 24 hours.)

SCALE MODEL OF THE SOLAR SYSTEM

Most drawings of the solar system are badly out of scale, because it is impossible to show both the sizes of the sun and planets and their relative distances on an ordinary-sized piece of paper. Constructing a simple scale model will help you develop a better picture of the real dimensions of the solar system.

Let a tennis ball about 7 cm in diameter represent the sun. The distance of the earth from the sun is 107 times the sun's diameter or, for this model, about 7.5 m. (You can confirm this easily. In the sky the sun has a diameter of 0.5°, about half the width of your thumb when held upright at arm's length in front of your nose. Check this, if you wish, by comparing your thumb to the angular diameter of the moon, which is nearly equal to that of the sun; both have diameters of 0.5°. Now hold your thumb in the same upright position and walk away from the tennis ball until its diameter is about half the width of your thumb. You will be between 7 and 8 m from the ball!) Since the diameter of the sun is about 1,400,000 km, in the model 1 cm represents about 200,000 km. From this scale, the proper scaled distances and sizes of all the other planets can be derived.

The moon has an average distance of 384,000 km from the earth and has a diameter of 3,476 km. Where is it on the scale model? How large is it? Completion of the column for the scale-model distances in Table 2-10) will yield some surprising results.

TABLE 2-10. A SCALE MODEL OF THE SOLAR SYSTEM.

Object	Solar Distance		Diameter		Sample Object
	AU	Model (cm)	km (approx.)	Model (cm)	
Sun	-------	--------	1,400,000	7	tennis ball
Mercury	0.39		4,600		
Venus	0.72		12,000		
Earth	1.00	750	13,000		pinhead
Mars	1.52		6.600		
Jupiter	5.20		140,000		
Saturn	9.45		120,000		
Uranus	19.2		48,000		
Neptune	30.0		45,000		
Pluto	39.5		6,000		
Nearest star	2.7×10^5				

The average distance between the earth and sun is called the *astronomical unit* (AU). This unit is used to describe distance within the solar system.

BUILD A SUNDIAL

If you are interested in building a sundial, there are numerous articles in the "Amateur Scientist" section of *Scientific American* that you will find helpful. See particularly the article in the issue of August 1959. Also see the issues of September 1953, October 1954, October 1959, or March 1964. The book *Sundials* by Mayall and Mayall (Charles T. Branford Co., publishers, Boston) gives theory and building instructions for a wide variety of sundials. Encyclopedias also have helpful articles.

PLOT AN ANALEMMA

Have you seen an analemma? Examine a globe of the earth, and you will usually find a graduated scale in the shape of a figure 8, with dates on it. This figure is called an *analemma.* It is used to summarize the changing positions of the sun during the year.

You can plot your own analemma. Place a small mirror on a horizontal surface so that the reflection of the sun at noon falls on a south-facing wall. Make observations each day at exactly the same time, such as noon, and mark the position of the reflection on a sheet of paper fastened to the wall. If you remove the paper each day, you must be sure to replace it in exactly the same position. Record the date beside the point. The north–south motion is most evident during September–October and March–April. You can find more about the east–west migration of the marks in astronomy texts and encyclopedias under the subject "equation of time."

STONEHENGE

Stonehenge (see Text p. 130) has been a mystery for centuries. Some scientists have thought that it was a pagan temple, others that it was a monument to slaughtered chieftains. Legends invoked the power of Merlin to explain how the stones were brought to their present location. Recent studies indicate that Stonehenge may have been an astronomical observatory and eclipse computer.

Read "Stonehenge Physics," in the April, 1966 issue of *Physics Today; Stonehenge Decoded*, by G.S. Hawkins and J.B. White (Dell, 1966); or see *Scientific American*, June, 1953. Make a report and/or a model of Stonehenge for your class.

MOON CRATER NAMES

Prepare a report about how some of the moon craters were named. See Isaac Asimov's *Biographical Encyclopedia of Science and Technology* for material about some of the scientists whose names were used for craters.

LITERATURE

The astronomical models that you read about in Chapters 5 and 6, Unit 2, of the Text strongly influenced the Elizabethan view of the world and the universe. In spite of the ideas of Galileo and Copernicus, writers, philosophers, and theologians continued to use Aristotelian and Ptolemaic ideas in their works. In fact, there are many references to the crystal-sphere model of the universe in the writings of Shakespeare, Donne, and Milton. The references often are subtle because the ideas were commonly accepted by the people for whom the works were written.

For a quick overview of this idea, with reference to many authors of the period, read the paperbacks *The Elizabethan World Picture*, by E.M.W. Tillyard (Vintage Press) or Basil Willey, *Seventeenth Century Background* (Doubleday).

An interesting specific example of the prevailing view, as expressed in literature, is found in Christopher Marlowe's *Doctor Faustus*, when Faustus sells his soul in return for the secrets of the universe. Speaking to the devil, Faustus says:

> "...Come, Mephistophilis, let us
> dispute again
> And argue of divine astrology.
> Tell me, are there many heavens
> above the moon?
> Are all celestial bodies but one
> globe
> As is the substance of this centric
> earth?..."

FRAMES OF REFERENCE

1. Two students, A and B, take hold of opposite ends of a meter stick or a piece of string one or two meters long. If A rotates about on one fixed spot so that A is always facing B while B walks around A in a circle, A will see B against a background of walls and furniture. How does A appear to B? Ask B to describe how A appears against the background of walls and furniture. How do the reports compare? In what direction did A see B move, toward the left or right? In which direction did B see A move, toward the left or right?

2. A second demonstration involves a camera, tripod, blinky, and turntable. Mount the camera on the tripod (using a motor-strobe bracket if the camera has no tripod connection) and put the blinky on a turntable. Aim the camera straight down (Fig. 2-74).

Fig. 2-74

Take a time exposure with the camera at rest and the blinky moving one revolution in a circle (Fig. 2-75). If you do not use the turntable, move

Fig. 2-75

the blinky by hand around a circle drawn faintly on the background. Then take a second print with the blinky at rest and the camera, on time exposure, moved steadily by hand about the axis of the tripod. Try to move the camera at the same rotational speed as the blinky moved in the first photo.

Can you tell, just by looking at the photos, whether the camera or the blinky was moving?

DEMONSTRATING SATELLITE ORBITS

A piece of thin plastic or a rubber sheet can be stretched tight and clamped in an embroidery hoop about 55 cm in diameter. Place the hoop on some books and put a heavy ball, for example, a 5-cm-diameter steel ball bearing, in the middle of the plastic (Fig. 2-76). The plastic will sag so that there is a greater force toward the center on the ball when it is closer to the center than when it is farther away.

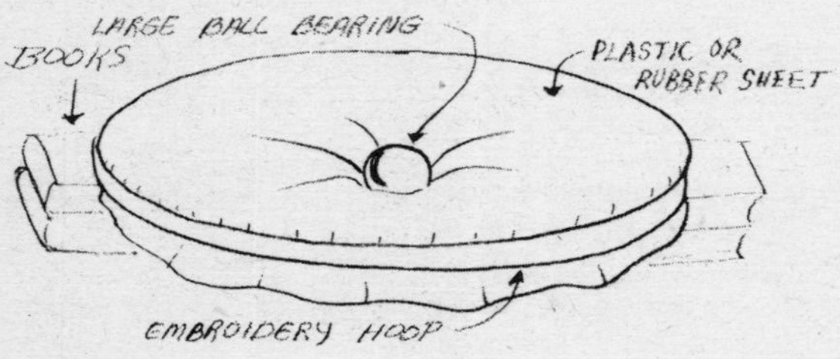

Fig. 2-76

You can use a smaller hoop on the stage of an overhead projector. Use small ball bearings, marbles, or beads as "satellites." Then you will have a shadow projection of the large central mass, with the small satellites racing around it. Be careful not to drop the ball through the glass.

If you take strobe photos of the motion, you can check whether Kepler's three laws are satisfied; you can see where satellites travel fastest in their orbits, and how the orbits themselves turn in space. To take the picture, set up the hoop on the floor with black paper under it.

You can use either the electronic strobe light or the slotted disk stroboscope to take the pictures. In either case, place the camera directly over the hoop and the light source at the side, slightly enough above the plane of the hoop so that the floor under the hoop is not well lighted (Fig. 2-77). A ball bearing or marble will make the best pictures.

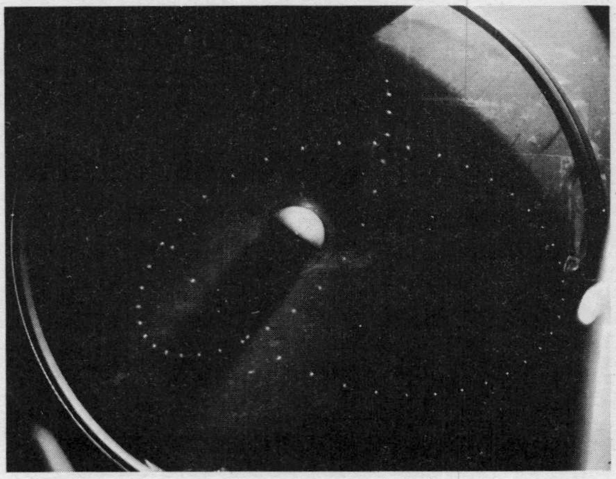

Fig. 2-77

Here are some questions to think about:

1. Does your model give a true representation of the gravitational force around the earth? In what ways does the model fail?

2. Is it more difficult to put a satellite into a perfectly circular orbit than into an elliptical one? What conditions must be satisfied for a circular orbit?

3. Are Kepler's three laws really verified? Should they be?

For additional detail and ideas see "Satellite Orbit Simulator," *Scientific American*, October, 1958.

GALILEO

Read Bertolt Brecht's play, *Galileo*, and present a part of it for the class. There is some controversy about whether the play truly reflects what historians believe were Galileo's feelings. For comparison, you could read *The Crime of Galileo*, by Giorgio de Santillana; *Galileo and the Scientific Revolution*, by Laura Fermi; The Galileo Quadricentennial Supplement in *Sky and Telescope*, February 1964; or articles in the April, 1966 issue of *The Physics Teacher*, "Galileo: Antagonist," and "Galileo Galilei: An Outline of His Life."

CONIC SECTIONS MODELS

Obtain from a mathematics teacher a demonstration cone thas has been cut along several different planes so that when it is taken apart the planes form the four conic sections.

If such a cone is not available, tape a cone of paper to the front of a small light source, such as a flashlight bulb. Shine the light on the wall and tilt the cone at different angles with respect to the wall. You can make all the conic sections shown in Sec. 7.3 of the Text.

If you have a wall lamp with a circular shade, the shadows cast on the wall above and below the lamp are usually hyperbolas. You can check this by tracing the curve on a large piece of paper, and seeing whether the points satisfy the definition of a hyperbola.

Fig. 2-78 *M. Babinet prévenu par sa portière de la visite de la comète.* A lithograph by the French artist Honoré Daumier (1808-1879), Museum of Fine Arts, Boston.

CHALLENGING PROBLEM: FINDING EARTH – SUN DISTANCE FROM VENUS PHOTOS

Here's a teaser: Assume that Venus has the same diameter as the earth. Also assume that the scale of the pictures on page 196 of the Text is 1.5 seconds of arc per millimeter.

Determine the distance from the earth to the sun in kilometers.

MEASURING IRREGULAR AREAS

Are you tired of counting squares to measure the area of irregular figures? A device called a *planimeter* can save you much drudgery. There are several styles, ranging from a simple pocket knife to a complex arrangement of worm gears and pivoted arms. See the "Amateur Scientist" section of *Scientific American*, August, 1958 and February, 1959.

OTHER COMET ORBITS

If you enjoyed making a model of the orbit of Halley's comet, you may want to make models of some other comet orbits. Data are given below for several other comets of interest.

Encke's comet is interesting because it has the shortest period known for a comet, only 3.3 years. In many ways, it is representative of all short-period comets that have orbits of low inclination and pass near the orbit of Jupiter, where they are often strongly deviated. The full ellipse can be drawn at the scale of 10 cm for 1 AU. The orbital elements for Encke's comet are:

$a = 2.22$ AU	$\Omega = 335°$
$e = 0.85$	$\omega = 185°$
$i = 15°$	

From these data you can calculate that the perihelion distance R_p is 0.33 AU and the aphelion distance R_a is 4.11 AU.

The comet of 1680 is discussed extensively in Newton's *Principia*, where approximate orbital elements are given. The best parabolic orbital elements known are:

$T =$ Dec. 18, 1680	$i = 60.16°$
$\omega = 350.7°$	$R_p = 0.00626$ AU
$\Omega = 272.2°$	

Note that this comet passed very close to the sun. At perihelion, it must have been exposed to intense destructive forces like the comet of 1965.

Comet Candy had the following parabolic orbital elements:

$T =$ Feb. 8, 1961	$i = 150.9°$
$\omega = 136.3°$	$R_p = 1.06$ AU
$\Omega = 176.6°$	

DRAWING A PARABOLIC ORBIT

The parabola is an unusual conic section whose eccentricity is exactly 1. Geometrically, it has the interesting property that all points on a parabola are equidistant both from the focus and from a line perpendicular to the major axis and twice the perihelion distance from the focus. This construction line is called the *directrix*. The geometrical property permits a quick development of points on a parabola, as Fig. 2-79 indicates.

Along the major axis, locate a point that is twice the distance to the perihelion. At that point draw a line perpendicular to the major axis. Then with a drawing compass swing an arc of any length from the focus. Without

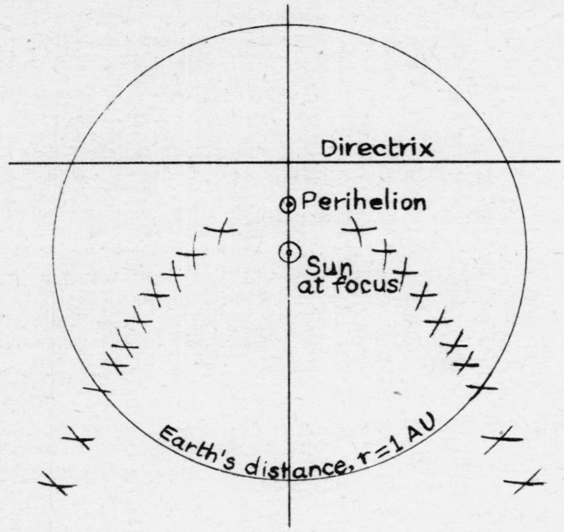

Fig. 2-79 Parabola for orbit with perihelion distance q = 0.20 AU.

changing the size of the arc, locate on the directrix a point such that an arc drawn from there will intersect the first arc as far as possible from the directrix; the line from the directrix to that intersection will be parallel to the major axis. By changing the size of the arc, you can establish a series of points on the parabola. Then draw a smooth curve through the points.

The number of days for a body moving around the sun in a parabolic orbit to move from a given solar distance to perihelion is given in the accompanying table (Table 2-11). With it, and the date of perihelion, you can establish the dates at which a comet was at any point on its parabolic orbit.

TABLE 2-11

Solar Distance	Perihelion Distance q (AU)						
× AU	0.0	0.2	0.4	0.6	0.8	1.0	1.2
2.0	77.8	88.1	97.1	105.8	108.0	109.6	107.8
1.8	66.1	78.2	84.3	98.0	93.2	93.0	88.6
1.6	56.1	64.8	72.0	76.7	78.2	76.0	69.4
1.4	45.4	56.0	60.3	63.8	63.4	59.0	46.8
1.2	38.2	43.9	48.9	50.7	49.5	50.0	0.0
1.0	29.4	36.3	38.0	36.2	31.9	0.0	
0.8	19.6	23.4	27.8	24.5	0.0		
0.7	16.7	21.3	22.5	14.5			
0.6	18.8	17.5	17.2	0.0			
0.5	9.7	15.5	11.3				
0.4	6.9	9.8	0.0				
0.3	4.5	6.1					
0.2	2.6	0.0					
0.1	0.9						

FORCES ON A PENDULUM

If a pendulum is drawn aside and released with a small sideways push, it will move in an almost elliptical path. This looks vaguely like the motion of a planet about the sun, but there are some differences.

To investigate the shape of the pendulum orbit and see whether the motion follows the law of areas, you can make a strobe photo with the setup shown in Fig. 2-80. Use either an electronic strobe flashing from the side, or use a small light and AA battery cell on the pendulum and a motor strobe disk in front of the lens. If you put tape over one slot of a 12-slot disk to make it half as wide as the rest, it will make every twelfth dot fainter, giving a handy time marker, as shown in Fig. 2-81. You can also set the camera on its back on the floor with the motor strobe above it, and suspend the pendulum overhead.

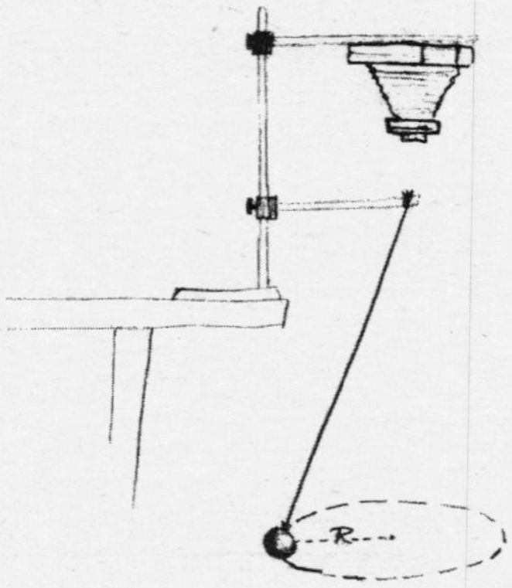

Fig. 2-80

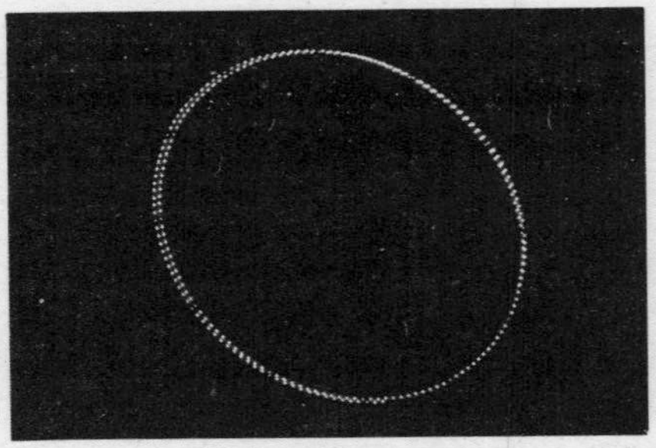

Fig. 2-81

Are the motions and the forces similar for the pendulum and the planets? The center of force for planets is located at one focus of the ellipse. Where is the center of force for the pendulum? Measure your photos to determine whether the pendulum bob follows the law of areas for motion under a central force.

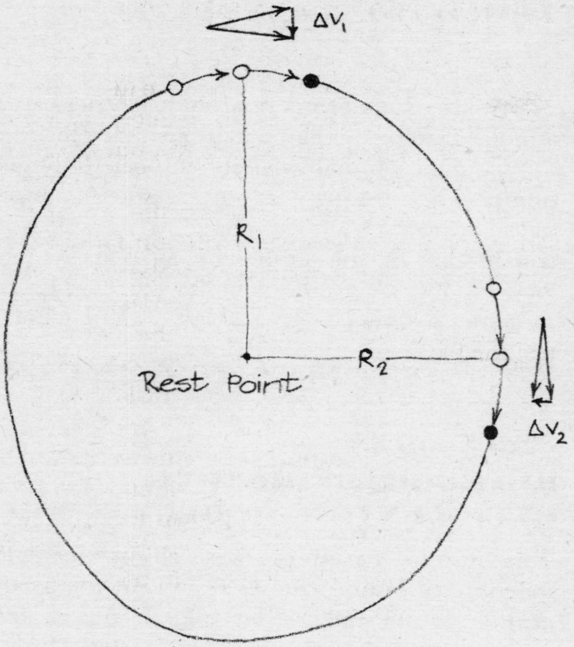

Fig. 2-82

In the case of the planets, the force varies inversely with the square of the distance between the sun and the planet. From your photograph, you can find how the restoring force on the pendulum changes with distance R from the rest point (Fig. 2-82). Find Δv between strobe flashes for two sections of the orbit, one near and one far from the rest point. How do the accelerations as indicated by the Δv's compare with the distances R? Does the restoring force depend on distance in the same way as it does for a planet? If you have a copy of Newton's *Principia* available, read Proposition X.

TRIAL OF COPERNICUS

Hold a mock trial for Copernicus. Have two groups of students represent the prosecution and the defense. If possible, have English, social studies, and language instructors serve as the jury for your trial.

FILM LOOP NOTES

Film Loop 10
RETROGRADE MOTION: GEOCENTRIC MODEL

The film illustrates the motion of a planet such as Mars, as seen from the earth. It was made using a large "epicycle machine" as a model of the Ptolemaic system (Fig. 2-83).

First, from above, you see the characteristic retrograde motion during the "loop" when the planet is closest to the earth. Then the studio lights go up and you see that the motion is due to the combination of two circular motions. One arm of the model rotates at the end of the other.

The earth, at the center of the model, is then replaced by a camera that points in a fixed direction in space. The camera views the motion of the planet relative to the fixed stars (the rotation of the earth on its axis is being ignored). This is the same as if you were looking at the stars and planets from the earth toward one constellation of the zodiac, such as Sagittarius.

The planet, represented by a white globe, is seen along the plane of motion. The direct motion of the planet, relative to the fixed stars, is eastward, toward the left (as it would be if you were facing south). A planet's retrograde motion does not always occur at the same place in the sky, so some retrograde motions are not visible in the chosen direction of observation. To simulate observations of planets better, an additional three retrograde loops were photographed using smaller bulbs and slower speeds.

Note the changes in apparent brightness and angular size of the globe as it sweeps close to the camera. Actual planets appear only as points of light to the eye, but a marked change in brightness can be observed. This was not considered in the Ptolemaic system, which focused only on positions in the sky. Film Loop 11 shows a similar model based on a heliocentric theory.

Film Loop 11
RETROGRADE MOTION: HELIOCENTRIC MODEL

This film is based on a large heliocentric mechanical model. Globes represent the earth and a planet moving in concentric circles around the sun (represented by a yellow globe). The earth (represented by a light blue globe) passes inside a slower moving outer planet such as Mars (represented by an orange globe).

Then the earth is replaced by a camera having a 25° field of view. The camera points in a fixed direction in space, indicated by an arrow, thus ignoring the daily rotation of the earth and concentrating on the motion of the earth relative to the sun.

The view from the moving earth is shown for more than one year. First the sun is seen in direct motion, then Mars comes to opposition and undergoes a retrograde motion loop, and finally you see the sun again in direct motion.

Scenes are viewed from above and along the plane of motion. Retrograde motion occurs whenever Mars is in opposition, that is,

Fig. 2-83

whenever Mars is opposite the sun as viewed from the earth. Not all these oppositions take place when Mars is in the sector the camera sees. The time between oppositions averages about 2.1 years. The film shows that the earth moves about 2.1 times around its orbit between oppositions of Mars.

You can calculate this value. The earth makes one cycle around the sun per year and Mars makes one cycle around the sun every 1.88 years. So the frequencies of orbital motion are:

$$f_{\text{earth}} = 1 \text{ cyc/yr and } f_{\text{mars}} = 1 \text{ cyc/1.88 yr} = 0.532 \text{ cyc/yr}$$

The frequency of the earth relative to Mars is $f_{\text{earth}} - f_{\text{mars}}$:

$$f_{\text{earth}} - f_{\text{mars}} = 1.00 \text{ cyc/yr} - 0.532 \text{ cyc/yr} = 0.468 \text{ cyc/yr}$$

That is, the earth catches up with and passes Mars once every

$$\frac{1}{0.468} = 2.14 \text{ years.}$$

Note the increase in apparent size and brightness of the globe representing Mars when it is nearest the earth. Viewed with the naked eye, Mars shows a large variation in brightness (ratio of about 50:1) but always appears to be only a point of light. Through a telescope, the angular size also varies as predicted by the model.

The heliocentric model is in some ways simpler than the geocentric model of Ptolemy, and gives the general features observed for the planets: angular position, retrograde motion, and variation in brightness. However, detailed numerical agreement between theory and observation cannot be obtained using circular orbits.

Film Loop 12
JUPITER SATELLITE ORBIT

This time-lapse study of the orbit of Jupiter's satellite, Io, was filmed at the Lowell Observatory in Flagstaff, Arizona, using a 60-cm refractor telescope.

Exposures were made at 1-min intervals during seven nights. An almost complete orbit of Io is reconstructed using all these exposures.

The film first shows a segment of the orbit as photographed at the telescope; a clock shows the passage of time. Due to small errors in guiding the telescope, and atmospheric turbulence, the highly magnified images of Jupiter and its satellites dance about. To remove this unsteadiness, each image (over 2100) was optically centered in the frame. The stabilized images were joined to give a continuous record of the motion of Io. Some variation in brightness was caused by haze or cloudiness.

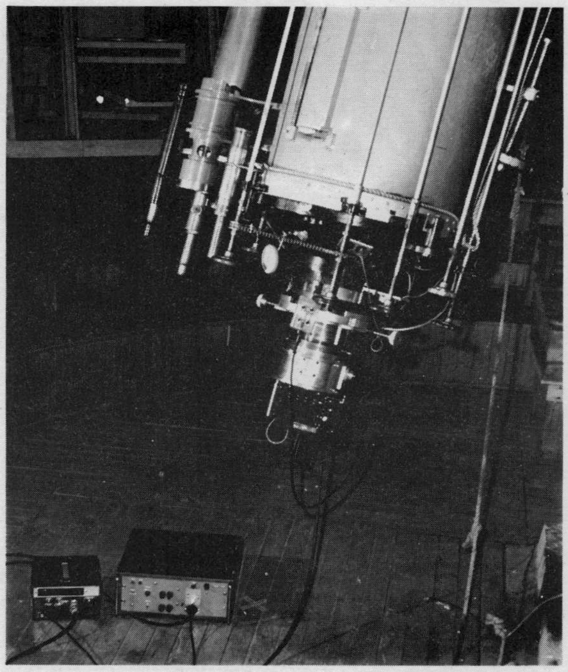

Fig. 2-84 Business end of the 60-cm refractor at Lowell Observatory.

The four Galilean satellites are listed in Table 2-12. On Feb. 3, 1967, they had the configuration shown in Fig. 2-85. The satellites move nearly in a plane viewed almost edge-on; thus, they seem

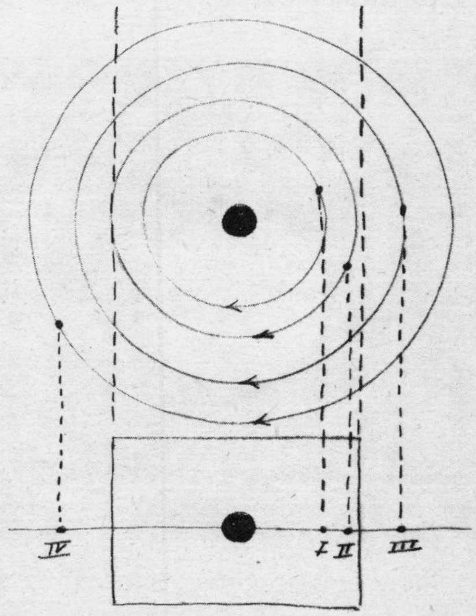

Fig. 2-85

TABLE 2-12. SATELLITES OF JUPITER

	Name	Period	Radius of Orbit (kilometers)	Eccentricity of Orbit	Diameter (kilometers)
I	Io	$1^d\ 18^h\ 28^m$	419,200	0.0000	3,200
II	Europa	$3^d\ 13^h\ 14^m$	667,200	0.0003	2,880
III	Ganymede	$7^d\ 3^h\ 43^m$	1,065,600	0.0015	4,960
IV	Callisto	$16^d\ 16^h\ 32^m$	1,873,600	0.0075	4,480

to move back and forth along a line. The field of view is large enough to include the entire orbits of I and II, but III and IV are outside the camera field when they are farthest from Jupiter.

The position of Io in the last frame of the Jan. 29 segment matches the position in the first frame of the Feb. 7 segment. However, since these frames were photographed nine days apart, the other three satellites had moved, so you can see them pop in and out while the image of Io is continuous. Lines identify Io in each section. Fix your attention on the steady motion of Io and ignore the comings and goings of the other satellites.

Interesting Features of the Film

1. At the start, Io appears almost stationary at the right, at its greatest elongation; another satellite is moving toward the left and overtakes it.

2. As Io moves toward the left (Fig. 2-86), it passes in front of Jupiter, a *transit*. Another satellite, Ganymede, has a transit at about the same time. Another satellite moves toward the right and disappears behind Jupiter, an *occulation*. It is a very active scene! If you look closely during the transit, you may see the shadow of Ganymede and perhaps that of Io, on the left part of Jupiter's surface.

Fig. 2-86 Still photograph from Film Loop 12 showing the positions of three satellites of Jupiter at the start of the transit and occultation sequence. Satellite IV is out of the picture, far to the right of Jupiter.

3. Near the end of the film, Io (moving toward the right) disappears; an occulation begins. Look for Io's reappearance; it emerges from an eclipse and appears to the right of Jupiter. Note that Io is out of sight part of the time because it is behind Jupiter as viewed from the earth, and part of the time because it is in Jupiter's shadow. It cannot be seen as it moves from D to E in Fig. 2-87.

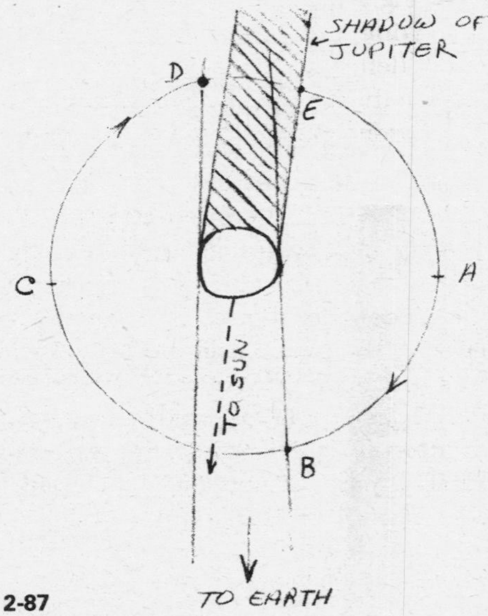

Fig. 2-87

4. Jupiter is seen as a flattened circle because its rapid rotation period ($9^h\ 55^m$) has caused it to flatten at the poles and bulge at the equator. The effect is quite noticeable: The equatorial diameter is 142,720 km and the polar diameter is 133,440 km.

Measurements

1. *Period of orbit.* Time the motion between transit and occulation (from B to D in Fig. 2-87), half a revolution, to find the period. The film is projected at about 18 frames/sec, so that the speed-up factor is 18 × 60, or 1,080. How can you calibrate your projector more accurately?

(There are 3,969 frames in the loop.) How does your result for the period compare with the value given in Table 2-12?

2. *Radius of orbit.* Project on paper and mark the two extreme positions of the satellite, farthest to the right (at A) and farthest to the left (at C). To find the radius in kilometers, use Jupiter's equatorial diameter for a scale.

3. *Mass of Jupiter.* You can use your values for the orbit radius and period to calculate the mass of Jupiter relative to that of the sun (a similar calculation based on the satellite Callisto is given in the Text). How does your experimental result compare with the accepted value, which is $m_j/m_s = 1/1{,}048$?

Film Loop 13
PROGRAM ORBIT I

A student (Fig. 2-88, right) is plotting the orbit of a planet, using a stepwise approximation. His instructor (left) is preparing the computer program for the same problem. The computer and the student follow a similar procedure.

Fig. 2-88

The computer "language" used was FORTRAN. The FORTRAN program (on a stack of punched cards) consists of the "rules of the game": the laws of motion and of gravitation. These rules describe precisely how the calculation is to be done. The program is translated and stored in the computer's memory.

The calculation begins with the choice of initial position and velocity of the planet. The initial position values of X and Y are selected and also the initial components of velocity XVEL and YVEL (XVEL is the name of a single variable, not a product of four variables).

Then the program instructs the computer to calculate the force on the planet from the sun from the inverse-square law of gravitation. Newton's laws of motion are used to calculate how far and in what direction the planet moves after each blow.

The computer's calculations can be displayed in several ways. A table of X and Y values can be typed or printed. An X– Y plotter can draw a graph from the values, similar to the hand-constructed graph made by the student. The computer results can also be shown on a cathode-ray tube (CRT), similar to that in a television set, in the form of a visual trace. In this film, the X– Y plotter was the mode of display used.

The dialogue between the computer and the operator for Trial 1 is as follows. The numerical values are entered at the computer typewriter by the operator after the computer types the message requesting them.

Computer:	GIVE ME INITIAL POSITION IN AU . . .
Operator:	X = 4 Y = 0
Computer:	GIVE ME INITIAL VELOCITY IN AU/YR . . .
Operator:	XVEL = 0 YVEL = 2
Computer:	GIVE ME CALCULATION STEP IN DAYS . . .
Operator:	60
Computer:	GIVE ME NUMBER OF STEPS FOR EACH POINT PLOTTED . . .
Operator:	1
Computer:	GIVE ME DISPLAY MODE . . .
Operator:	X– Y PLOTTER

You can see that the orbit displayed on the X– Y plotter, like the student's graph, does not close. This is surprising, as you know that the orbits of planets are closed. Both orbits fail to close exactly. Perhaps too much error is introduced by using such large steps in the step-by-step approximation. The blows may be too infrequent near perihelion, where the force is largest, to be a good approximation to a continuously acting force. In Film Loop 14, "Program Orbit II," the calculations are based upon smaller steps, and you can see if this explanation is reasonable.

Film Loop 14
PROGRAM ORBIT II

In this continuation of the film "Program Orbit I," a computer is again used to plot a planetary orbit with a force inversely proportional to the square of the distance. The computer program adopts Newton's laws of motion. At equal intervals, blows act on the body. The orbit calculated in the previous film probably failed to close because the blows were spaced too far apart. You could calculate the orbit using many more blows, but to do this by hand would require much more time and effort. In the computer calculation, you need only specify a smaller time interval between the calculated points. The laws of motion are the same as before, so the same program is used.

A portion of the "dialogue" between the computer and the operator for Trial 2 is as follows:

Computer:	GIVE ME CALCULATION STEP IN DAYS . . .
Operator:	3
Computer:	GIVE ME NUMBER OF STEPS FOR EACH POINT PLOTTED . . .
Operator:	7
Computer:	GIVE ME DISPLAY MODE . . .
Operator:	X – Y PLOTTER.

Points are now calculated every three days (20 times as many calculations as for Trial 1 on the "Program Orbit I" film), but to avoid a graph with too many points, only one out of seven of the calculated points is plotted.

The computer output in this film can also be displayed on the face of a cathode-ray tube (CRT). The CRT display has the advantage of speed and flexibility and is used in the other loops in this series, Film Loops 15, 16, and 17. On the other hand, the permanent record produced by the X – Y plotter is sometimes very convenient.

Orbit Program

The computer program for orbits is written in FORTRAN II and includes "ACCEPT" (data) statements used on an IBM 1620 input typewriter (Fig. 2-89).

With slight modification, the program worked on a CDC 3100 and CDC 3200, as shown in Film Loops 13 and 14, "Program Orbit I" and "Program Orbit II." With additional slight modifications (in statement 16 and the three succeeding statements in Fig. 2-89) it can be

```
      PROGRAM ORBIT
C
C        HARVARD PROJECT PHYSICS ORBIT PROGRAM.
C        EMPIRICAL VERIFICATION OF KEPLERS LAWS
C        FROM NEWTONS LAW OF UNIVERSAL GRAVITATION.
C
      G=40.
    4 CALL MARKF(0.,0.)
    6 PRINT 7
    7 FORMAT(9HGIVE ME Y )
      X=0.
      ACCEPT 5,Y
      PRINT 8
    8 FORMAT(12HGIVE ME XVEL)
    5 FORMAT(F10.6)
      ACCEPT 5,XVEL
      YVEL=0.
      PRINT 9
    9 FORMAT(49HGIVE ME DELTA IN DAYS,  AND NUMBER BETWEEN PRINTS)
      ACCEPT 5,DELTA
      DELTA=DELTA/365.25
      ACCEPT 5,PRINT
      IPRINT = PRINT
      INDEX = 0
      NFALLS = 0
   13 CALL MARKF(X,Y)
      PRINT 10,X,Y
   15 IF(SENSE SWITCH 3) 20,16
   20 PRINT 21
   10 FORMAT(2F7.3)
      NFALLS = NFALLS + IPRINT
   21 FORMAT(23HTURN OFF SENSE SWITCH 3 )
   22 CONTINUE
      IF(SENSE SWITCH 3) 22,4
   16 RADIUS = SQRTF(X*X + Y*Y)
      ACCEL = -G/(RADIUS*RADIUS)
      XACCEL = (X/RADIUS)*ACCEL
      YACCEL = (Y/RADIUS)*ACCEL
C  FIRST TIME THROUGH WE WANT TO GO ONLY 1/2 DELTA
      IF(INDEX) 17,17,18
   17 XVEL = XVEL + 0.5 * XACCEL * DELTA
      YVEL = YVEL + 0.5 * YACCEL * DELTA
      GO TO 19
C  DELTA V = ACCELERATION TIMES DELTA T
   18 XVEL = XVEL + XACCEL * DELTA
      YVEL = YVEL + YACCEL * DELTA
C  DELTA X = XVELOCITY TIMES DELTA T
   19 X = X + XVEL * DELTA
      Y = Y + YVEL * DELTA
      INDEX = INDEX + 1
      IF(INDEX - NFALLS) 15,15,13
      END
```

Fig. 2-89

used for other force laws. A similar program is presented and explained in *FORTRAN for Physics* by Alfred M. Bork (Addison-Wesley, 1967).

Note that it is necessary to have a subroutine MARK. In this case, it is used to plot the points on an X – Y plotter, but MARK could be replaced by a PRINT statement to print the X and Y coordinates.

Film Loop 15
CENTRAL FORCES: ITERATED BLOWS

In Chapter 8 and in Experiment 2-11 and Film Loop 13 on the stepwise approximation of orbits, Kepler's law of areas applies to objects acted on by a central force. The force in each case was attractive and was either constant or varied smoothly according to some pattern. But suppose the central force is repulsive; that is, directed *away* from the center? Suppose it is sometimes attractive and sometimes repulsive? What if the amount of force applied each time varies unsystematically? Under these circumstances, would the law of areas still hold? You can use this film to find out.

The film was made by photographing the face of a cathode-ray tube (CRT) that displayed the output of a computer. It is important to realize the role of the computer program in this film. It controlled the change in direction and change in speed of the "object" as a result of a "blow." This is how the computer program uses Newton's laws of motion to predict the result of applying a brief impulsive force, or blow. The program remained the same for all parts of the loop, just as Newton's laws remain the same during all experiments in a laboratory. However, at one place in the program, the operator had to specify how he wanted the force to vary.

Random Blows

Figure 2-90 shows part of the motion of the body as blows are repeatedly applied at equal time intervals. No one decided in advance how great each blow was to be. The computer was programmed to select a number at random to represent the magnitude of the blow. The directions toward or away from the center were also selected at random, although a slight preference for attractive blows was built in so that the pattern would be likely to stay on the face of the CRT. The dots appear at equal time intervals. The intensity and direction of each blow is represented by the length of line at the point of the blow.

Study the photograph. How many blows were attractive? How many were repulsive? Were any blows so small as to be negligible?

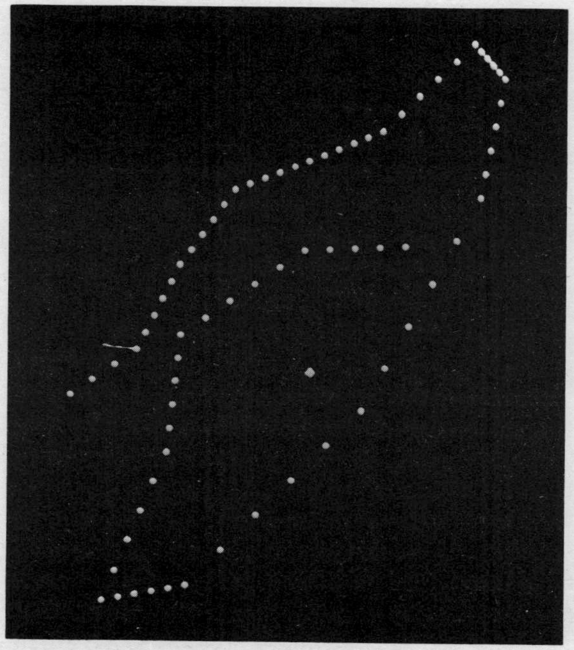

Fig. 2-90

You can see if the law of areas applies to this random motion. Project the film on a piece of paper, mark the center, and mark the points where the blows were applied. Now measure the areas of the triangles. Does the moving body sweep over equal areas in equal time intervals?

Force Proportional to Distance

If a weight on a string is pulled back and released with a sideways shove, it moves in an elliptical orbit with the force center (lowest point) at the center of the ellipse. A similar path is traced on the CRT in this segment of the film. Notice how the force varies at different distances from the center. A smooth orbit is approximated by the computer by having the blows come at shorter time intervals. In frame 2(a), four blows are used for a full orbit; in 2(b) there are nine blows, and in 2(c), 20 blows, which give a good approximation of the ellipse that is observed with this force. Geometrically, how does this orbit differ from planetary orbits? How is it different physically?

Inverse-Square Force

A similar program is used with two planets simultaneously, but with a force on each varying *inversely* as the *square* of the distance from a force center. Unlike the real situation, the program assumes that the planets do not exert forces on one another. For the resulting ellipses, the force center is at one *focus* (Kepler's first law), not at the center of the ellipse as in the previous case.

In this film, the computer has done thousands of times faster what you could do if you had enormous patience and time. With the computer you can change conditions easily, and thus investigate many different cases and display the results. Once told what to do, the computer makes fewer calculation errors than a person.

Film Loop 16
KEPLER'S LAWS

A computer program similar to that used in the film "Central Forces: Iterated Blows" causes the computer to display the motion of two planets. Blows directed toward a center (the sun), act on each planet in equal time intervals. The force exerted by the planets on one another is ignored in the program; each is attracted only by the sun, by a force that varies inversely as the square of the distance from the sun.

Initial positions and initial velocities for the planets were selected. The positions of the planets are shown as dots on the face of the cathode-ray tube at regular intervals. (Many more points were calculated between those displayed.)

You can check Kepler's three laws by projecting on paper and marking successive positions of the planets. The law of areas can be verified by drawing triangles and measuring areas. Find the areas swept out in at least three places: near perihelion, near aphelion, and at a point approximately midway between perihelion and aphelion.

Kepler's third law holds that in any given planetary system the squares of the periods of the planets are proportional to the cubes of their average distances from the object around which they are orbiting. In symbols,

$$T^2 \propto R_{av}^3$$

where T is the period and R_{av} is the average distance. Thus, in any one system, the value of T^2/R_{av}^3 ought to be the same for all planets.

You can use this film to check Kepler's law of periods by measuring T for each of the two orbits shown, and then computing T^2/R_{av}^3 for each. To measure the periods of revolution, use a clock or watch with a sweep second hand. Another way is to count the number of plotted points in each orbit. To find R_{av} for each orbit, measure the perihelion and aphelion distances (R_p and R_a) and take their average (Fig. 2-91).

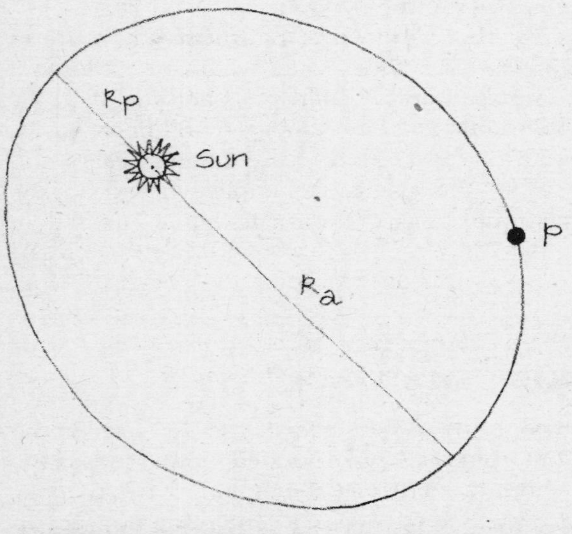

Fig. 2-91 The mean distance R_{av} of a planet P orbiting about the sun is $(R_p + R_a)/2$.

How close is the agreement between your two values of T^2/R_{av}^3? Which is the greater source of error, the measurement of T or of R_{av}?

To check Kepler's first law, see if the orbit is an ellipse with the sun at a focus. You can use string and thumbtacks to draw an ellipse. Locate the empty focus, symmetrical with respect to the sun's position. Place tacks in a board at these two points. Make a loop of string as shown in Fig. 2-92.

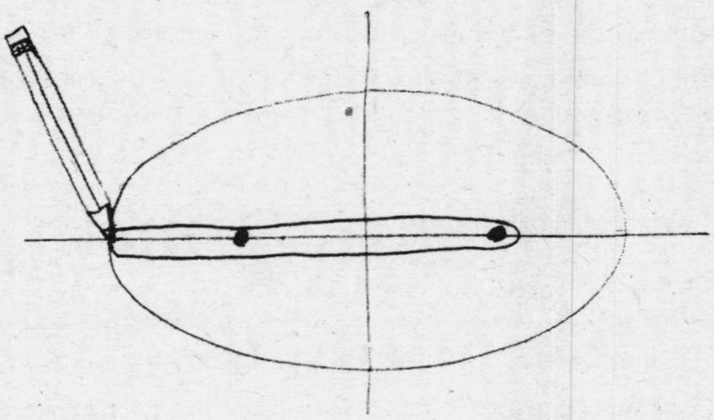

Fig. 2-92

Put your pencil in the string loop and draw the ellipse, keeping the string taut. Does the ellipse match the observed orbit of the planet? What other methods can be used to find whether a curve is a good approximation of an ellipse?

You might ask whether checking Kepler's laws for these orbits is just busy-work, since the computer already "knew" Kepler's laws and used them in calculating the orbits. But the computer was *not* given instructions for Kepler's laws. What you are checking is whether Newton's laws lead to motions that fit Kepler's descriptive laws. The computer "knew" (through the program given it) only Newton's laws of motion and the inverse-square law of gravitation. This computation is exactly what Newton did, but without the aid of a computer to do the routine work.

Film Loop 17
UNUSUAL ORBITS

In this film, a modification of the computer program described in "Central Forces: Iterated Blows" is used. There are two sequences: The first shows the effect of a disturbing force on an orbit produced by a central inverse-square force; the second shows an orbit produced by an inverse-cube force.

The word *perturbation* refers to a small variation in the motion of a celestial body

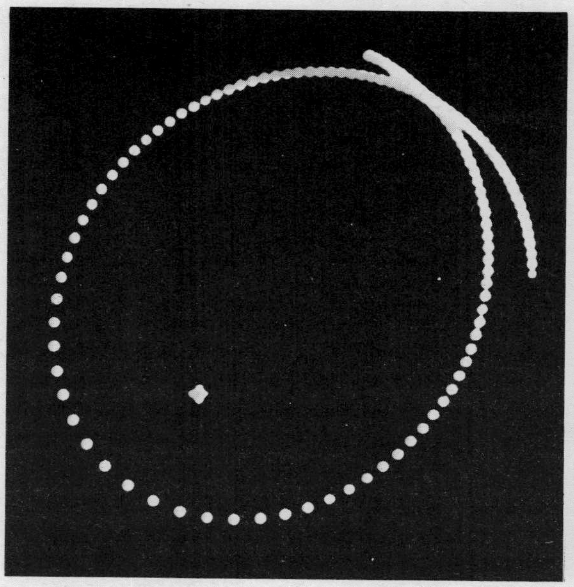

Fig. 2-93

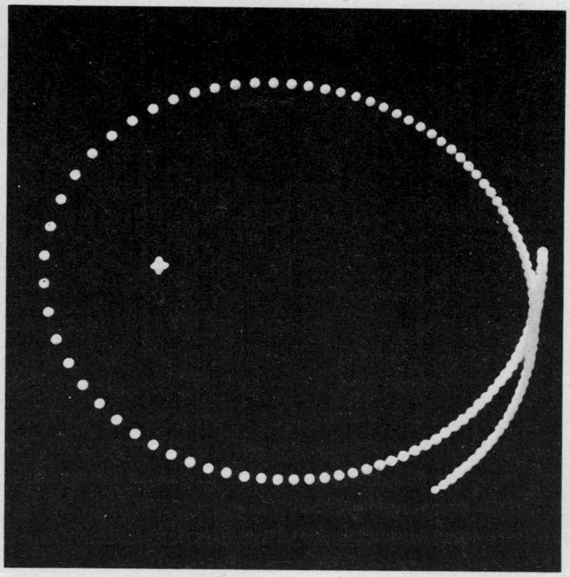

Fig. 2-94

caused by the gravitational attraction of another body. For example, the planet Neptune was discovered because of the perturbation it caused in the orbit of Uranus. The main force on Uranus is the gravitational pull of the sun, and the force exerted on it by Neptune causes a perturbation that changes the orbit of Uranus very slightly. By working backward, astronomers were able to predict the position and mass of the unknown planet from its small effect on the orbit of Uranus. This spectacular "astronomy of the invisible" was rightly regarded as a triumph for the Newtonian law of universal gravitation.

Typically, a planet's entire orbit rotates slowly, because of the small pulls of other planets and the retarding force of friction due to dust in space. This effect is called *advance of perihelion*. Mercury's perihelion advances about 500 seconds of arc (0.14°) per century. Most of this was explained by perturbations due to the other planets. However, about 43 seconds per century remained unexplained. When Einstein reexamined the nature of space and time in developing the theory of relativity, he developed a new gravitational theory that modified Newton's theory in crucial ways. Relativity theory is important for bodies moving at high speeds or near massive bodies. Mercury's orbit is closest to the sun and therefore most affected by Einstein's extension of the law of gravitation. Relativity was successful in explaining the extra 43 seconds per century of advance of Mercury's perihelion.

The first sequence shows the advance of perihelion due to a small force proportional to the distance R, added to the usual inverse-square force. The "dialogue" between operator and computer starts as follows:

```
PRECESSION PROGRAM WILL USE
  ACCEL = G/(R*R) + P*R
GIVE ME PERTURBATION P
  P = 0.66666
GIVE ME INITIAL POSITION IN AU
  X = 2
  Y = 0
GIVE ME INITIAL VELOCITY IN AU/YR
  XVEL = 0
  YVEL = 3
```

The symbol * means multiplication in the FORTRAN language used in the program. Thus, $G/(R*R)$ is the inverse-square force, and $P*R$ is the perturbing force, proportional to R.

In the second part of the film, the force is an inverse-cube force. The orbit resulting from the inverse-cube attractive force, as from most force laws, is not closed. The planet spirals into the sun in a "catastrophic" orbit. As the planet approaches the sun, it speeds up, so points are separated by a large fraction of a revolution. Different initial positions and velocities would lead to quite different orbits.

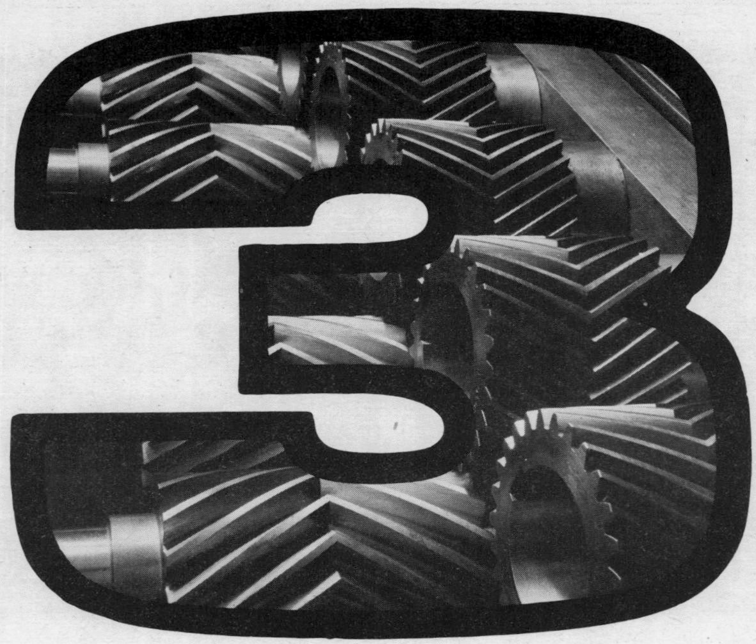

The Triumph of Mechanics

EXPERIMENTS

Experiment 3-1 COLLISIONS IN ONE DIMENSION. I

In this experiment, you will investigate the motion of two objects interacting in one dimension. The interactions (explosions and collisions in the cases treated here) are called one-dimensional because the objects move along a single straight line. Your purpose is to look for quantities or combinations of quantities that remain unchanged before and after the interaction, that is, quantities that are conserved.

Your experimental explosions and collisions may seem not only tame but also artificial and unlike the ones you see around you in everyday life. But this is typical of many scientific experiments, which simplify the situation so as to make it easier to make meaningful measurements and to discover patterns in the observed behavior. The underlying laws are the same for all phenomena, whether or not they are in a laboratory.

Two different ways of observing interactions are described here (and two others in Experiment 3-2). You will probably use only one of them. In each method, the friction between the interacting objects and their surroundings is kept as small as possible, so that the objects are a nearly isolated system. Whichever method you do follow, you should handle your results in the way described in the final section, "Analysis of Data."

METHOD A. Dynamics Carts

"Explosions" are easily studied using the low-friction dynamics carts. Squeeze the loop of spring steel flat and slip a loop of thread over it to hold it compressed. Put the compressed loop between two carts on the floor or on a smooth table (Fig. 3-1). When you release the spring by burning the thread, the carts fly apart with velocities that you can measure from a strobe photograph or by any of the techniques you learned in earlier experiments.

Load the carts with a variety of weights to create simple ratios of masses, for example, 2 to 1 or 3 to 2. Take data for as great a variety of mass ratios as time permits. Because friction will gradually slow the carts down, you should

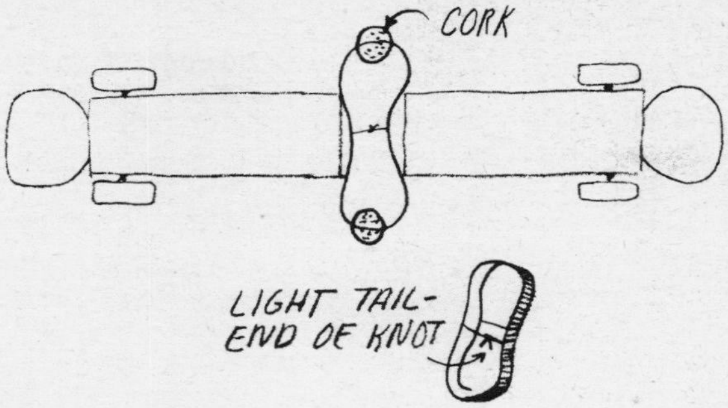

Fig. 3-1

make measurements on the speeds immediately after the explosion is over (that is, when the spring stops pushing).

Since you are interested only in comparing the speeds of the two carts, you can express those speeds in any units you wish, without worrying about the exact scale of the photograph and the exact strobe rate. For example, you can use distance units measured directly from the photograph (in millimeters) and use time units equal to the time interval between strobe images. If you follow that procedure, the speeds recorded in your notes will be in millimeters per interval.

Remember that you can get data from the negative of a Polaroid picture as well as from the positive print.

METHOD B. Air Track

The air track allows you to observe collisions between objects, "gliders," that move with almost no friction. You can take stroboscopic photographs of the gliders either with the xenon strobe or by using a rotating slotted disk in front of the camera.

The air track has three gliders: two small ones with the same mass, and a larger one with just twice the mass of a small one. A small and a large glider can be coupled together to make one glider so that you can have collisions between gliders whose masses are in the ratio of 1:1, 2:1, and 3:1. (If you add light sources to the gliders, their masses will no longer be in the same simple ratios. You can find the masses from the measured weights of the glider and light source.)

You can arrange to have the gliders bounce apart after they collide (elastic collision) or stick together (inelastic collision). Good technique is important if you are to get consistent results. Before taking any pictures, try both elastic and inelastic collisions with a variety of mass ratios. Then, when you have chosen one type to analyze, rehearse each step of your procedure with your partners before you go ahead.

You can use a good photograph to find the speeds of both carts, before and after they collide. Since you are interested only in comparing the speeds before and after each collision, you can express speeds in any unit you wish, without worrying about the exact scale of the photograph or the exact strobe rate. For example, if you use distance units measured directly from the photograph (in millimeters) and time units equal to the time interval between strobe images, the speeds recorded in your notes will be in millimeters per interval.

Remember that you can get data from the negative of your Polaroid picture as well as from your positive print.

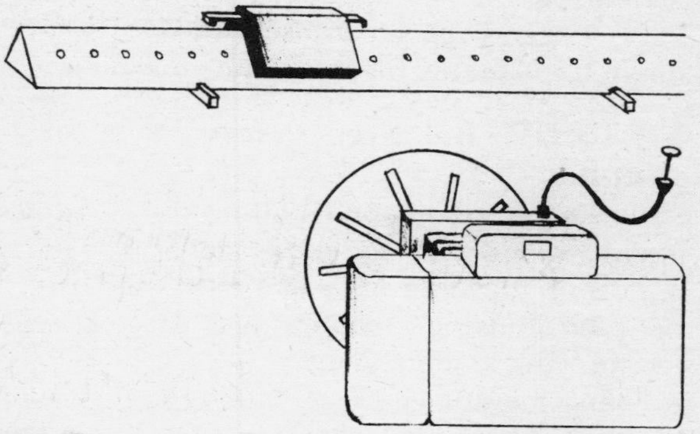

Fig. 3-2

Analysis of Data

Assemble all your data in a table having column headings for the mass of each object, m_A and m_B, the speeds before the interaction, v_A and v_B (for explosions, $v_A = v_B = 0$), and the speeds after the collision, v_A' and v_B'.

Examine your table carefully. Search for quantities or combinations of quantities that remain unchanged before and after the interaction.

?

1. Is *speed* a conserved quantity? That is, does the quantity $(v_A + v_B)$ equal the quantity $(v_A' + v_B')$?

2. Consider the *direction* as well as the speed. Define velocity to the right as positive and velocity to the left as negative. Is *velocity* a conserved quantity?

3. If neither speed nor velocity is conserved, try a quantity that combines the mass and velocity of each cart. Compare $(m_A v_A + m_B v_B)$ with $(m_A v_A' + m_B v_B')$ for each interaction. In the same way compare m/v, $\vec{mv}$, m^2v, or any other likely combinations you can think of, before and after interaction. What conclusions do you reach?

Experiment 3-2 COLLISIONS IN ONE DIMENSION. II

METHOD A. Film Loops

Film Loops 18, 19, and 20 show one-dimensional collisions that you cannot easily perform in your own laboratory. They were filmed with a very high-speed camera, producing the effect of slow motion when projected at the standard rate. You can make measurements directly from the pictures projected onto graph paper. Since you are interested only in comparing speeds before and after a collision, you can express speeds in any unit you wish, that is, you can make measurements in any convenient distance and time units.

Notes for these film loops are located on pages 156 and 157. If you use these loops, read the notes carefully before taking your data.

METHOD B. Stroboscopic Photographs

Stroboscopic photographs showing seven different examples of one-dimensional collisions appear on the following pages.* They are useful here for studying momentum and again later for studying kinetic energy.

For each event, you should find the speeds of the balls before and after collision. From the values for the mass and speed of each ball, calculate the total momentum before and after collision. Use the same values to calculate the total kinetic energy before and after collision.

You should read Section I before analyzing any of the events, in order to find out what measurements to make and how the collisions were produced. After you have made your measurements, turn to Section II for questions to answer about each event.

I. The Measurements You Will Make

To make the necessary measurements, you will need a metric ruler marked in millimeters, preferably of transparent plastic with sharp scale markings. Before starting your work, consult Fig. 3-2 for suggestions on improving your measuring technique.

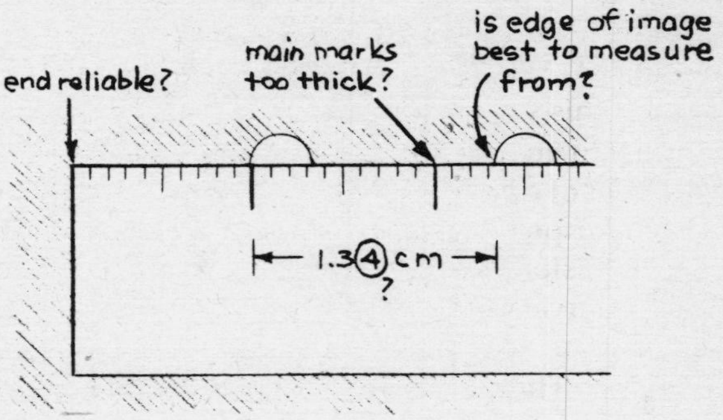

Figure 3-3 shows schematically that the colliding balls were hung from very long wires. The balls were released from rest, and their double-wire (bifilar) suspensions guided them to a squarely head-on collision. Stroboscopes illuminated the 90 cm × 120 cm rectangle that was the field of view of the camera. The stroboscopes are not shown in Fig. 3-3.

Notice the two rods whose tops reach into the field of view. These rods were 1 m (±2 mm) apart, measured from top center of one rod to top center of the other. The tops of these rods are visible in the photographs on which you will make your measurements. This enables you to convert your measurements to actual distances if you wish. However, it is easier to use the lengths in millimeters measured directly on the photograph if you are merely going to compare momenta.

The balls speed up as they move into the field of view. Likewise, as they leave the field of view, they slow down. Therefore successive

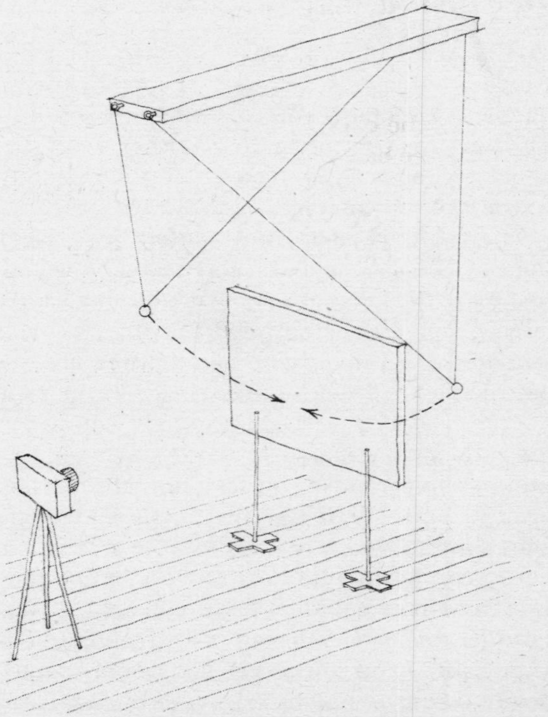

Fig. 3-3 Setup for photographing one-dimensional collisions.

*Reproduced by permission of National Film Board of Canada.

displacements on the stroboscopic photograph, each of which took exactly the same time, will not necessarily be equal in length. Check this with your ruler.

As you measure a photograph, number the position of each ball at successive flashes of the stroboscope. Note the interval during which the collision occurred. Identify the clearest time interval for finding the velocity of each ball (a) before the collision and (b) after the collision. Then mark this information near each side of the interval.

II. Questions To Be Answered About Each Event

After you have recorded the masses (or relative masses) given for each ball and have recorded the necessary measurements of velocities, answer the following questions.

1. What is the total momentum of the system of two balls before the collision? Keep in mind here that velocity and, therefore, momentum are vector quantities.

2. What is the total momentum of the system of two balls after the collision?

3. Was momentum conserved within the limits of precision of your measurements?

Event 1

The photographs of this Event 1 and all the following events appear as Figs. 3-11 to 3-17. This event is also shown as the first example in Film Loop 18, "One-Dimensional Collisions. I."

Figure 3-4 shows that ball B was initially at rest. After the collision, both balls moved off to the left. (The balls are made of hardened steel.)

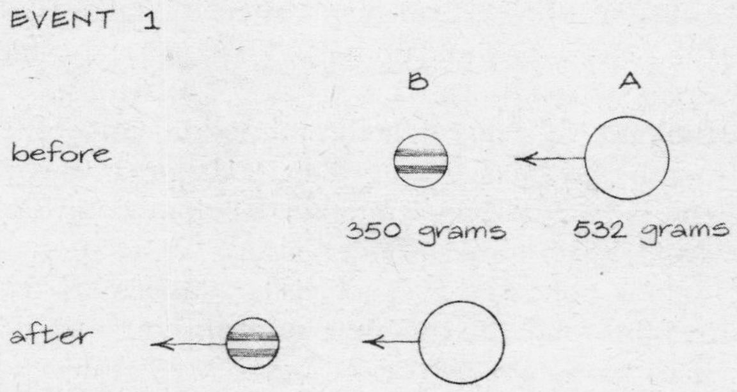

Fig. 3-4

Event 2

This event, the reverse of Event 1, is shown as the second example in Film Loop 18.

Figure 3-5 shows that ball B came in from the left and that ball A was initially at rest. The collision reversed the direction of motion of ball B and sent ball A off to the right. (The balls are of hardened steel.)

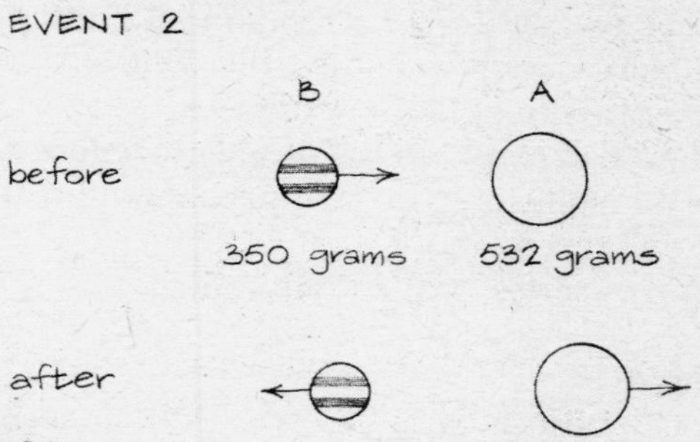

Fig. 3-5

As you can tell by inspection, ball B moved slowly after collision, and thus you may have difficulty getting a precise value for its speed. This means that your value for this speed is the least reliable of your four measurements of speed. Nevertheless, this fact has only a small influence on the reliability of your value for the total momentum after collision. Can you explain why this should be so?

Why was the direction of motion of ball B reversed by the collision?

If you have already studied Event 1, you will notice that the same balls were used in Events 1 and 2. Check your velocity data, and you will find that the *initial* speeds were nearly equal. Thus, Event 2 was truly the reverse of Event 1. Why, then, was the direction of motion of ball A in Event 1 not reversed although the direction of ball B in Event 2 was reversed?

Event 3

This event is shown as the first example in Film Loop 19, "One-Dimensional Collisions. II." Event 3 is not recommended until you have studied one of the other events. Event 3 is especially recommended as a companion to Event 4.

Figure 3-6 shows that a massive ball A entered from the left. A less massive ball B came

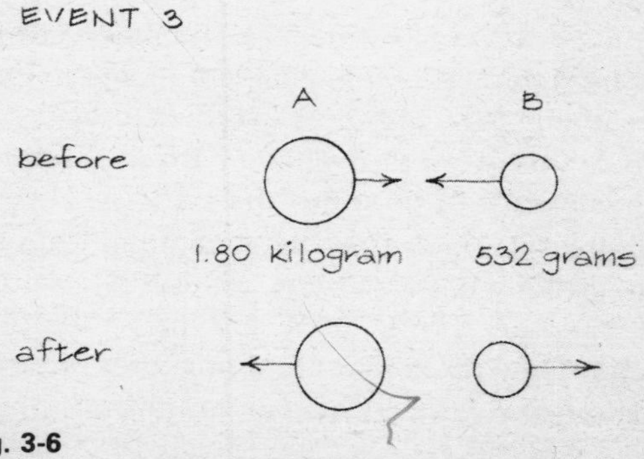

Fig. 3-6

in from the right. The directions of motion of both balls were reversed by the collision. (The balls are made of hardened steel.)

When you compare the momenta before and after the collision, you will probably find that they differed by more than any other event so far in this series. Explain why this is so.

Event 4

This event is also shown as the second example in Film Loop 19.

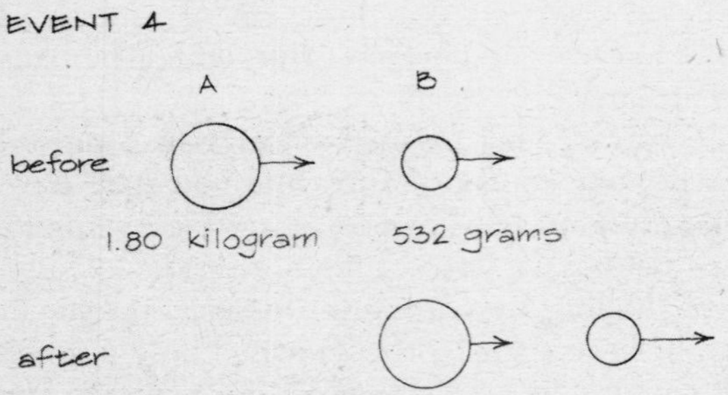

Fig. 3-7

Figure 3-7 shows that two balls came in from the left, that ball A was far more massive than ball B, and that ball A was moving faster than ball B before collision. The collision occurred when A caught up with B, increasing B's speed at some expense to its own speed. (The balls are made of hardened steel.)

Each ball moved across the camera's field from left to right on the same line. In order to be able to tell successive positions apart on a stroboscopic photograph, the picture was taken twice. The first photograph shows only the progress of the large ball A because ball B had been given a thin coat of black paint (of negligible mass). Ball A was painted black when the second picture was taken. It will help you to analyze the collision if you number white-ball positions at successive stroboscope flashes in each picture.

Event 5

This event is also shown as the first example in Film Loop 20, "Inelastic One-Dimensional Collisions." You should find it interesting to analyze this event or Event 6 or Event 7, but it is not necessary to do more than one.

Figure 3-8 shows that ball A came in from the right, striking ball B which was initially at rest. The balls are made of a soft material (plasticene). They remained stuck together after the collision and moved off to the left as one. A collision of this type is called *perfectly inelastic*.

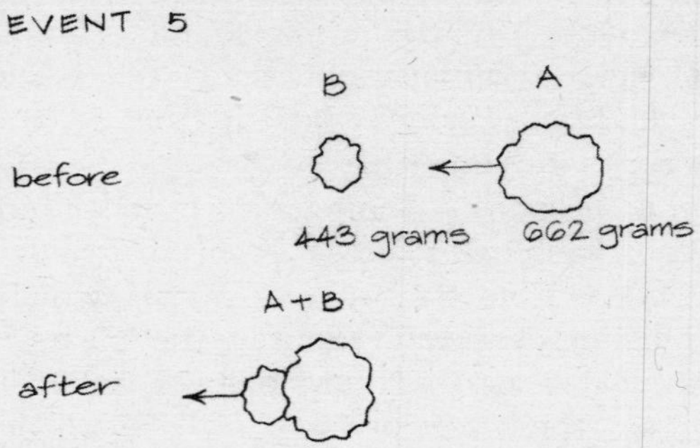

Fig. 3-8

Event 6

This event is shown as the second example in Film Loop 20.

Figure 3-9 shows that balls A and B moved in from the right and left, respectively, before collision. The balls are made of a soft material (plasticene). They remained stuck together after the collision and moved off together to the left. This is another perfectly inelastic collision, like that in Event 5.

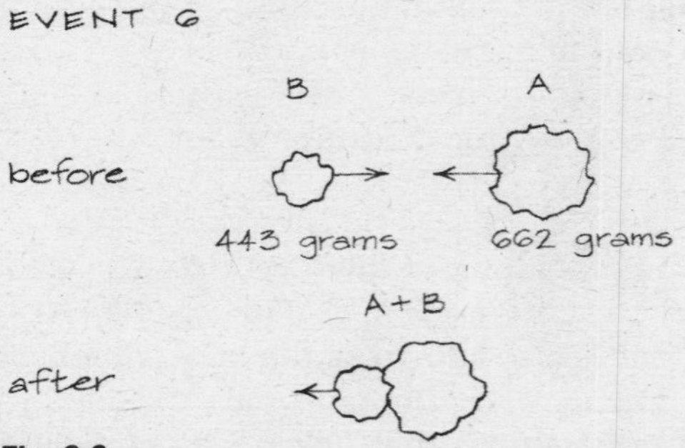

Fig. 3-9

This event was photographed in two parts. The first print shows the conditions before collision, the second print, after collision. Had the picture been taken with the camera shutter open throughout the motion, it would be difficult to take measurements because the combined balls (A + B) after collision retraced the path that ball B followed before collision. You can number the positions of each ball before collision at successive flashes of the stroboscope (in the first photo); you can do likewise for the combined balls (A + B) after the collision in the second photo.

Event 7

Figure 3-10 shows that balls A and B moved in from opposite directions before collision. The balls are made of a soft material (plasticene). They remained stuck together after collision and moved off together to the right. This is another perfectly inelastic collision.

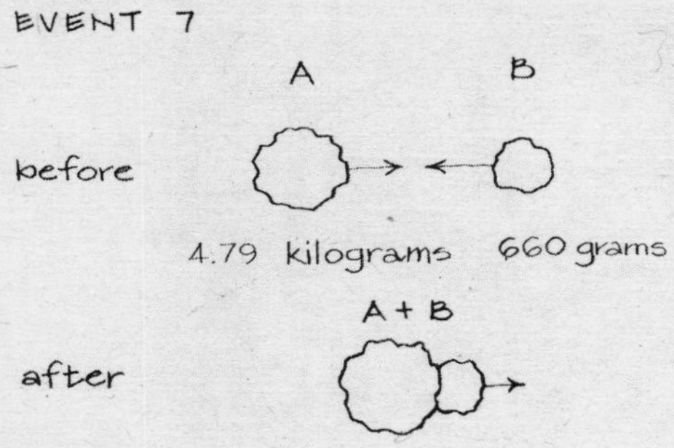

Fig. 3-10

This event was photographed in two parts. The first print shows the conditions before collision, the second print, after collision. Had the picture been made with the camera shutter open throughout the motion, it would be difficult to take measurements because the combined balls (A + B) return along the same path as incoming ball B. You can number the positions of each ball before collision at successive flashes of the stroboscope (in the first photograph); you can do likewise for the combined balls (A + B) after collision in the second photograph.

Photographs of the Events

The photographs of the events are shown in Figs. 3-11 through 3-17.

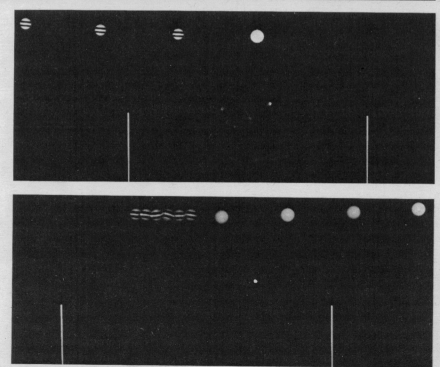

Fig. 3-11 Event 1, 10 flashes/sec.

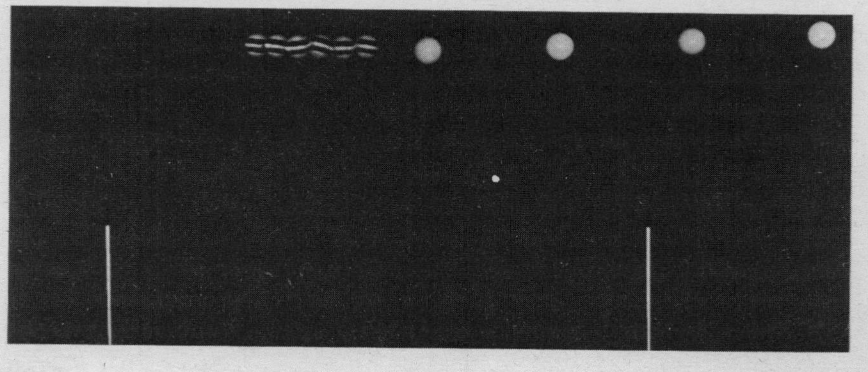

Fig. 3-12 Event 2, 10 flashes/sec.

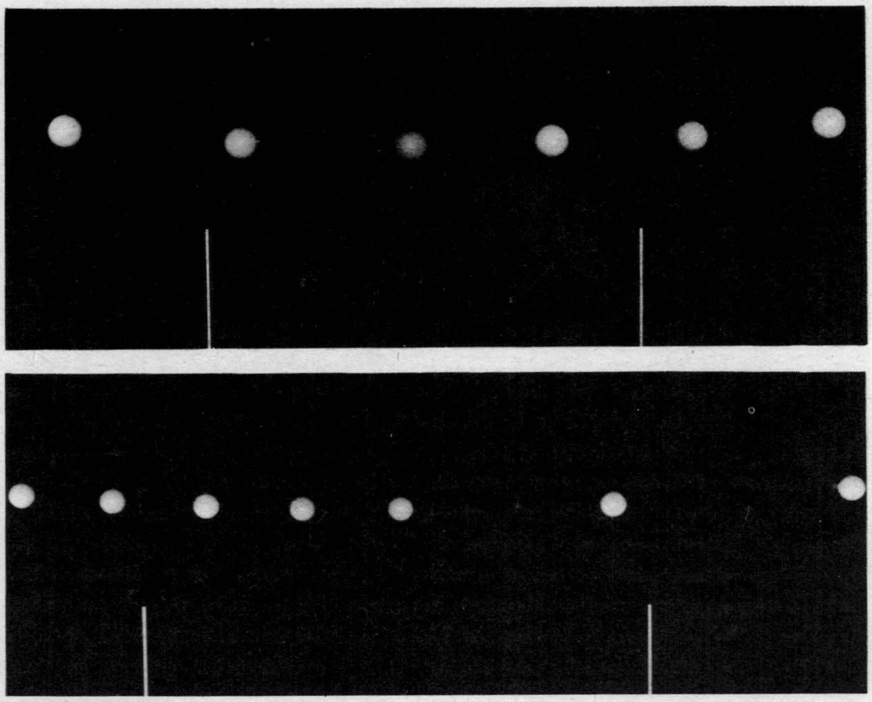

Fig. 3-13 Event 3, 10 flashes/sec.

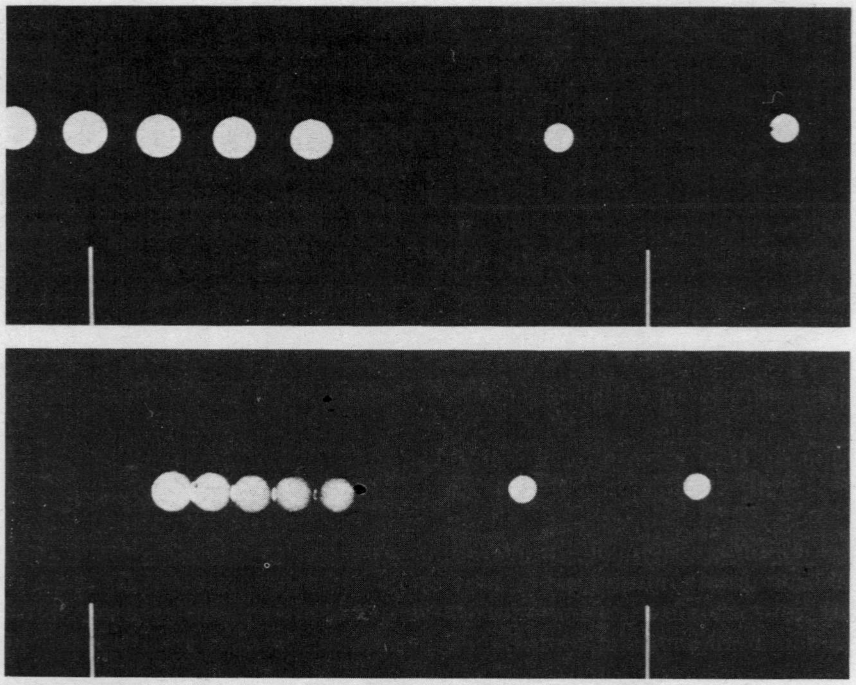

Fig. 3-14 Event 4, 10 flashes/sec.

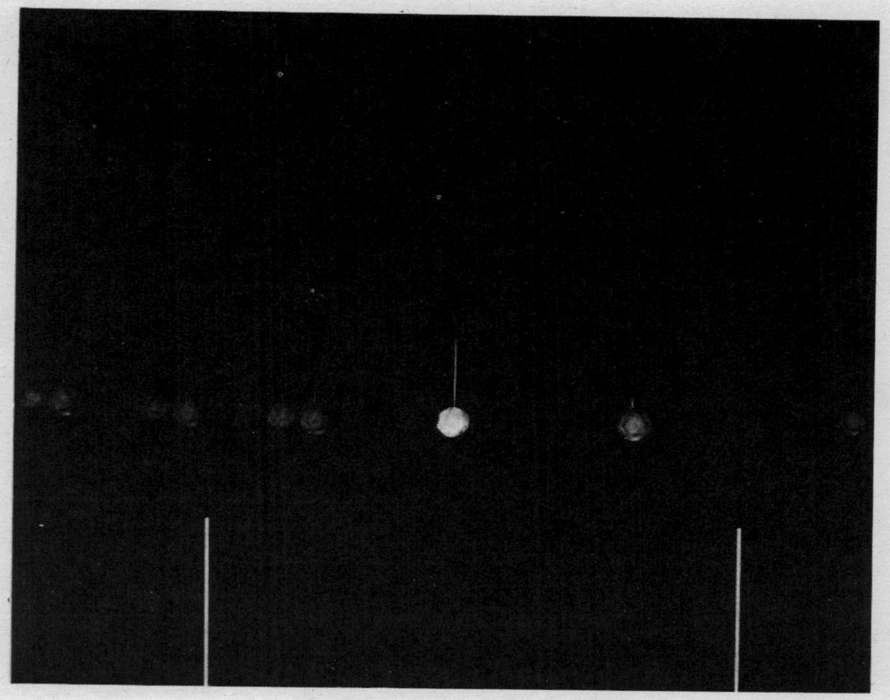

Fig. 3-15 Event 5, 10 flashes/sec.

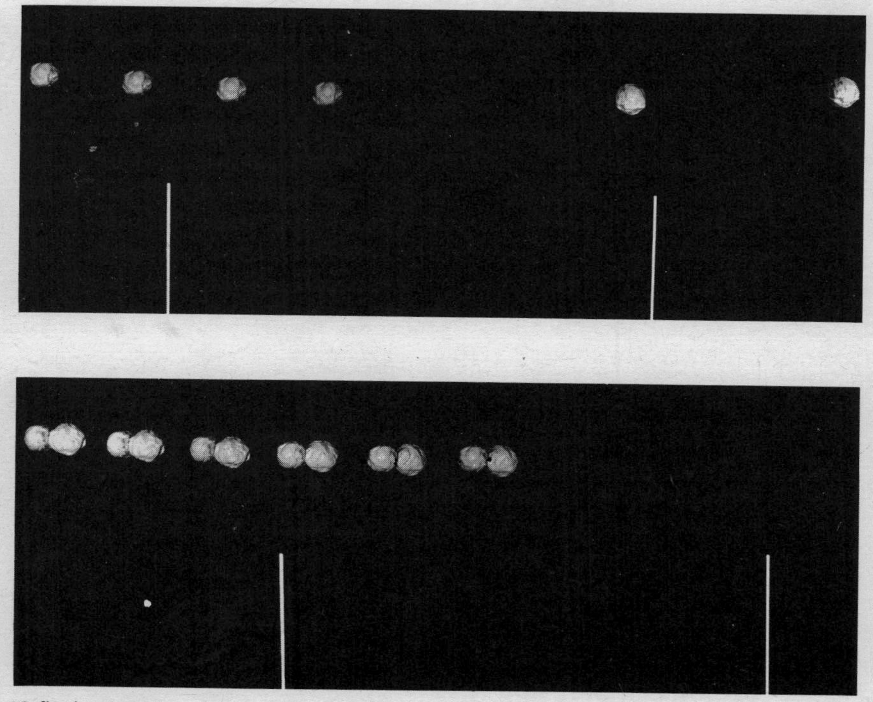

Fig. 3-16 Event 6, 10 flashes/sec.

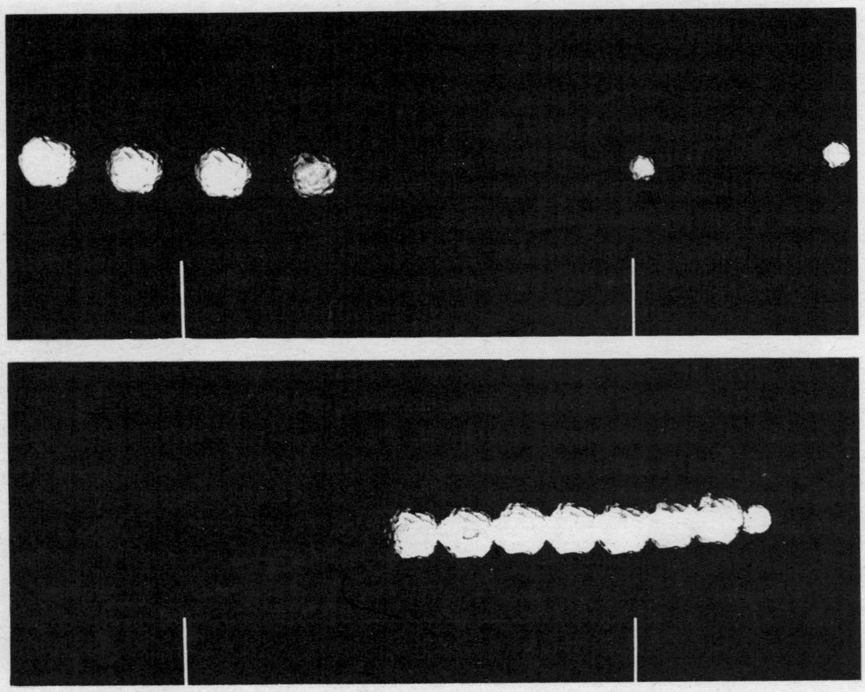

Fig. 3-17 Event 7, 10 flashes/sec.

Experiment 3-3 COLLISIONS IN TWO DIMENSIONS. I

Collisions rarely occur in only one dimension, that is, along a straight line. In billiards, basketball, and tennis, the ball usually rebounds at an angle to its original direction. Ordinary explosions (which can be thought of as collisions in which initial velocities are all zero) send pieces flying off in all directions.

This experiment deals with collisions that occur in two dimensions, that is, in a single plane, instead of along a single straight line. It assumes that you know what momentum is and understand what is meant by "conservation of momentum" in one dimension. In this experiment, you will discover a general form of the rule for one dimension that applies also to the conservation of momentum in cases where the parts of the system move in two (or three) dimensions.

Two methods of getting data on two-dimensional collisions are described in Methods A and B (and two others in Experiment 3-4), but you will probably want to follow only one method. Whichever method you use, handle your results in the way described in "Analysis of Data."

METHOD A. Colliding Pucks

On a carefully leveled glass tray covered with a sprinkling of Dylite spheres, you can make pucks coast with almost uniform speed in any direction. Set one puck motionless in the center of the table and push a second similar one toward it, a little off-center. You can make excellent pictures of the resulting two-dimensional glancing collision with a camera mounted directly above the surface.

To reduce reflection from the glass tray, the photograph should be taken using the xenon stroboscope with the light on one side and almost level with the glass tray. To make each puck's location clearly visible in the photograph, attach a steel ball or a small white Styrofoam hemisphere to its center.

You can get a great variety of masses by stacking pucks one on top of the other and fastening them together with tape (avoid having the collisions cushioned by the tape).

Two people are needed to do the experiment. One experimenter, after some preliminary practice shots, launches the projectile puck while the other experimenter operates the camera. The resulting picture should consist of a series of white dots in a rough "Y" pattern.

Using your picture, measure and record all the speeds before and after collision. Record the masses in each case too. Since you are interested only in comparing speeds, you can use any convenient speed units. You can simplify your work if you record speeds in millimeters per dot instead of trying to work them out in centimeters per second. Because friction does slow the pucks down, find speeds as close to the impact as you can. You can also use the "puck" instead of the kilogram as your unit of mass.

METHOD B. Colliding Disk Magnets

Disk magnets will also slide freely on Dylite spheres as described in Method A.

The difference here is that the magnets need never touch during the "collision." Since the interaction forces are not really instantaneous as they are for the pucks, the magnets follow *curving* paths during the interaction. Consequently the "before" velocity should be determined as early as possible and the "after" velocities should be measured as late as possible.

Following the procedure described above for pucks, photograph one of these "collisions." Again, small Styrofoam hemispheres or steel balls attached to the magnets should show up in the strobe picture as a series of white dots. Be sure the paths you photograph are long enough so that the dots near the ends are along straight lines rather than curves (see Fig. 3-18).

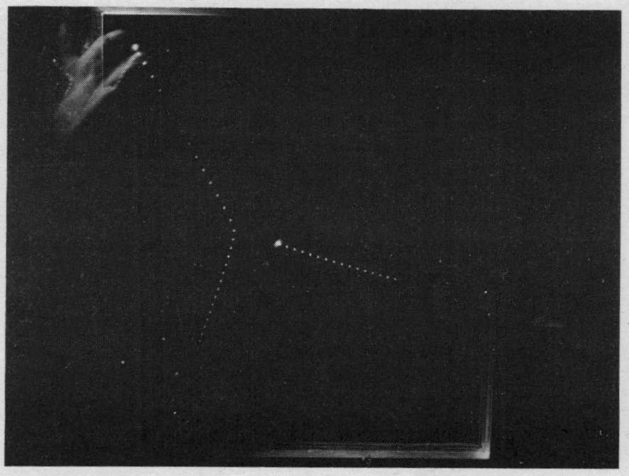

Fig. 3-18

Using your photograph, measure and record the speeds and record the masses. You can simplify your work if you record speeds in millimeters per dot instead of working them out in centimeters per second. You can use the disk instead of the kilogram as your unit of mass.

Analysis of Data

Whichever procedure you used, you should analyze your results in the same way.

?

1. Multiply the mass of each object by its before-the-collision speed, and add the products.
2. Do the same thing for each of the objects in the system after the collision, and add the after-the-collision products together. Does the sum before the collision equal the sum after the collision?

Imagine the collision you observed was an explosion of a cluster of objects at rest; the total quantity mass-times-speed before the explosion will be zero. But surely, the mass-times-speed of each of the flying fragments after the explosion is more than zero! "Mass-times-speed" is obviously *not* conserved in an explosion. You probably found it was not conserved in the experiments with pucks and magnets, either. You may already have suspected that you ought to be taking into account the *directions* of motion.

To see what *is* conserved, proceed as follows.

Use your measurements to construct a drawing like Fig. 3-19, in which you show the directions of motion of all the objects both before and after the collision.

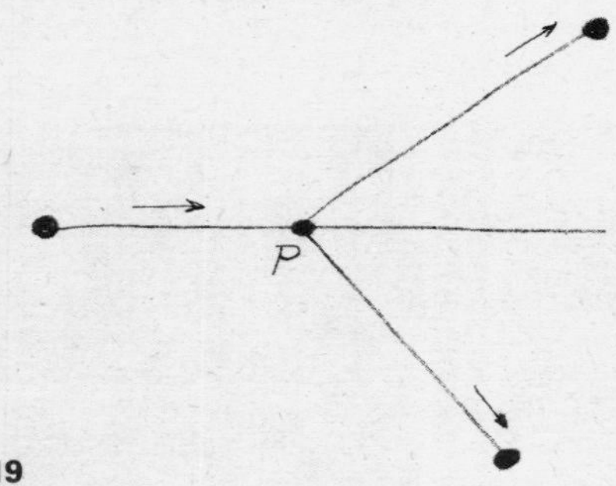

Fig. 3-19

Have all the direction lines *meet at a single point* in your diagram. The actual paths in your photographs will not do so, because the pucks and magnets are large objects instead of points, but you can still draw the *directions* of motion as lines through the single point P.

On this diagram draw a vector arrow whose magnitude (length) is proportional to the mass times the speed of the projectile *before* the collision. (You can use any convenient scale.) In Fig. 3-20, this vector is marked $m_A\vec{v}_A$.

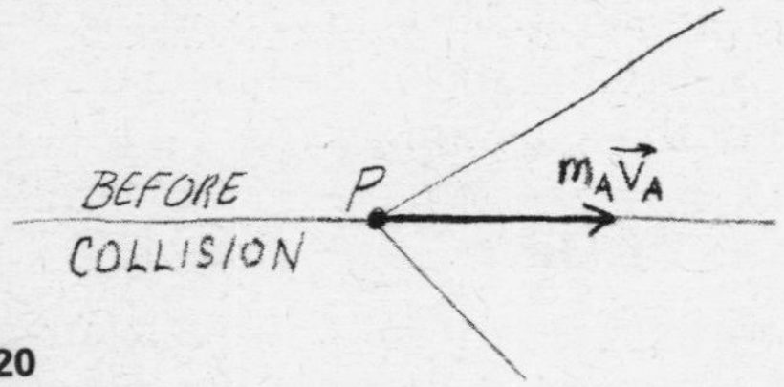

Fig. 3-20

Below your first diagram draw a second one in which you once more draw the directions of motion of all the objects exactly as before. On this second diagram, construct the vectors for mass-times-speed for each of the objects leaving P *after* the collision. For the collisions of pucks and magnets, your diagram will resemble Fig. 3-21. Now construct the "after-the-collision" vector sum.

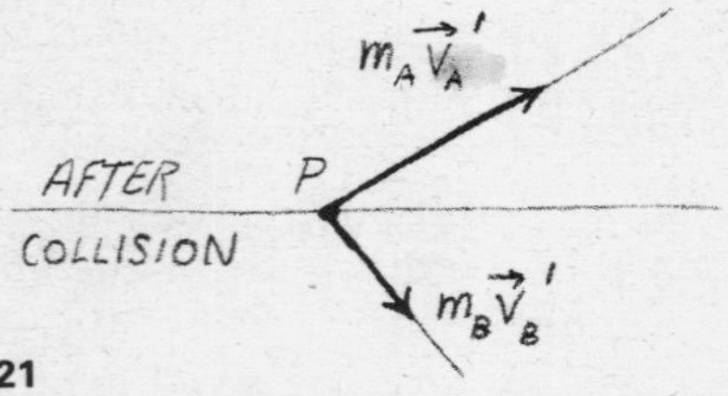

Fig. 3-21

The length of each of your arrows is given by the product of mass and speed. Since each arrow is drawn in the *direction* of the speed, the arrows represent the product of mass and velocity $m\vec{v}$ which is called *momentum*. The vector sums "before" and "after" collision, therefore, represent the total momentum of the system of objects before and after the collision. If the "before" and "after" arrows are equal, then the total momentum of the system of interacting objects is conserved.

?

3. How does this vector sum compare with the vector sum on your before-the-collision figure? Are they equal within the uncertainty?

4. Is the principle of conservation of momentum for one dimension different from that for two, or merely a special case of it? How can the principle of conservation of momentum be extended to three dimensions? Sketch at least one example.

5. Write an equation that would express the principle of conservation of momentum for collisions of (a) three objects in two dimensions, (b) two objects in three dimensions, (c) three objects in three dimensions.

Fig. 3-22 A 1,350-kg steel ball swung by a crane against the walls of a condemned building. What happens to the momentum of the ball?

Experiment 3-4 COLLISIONS IN TWO DIMENSIONS. II

METHOD A. Film Loops

Several Film Loops (21, 22, 23, 24, and 25) show two-dimensional collisions that you cannot conveniently reproduce in the laboratory. Notes on these films appear on pages 77–79. Project one of the loops on the chalkboard or on a sheet of graph paper. Trace the paths of the moving objects and record their masses and measure their speeds. Then go on to the analysis described in notes for Film Loop 21.

METHOD B. Stroboscopic Photographs

Stroboscopic photographs* of seven different two-dimensional collisions in a plane are used in this experiment. The photographs (Figs. 3-27 to 3-34) are shown on the pages immediately following the description of these events. They were photographed during the making of Film Loops 21–25.

I. Material Needed

1. A transparent plastic ruler, marked in millimeters.

2. A large sheet of paper for making vector diagrams. Graph paper is convenient.

3. A protractor and two large drawing triangles are useful for transferring directional vectors from the photographs to the vector diagrams.

II. How the Collisions Were Produced

Balls were hung on 10-m wires, as shown schematically in Fig. 3-23. They were released so as to collide directly above the camera, which was facing upward. Electronic strobe lights (shown in Fig. 3-26) illuminated the rectangle shown in each frame.

Two white bars are visible at the bottom of each photograph. These are rods that had their tips 1 m (±2 mm) apart in the actual situation. The rods make it possible for you to convert your measurements to the actual distance. It is not necessary to do so, if you choose instead to use actual on-the-photograph distances in millimeters (as you may have done in your study of one-dimensional collisions).

Since the balls are pendulum bobs, they move faster near the center of the photographs than near the edge. Your measurements, therefore, should be made near the center.

*Reproduced by permission of National Film Board of Canada.

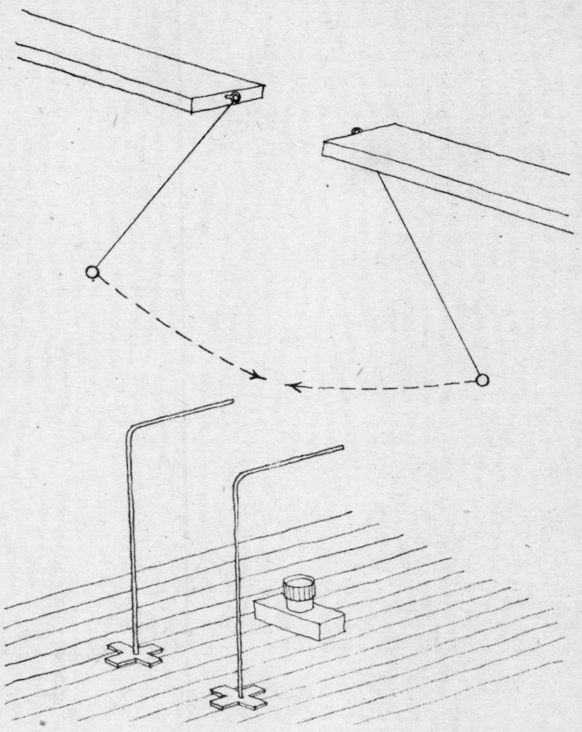

Fig. 3-23 Setup for photographing two-dimensional collisions.

III. A Sample Procedure

The purpose of your study is to see to what extent momentum seems to be conserved in two-dimensional collisions. For this purpose you need to construct vector diagrams.

Consider an example: In Fig. 3-24, a 450-g and a 500-g ball are moving toward each other. Ball A has a momentum of 1.8 kg-m/sec, in the direction of the ball's motion. Using the scale shown, draw a vector 1.8 units long, parallel to the direction of motion of A. Similarly, for ball B draw a momentum vector 2.4 units long, parallel to the direction of motion of B.

The system of two balls has a total momentum before the collision equal to the vector sum of the two momentum vectors for A and B.

The total momentum after the collision is also found the same way, by adding the momentum vector for A after the collision to that for B after the collision (see Fig. 3-25).

This same procedure is used for any event you analyze. Determine the momentum (magnitude *and* direction) for each object in the system before the collision, graphically add them, and then do the same thing for each object after the collision.

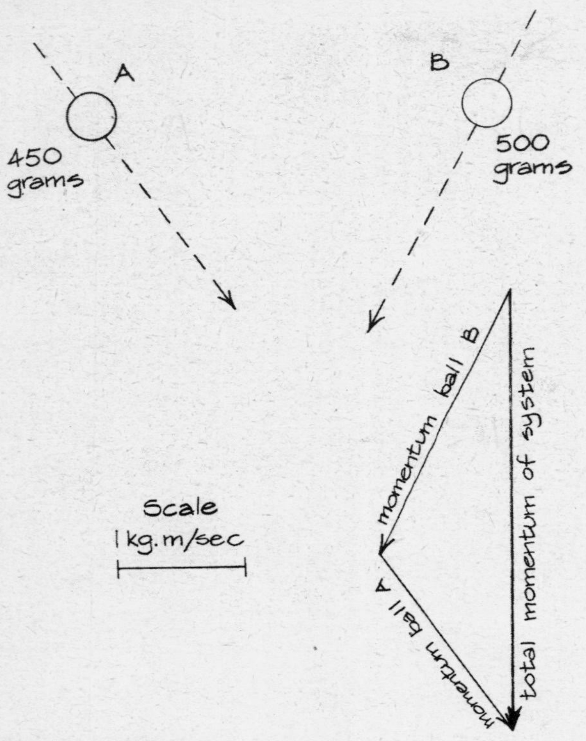

Fig. 3-24 Two balls moving in a plane. Their individual momenta, which are vectors, are added together vectorially in the diagram on the lower right. The vector sum is the total momentum of the system of two balls. (Your own vector drawings should be at least twice this size.)

For each event that you analyze, consider whether momentum is conserved.

Events 8, 9, 10, and 11

Event 8 is shown as the first example in Film Loop 22, "Two-Dimensional Collisions. II."

Event 10 is shown as the second example in Film Loop 22.

Event 11 is also shown in Film Loop 21, "Two-Dimensional Collisions. I."

These are all elastic collisions. Events 8 and 10 are simplest to analyze because each shows a collision of equal masses. In Events 8 and 9, one ball is initially at rest.

A small sketch next to each photograph indicates the direction of motion of each ball. The mass of each ball and the strobe rate are also given.

Events 12 and 13

Event 12 is shown as the first example in Film Loop 23, "Inelastic Two-Dimensional Collisions."

Event 13 is shown as the second example in Film Loop 23.

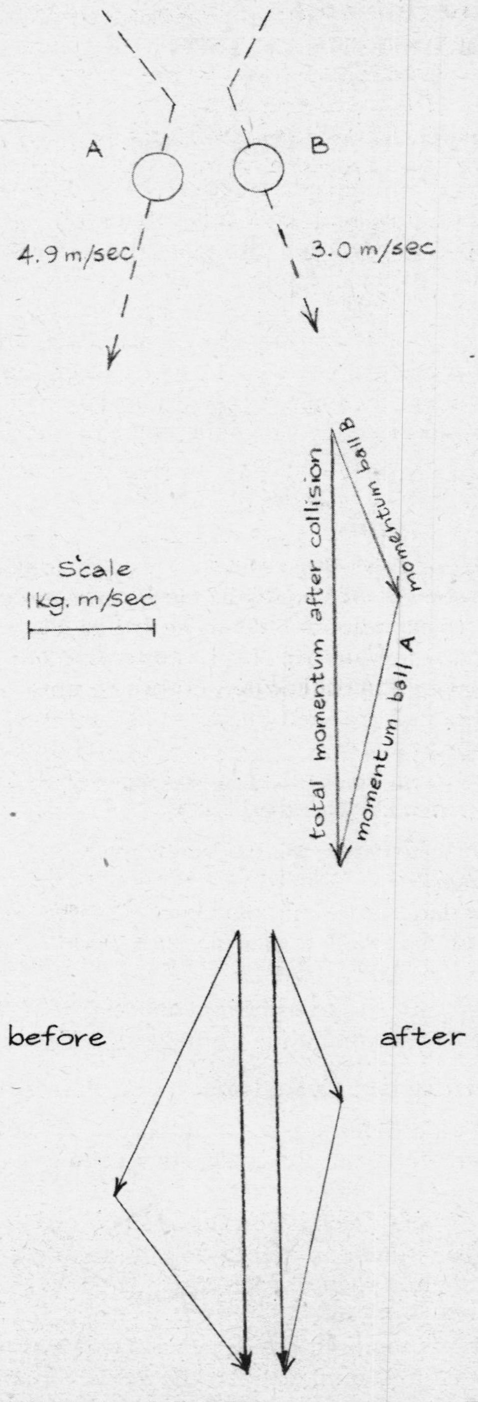

Fig. 3-25 The two balls collide and move away. Their individual momenta after collision are added vectorially. The resultant vector is the total momentum of the system after collision.

Since Events 12 and 13 are similar, there is no need to do both.

Events 12 and 13 show inelastic collisions between two plasticene balls that stick together and move off as one compound object after the collision. In Event 13 the masses are equal; in 12 they are unequal.

CAUTION: You may find that the two objects rotate slightly about a common center after the collision. For each image after the collision, you should make marks halfway between the centers of the two objects. Then determine the velocity of this center of mass, and multiply it by the combined mass to get the total momentum after the collision.

Event 14

Do *not* try to analyze Event 14 unless you have done at least one of the simpler Events 8 through 13.

Event 14 is shown on Film Loop 24 "Scattering of a Cluster of Objects."

Figure 3-26 shows the setup used in photographing the scattering of a cluster of balls. The photographer and camera are on the floor; and four electronic stroboscope lights are on tripods in the lower center of the picture.

Use the same graphical methods as you used for Events 8–13 to see if the conservation of momentum holds for more than two objects. Event 14 is much more complex because you must add seven vectors, rather than two, to get the total momentum after the collision.

In Event 14, one ball comes in and strikes a cluster of six balls of various masses, which were initially at rest. Two photographs are included: print 1 shows only the motion of ball A before the event; print 2 shows the positions of all seven balls just before the collision and the motion of each of the seven balls after the collision.

Fig. 3-26 Catching the seven scattered balls to avoid tangling the wires from which they hang. The photographer and the camera are on the floor. The four stroboscopes are on tripods in the lower center of the picture.

You can analyze this event in two different ways. One way is to determine the initial momentum of ball A from measurements taken on print 1 and then compare it to the total final momentum of the system of seven balls from measurements taken on print 2. The second method is to determine the total final momentum of the system of seven balls on print 2, predict the momentum of ball A, and then take measurements of print 1 to see whether ball A had the predicted momentum. Choose one method.

The tops of prints 1 and 2 lie in identical positions. To relate measurements on one print

Fig. 3-27 Event 8, 20 flashes/sec.

A ◯ 367g

B ◯ 367g

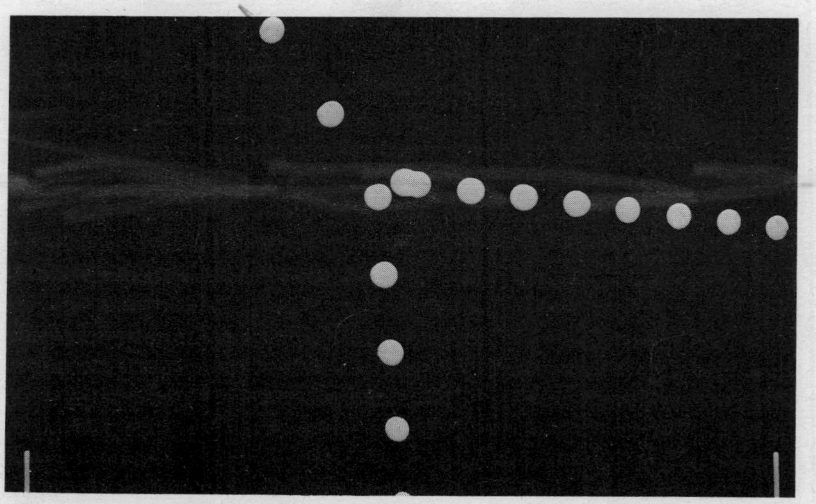

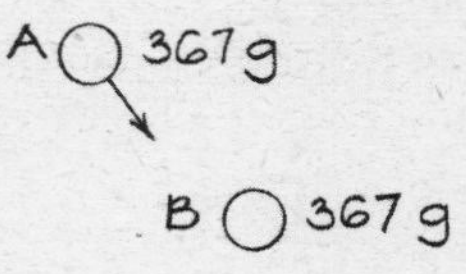

to measurements on the other, measure a ball's distance relative to the top of one picture with a ruler; the ball would lie in precisely the same position in the other picture if the two pictures could be superimposed.

There are two other matters you must consider. First, the time scales are different on the two prints. Print 1 was taken at a rate of 5 flashes/sec, and print 2 was taken at a rate of 20 flashes/sec. Second, the distance scale may not be exactly the same for both prints. Remember that the distance from the center of the tip of one of the white bars to the center of the tip of the other is 1 m (±2 mm) in real space. Check this scale carefully on both prints to determine the conversion factor.

The stroboscopic photographs for Events 8–14 appear in Figs. 3-27 to 3-34.

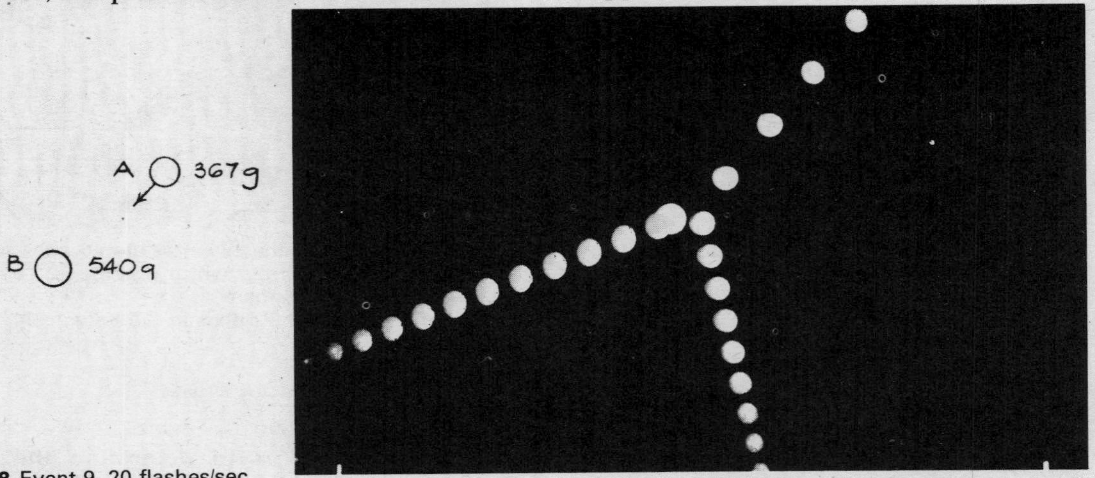

Fig. 3-28 Event 9, 20 flashes/sec.

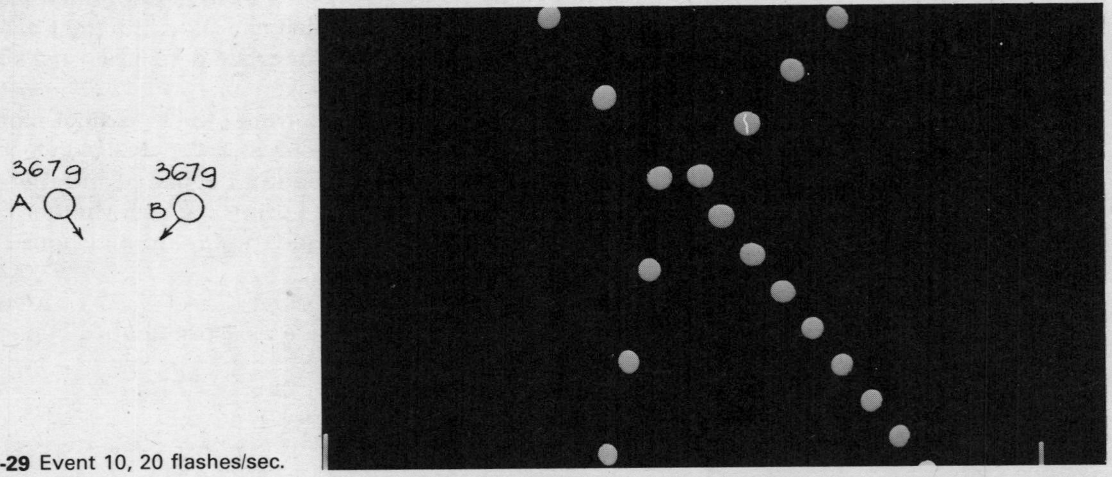

Fig. 3-29 Event 10, 20 flashes/sec.

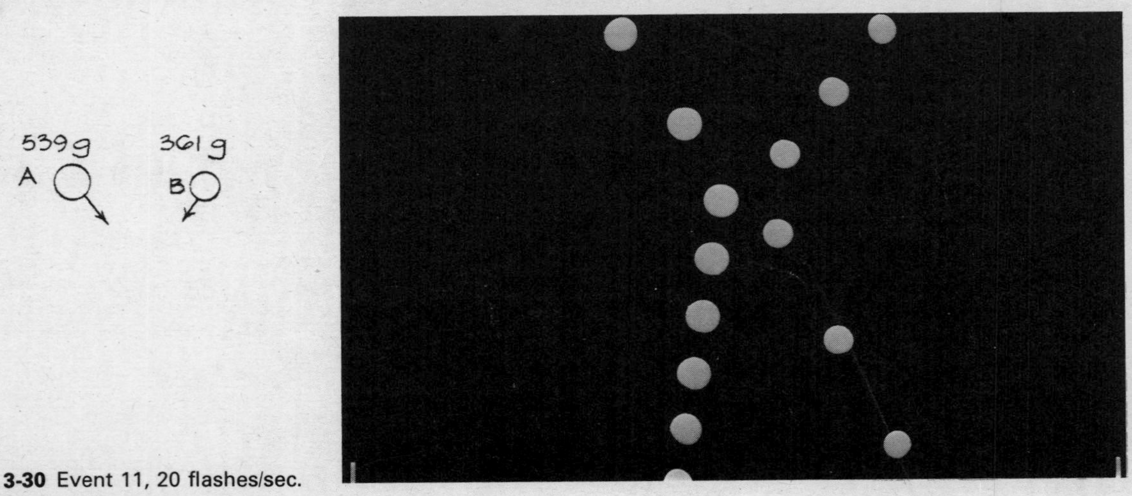

Fig. 3-30 Event 11, 20 flashes/sec.

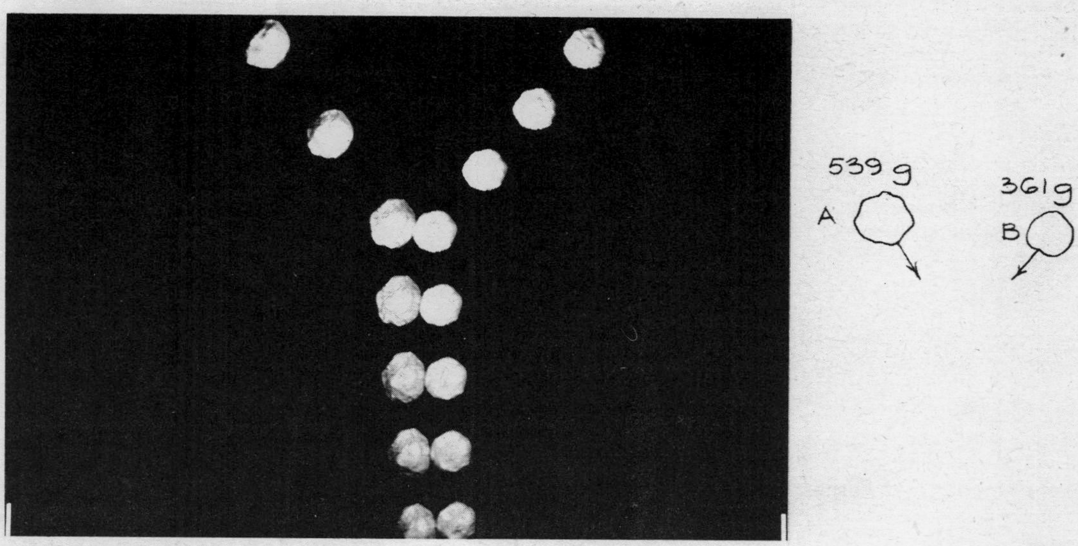

Fig. 3-31 Event 12, 20 flashes/sec.

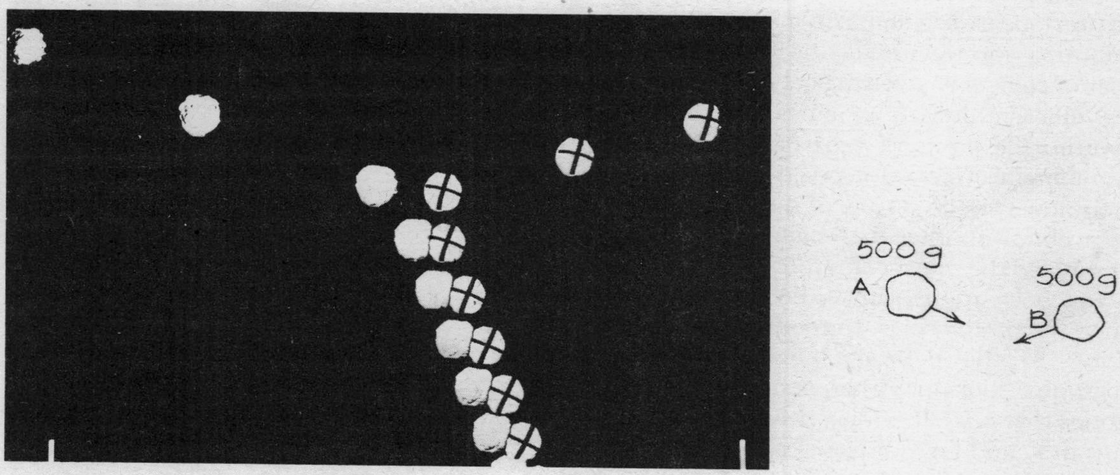

Fig. 3-32 Event 13, 10 flashes/sec.

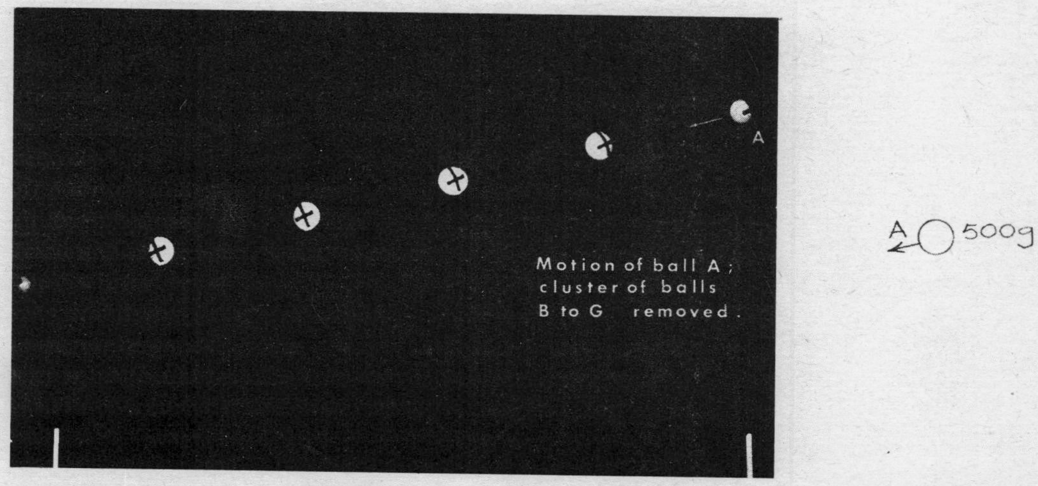

Fig. 3-33 Event 14, print 1, 20 flashes/sec.

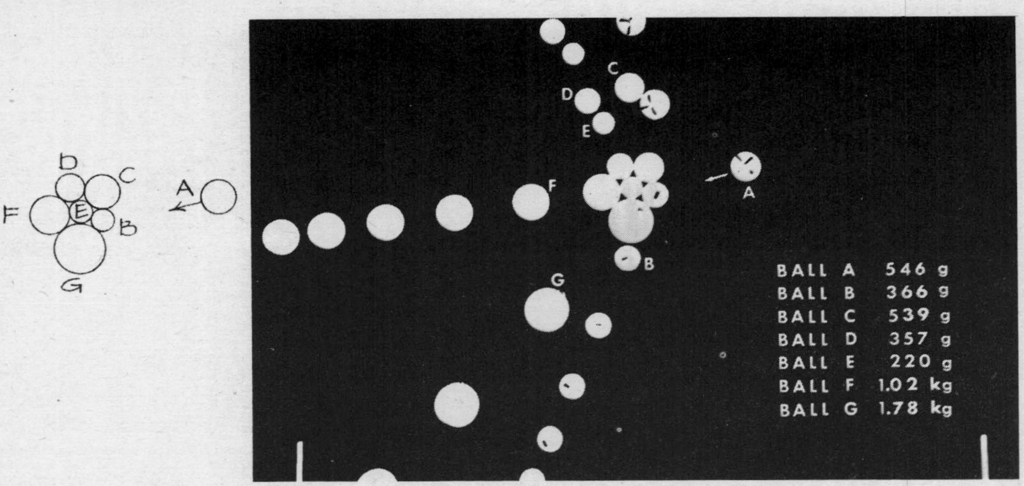

Fig. 3-34 Event 14, print 2, 20 flashes/sec.

Experiment 3-5
CONSERVATION OF ENERGY. I

In the previous experiments on conservation of momentum, you recorded the results of a number of collisions involving carts and gliders having different initial velocities. You found that within the limits of experimental uncertainty, momentum was conserved in each case. You can now use the results of these collisions to learn about another extremely useful conservation law, the conservation of energy.

Do you have any reason to believe that the product of m and v is the only conserved quantity? In the data obtained from your photographs, look for other combinations of quantities that might be conserved. Find values for m/v, m^2v, and mv^2 for each cart before and after collision, to see if the sum of these quantities for both carts is conserved. Compare the results of the elastic collisions with the inelastic ones. Consider the "explosion," too.

Is there a quantity that is conserved for one type of collision but not for the other?

There are several alternative methods to explore further the answer to this question; you will probably wish to do just one. Check your results against those of classmates who use other methods.

METHOD A. Dynamics Carts

To take a closer look at the details of an elastic collision, photograph two dynamics carts as you may have done in a previous experiment. Set the carts up as shown in Fig. 3-35.

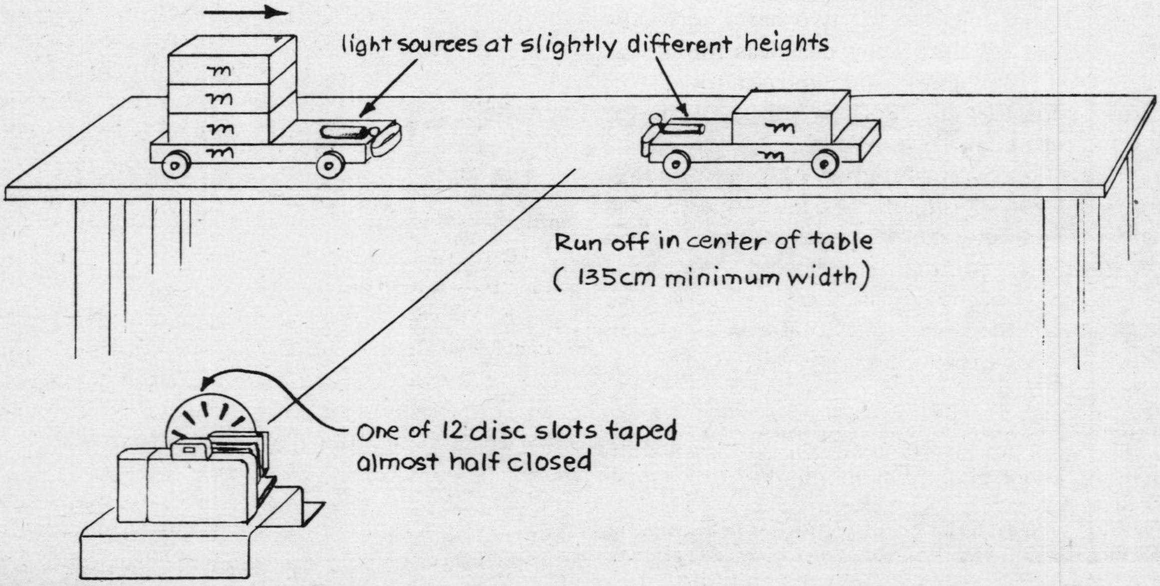

Fig. 3-35

The mass of each cart is 1 kg. Extra mass is added to make the total masses 2 kg and 4 kg. Tape a light source on each cart. So that you can distinguish between the images formed by the two lights, make sure that one of the bulbs is slightly higher than the other.

Place the 2-kg cart at the center of the table and push the other cart toward it from the left. If you use the 12-slot disk on the stroboscope, you should get several images during the time that the spring bumpers are touching. You will need to know which image of the right-hand cart was made at the same instant as a given image of the left-hand cart. Matching images will be easier if one of the 12 slots on the stroboscope disk is half-covered with tape (Fig. 3-36). Images formed when that slot is in front of the lens will be fainter than the others.

Fig. 3-36

Compute the values for the momentum mv for each cart for each time interval while the springs were touching, plus at least three intervals before and after the springs touched. List the values in a table, making sure that you pair off the values for the two carts correctly. Remember that the lighter cart was initially at rest while the heavier one moved toward it. This means that the first few values of mv for the lighter cart will be zero.

On a sheet of graph paper, plot the momentum of each cart as a function of time, using the same coordinate axes for both. Connect each set of values with a smooth curve.

Now draw a third curve that shows the sum of the two values of mv (the total momentum of the system) for each time interval.

?

1. Compare the final value of mv for the system with the initial value. Was momentum conserved in the collision?
2. What happened to the momentum of the system while the springs were touching; was momentum conserved *during* the collision?

Now compute values for the scalar quantity mv^2 for each cart for each time interval, and add them to your table. On another sheet of graph paper, plot the values of mv^2 for each cart for each time interval. Connect each set of values with a smooth curve.

Now draw a third curve that shows the sum of the two values of mv^2 for each time interval.

?

3. Compare the final value of mv^2 for the system with the initial value. Is mv^2 a conserved quantity?
4. How would the appearance of your graph change if you multiplied each quantity by ½? (The quantity $\frac{1}{2}mv^2$ is called the *kinetic energy* of the object of mass m and speed v.)

Compute values for the scalar quantity $\frac{1}{2}mv^2$ for each cart for each time interval. On a sheet of graph paper, plot the kinetic energy of each cart as a function of time, using the same coordinate axes for both.

Now draw a third curve that shows the sum of the two values of $\frac{1}{2}mv^2$ for each time interval.

?

5. Does the total amount of kinetic energy vary during the collision? If you found a change in the total kinetic energy, how do you explain it?

METHOD B. Magnets

Spread some Dylite spheres (tiny plastic beads) on a glass tray or other hard, flat surface. A disk magnet will slide freely on this low-friction surface. Level the surface carefully.

Put one disk magnet at the center and push a second one toward it, slightly off center. You want the magnets to repel each other without actually touching. Try varying the speed and direction of the pushed magnet until you find conditions that make both magnets move off after the collision with about equal speeds.

To record the interaction, set up a camera directly above the glass tray (using the motor-strobe mount if your camera does not attach directly to the tripod) and a xenon stroboscope to one side as in Fig. 3-37. Mount a steel ball or a Styrofoam hemisphere on the center of each disk with a small piece of clay. The ball will give a sharp reflection of the strobe light.

Take strobe photographs of several interactions. There must be several images before and after the interaction, but you can vary the initial speed and direction of the moving magnet to

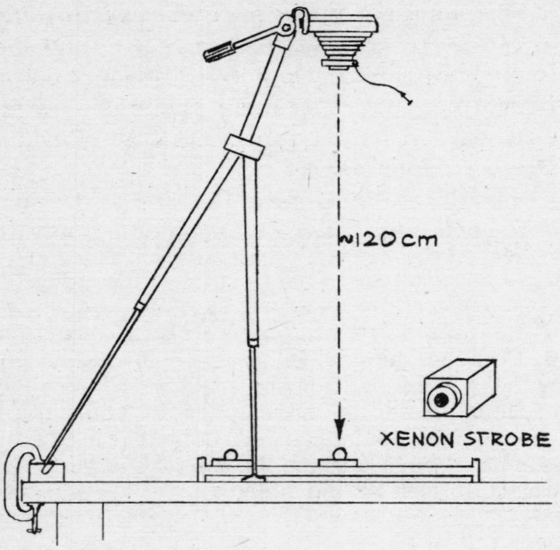

Fig. 3-37

get a variety of interactions. Using your photograph, calculate the "before" and "after" speeds of each disk. Since you are interested only in comparing speeds, you can use any convenient units of measurement for speed.

?

1. Is mv^2 a conserved quantity? Is $\frac{1}{2}mv^2$ a conserved quantity?

If you find that there has been a decrease in the total kinetic energy of the system of interacting magnets, consider the following: The surface is not perfectly frictionless and a single magnet disk pushed across it will slow down a bit. Make a plot of $\frac{1}{2}mv^2$ against time for a moving disk to estimate the rate at which kinetic energy is lost in this way.

?

2. How much of the loss in $\frac{1}{2}mv^2$ that you observed in the interaction can be due to friction?
3. What happens to your results if you consider kinetic energy to be a *vector* quantity?

When the two disks are close together (but not touching) there is quite a strong force between them pushing them apart. If you put the two disks down on the surface close together and release them, they will fly apart; the kinetic energy of the system has increased.

If you have time to go on, you should try to find out what happens to the total quantity $\frac{1}{2}mv^2$ of the disks while they are close together during the interaction. To do this you will use a fairly high strobe rate, and push the projectile magnet at fairly high speed, without letting the two magnets actually touch, of course. Close the camera shutter before the disks are out of the field of view so that you can match images by counting backward from the last images.

Now, working backward from the last interval, measure v and calculate $\frac{1}{2}mv^2$ for each disk. Make a graph in which you plot $\frac{1}{2}mv^2$ for each disk against time. Draw smooth curves through the two plots.

Now draw a third curve that shows the sum of the two $\frac{1}{2}mv^2$ values for each time interval.

?

4. Is the quantity $\frac{1}{2}mv^2$ conserved *during* the interaction, that is, while the repelling magnets approach very closely?
Try to explain your observations.

METHOD C. Inclined Air Tracks

Suppose you give the glider a push at the bottom of an inclined air track. As it moves up the slope it slows down, stops momentarily, and then begins to come back down the track.

Clearly the bigger the push you give the glider (the greater its initial velocity v_i), the higher up the track it will climb before stopping. From experience you know that there is some connection between v_i and d, the distance the glider moves along the track.

According to physics texts, when a stone is thrown upward, the kinetic energy that it has initially ($\frac{1}{2}mv_i^2$) is transformed into gravitational potential energy (ma_gh) as the stone moves up. In this experiment, you will test to see whether the same relationship applies to the behavior of the glider on the inclined air track. In particular, your task is to find the initial kinetic energy and the increase in potential energy of the air track glider and to compare them.

The purpose of the first set of measurements is to find the initial kinetic energy $\frac{1}{2}mv_i^2$. You cannot measure v_i directly, but you can find it from your calculation of the *average* velocity v_{av} as follows. In the case of uniform acceleration $v_{av} = \frac{1}{2}(v_i + v_f)$. Since final velocity $v_f = 0$ at the top of the track, $v_{av} = \frac{1}{2}v_i$ or $v_i = 2v_{av}$. Remember that $v_{av} = \Delta d/\Delta t$, so $v_i = 2(\Delta d/\Delta t)$; Δd and Δt are easy to measure with your apparatus.

To measure Δd and Δt, three people are needed: one gives the glider the initial push, another marks the highest point on the track reached by the glider, and the third uses a stopwatch to time the motion from push to rest.

Raise one end of the track a few centimeters above the tabletop. The launcher should practice pushing to produce a push that will send the glider nearly to the raised end of the track.

Record the distance traveled and time taken for several trials, and weigh the glider. Determine and record the initial kinetic energy.

To calculate the increase in gravitational potential energy, you must measure the vertical height h through which the glider moves for each push. You will probably find that you need to measure from the tabletop to the track at the initial h_i and final h_f points of the glider's motion (see Fig. 3-38), since $h = h_f - h_i$. Calculate the potential energy increase, the quantity ma_gh, for each of your trials.

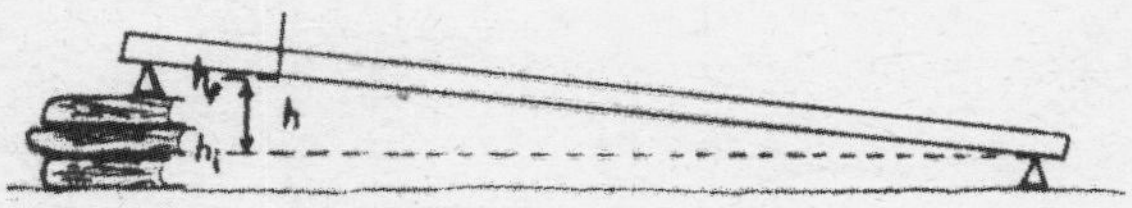

Fig. 3-38

For each trial, compare the kinetic energy loss with the potential energy increase. Be sure that you use consistent units: m in kilograms, v in meters/second, a_g in meters/second², h in meters.

?

1. Are the kinetic energy loss and the potential energy increase equal within your experimental uncertainty?
2. Explain the significance of your result.

Here are more things to do if you have time to go on:

(a) See if your answer to Question 1 continues to be true as you make the track steeper and steeper.

(b) When the glider rebounds from the rubber band at the bottom of the track, it is momentarily stationary; its kinetic energy is zero. The same is true of its gravitational potential energy, if you use the bottom of the track as the zero level. Yet the glider will rebound from the rubber band (regain its kinetic energy) and go quite a way up the track (gaining gravitational potential energy) before it stops. See if you can explain what happens at the rebound in terms of the conversation of mechanical energy.

(c) The glider does not get quite so far up the track on the second rebound as it did on the first. There is evidently a loss of energy. See if you can measure how much energy is lost each time.

Experiment 3-6 CONSERVATION OF ENERGY. II

METHOD A. Film Loops

You may have used one or more of Film Loops 18–25 in your study of momentum. You will find it helpful to view these slow-motion films of one- and two-dimensional collisions again, but this time in the context of the study of energy. The data you collected previously will be sufficient for you to calculate the kinetic energy of each ball before and after the collision. Remember that kinetic energy $\frac{1}{2}mv^2$ is *not* a vector quantity, and therefore you need only use the magnitude of the velocities in your calculations.

On the basis of your analysis you may wish to try to answer such questions as: Is kinetic energy conserved in such interactions? If not, what happened to it? Is the loss in kinetic energy related to such factors as relative speed, angle of impact, or relative masses of the colliding balls? Is there a difference in the kinetic energy lost in elastic and inelastic collisions?

The Film Loops were made in a highly controlled laboratory situation. After you have developed the technique of measurement and analysis from Film Loops, you may want to turn to one or more loops dealing with things outside the laboratory setting. Film Loops 26–33 involve freight cars, billiard balls, pole vaulters, and the like. Suggestions for using these loops can be found on pages 81–87.

METHOD B. Stroboscopic Photographs of Collisions

When studying momentum, you may have taken measurements on the one-dimensional and two-dimensional collisions shown in stroboscopic photographs on pages 107–110 and 115–118. If so, you can now easily reexamine your data and compute the kinetic energy $\frac{1}{2}mv^2$ for each ball before and after the interaction. Remember that kinetic energy is a scalar quantity, and so you will use the magnitude of the velocity but not the direction in making your computations. You would do well to study one or more of the simpler events (for example, Events 1, 2, 3, 8, 9, or 10) before attempting the more complex ones involving inelastic collisions or several balls. Also, you may wish to review the discussions given earlier for each event.

If you find there is a loss of kinetic energy beyond what you would expect from measurement error, try to explain your results. Some questions you might try to answer are: How does kinetic energy change as a function of the distance from impact? Is it the same before and after impact? How is energy conservation influenced by the relative speed at the time of collision? How is energy conservation influenced by the angle of impact? Is there a difference between elastic and inelastic interactions in the fraction of energy conserved?

Experiment 3-7 MEASURING THE SPEED OF A BULLET

In this experiment you will use the principle of the conservation of momentum to find the speed of a bullet. Sections 9.2 and 9.3 in the Text discuss collisions and define momentum.

You will use the general equation of the principle of conservation of momentum for two-body collisions: $m_A\vec{v}_A + m_B\vec{v}_B = m_A\vec{v}'_A + m_B\vec{v}'_B$.

The experiment consists of firing a projectile into a can packed with cotton or into a heavy block that is free to move horizontally. Since all velocities before and after the collision are in the same direction, you may neglect the vector nature of the equation above and work only with speeds. To avoid subscripts, call the mass of the target M and the much smaller mass of the projectile m. Before impact, the target is at rest, so you have only the speed v of the projectile to consider. After impact, both the target and the added projectile move with a common speed v'. Thus, the general equation becomes

$$mv = (M + m)v'$$

or

$$v = \frac{(M + m)v'}{m}$$

Both masses are easy to measure. Therefore, if the comparatively slow speed v' can be found after impact, you can compute the high speed v of the projectile before impact. There are at least two ways to find v'.

METHOD A. Air Track

The most direct way to find v' is to mount the target on the air track and to time its motion after the impact. (See Fig. 3-39.) Mount a small can, lightly packed with cotton, on an air-track glider. Make sure that the glider will still ride freely with this extra load. Fire a "bullet" (a pellet from a toy gun that has been checked for safety by your instructor) horizontally, parallel to the length of the air track. If M is large enough, compared to m, the glider's speed will be low enough so that you can use a stopwatch to time it over a 1-m distance. Repeat the measurement a few times until you get consistent results.

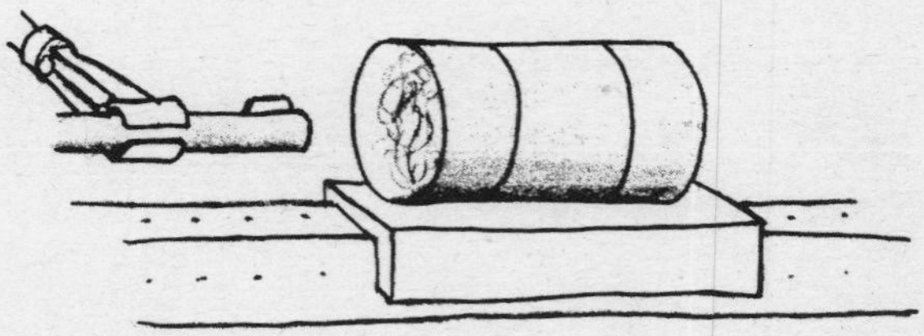

Fig. 3-39

?

1. What is your value for the bullet's speed?
2. Suppose the collision between bullet and can was not completely inelastic, so that the bullet bounced back a little after impact. Would this increase or decrease your value for the speed of the bullet?
3. Can you think of an independent way to measure the speed of the bullet? If you can, make the independent measurement. Then see if you can account for any differences between the two results.

METHOD B. Ballistic Pendulum

This was the original method of determining the speed of bullets, invented in 1742 and still used in some ordnance laboratories. A movable block is suspended as a freely swinging pendulum whose motion reveals the bullet's speed.

Obtaining the Speed Equation

The collision is inelastic, so kinetic energy is not conserved in the impact. But during the nearly frictionless swing of the pendulum after the impact, mechanical energy is conserved, that is, the increase in gravitational potential energy of the pendulum at the end of its upward swing is equal to its kinetic energy immediately after impact. Written as an equation, this becomes

$$(M+m)a_g h = \frac{(M+m)v'^2}{2}$$

where h is the increase in height of the pendulum bob.

Solving this equation for v' gives:

$$v' = \sqrt{2a_g h}$$

Substituting this expression for v' in the momentum equation above leads to

$$v = \frac{M+m}{m}\sqrt{2a_g h}$$

Now you have an equation for the speed v of the bullet in terms of quantities that are known or can be measured.

A Useful Approximation

The change h in vertical height is difficult to measure accurately, but the horizontal displacement d may be 10 cm or more and can be found easily. Can h be replaced by an equivalent expression involving d? The relation between h and d can be found by using a little plane geometry.

In Fig. 3-40, the center of the circle, O, represents the point from which the pendulum is hung. The length of the cords is l.

In the triangle OBC,

$$l^2 = d^2 + (l-h)^2$$

so $$l^2 = d^2 + l^2 - 2lh + h^2$$

and $$2lh = d^2 + h^2$$

For small swings, h is small compared with l and d, so you may neglect h^2 in comparison with d^2, and write the close approximation

$$2lh = d^2$$

or $$h = d^2/2l$$

Putting this value of h into your last equation for v above and simplifying gives:

$$v = \frac{(M+m)d}{m}\sqrt{\frac{a_g}{l}}$$

If the mass of the projectile is small compared with that of the pendulum, this equation can be simplified to another good approximation. How?

Finding the Projectile's Speed

Now you are ready to begin the experiment. The kind of pendulum you use will depend on the nature and speed of the projectile. If you use pellets from a toy gun, a cylindrical cardboard carton stuffed lightly with cotton and suspended by threads from a laboratory stand will do. If you use a good bow and arrow, stuff straw into a fairly stiff corrugated box and hang it from the ceiling. To prevent the target pendulum from twisting, hang it by parallel cords connecting four points on the pendulum to four points directly above them, as in Fig. 3-41.

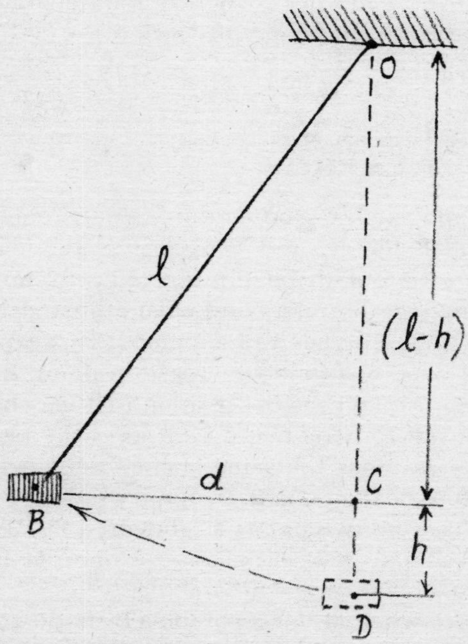

Fig. 3-40

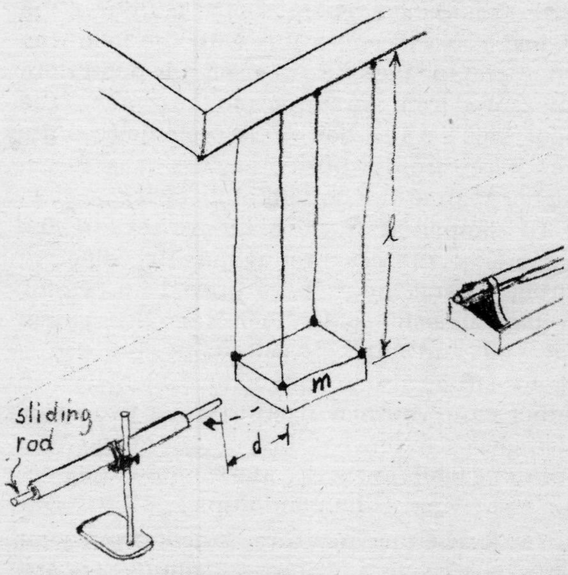

Fig. 3-41

To measure d, a light rod (a pencil or a soda straw) is placed in a tube clamped to a stand. The rod extends out of the tube on the side toward the pendulum. As the pendulum swings, it shoves the rod back into the tube so that the rod's final position marks the end of the swing of the pendulum. Of course the pendulum must not hit the tube and there must be sufficient friction between rod and tube so that the rod stops when the pendulum stops. The original rest position of the pendulum is readily found so that the displacement d can be measured.

Repeat the experiment a few times to get an idea of how precise your value for d is. Then substitute your data in the equation for v, the bullet's speed.

?

1. What is your value for the bullet's speed?
2. From your results, compare the kinetic energy of the bullet before impact with that of the pendulum after impact. Why is there such a large difference in kinetic energy?
3. Can you describe an independent method for finding v? If you have time, try it, and explain any difference betwen the two values of v.

Experiment 3-8
ENERGY ANALYSIS OF A PENDULUM SWING

According to the law of conservation of energy, the loss in gravitational potential energy of a simple pendulum as it swings from the top of its swing to the bottom is completely transferred into kinetic energy at the bottom of the swing. You can check this with the following photographic method. A 1-m simple pendulum (measured from the support to the *center* of the bob) with a 0.5-kg bob works well. Release the pendulum from a position where it is 10 cm higher than at the bottom of its swing.

To simplify the calculations, set up the camera for 10:1 scale reduction. Two different strobe arrangements have proved successful: (1) tape an ac blinky to the bob, or (2) attach an AA cell and bulb to the bob and use a motor-strobe disk in front of the camera lens. In either case, you may need to use a two-string suspension to prevent the pendulum bob from spinning while swinging. Make a time exposure for one swing of the pendulum.

You can either measure directly from your print (which should look something like the one in Fig. 3-42) or make pinholes at the center of each image on the photograph and project the hole images onto a larger sheet of paper. Calculate the instantaneous speed v at the bottom of the swing by dividing the distance traveled between the images nearest the bottom of the swing by the interval between the images. The kinetic energy at the bottom of the swing, $\frac{1}{2}mv^2$, should equal the change in potential energy from the top of the swing to the bottom. If Δh is the difference in vertical height between the bottom of the swing and the top, then

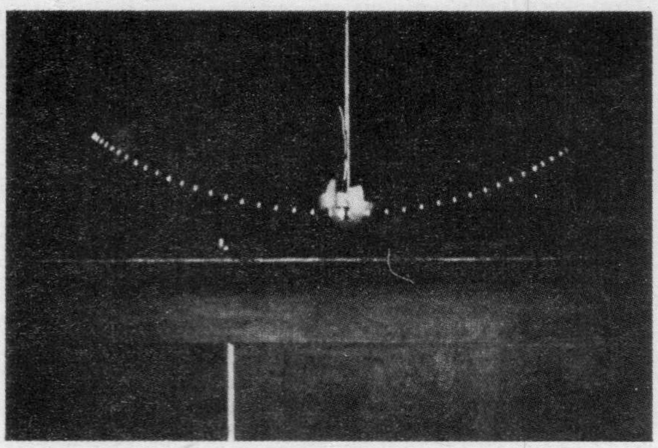

Fig. 3-42

$$v = \sqrt{2a_g \Delta h}$$

If you plot both the kinetic and potential energy on the same graph (using the bottom-most point as a zero level for gravitational potential energy), and then plot the sum of $KE + PE$, you can check whether total energy is conserved during the entire swing.

Experiment 3-9
LEAST ENERGY

Concepts such as momentum, kinetic energy, potential energy, and the conservation laws often turn out to be unexpectedly useful in helping you to understand what at first glance seem to be unrelated phenomena. This experiment offers just one such case in point: How can you explain the observation that if a chain is allowed to hang freely from its two ends, it always assumes the same shape? Hang a 1-m length of beaded chain, the type used on light sockets, from two points as shown in Fig. 3-43. What shape does the chain assume? At first glance it seems to be a parabola.

Check whether it is a parabola by finding the equation for the parabola that would go through the vertex and the two fixed points.

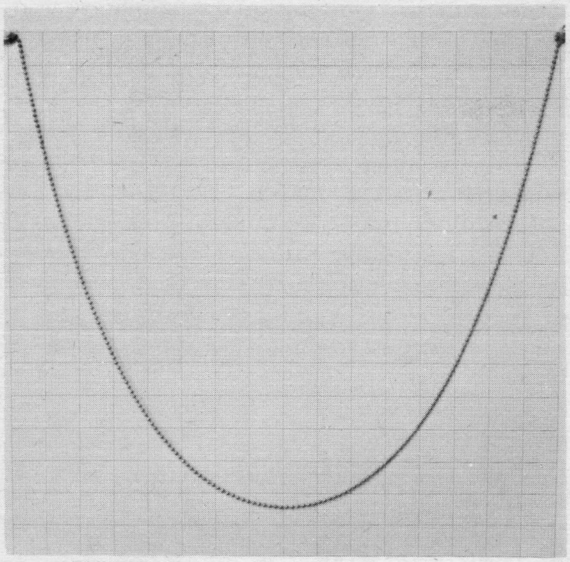

Fig. 3-43

Determine other points on the parabola by using the equation. Plot them and see whether they match the shape of the chain.

One way to plot the parabola is as follows. The vertex in Fig. 3-43 is at (0,0) and the two fixed points are at (−8, 14.5) and (8, 14.5). All parabolas symmetric to the y axis have the formula $y = kx^2$, where k is a constant. For this example, you must have $14.5 = k(8)^2$, or $14.5 = 64k$. Therefore, $k = 0.227$, and the equation for the parabola going through the given vertex and two points is $y = 0.227x^2$. Now substitute values for x, producing a table of x and y values for the parabola. When you plot these values on the graph paper behind the chain, do the chain and the plotted points coincide?

A more interesting question is why the chain assumes the particular shape it does, which is called a *catenary curve*. Recall that the gravitational potential energy of a body mass m is defined as ma_gh, where a_g is the acceleration due to gravity, and h is the height of the body above the reference level chosen. Remember that only a *difference* in energy level is meaningful; a different reference level only adds a constant to each value associated with the original reference level. In theory, you could measure the mass of one bead on the chain, measure the height of each bead above the reference level, and total the potential energies for all the beads to get the total potential energy for the whole chain.

In practice, that would be quite tedious, so you will use an approximation that will still allow you to get a reasonably good result. (This would be an excellent computer problem.) Draw vertical parallel lines about 2 cm apart on the paper behind the chain (or use graph paper). In each vertical section, make a mark beside the chain in the middle of that section (see Fig. 3-44).

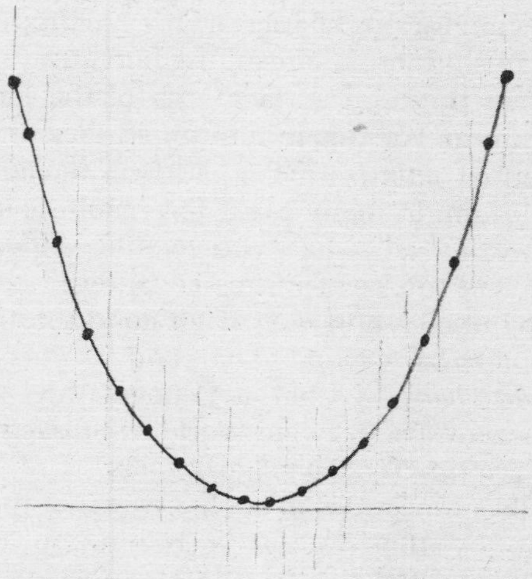

Fig. 3-44

The total potential energy for that section of the chain will be approximately Ma_gh_{av}, where h_{av} is the average height marked, and M is the total mass in that section of chain. Notice that near the ends of the chain there are more beads in one horizontal interval than there are near the center of the chain. To simplify the solution further, assume that M is always an integral number of beads that you can count.

In summary, for each interval multiply the number of beads by the average height for that interval. Total all these products. This total is a good approximation of the gravitational potential energy of the chain.

After doing this for the freely hanging chain, pull the chain with thumbtacks into different shapes such as those shown in Fig. 3-45.

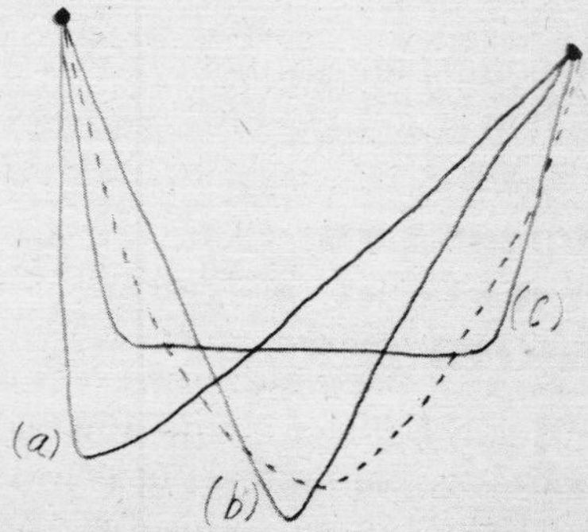

Fig. 3-45

Calculate the total potential energy for each shape. Does the catenary curve (the freely formed shape) or one of these others have the minimum total potential energy?

If you would like to explore other instances of the minimization principles, try the following:

1. When various shapes of wire are dipped into a soap solution, the resulting film always forms so that the total surface area of the film is a minimum. For this minimum surface, the total potential energy due to surface tension is a minimum. In many cases the resulting surface is not at all what you would expect. An excellent source of suggested experiments with soap bubbles, and recipes for good solutions, is the paperback *Soap Bubbles and the Forces that Mould Them*, by Charles V. Boys, (Dover, 1959). Also, see "The Strange World of Surface Film," *The Physics Teacher* (Sept., 1966).

2. Rivers meander in such a way that the work done by the river is a minimum. For an explanation of this, see "A Meandering River," in the June, 1966 issue of *Scientific American*.

3. Suppose that points A and B are placed in a vertical plane as shown in Fig. 3-46. You want to build a track between the two points so that a ball will roll from A to B in the least possible time. Should the track be straight or in the shape of a circle, parabola, cycloid, catenary, or some other shape? An interesting property of a cycloid is that no matter where on a cycloidal track you release a ball, it will take the same amount of time to reach the bottom of the track. You may want to build a cycloidal track in order to check this. Do not make the track so steep that the ball slips instead of rolling.

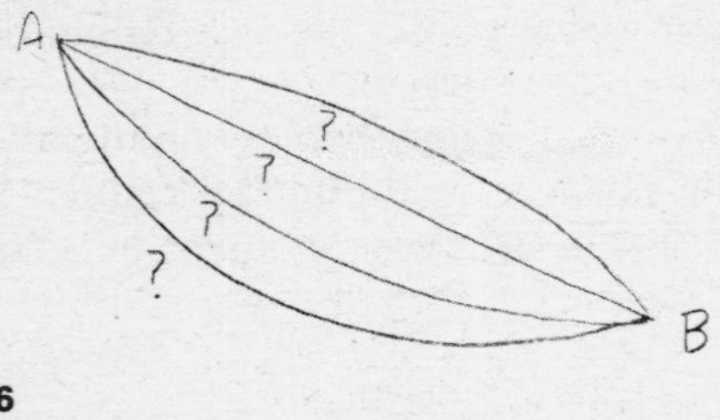

Fig. 3-46

A more complete treatment of this principle of least action is given in the *Feynman Lectures on Physics*, Vol. 2.

Experiment 3-10 TEMPERATURE AND THERMOMETERS

You can usually tell just by touch which of two similar bodies is the hotter. But if you want to tell exactly *how* hot something is or to communicate such information to somebody else, you have to find some way of assigning a *number* to "hotness." This number is called *temperature*, and the instrument used to measure this number is called a *thermometer*.

Standard units for measuring intervals of time and distance, the day and the meter, are both familiar. But try to imagine yourself living in an era before the invention of thermometers and temperature scales, that is, before the time of Galileo. How would you describe, and if possible give a number to, the "degree of hotness" of an object?

Any property (such as length, volume, density, pressure, or electrical resistance) that changes with hotness and that can be measured could be used as an indication of temperature; any device that measures this property could be used as a thermometer.

In this experiment you will be using thermometers based on properties of liquid expansion, gas expansion, and electrical resistance. (Other common kinds of thermometers are based on electrical voltages, color, or gas pressure.) Each of these devices has its own particular merits that make it suitable, from a practical point of view, for some applications, and difficult or impossible to use in others.

Of course, temperature estimates given by two different types of thermometers must agree over the range that they are to be used in common. In this experiment you will make your own thermometers, put temperature scales on them, and then compare them to see how well they agree with each other.

Defining a Temperature Scale

How do you make a thermometer? First, you decide what *property* (length, volume, etc.) of what *substance* (mercury, air, etc.) to use in your thermometer. Then you must decide on two fixed points in order to arrive at the size of a degree. A fixed point is based on a physical phenomenon that always occurs at the same degree of hotness. Two convenient fixed points to use are the melting point of ice and the boiling point of water. On the Celsius (centigrade) scale they are assigned the values 0°C and 100°C at ordinary atmospheric pressure.

When you are making a thermometer of any sort, you have to put a scale on it against which you can read the hotness-sensitive quantity. Often a piece of centimeter-marked tape or a short piece of ruler will do. Submit your thermometer to two fixed points of hotness (for example, a bath of boiling water and a bath of

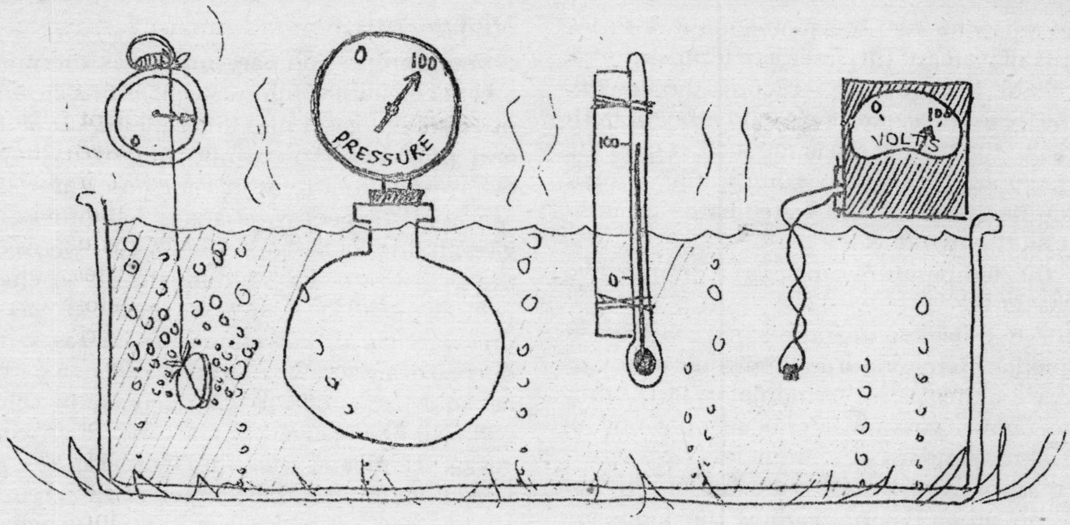

Fig. 3-47 Any quantity that varies with hotness can be used to establish a temperature scale (even the time it takes for an antacid tablet to dissolve in water!). Two "fixed points" (such as the freezing and boiling points of water) are needed to define the size of a degree.

ice water) and mark the positions on the indicator.

The length of the column can now be used to define a temperature scale by assuming that equal temperature changes cause equal changes along the scale between the two fixed-point positions. Suppose you marked the length of a column of liquid at the freezing point and again at the boiling point of water. You can now divide the total increase in length into equal parts and call each of these parts "one degree" change in temperature.

On the Celsius scale, the degree is 1/100 of the temperature range between the boiling and freezing points of water.

To identify temperatures between the fixed points on a thermometer scale, mark off the actual distance between the two fixed points on the vertical axis of a graph and equal intervals for degrees of temperature on the horizontal axis, as in Fig. 3-48. Then plot the fixed points, X, on the graph and draw a straight line between them.

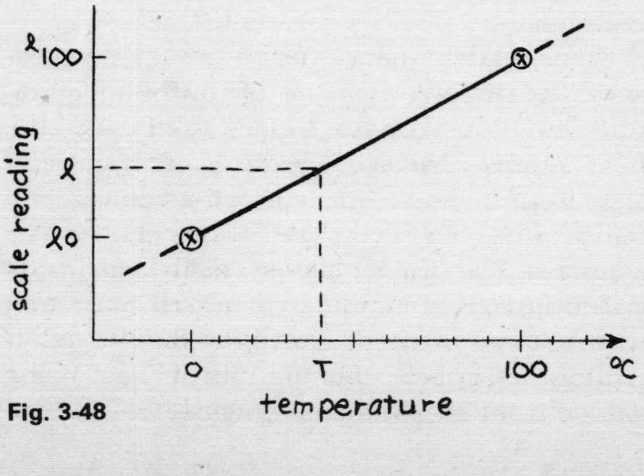

Fig. 3-48

The temperature on this scale, corresponding to any intermediate position l, can be read from the graph.

Other properties and other substances can be used (the volume of different gases, the electrical resistance of different metals, and so on), and the temperature scale defined in the same way. All such thermometers will have to agree at the two fixed points, but do they agree at intermediate temperatures?

If different physical properties do not change in the same way with hotness, then the temperature values you read from thermometers using these properties will not agree. Do similar temperature scales defined by different physical properties agree anywhere besides at the fixed points? That is a question that you can answer from this lab experience.

Comparing Thermometers

You will make or be given two "thermometers" to compare. Take readings of the appropriate quantities, such as length of liquid column, volume of gas, electrical resistance, or thermocouple voltage, when the devices are placed in an ice bath and again when they are placed in a boiling-water bath. Record these values. Define these two temperatures as 0° and 100° and draw the straight-line graphs that define intermediate temperatures as described above.

Now put your two thermometers in a series of baths of water at intermediate temperatures, and again measure and record the length, volume, resistance, etc., for each bath. Put both devices in the bath at the same time in case the bath is cooling down. Use your graphs to read the temperatures of the water baths as indicated by the two devices.

Do the temperatures measured by the two devices agree?

If the two devices give the same readings at intermediate temperatures, then you can apparently use either as a thermometer. But if they do not agree, you must choose only one of them as a standard thermometer. Give whatever reasons you can for choosing one rather than the other before reading the following discussion. If possible, compare your results with those of classmates using the same or different kinds of thermometers.

There will, of course, be some uncertainty in your measurements, and you must decide whether the differences you observe between the two thermometers might be due only to this uncertainty.

The relationship between the readings from two different thermometers can be displayed on another graph, where one axis is the reading on one thermometer and the other axis is the reading on the other thermometer. Each bath will give a plot on this graph. If the points fall along a straight line, then the two thermometer properties must change in the same way. If, however, a fairly regular smooth curve can be drawn through the points, then the two thermometer properties probably change with hotness in different ways. (Figure 3-49 shows possible results for two thermometers.)

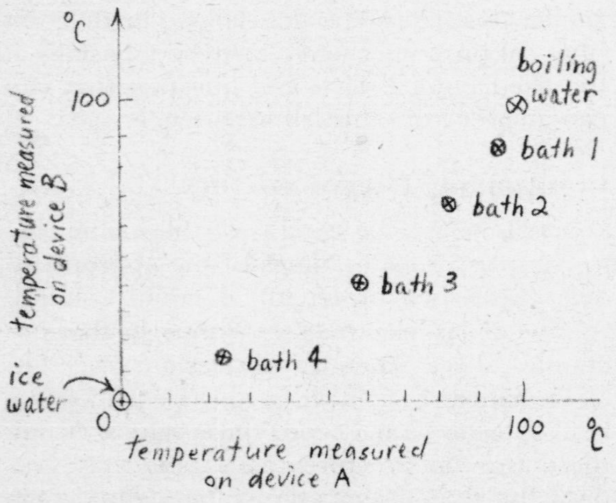

Fig. 3-49

Discussion

If you were to compare many gas thermometers at constant volume as well as pressure, and use different gases and different initial volumes and pressures, you would find that they all behave quantitatively in very much the same way with respect to changes in hotness. If a given hotness change causes a 10% increase in the pressure of gas A, then the same change will also cause a 10% increase in gas B's pressure. Or, if the volume of one gas sample decreases by 20% when transferred to a particular cold bath, then a 20% decrease in volume will also be observed in a sample of any other gas. This means that the temperatures read from different gas thermometers all agree.

This sort of close similarity of behavior between different substances is not found as consistently in the expansion of liquids or solids, or in their other properties, and so these thermometers do not agree, as you may have just discovered.

This suggests two things. First, there is quite a strong case for using the change in pressure (or volume) of a gas to *define* the temperature change. Second, the fact that, in such experiments, all gases do behave quantitatively in the same way suggests that there may be some underlying simplicity in the behavior of gases not found in liquids and solids, and that if one wants to learn more about the way matter changes with temperature, one would do well to start with gases.

Experiment 3-11
CALORIMETRY

Speedometers measure speed, voltmeters measure voltage, and accelerometers measure acceleration. In this experiment, you will use a device called a *calorimeter*. As the name suggests, it measures a quantity connected with heat.

Unfortunately, heat energy cannot be measured as directly as some of the other quantities mentioned above. In fact, to measure the heat energy absorbed or given off by a substance you must measure the change in temperature of a second substance chosen as a standard. The heat exchange takes place inside a calorimeter, a container in which measured quantities of materials can be mixed together without an appreciable amount of heat being gained from or lost to the outside.

Fig. 3-50

A Preliminary Experiment

The first experiment will give you an idea of how good a calorimeter's insulating ability really is.

Fill a calorimeter cup (a Styrofoam coffee cup does nicely) about half full of ice water. Put the same amount of ice water with one or two ice cubes floating in it in a second cup. Into a third cup, pour the same amount of water that has been heated to nearly boiling. Measure the temperature of the water in each cup, and record the temperature and the time of observation. (See Fig. 3-50.)

Repeat the observations at about 5-min intervals throughout the period. Between observations, prepare a sheet of graph paper with coordinate axes so that you can plot temperature as a function of time.

Mixing Hot and Cold Liquids

(You can do this experiment while continuing to take readings of the temperature of the water in your three cups.) You will make several assumptions about the nature of heat. Then you will use these assumptions to predict what will happen when you mix two samples that are initially at different temperatures. If your prediction is correct, you can feel some confidence in your assumptions; at least, you can continue to use the assumptions until they lead to a prediction that turns out to be wrong.

First, assume that, in your calorimeter, heat behaves like a fluid that is conserved. Assume that it can flow from one substance to another, but that the total quantity of heat H present in the calorimeter in any given experiment is constant. Then the heat lost by a warm object should just equal the heat gained by a cold object. In symbols,

$$-\Delta H_1 = \Delta H_2$$

Next, assume that if two objects at different temperatures are brought together, heat will flow from the warmer to the cooler object until they reach the same temperature.

Finally, assume that the amount of heat fluid ΔH that enters or leaves an object is proportional to the change in temperature ΔT and to the mass of the object, m. In symbols,

$$\Delta H = cm\,\Delta T$$

where c is a constant of proportionality that depends on the units, and is different for different substances.

The units in which heat is measured have been defined so that they are convenient for calorimeter experiments. The *Calorie* (cal) is defined as the quantity of heat necessary to change the temperature of 1 kg of water by one Celsius degree. (This definition has to be refined somewhat for very precise work, but it is adequate for your purpose.) In the expression

$$\Delta H = cm\,\Delta T$$

when m is measured in kilograms of water and T in Celsius degrees, H will be the number of Calories. Because the Calorie was defined this way, the proportionality constant c has the value 1 Cal/kg C° when water is the only substance in the calorimeter. In metric units, 1 Cal = 4.19 kJ. Therefore, you could also measure directly in joules: 1 J of energy heats 1 g of water by 1/4.19, or 0.240 Celsius degree.

Checking the Assumptions

Measure and record the mass of two empty plastic cups. Then put about one-half cup of cold water in one and about the same amount of hot water in the other, and record the mass and temperature of each. (Subtract the mass of the empty cup.) Now mix the two together in one of the cups, stir *gently* with a thermometer, and record the final temperature of the mixture.

Multiply the change in temperature of the cold water by its mass. Do the same for the hot water.

?

1. What is the product (mass × temperature change) for the cold water?
2. What is this product for the hot water?
3. Are your assumptions confirmed, or is the difference between the two products greater than can be accounted for by uncertainties in your measurement?

Predicting from the Assumptions

Try another mixture using different quantities of water, for example one-quarter cup of hot water and one-half cup of cold. Before you mix the two, try to predict the final temperature.

?

4. What do you predict the temperature of the mixture will be?
5. What final temperature do you observe?
6. Estimate the uncertainty of your thermometer readings and your mass measurements. Is this uncertainty enough to account for the difference between your predicted and observed values?
7. Do your results support the assumptions?

Melting

The cups you filled with hot and cold water at the beginning of the period should show a measurable change in temperature by this time. If you are to hold to your assumption of conservation of heat fluid, then it must be that some heat has gone from the hot water into the room and from the room to the cold water.

?

8. How much has the temperature of the cold water changed?
9. How much has the temperature of the water that had ice in it changed?

The heat that must have gone from the room to the water–ice mixture evidently did not change the temperature of the water as long as the ice was present. But some of the ice melted, so apparently the heat that leaked in melted the ice. Evidently, heat was needed to cause a "change of state" (in this case, to change ice to water) even if there was no change in temperature. The additional heat required to melt 1 g of ice is called *latent heat of melting*. *Latent* means hidden or dormant. The units are Calories per gram; there is no temperature unit here because the temperature does not change.

Next, you will do an experiment mixing materials other than liquid water in the calorimeter to see if your assumptions about heat as a fluid can still be used. Two such experiments are described below, "Measuring Heat Capacity" and "Measuring Latent Heat." If you have time for only one of them, choose either one. Finally, do "Rate of Cooling" to complete your preliminary experiment.

Measuring Heat Capacity

(While you are doing this experiment, continue to take readings of the temperature of the water in your three test cups.) Measure the mass of a small metal sample. Put just enough cold water in a calorimeter to cover the sample. Tie a thread to the sample and suspend it in a beaker of boiling water. Measure the temperature of the boiling water.

Record the mass and temperature of the water in the calorimeter.

When the sample has been immersed in the boiling water long enough to be heated uniformly (2 or 3 min), lift it out and hold it just above the surface for a few seconds to let the water drip off, then transfer it quickly to the calorimeter cup. Stir gently with a thermometer and record the temperature when it reaches a steady value.

?

10. Is the product of mass and temperature change the same for the metal sample and for the water?
11. If not, must you modify the assumptions about heat that you made earlier in the experiment?

In the expression $\Delta H = cm\,\Delta T$, the constant of proportionality c (called the *specific heat capacity*) may be different for different materials. For water the constant has the value 1 cal/kgC°, or 0.240 J/gC°. You can find a value of c for the metal by using the assumption that heat gained by the water equals the heat lost by the sample. Writing subscripts for water H_w and metal sample H_s, $\Delta H_w = -\Delta H_s$.

Then $$c_w m_w \Delta t_w = -c_s m_s \Delta t_s$$

and $$c_s = \frac{-c_w m_w \Delta t_w}{m_s \Delta t_s}$$

?

12. What is your calculated value for the specific heat capacity c_1 for the metal sample you used?

If your assumptions about heat being a fluid are valid, you now ought to be able to predict the final temperature of *any* mixture of water and your material.

Try to verify the usefulness of your value. Predict the final temperature of a mixture of water and a heated piece of your material, using different masses and different initial temperatures.

?

13. Does your result support the fluid model of heat?

Measuring Latent Heat

Use your calorimeter to find the latent heat of melting of ice. Start with about one-half cup of water that is a little above room temperature, and record its mass and temperature. Take a small piece of ice from a mixture of ice and water that has been standing for some time; this will assure that the ice is at 0°C and will not have to be warmed up before it can melt. Place the small piece of ice on paper toweling for a moment to dry off water on its surface, and then transfer it quickly to the calorimeter.

Stir gently with a thermometer until the ice is melted and the mixture reaches an equilibrium temperature. Record this temperature and the mass of the water plus melted ice.

?

14. What was the mass of the ice that you added?

The heat given up by the warm water is

$$\Delta H_w = c_w m_w \Delta t_w$$

The heat gained by the water formed by the melted ice is

$$H_i = c_w m_i \Delta t_i$$

The specific heat capacity c_w is the same in both cases; that is, the specific heat of water.

The heat given up by the warm water first melts the ice, and then heats the water formed by the melted ice. Using the symbol ΔH_L for the heat energy required to melt the ice,

$$-\Delta H_w = \Delta H_L + \Delta H_i$$

So the heat energy needed to melt the ice is

$$\Delta H_L = -\Delta H_w - \Delta H_i$$

The latent heat of melting is the heat energy needed *per gram* of ice, so

$$\text{latent heat of melting} = \frac{\Delta H_L}{m}$$

?

15. What is your value for the latent heat of melting of ice?

When this experiment is done with ice made from distilled water with no inclusions of liquid water, the latent heat is found to be 80 cal/g, or 335 J/g. How does your result compare with the accepted value?

Rate of Cooling

If you have been measuring the temperature of the water in your three test cups, you should have enough data by now to plot three curves of temperature against time. Mark the temperature of the air in the room on your graph too.

?

16. How does the rate at which the hot water cools depend on its temperature?
17. How does the rate at which the cold water heats up depend on its temperature?

Weigh the amount of water in the cups. From the rates of temperature change (degrees/minute) and the masses of water, calculate the rates at which heat leaves or enters the cups at various temperatures. Use this information to estimate the error in your earlier results for latent or specific heat.

Experiment 3-12
ICE CALORIMETRY

A simple apparatus made up of thermally insulating Styrofoam cups can be used for doing some ice calorimetry experiments. Although the apparatus is simple, careful use will give you excellent results. To determine the heat transferred in processes in which heat energy is given off, you will be measuring either the volume of water or the mass of water from a melted sample of ice.

You will need either three cups the same size, or two large and one slightly smaller cup. Also have some extra cups ready. One large cup serves as the collector, A (Fig. 3-51), the second cup as the ice container, I, and the smaller cup (or one of the same size cut back to fit inside the ice container as shown) as the cover, K.

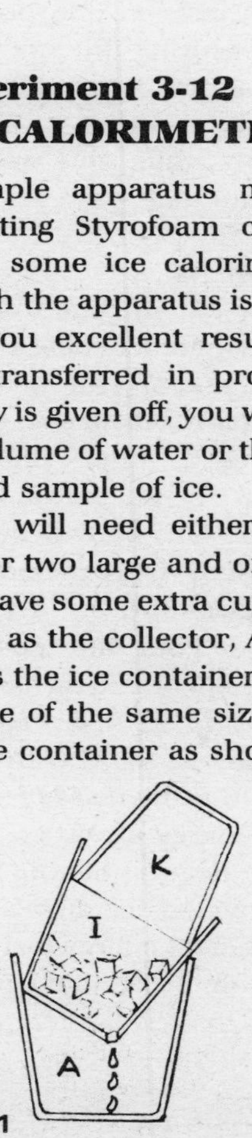

Fig. 3-51

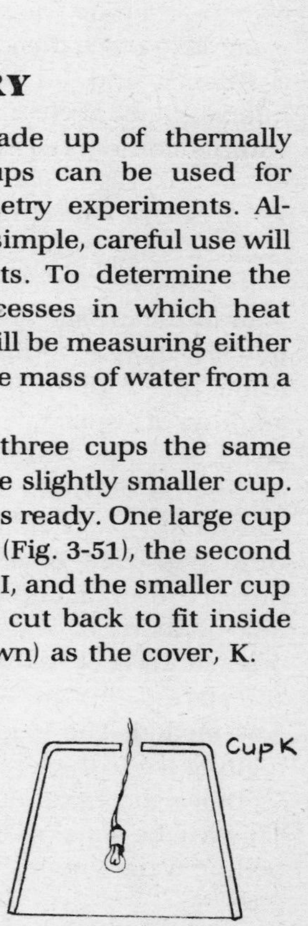

Fig. 3-52

Cut a hole about 0.5 cm in diameter in the bottom of cup I so that melted water can drain out into cup A. To keep the hole from becoming clogged by ice, place a bit of window screening in the bottom of I.

In each experiment, ice is placed in cup I. This ice should be carefully prepared, free of bubbles, and dry, if you plan to use the known value of the heat of fusion of ice. However, you can use ordinary crushed ice, and, before doing any of the experiments, determine experimentally the effective heat of melting of this nonideal ice. (Why should these two values differ?)

In some experiments that require some time to complete (such as Experiment b), you should set up two identical sets of apparatus (same quantity of ice, etc.), except that one does not contain a source of heat. One will serve as a fair measure of the background effect. Measure the amount of water collected in this control apparatus during the same time, and subtract this amount from the total amount of water collected in the experimental apparatus, thereby correcting for the amount of ice melted just by the heat leaking in from the room. An efficient method for measuring the amount of water is to place the arrangement on the pan of a balance and lift cups I and K at regular intervals (about 10 min) while you weight A with its contents of melted ice water.

(a) Heat of melting of ice

Fill a cup about one-half to one-third full with crushed ice. (Crushed ice has a larger amount of surface area, and so will melt more quickly, thereby minimizing errors due to heat from the room.) Bring a small measured amount of water (about 20 mL) to a boil in a beaker or large test tube and pour it over the ice in the cup. Stir briefly with a poor heat conductor, such as a glass rod, until equilibrium has been reached. Pour the ice–water mixture through cup I. Collect and measure the final amount of water (m_f) in A. If m_o is the original mass of hot water at 100°C with which you started, then $m_f - m_o$ is the mass of ice that was melted. The heat energy absorbed by the melting ice is the latent heat of melting for ice, L_i, times the mass of melted ice: $L_i(m_f - m_o)$. This will be equal to the heat energy lost by the boiling water cooling from 100°C to 0°C. Therefore,

$$L_i(m_f - m_o) = m_o \Delta T$$

and

$$L_i = \frac{m_o}{m_f - m_o} 100\text{C}^\circ$$

Note: This derivation is correct only if there is still some ice in the cup afterwards. If you start with too little ice, the water will come out at a higher temperature.

For crushed ice that has been standing for some time, the value of L_i will vary between 70 and 75 Cal/g.

(b) Heat exchange and transfer by conduction and radiation

For several possible experiments you will need the following additional apparatus. Make a small hole in the bottom of cup K and thread two wires, soldered to a lightbulb, through the hole. A flashlight bulb that operates with an electric current between 300 and 600 milliamperes (mA) is preferable; but even a GE #1130 6-volt automobile headlight bulb (which draws 2.4 A) has been used with success. (See Fig. 3-52.) In each experiment, you are to observe how the different apparatus affects heat transfer into or out of the system.

1. Place the bulb in the ice and turn it on for 5 min. Measure the ice melted.

2. Repeat 1, but place the bulb above the ice for 5 min.

3. and **4.** Repeat 1 and 2, but cover the inside of cup K with aluminum foil.

5. and **6.** Repeat 3 and 4, but, in addition, cover the inside of cup I with aluminum foil.

7. Prepare "heat-absorbing" ice by freezing water to which you have added a small amount of dye, such as India ink. Repeat any or all of experiments 1 through 6 using this specially prepared ice.

Some questions to guide your observations: Does any heat escape when the bulb is immersed in the ice? What arrangement keeps in as much heat as possible?

Experiment 3-13
MONTE CARLO EXPERIMENT ON MOLECULAR COLLISIONS

A model for a gas consisting of a large number of very small particles in rapid random motion has many advantages. One of these is that it makes it possible to estimate the properties of a gas as a whole from the behavior of a comparatively small random sample of its molecules. In this experiment, you will not use actual gas particles, but instead employ analogs of molecular collisions. The technique is named the *Monte Carlo method* after that famous gambling casino in Monaco. The experiment consists of two games, both of which involve the concept of randomness. You will probably have time to play only one game.

Game I: Collision Probability for a Gas of Marbles

In this part of the experiment, you will try to find the diameter of marbles by rolling a "bombarding marble" into an array of "target marbles" placed at random positions on a level sheet of graph paper. The computation of the marble diameter will be based on the proportion of hits and misses. In order to assure randomness in the motion of the bombarding marble, release each marble from the top of an inclined board studded with nails spaced about 2.5 cm apart: a sort of pinball machine (Fig. 3-53). To get a fairly even, yet random, distribution of the bombarding marble's motion, move its release position over one space for each release in the series. The launching board should be about 0.5 m from the target board; from this distance, the bombarding marbles will move in nearly parallel paths through the target board.

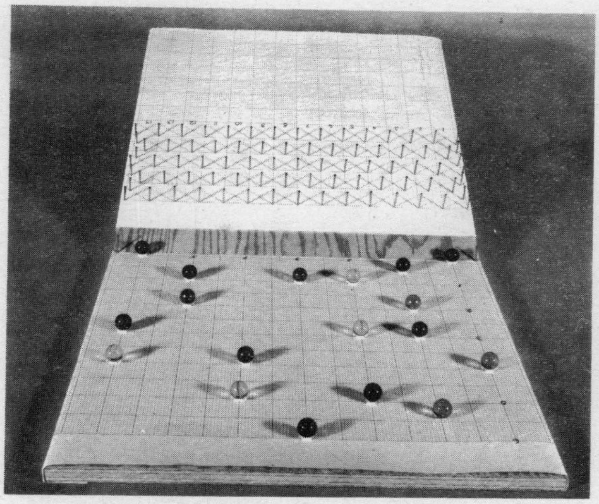

Fig. 3-53

First you need to place the target marbles at random. Then draw a network of crossed-grid lines spaced at least two marble diameters apart on your graph paper. (If you are using marbles whose diameters are 1 cm, these grid lines should be spaced 3–4 cm apart.) Number the grid lines as shown in Fig. 3-54.

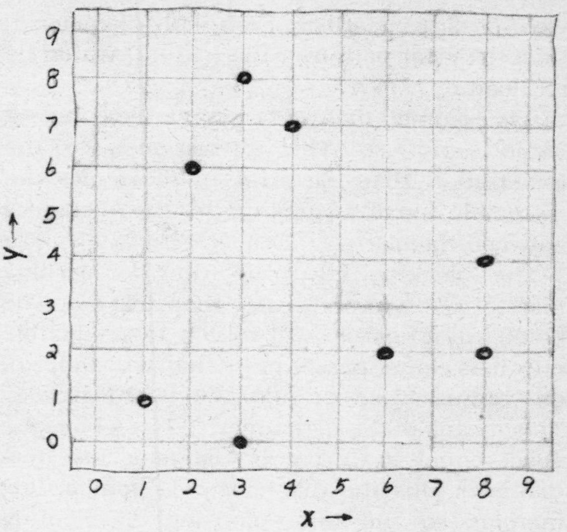

Fig. 3-54 Eight consecutive two-digit numbers in a table of random numbers were used to place the marbles.

One way of placing the marbles at random is to turn to the table of random numbers at the end of this experiment (Table 3-1). Each student should start at a different place in the table and then select the next eight numbers. Use the first two digits of these numbers to locate positions on the grid. The first digit of each number gives the x coordinate, the second gives the y coordinate, or vice versa. Place the target marbles in these positions. Books may be placed around the sides of the graph paper to serve as containing walls.

With your array of marbles in place, make about 50 trials with the bombarding marble. From your record of hits and misses compute R, the ratio between the number of runs in which there are one or more hits to the total number of runs. Remember that you are counting "runs with hits," not hits. Therefore, several hits in a single run are still counted as "one."

Inferring the size of the marbles

How does the ratio R lead to the diameter of the target object? The theory applies just as well to determining the size of molecules as it does to marbles, although there would be 10^{20} or so molecules instead of eight "marble molecules."

If there were no target marbles, the bombarding marble would get a clear view of the full width, D, of the target area. There could be no hit. If, however, there were target marbles, the 100% clear view would be cut down. If there were N target marbles, each with diameter d, then the clear path over the width D would be reduced by $N \times d$.

It is assumed that no target marble is hiding behind another. (This corresponds to the assumption that the sizes of molecules are extremely small compared to the distances between them.)

The blocking effect on the bombarding marble is, however, greater than just Nd. The bombarding marble will miss a target marble only if its center passes more than a distance of one radius on either side of it. (See Fig. 3-55.) This means that a target marble has a blocking effect equal to twice its diameter (its own diameter plus the diameter of a bombarding marble), so the total blocking effect of N marbles is $2Nd$. Therefore, the expected ratio R of hits to total trials is $2Nd/D$ (total blocked width to total width). Thus,

$$R = \frac{2Nd}{D},$$

which can be rearranged to give an expression for d:

$$d = \frac{RD}{2N}$$

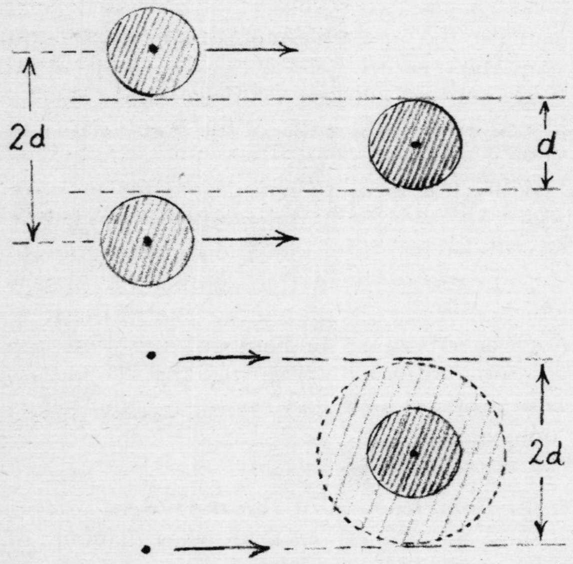

Fig. 3-55 A projectile will clear a target only if it passes outside a center-to-center distance d on either side of it. Therefore, thinking of the projectiles as points, the effective blocking width of the target is $2d$.

To check the accuracy of the Monte Carlo method, compare the value for d obtained from the formula above with that obtained by direct measurement of the marbles. (For example, line up your eight marbles against a book. Measure the total length of all of them together and divide by eight to find the diameter d of one marble.)

?

1. What value to you calculate for the marble diameter?
2. How well does your experiment prediction agree with direct measurement?

Game II: Mean Free Path Between Collision Squares

In this part of the experiment you play with blacked-in squares as target molecules in place of marble molecules in a pinball game. On a sheet of graph paper 50 units on a side (2,500 squares), you will locate, by the Monte Carlo method, between 40 and 100 molecules. Each student should choose a different number of molecules.

You will find a table of random numbers (from 0 to 50) at the end of this experiment (Table 3-1). Begin anywhere you wish in the table, but then proceed in a regular sequence. Let each pair of numbers be the x and y coordinates of a point on your graph. (If one of the pair is greater than 49, you cannot use it. Ignore it and take the next pair.) Then shade in the squares for which these points are the lower left-hand corners (Fig. 3-56). You now have a random array of square target "molecules."

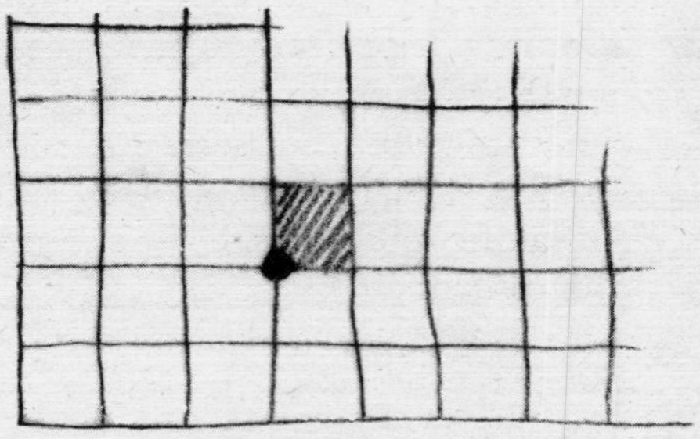

Fig. 3-56

Rules of the game

The way a bombarding particle passes through this array, it is bound to collide with some of the target particles. There are five rules for this game of collision. All of them are illustrated in Fig. 3-57.

(a) The particle can travel only along lines of the graph paper, up or down, left or right. A particle starts at some point (chosen at random) on the left-hand edge of the graph paper. The particle initially moves horizontally from the starting point until it collides with a blackened square or another edge of the graph paper.

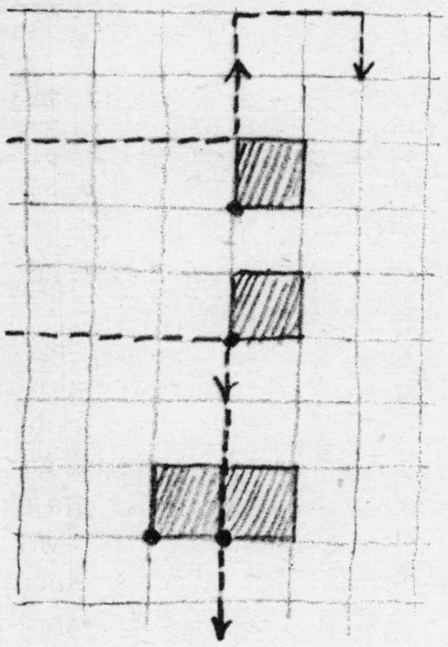

Fig. 3-57

(b) If the particle strikes the upper left-hand corner of a target square, it is diverted upward through a right angle. If it should strike a lower left-hand corner it is diverted downward, again through an angle of 90°.

(c) When the path of the particle meets an edge or boundary of the graph paper, the particle is *not* reflected directly back. (Such a reversal of path would make the particle retrace its previous path.) Rather it moves two spaces to its *right* along the boundary edge before reversing its direction.

(d) There is an exception to rule (c). Whenever the particle strikes the edge so near a corner that there is no room for it to move two spaces to the right without meeting another edge of the graph paper, it moves two spaces to the *left* along the boundary.

(e) Occasionally two target molecules may occupy adjacent squares and the particle may hit touching corners of the two target molecules at the same time. The rule is that this counts as two hits and the particle goes straight through without changing its direction.

Finding the mean free path

With these collision rules in mind, trace the path of the particle as it bounces about among the random array of target squares. Count the number of collisions with targets. Follow the path of the particle until you get 50 hits with target squares (collisions with the edge do not count). Next, record the 50 lengths of the paths of the particle between collisions. Distances to and from a boundary should be included, but *not* distances *along* a boundary (the two spaces introduced to avoid backtracking). These 50 lengths are the free paths of the particle. Total them and divide by 50 to obtain the *mean free path*, L for your random two-dimensional array of square molecules.

In this game, your molecule analogs were pure points, that is, dimensionless. In his investigations, Clausius modified this model by giving the particles a definite size. Clausius showed that the average distance L a molecule travels between collisions, the so-called *mean free path*, is given by

$$L = \frac{V}{Na}$$

where V is the volume of the gas, N is the number of molecules in that volume, and a is the cross-sectional area of an individual molecule. In this two-dimensional game, the particle was moving over an area A, instead of through a volume V, and was obstructed by targets of width d, instead of cross-sectional area a. A two-dimensional version of Clausius's equation is

$$L = \frac{A}{2Nd}$$

where N is the number of blackened square "molecules."

?

3. What value of L do you get from the data for your runs?

4. Using the two-dimensional version of Clausius's equation, what value do you estimate for d (the width of a square)?

5. How does your calculated value of d compare with the actual value? How do you explain the difference?

TABLE 3-1

TABLE OF 1000 RANDOM TWO-DIGIT NUMBERS (FROM 0 to 50)

03 47	44 22	30 30	22 00	00 49	22 17	38 30	23 21	20 11	24 33
16 22	36 10	44 39	46 40	24 02	19 36	38 21	45 33	14 23	01 31
33 21	03 29	08 02	20 31	37 07	03 28	47 24	11 29	49 08	10 39
34 29	34 02	43 28	03 43	43 40	26 08	28 06	50 14	21 44	47 21
32 44	11 05	05 05	05 50	23 29	26 00	09 05	27 31	08 43	04 14
18 18	04 02	48 39	48 22	38 18	15 39	48 34	50 28	37 21	15 09
23 42	31 08	19 30	06 00	20 18	30 24	15 33	10 07	14 29	05 24
35 12	11 12	11 04	01 10	25 39	48 50	24 44	03 47	34 04	44 07
12 13	42 10	40 48	45 44	42 35	41 26	41 10	23 05	06 36	08 43
37 35	12 41	02 02	19 11	06 07	42 31	23 47	47 25	10 43	12 38
16 08	18 39	03 31	49 26	07 12	17 31	17 31	35 07	44 38	40 35
31 16	10 47	38 45	28 40	33 34	24 16	42 38	19 09	41 47	50 41
32 43	45 37	30 38	22 01	30 14	02 17	45 18	29 06	13 27	46 24
27 42	03 09	08 32	24 02	05 49	18 05	22 00	23 02	44 43	43 20
00 39	05 03	49 37	23 22	33 42	26 29	00 20	12 03	10 05	02 39
11 27	39 32	13 30	36 45	09 03	46 40	22 07	03 03	05 39	03 46
35 24	22 49	17 33	35 01	01 32	18 09	47 03	39 41	36 23	19 41
16 20	38 36	29 48	07 27	48 14	34 13	07 48	39 12	20 18	19 42
38 23	33 26	15 29	20 02	21 45	04 31	48 13	23 32	37 30	09 24
45 11	27 07	39 43	13 05	47 45	47 45	00 06	41 18	05 02	03 09
18 00	14 21	49 17	30 37	25 15	04 49	24 19	40 23	24 17	17 16
20 46	06 18	45 07	06 28	49 44	10 08	43 00	38 26	34 41	11 16
05 26	50 25	38 47	39 38	42 45	10 08	16 06	43 18	34 48	27 03
21 19	13 42	16 04	00 18	16 46	13 13	16 29	44 10	29 18	22 45
41 23	03 10	35 30	24 36	38 09	25 21	08 40	20 46	39 14	37 31
34 50	20 14	21 46	38 46	12 27	20 44	46 06	01 41	30 49	18 48
39 43	13 04	24 15	08 22	13 29	04 05	42 29	50 47	01 50	01 48
18 14	04 43	27 46	23 07	19 28	07 10	23 19	41 45	25 27	19 10
09 47	34 45	08 45	25 21	49 21	18 46	16 40	35 14	41 28	41 15
44 17	04 33	15 22	12 45	39 07	34 27	14 47	35 33	42 29	47 47
40 33	42 45	07 08	38 15	08 25	22 06	07 26	32 44	03 42	42 34
33 27	10 45	18 40	11 48	48 03	07 16	32 25	20 25	44 22	39 28
06 09	04 26	14 35	36 03	15 22	02 07	46 48	45 12	47 11	30 19
33 32	34 25	45 17	13 26	03 37	33 35	08 13	15 26	09 18	34 25
42 38	40 01	43 31	30 33	39 11	49 41	27 44	11 39	06 19	47 23
15 06	22 08	50 44	50 11	18 16	00 41	07 47	34 25	28 10	50 03
22 35	49 36	44 21	25 12	19 44	31 51	49 18	40 36	00 27	22 12
31 04	32 17	08 23	38 32	01 47	43 53	44 04	10 27	16 00	16 33
39 00	01 50	07 28	35 02	38 00	46 47	33 29	28 41	09 23	47 48
37 32	07 02	07 48	07 41	22 13	37 27	27 12	34 21	07 04	49 34
05 03	36 07	10 15	21 48	14 44	39 39	15 09	23 23	37 31	00 25
17 37	13 41	13 39	40 14	19 48	34 18	08 18	08 06	44 26	12 45
32 24	24 30	29 13	34 39	27 44	11 20	37 40	36 46	35 22	09 09
07 45	29 12	48 35	05 38	43 11	45 18	28 14	04 37	48 38	43 12
14 08	04 04	18 17	10 33	04 32	27 37	33 42	34 41	07 41	49 14
31 38	08 31	38 30	42 10	08 09	17 32	46 15	15 43	15 31	46 45
42 34	46 31	29 03	08 32	11 06	20 21	24 16	13 17	29 34	42 31
16 00	02 48	10 34	32 14	25 39	29 31	18 37	28 50	07 28	08 24
20 15	60 11	21 31	20 49	07 35	41 16	16 17	43 36	20 26	39 38
00 49	14 10	29 01	49 28	21 30	40 15	01 07	16 04	19 09	36 12

Experiment 3-14
BEHAVIOR OF GASES

Air is elastic or springy. You can feel this when you put your finger over the outlet of a bicycle pump and push down on the pump plunger. You can tell that there is some connection between the volume of the air in the pump and the force you exert in pumping, but the exact relationship is not obvious. About 1660, Robert Boyle performed an experiment that disclosed a very simple relationship between gas pressure and volume, but not until two centuries later was the kinetic theory of gases developed, which satisfactorily accounted for Boyle's law.

The purpose of these experiments is not simply to show that Boyle's law and Gay Lussac's law (which relates temperature and volume) are "true." The purpose is also to show some techniques for analyzing data that can lead to such laws.

I. Volume and Pressure

Boyle used a long glass tube in the form of a J to investigate the "spring of the air." The short arm of the J was sealed, and air was trapped in it by pouring mercury into the top of the long arm.

A simpler method requires only a small plastic syringe, calibrated in milliliters, and mounted so that you can push down the piston by piling weights on it (see Fig. 3-58). The volume of the air in the syringe can be read directly from the calibrations on the side. The pressure on the air due to the weights on the piston is equal to the force exerted by the weights divided by the area of the face of the piston:

$$P_w = \frac{F_w}{A}$$

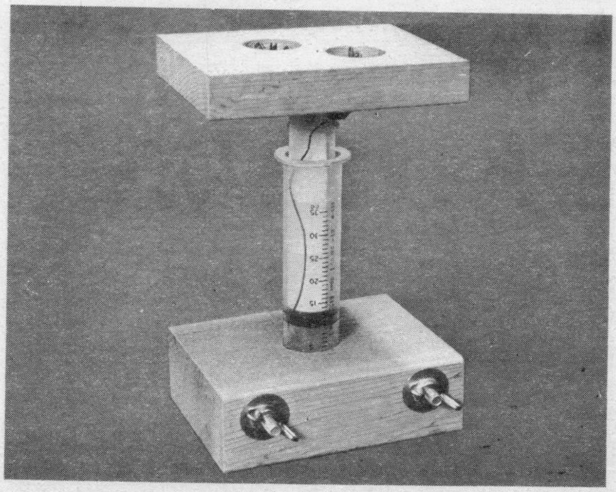

Fig. 3-58

Because "weights" are usually marked with the value of their *mass*, you will have to compute the force from the relation $F_{grav} = ma_{grav}$. (It will help if you answer this question before going on: What is the weight, in newtons, of a 0.1-kg mass?)

To find the area of the piston, remove it from the syringe. Measure the diameter ($2R$) of the piston face, and compute its area from the familiar formula $A = \pi R^2$.

You will want to both decrease and increase the volume of the air, so insert the piston about halfway down the barrel of the syringe. The piston may tend to stick slightly. Give it a twist to free it and help it come to its equilibrium position. Then record this position.

Add weights to the top of the piston and each time record the equilibrium position, after you have given the piston a twist to help overcome friction.

Record your data in a table with columns for volume, weight, and pressure. Then remove the weights one by one to see if the volumes are the same with the piston coming up as they were going down.

If your apparatus can be turned over so that the weights pull out on the plunger, obtain more readings this way, adding weights to increase the volume. Record these as negative forces. (Stop adding weights before the piston is pulled all the way out of the barrel!) Again remove the weights and record the values on returning.

Interpreting Your Results

You now have a set of numbers somewhat like the ones Boyle reported for his experiment. One way to look for a relationship between the pressure P_w and the volume V is to plot the data on graph paper. Plot volume V (vertical axis) as a function of pressure P_w (horizontal axis). Then draw a smooth curve that gives an overall "best fit." Because errors of measurement affect each plotted point, your smooth curve need not go through all the points.

Since V decreases as P_w increases, you can tell before you plot it that your curve represents an "inverse" relationship. As a first guess at the mathematical description of this curve, try the simplest possibility, that $1/V$ is proportional to P_w. That is, $1/V \propto P_w$. If $1/V$ *is* proportional to P_w, then a plot of $1/V$ against P_w will lie on a straight line.

Add another column to your data table for values of $1/V$ and plot this against P_w.

?

1. Does this curve pass through the origin?
2. If not, at what point does your curve cross the horizontal axis? (In other words, what is the value of P_w for which $1/V$ would be zero?) What is the physical significance of the value of P_w?

In Boyle's time, it was not understood that air is really a mixture of several gases. Do you believe you would find the same relationship between volume and pressure if you tried a variety of pure gases instead of air? If there are other gases available in your laboratory, flush out and refill your apparatus with one of them and try the experiment again.

?

3. Does the curve you plot have the same shape as the previous one?

II. Volume and Temperature

Boyle suspected that the temperature of his air sample had some influence on its volume, but he did not do a quantitative experiment to find the relationship between volume and temperature. It was not until about 1880, when there were better ways of measuring temperature, that this relationship was established.

You could use several kinds of equipment to investigate the way in which volume changes with temperature. Such a piece of equipment is a glass bulb with a J tube of mercury or the syringe described above. Make sure the gas inside is dry and at atmospheric pressure. Immerse the bulb or syringe in a beaker of cold water and record the volume of gas and temperature of the water (as measured on a suitable thermometer) periodically as you slowly heat the water.

Interpreting Your Results

?

4. With either of the methods mentioned here, the pressure of the gas remains constant. If the curve is a straight line, does this "prove" that the volume of a gas at constant pressure is proportional to its temperature?
5. Remember that the thermometer you used probably depended on the expansion of a liquid such as mercury or alcohol. Would your graph have been a straight line if a different type of thermometer had been used?
6. If you could continue to cool the air, would there be a lower limit to the volume it would occupy?

Draw a straight line as nearly as possible through the points on your $V-T$ graph and extend it to the left until it shows the approximate temperature at which the volume would be zero. Of course, you have no reason to assume that gases have this simple linear relationship all the way down to zero volume. (In fact, air would change to a liquid long before it reached the temperature indicated on your graph for zero volume.) However, some gases do show this linear behavior over a wide temperature range, and for these gases the straight line always crosses the T axis at the same point. Since the volume of a sample of gas cannot be less than zero, this point represents the lowest possible temperature of the gases, the *absolute zero* of temperature.

?

7. What value does your graph give for absolute zero?

III. Questions for Discussion

Both the pressure and the temperature of a gas sample affect its volume. In these experiments, you were asked to consider each of these factors separately.

?

8. Were you justified in assuming that the temperature remained constant in the first experiment as you varied the pressure? How could you check this? How would your results be affected if, in fact, the temperature went up each time you added weight to the plunger?
9. In the second experiment, the gas was at atmospheric pressure. Would you expect to find the same relationship between volume and temperature if you repeated the experiment with a different pressure acting on the sample?

Gases such as hydrogen, oxygen, nitrogen, and carbon dioxide are very different in their chemical behavior. Yet they all show the same simple relationships between volume, pressure, and temperature that you found in these experiments, over a fairly wide range of pressures and temperatures. This suggests that perhaps there is a simple physical model that will explain the behavior of all gases within these limits of temperature and pressure. Chapter 11 of the Text describes just such a simple model and its importance in the development of physics.

Experiment 3-15
WAVE PROPERTIES

In this laboratory exercise you will become familiar with a variety of wave properties in one- and two-dimensional situations.* Using ropes, springs, Slinkies, or a ripple tank, you can find out what determines the speed of waves, what happens when they collide, and how waves reflect and go around corners.

Waves in a Spring

Many waves move too fast or are too small to watch easily, but in a long "soft" metal spring you can make big waves that move slowly. With a partner to help you, pull the spring out on a smooth floor to a length of about 6–9 m. Now, with your free hand, grasp the stretched spring about 50 cm from the end. Pull the spring together toward the end and then release it, being careful *not* to let go of the fixed end with your other hand! Notice the single wave, called a *pulse*, that travels along the spring. In such a *longitudinal* pulse, the spring coils move back and forth along the same direction as the wave travels. The wave carries energy, and thus could be used to carry a message from one end of the spring to the other.

You can see a longitudinal wave more easily if you tie pieces of string to several of the loops of the spring and watch their motion when the spring is pulsed.

A *transverse* wave is easier to see. To make one, practice moving your hand very quickly back and forth at right angles to the stretched spring, until you can produce a pulse that travels down only one side of the spring. This pulse is called *transverse* because the individual coils of wire move at right angles to (transverse to) the length of the spring.

Perform experiments to answer the following questions about transverse pulses.

?

1. Does the size of the pulse change as it travels along the spring? If so, in what way?
2. Does the pulse reflected from the far end return to you on the same side of the spring as the original pulse, or on the opposite side?
3. Does a change in the tension of the spring have any effect on the speed of the pulses? When you stretch the spring farther, in effect you are changing the nature of the *medium* through which the pulses move.

*Adapted from R.F. Brinckerhoff and D.S. Taft, *Modern Laboratory Experiments in Physics*, by permission of Science Electronics, Nashua, N.H.

Next observe what happens when waves go from one material into another, an effect called *refraction*. To one end of your spring attach a length of rope or rubber tubing (or a different kind of spring) and have your partner hold this end.

?

4. The far end of your first spring is now free to move back and forth at the joint. What happens to a pulse (size, shape, speed, direction) when it reaches the boundary between the two media?

Have your partner detach the extra spring and once more grasp the far end of your original spring. Then you both send a pulse on the same side, at the same instant, so that the two pulses meet in the center. The interaction of the two pulses is called *interference*.

?

5. What happens (size, shape, speed, direction) when two pulses reach the center of the spring? (It will be easier to see what happens in the interaction if one pulse is larger than the other.)
6. What happens when two pulses on opposite sides of the spring meet?

As the two pulses pass on opposite sides of the spring, can you observe a point on the spring that does not move at all?
7. From these observations, what can you say about the displacement caused by the addition of two pulses at the same point?

By vibrating your hand steadily back and forth, you can produce a train of pulses, a *periodic wave*. The distance between any two neighboring crests on such a periodic wave is the *wavelength*. The rate at which you vibrate your hand will determine the *frequency* of the periodic wave. Use a long spring and produce short bursts of periodic waves so you can observe them without interference by reflections from the far end.

?

8. How does the wavelength seem to depend on the frequency?

You have now observed the reflection, refraction, and interference of single waves, or pulses, traveling through different materials. These waves, however, moved only along one dimension. So that you can make a more realistic comparison with other forms of traveling energy, in the next experiment you will examine these same wave properties spread out over a two-dimensional surface.

Experiment 3-16
WAVES IN A RIPPLE TANK

In the laboratory, one or more ripple tanks will have been set up. To the one you and your partner are going to use, add water (if necessary) to a depth of 6–8 mm. Check to see that the tank is level so that the water has equal depth at all four corners. Place a large sheet of white paper on the table below the ripple tank, and then switch on the overhead light source. Disturbances on the water surface are projected onto the paper as light and dark patterns, thus allowing you to "see" the shape of the disturbances in the horizontal plane.

To see what a single pulse looks like in a ripple tank, gently touch the water with your fingertip, or, better, let a drop of water fall into it from a medicine dropper held only a few millimeters above the surface.

For certain purposes, it is easier to study pulses in water if their crests are straight. To generate single straight pulses, place a dowel, or a section of a broom handle, along one edge of the tank and roll it backward a fraction of a centimeter. By rolling the dowel backward and forward with a uniform frequency, a periodic wave, a continuous train of pulses, can be formed.

Use straight pulses in the ripple tank to observe reflection, refraction, and diffraction, and circular pulses from point sources to observe interference.

Reflection

Generate a straight pulse and notice the direction of its motion. Now place a barrier in the water so that it intersects that path. Generate new pulses and observe what happens to the pulses when they strike the barrier. Try different angles between the barrier and the incoming pulse.

?

1. What is the relationship between the *direction* of the incoming pulse and the reflected one?
2. Replace the straight barrier with a curved one. What is the shape of the reflected pulse?
3. Find the point where the reflected pulses run together. What happens to the pulse after it converges at this point? At this point, called the *focus*, start a pulse with your finger or a drop of water. What is the shape of the pulse after reflection from the curved barrier?

Refraction

Lay a sheet of glass in the center of the tank, supported by coins if necessary, to make an area of very shallow water. Try varying the angle at which the pulse strikes the boundary between the deep and shallow water.

?

4. What happens to the wave speed at the boundary?
5. What happens to the wave direction at the boundary?
6. How is change in direction related to change in speed?

Interference

Arrange two point sources side by side a few centimeters apart. When tapped gently, they should produce two pulses. You will see the action of interference better if you vibrate the two point sources continuously with a motor and study the resulting pattern of waves.

?

7. How does changing the wave frequency affect the original waves?

Find regions in the interference pattern where the waves from the two sources cancel and leave the water undisturbed. Find the regions where the two waves add up to create a doubly great disturbance.

8. Make a sketch of the interference pattern indicating these regions.
9. How does the pattern change as you change the wavelength?

Diffraction

With two-dimensional waves you can observe a new phenomenon: the behavior of a wave when it passes around an obstacle or through an opening. The spreading of the wave into the "shadow" area is called *diffraction*. Generate a steady train of waves by using the motor driven straight-pulse source. Place a small barrier in the path of the waves so that it intercepts part, but not all, of the wave front. Observe what happens as the waves pass the edge of the barrier. Now vary the wavelength of the incoming wave train by changing the speed of the motor on the source.

?

10. How does the interaction with the obstacle vary with the wavelength?

Place two long barriers in the tank, leaving a small opening between them.

11. How does the angle by which the wave spreads out beyond the opening depend on the size of the opening?
12. In what way does the spread of the diffraction pattern depend on the length of the waves?

Experiment 3-17
MEASURING WAVELENGTH

There are three ways you can conveniently measure the wavelength of the waves generated in your ripple tank. You should try them all, if possible, and cross-check the results. If there are differences, indicate which method you believe is most accurate and explain why.

Fig. 3-59 An interference pattern in water. Two point sources vibrating in phase generate waves in a ripple tank. A and C are points of maximum disturbance (in opposite directions) and B is a point of minimum disturbance.

METHOD A: Direct

Set up a steady train of pulses using either a single-point source or a straight-line source. Observe the moving waves with a stroboscope, and then adjust the vibrator motor to the lowest frequency that will "freeze" the wave pattern. Place a meter stick across the ripple tank and measure the distance between the crests of a counted number of waves.

METHOD B: Standing Waves

Place a straight barrier across the center of the tank parallel to the advancing waves. When the distance of the barrier from the generator is properly adjusted, the superposition of the advancing waves and the waves reflected from the barrier will produce *standing waves*. In other words, the reflected waves are, at some points, reinforcing the original waves, while at other points there is always cancellation. The points of continual cancellation are called *nodes*. The distance between nodes is one-half wavelength.

METHOD C: Interference Pattern

Set up the ripple tank with two point sources. The two sources should strike the water at the same instant so that the two waves will be exactly in phase and of the same frequency as they leave the sources. Adjust the distance between the two sources and the frequency of vibration until a distinct pattern is obtained, such as in Fig. 3-59.

As you study the pattern of ripples, you will notice lines along which the waves cancel almost completely so that the amplitude of the disturbance is almost zero. These lines are called *nodal lines,* or *nodes.* You have already seen nodes in your earlier experiment with standing waves in the ripple tank.

At every point along a node the waves arriving from the two sources are half a wavelength out of step, or "out of phase." This means that for a point (such as B in Fig. 3-59) to be on a line of nodes it must be ½ or 1½ or 2½ ... wavelengths farther from one source than from the other.

Between the lines of nodes are regions of maximum disturbance. Points A and C in Fig. 3-59 are on lines down the center of such regions, called *antinodal lines*. Reinforcement of waves from the two sources is at a maximum along these lines.

For reinforcement to occur at a point, the two waves must arrive in step or "in phase." This means that any point on a line of antinodes is a whole number of wavelengths 0, 1, 2, ... farther from one source than from the other. The relationship between crests, troughs, nodes, and antinodes in this situation is summarized schematically in Fig. 3-60.

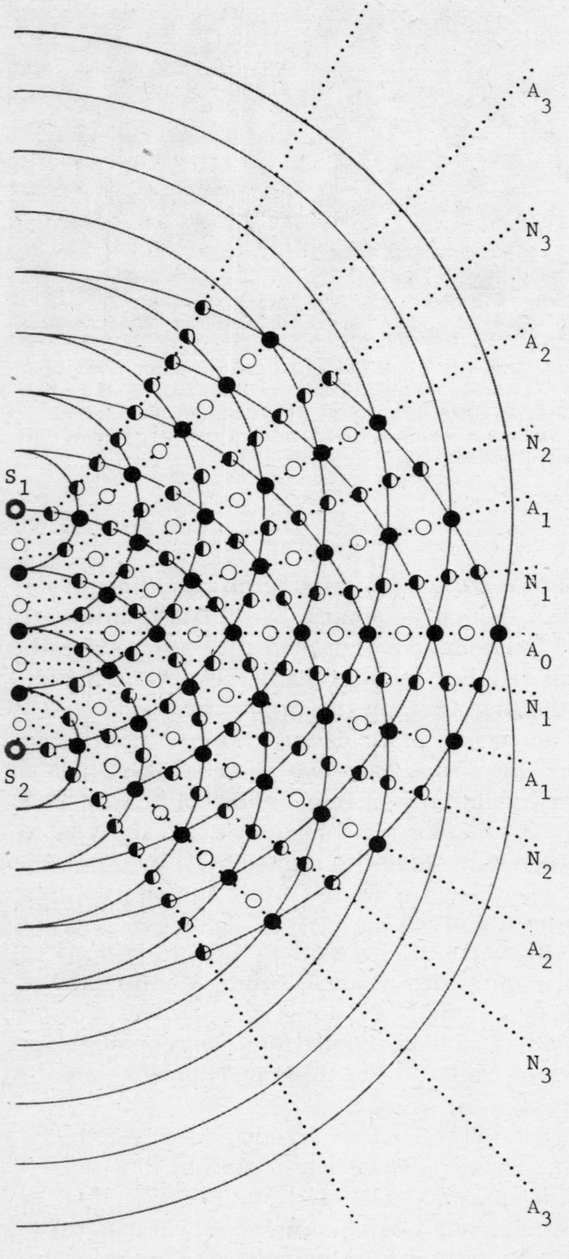

Fig. 3-60 Analysis of interference pattern similar to that of Fig. 3-59 set up by two in-phase periodic sources. (Here S_1 and S_2 are separated by four wavelengths.) The letters A and N designate antinodal and nodal lines. The dark circles indicate where crest is meeting crest, the blank circles where trough is meeting trough, and the half-dark circles where crest is meeting trough.

Most physics textbooks develop the mathematical argument of the relationship of wavelength to the geometry of the interference pattern. (See, for example, p. 119 in Unit 3 of the Text.) If the distance between the sources is d and the detector is at a comparatively greater distance L from the sources, then d, L, and λ are related by the equations

$$\frac{\lambda}{d} = \frac{x}{L}$$

or

$$\lambda = \frac{xd}{L}$$

where x is the distance between neighboring antinodes (or neighboring nodes).

You now have a method for computing the wavelength λ from the distances that you can measure precisely. Measure x, d, and L in your ripple tank and compute λ.

Experiment 3-18
SOUND

In previous experiments, you observed how waves of relatively low frequency behave in different media. In this experiment, you will try to determine to what extent audible sound exhibits similar properties.

At the laboratory station where you work, there should be the following: an oscillator, a power supply, two small loudspeakers, and a group of materials to be tested. A loudspeaker is the source of audible sound waves, and your ear is the detector. First connect one of the loudspeakers to the output of the oscillator and adjust the oscillator to a frequency of about 4000 Hz. Adjust the loudness so that the signal is just audible 1 m away from the speaker. The

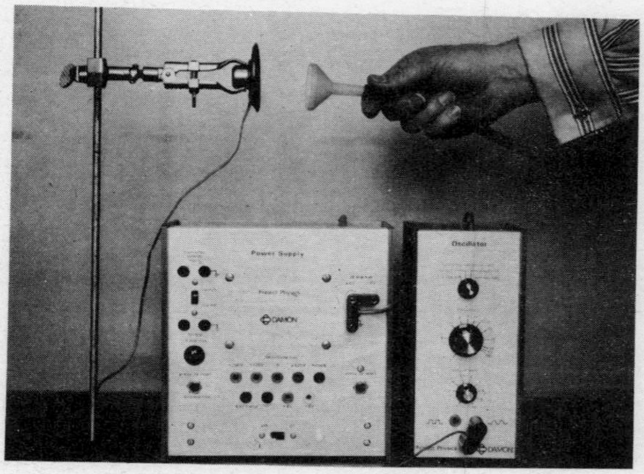

Fig. 3-61 Sound from the speaker can be detected by using a funnel and rubber hose, the end of which is placed to the ear. The oscillator's banana plug jacks must be inserted into the −8 V, +8 V, and ground holes of the power supply. Insert the speaker's plugs into the sine wave ground receptacles of the oscillator. Select the audio range by means of the top knob of the oscillator and then turn on the power supply.

gain-control setting should be low enough to produce a clear, pure tone. Since reflections from the floor, tabletop, and hard-surfaced walls may interefere with your observations, set the sources at the edge of a table. Put soft material over any close hard surface that could cause reflective interference.

You may find that you can localize sounds better if you make an "ear trumpet" or stethoscope from a small funnel or thistle tube and a short length of rubber tubing (Fig. 3-61). Cover the ear not in use to minimize confusion when you are hunting for nodes and maxima.

Transmission and Reflection

Place samples of various materials at your station between the speaker and the receiver to see how they transmit the sound wave. In a table, record your qualitative judgments as best, good, poor, and so on.

Test the same materials for their ability to reflect sound and record your results. Be sure that the sound is really being reflected and is not coming to your detector by some other path. You can check how the intensity varies at the detector when you move the reflector and rotate it about a vertical axis (see Fig. 3-62). If suitable materials are available to you, also test the reflection from curved surfaces.

Fig. 3-62

?

1. On the basis of your findings, what generalizations can you make relating transmission and reflection to the properties of the test materials?

Refraction

You have probably observed the refraction or "bending" of a wave front in a ripple tank as the wave slowed down in passing from water of one depth to shallower water.

You may observe the refraction of sound waves using a "lens" made of gas. Inflate a spherical balloon with carbon dioxide gas to a diameter of about 10–15 cm. Explore the area near the balloon on the side away from the source. Locate a point where the sound seems loudest, and then remove the balloon.

?

2. Do you notice any difference in loudness when the balloon is in place? Explain.

Diffraction

In front of a speaker set up as before; place a thick piece of hard material about 25 cm long, mounted vertically about 25 cm directly in front of the speaker. Slowly probe the area about 75 cm beyond the obstacle.

?

3. Do you hear changes in loudness? Is there sound in the "shadow" area? Are there regions of silence where you would expect to hear sound? Does there seem to be any pattern to the areas of minimum sound?

For another way to test for diffraction, use a large piece of board placed about 25 cm in front of the speaker with one edge aligned with the center of the source. Now explore the area inside the shadow zone and just outside it. (See Fig. 3-63.)

Describe the pattern of sound interference that you detect.

?

4. Is the pattern analogous to the pattern you observed in the ripple tank?

Wavelength

METHOD A: Standing Wave

Set your loudspeaker about 0.5 m above and facing toward a *hard* tabletop or floor, or about that distance from a hard, smooth plaster wall or other good sound reflector. Your ear is *most* sensitive to the changes in intensity of faint sounds, so be sure to keep the volume low.

Explore the space between the source and reflector, listening for changes in loudness. Record the positions of minimum loudness, or at least find the approximate distance between two consecutive minima. These minima are located one-half wavelength apart.

?

5. Does the spacing of the minima depend on the intensity of the wave?

Measure the wavelength of sound at several different frequencies.

6. How does the wavelength change when the frequency is changed?

METHOD B: Interference

Connect the two loudspeakers to the output of the oscillator and mount them at the edge of the table about 25 cm apart. Set the frequency at about 4,000 Hz to produce a high-pitched tone. Keep the gain setting low during the entire experiment to make sure the oscillator is producing a pure tone, and to reduce reflections that would interfere with the experiment.

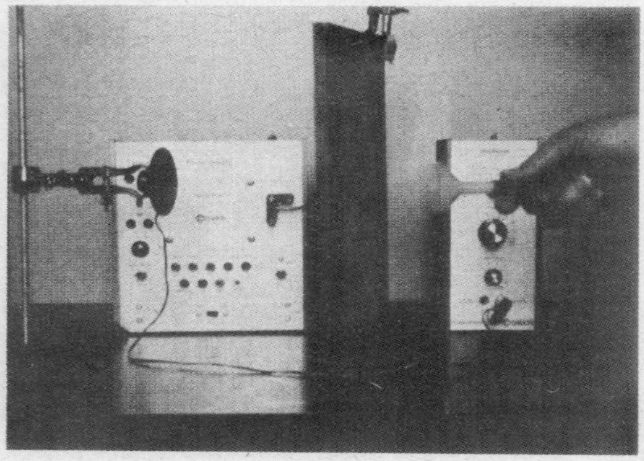

Fig. 3-63

Move your ear or "stethoscope" along a line parallel to, and about 50 cm from, the line joining the sources. Can you detect distinct maxima and minima? Move farther away from the sources; do you find any change in the pattern spacing?

?

7. What effect does a change in the source separation have on the spacing of the nodes?

8. What happens to the spacing of the nodes if you change the frequency of the sound? To make this experiment quantitative, work out for yourself a procedure similar to that used with the ripple tank.

Measure the separation d of the source centers and the distance x between nodes and use these data to calculate the wavelength λ.

?

9. Does the wavelength change with frequency? If so, does it change directly or inversely?

Calculating the Speed of Sound

The relationship between speed v, wavelength λ, and frequency f is $v = \lambda f$. The oscillator dial gives a rough indication of the frequency (and your instructor can advise you on how to use an oscilloscope to make precise frequency settings). Using your best estimate of λ, calculate the speed of sound. If you have time, extend your data to answer the following questions.

?

10. Does the speed of the sound waves depend on the intensity of the waves?

11. Does the speed depend on the frequency?

Experiment 3-19
ULTRASOUND

The equipment needed for this experiment is an oscillator, power supply, and three ultrasonic transducers (crystals that transform electrical impulses into sound waves, or vice versa), and several materials to be tested. The signal from the detecting transducer can be displayed with either an oscilloscope (as in Fig. 3-64) or an

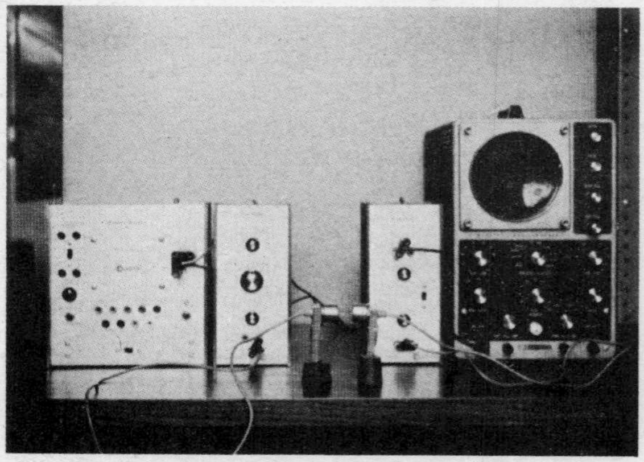

Fig. 3-64 Complete ultrasound equipment. Plug the +8 V, −8 V ground jacks from the amplifier and oscillator into the power supply. Plug the coaxial cable attached to the transducer to the sine wave output of the oscillator. Plug the coaxial cable attached to a second transducer into the input terminals of the amplifier. Be sure that the shield of the coaxial cable is attached to ground. Turn the oscillator range switch to the 5 kHz-50 position. Turn the horizontal frequency range switch of the oscilloscope to at least 10 kHz. Turn on the oscillator and power supply. Tune the oscillator for maximum reception, about 40 kHz.

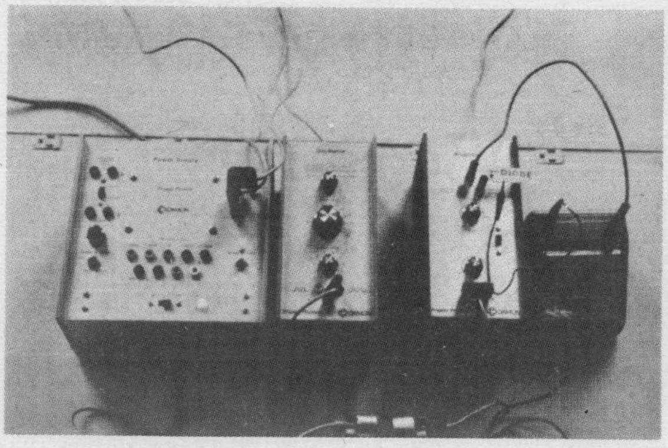

Fig. 3-65 *Above,* ultrasound transmitter and receiver. The signal strength is displayed on a microammeter connected to the receiver amplifier. *Below,* a diode connected between the amplifier and the meter, to rectify the output current. The amplifier selector switch should be turned to *ac*. The *gain* control on the amplifier should be adjusted so that the meter will deflect about full-scale for the loudest signal expected during the experiment. The *offset* control should be adjusted until the meter reads zero when there is no signal.

amplifier and meter (Fig. 3-65). One or two of the transducers, driven by the oscillator, are sources of the ultrasound, while the third transducer is a detector. Before you proceed, have the instructor check your setup and help you get a pattern on the oscilloscope screen or a reading on the meter.

The energy output of the transducer is highest at about 40,000 Hz, and the oscillator must be carefully "tuned" to that frequency. Place the detector a few centimeters directly in front of the source and set the oscillator range to the 5–50 kHz position. Tune the oscillator carefully around 40,000 Hz for maximum deflection of the meter or the scope track. If the signal output is too weak to detect beyond 25 cm, plug the detector transducer into an amplifier and connect the output of the amplifier to the oscilloscope or meter input.

Transmission and Reflection

Test the various samples at your station to see how they *transmit* the ultrasound. Record your judgments as best, good, poor, etc. Hold the sample of the material being tested close to the detector.

Test the same materials for their ability to *reflect* ultrasound. Be sure that the ultrasound is really being reflected and is not coming to your detector by some other path. You can check this by seeing how the intensity varies at the detector when you move the reflector.

Make a table of your observations.

?

1. What happens to the energy of ultrasonic waves in a material that neither reflects nor transmits well?

Diffraction

To observe diffraction around an obstacle, put a 3-cm wide piece of hard material about 8 or 10 cm in front of the source (see Fig. 3-66). Explore the region 5–10 cm behind the obstacle.

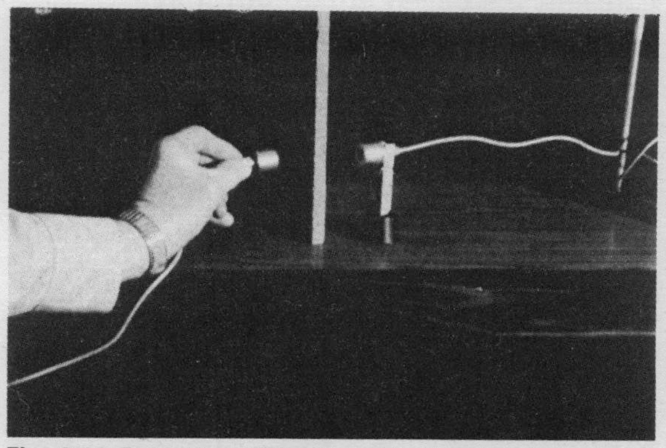

Fig. 3-66 Detecting diffraction of ultrasound around a barrier.

?

2. Do you find any signal in the "shadow" area? Do you find minima in the regions where you would expect a signal to be? Does there seem to be any pattern relating the places of minimum and maximum signals?

Put a larger sheet of absorbing material 10 cm in front of the source so that the edge obstructs about one-half of the source.

Again probe the "shadow" area and the area near the edge to see if a pattern of maxima and minima seems to appear.

Measuring Wavelength

METHOD A: Standing Wave

Investigate the standing waves set up between a source and a reflector, such as a hard tabletop or metal plate. Place the source about 10 to 15 cm from the reflector with the detector.

Find the approximate distance between two consecutive maxima or two consecutive minima. This distance is one-half the wavelength.

?

3. Does the spacing of nodes depend on the intensity of the waves?

METHOD B: Interference

For sources, connect two transducers to the output of the oscillator and set them about 5 cm apart. Set the oscillator switch to the 5–50 kHz position. For a detector, connect a third transducer to an oscilloscope or amplifier and meter as described in Method A of the experiment. Then tune the oscillator for maximum signal from the detector when it is held near one of the sources (about 40,000 Hz). Move the detector along a line parallel to and about 25 cm in front of a line connecting the sources. Do you find distinct maxima and minima? Move closer to the sources. Do you find any change in the pattern spacing?

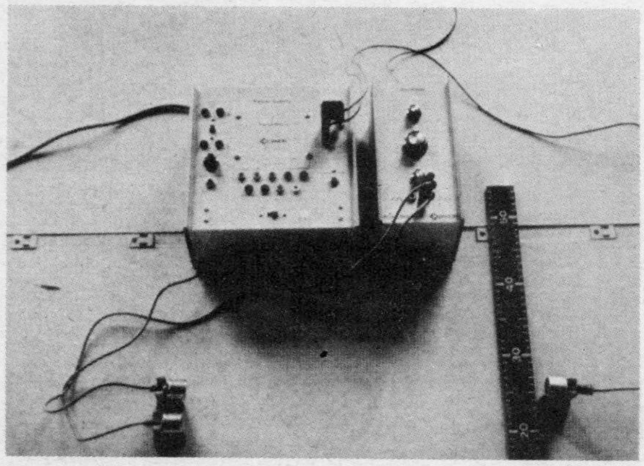

Fig. 3-67 Setup for determination of wavelength by the interference method.

?

4. What effect does a change in the separation of the sources have on the spacing of the nulls?

To make this experiment quantitative, work out a procedure for yourself similar to that used with the ripple tank. Measure the appropriate distances and then calculate the wavelength using the relationship

$$\lambda = \frac{xd}{L}$$

derived earlier for interference patterns in a ripple tank.

?

5. In using that equation, what assumptions are you making?

The Speed of Ultrasound

The relationship between speed v, wavelength λ, and frequency f is $v = \lambda f$. Using your best estimate of λ, calculate the speed of sound.

?

6. Does the speed of the ultrasound waves depend on the intensity of the wave?
7. How does the speed of sound in the inaudible range compare with the speed of audible sound?

ACTIVITIES

IS MASS CONSERVED?

You have read about some of the difficulties in establishing the law of conservation of mass. You can do several different experiments to check this law.

Antacid Tablet

You will need the following equipment: antacid tablets; 2-L flask, or plastic jug (such as is used for bleach, distilled water, or duplicating fluid); stopper for flask or jug; warm water; balance (sensitivity better than 0.1 g); spring scale (sensitivity better than 0.5 g).

Balance a tablet and 2-L flask containing 200–300 mL of water on a sensitive balance. Drop the tablet in the flask. When the tablet disappears and no more bubbles appear, readjust the balance. Record any change in mass. If there is a change, what caused it?

Repeat the procedure above, but include the rubber stopper in the initial balancing. Immediately after dropping in the tablet, place the stopper tightly in the mouth of the flask. (The pressure in a 2-L flask goes up by no more than 20%, so it is not necessary to tape or wire the stopper to the flask. Do not use smaller flasks in which proportionately higher pressure would be built up.) Is there a change in mass? Remove the stopper after all reaction has ceased. What happens? Discuss the difference between the two procedures.

Brightly Colored Precipitate

You will need: 20 g lead nitrate; 11 g potassium iodide; Erlenmeyer flask, 1000 mL with stopper; test tube, 25 × 150 mm; balance.

Place 400 mL of water in the Erlenmeyer flask, add the lead nitrate, and stir until dissolved. Place the potassium iodide in the test tube, add 30 mL of water, and shake until dissolved. Place the test tube, open and upward, carefully inside the flask and seal the flask with the stopper. Place the flask on the balance and bring the balance to equilibrium. Tip the flask to mix the solutions. Replace the flask on the balance. Does the total mass remain conserved? What *does* change in this experiment?

Magnesium Flashbulb

On the most sensitive balance you have available, measure the mass of an unflashed magnesium flashbulb. Repeat the measurement several times to make an estimate of the precision of the measurement.

Flash the bulb by connecting it to a battery. Be careful to touch the bulb as little as possible, so as not to wear away any material or leave any fingerprints. Measure the mass of the bulb several times, as before. You can get a feeling for how small a mass change your balance could have detected by seeing how large a piece of tissue paper you have to put on the balance to produce a detectable difference.

EXCHANGE OF MOMENTUM DEVICES

The four situations described below are more complex tests for conversation of momentum, to give you a deeper understanding of the generality of the conservation law and of the importance of your frame of reference.

(a) Fasten a section of HO-gauge model railroad track to two rings stands as shown in Fig. 3-68. Set one truck of wheels, removed from a car, on the track and from it suspend an object with mass roughly equal to that of the truck. Hold the truck, pull the object to one side, parallel to the track, and release both at the same instant. What happens?

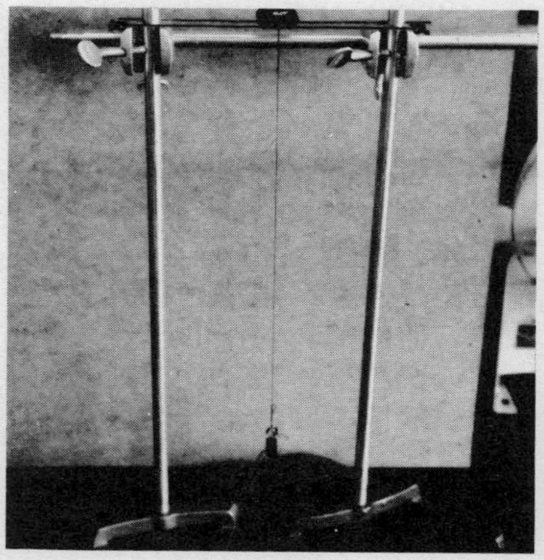

Fig. 3-68

Predict what you expect to see happen if you released the truck an instant after releasing the object. Try it. Also, try increasing the suspended mass.

(b) An air track supported on ring stands can also be used. An object of 20 g mass was suspended by a 50-cm string from one of the small air-track gliders. (One student trial continued for 166 swings.)

(c) Fasten two dynamics carts together with four hacksaw blades as shown in Fig. 3-69. Push the top one to the right, the bottom to the left, and release them. Try giving the bottom cart a push across the room at the same instant you release them.

Fig. 3-69

What would happen when you released the two if there were 10 or 20 ball bearings or small wooden balls hung as pendula from the top cart?

(d) Push two large rubber stoppers onto a short piece of glass tubing or wood (Fig. 3-70). Let the "dumbbell" roll down a wooden wedge so that the stoppers do not touch the table until the dumbbell is almost to the bottom. When the dumbbell touches the table, it suddenly increases its linear momentum as it moves along the table. Principles of rotational momentum and energy are involved here that are not covered in the Text, but even without extending the Text, you can deal with the "mysterious" increase in linear momentum when the stoppers touch the table.

Fig. 3-70

Using what you have learned about conservation of momentum, what do you think could account for this increase? (*Hint:* Set the wedge on a piece of cardboard supported on plastic beads and try the experiment.)

STUDENT HORSEPOWER

When you walk up a flight of stairs, the work you do goes into frictional heating and increasing gravitational potential energy. The $\Delta(PE)_{grav}$, in joules, is the product of your weight in newtons and the height of the stairs in meters.

Your useful power output is the average rate at which you did the lifting work, that is, the total change in $(PE)_{grav}$, divided by the time it took to do the work.

Walk or run up a flight of stairs and have someone time how long it takes. Determine the total vertical height that you lifted yourself by measuring one step and multiplying by the number of steps.

Calculate your useful work output and your power, in both units of watts and in horsepower. (Take 1 horsepower to be equal to 746 watts.)

DRINKING DUCK

A toy called a Drinking Duck demonstrates very well the conversion of heat energy into energy of gross motion by the processes of evaporation and condensation. The duck will continue to bob up and down as long as there is enough water in the cup to wet the beak (see Fig. 3-71).

Fig. 3-71

Rather than dampen your spirit of adventure, we will not tell you how it works. If you cannot figure out a possible mechanism for yourself,

George Gamow's book, *The Biography of Physics,* has a very good explanation. Gamow also calculates how far the duck could raise water in order to feed itself. An interesting extension is to replace the water with rubbing alcohol. What do you think will happen?

MECHANICAL EQUIVALENT OF HEAT

By dropping a quantity of lead shot from a measured height and measuring the resulting change in temperature of the lead, you can get a value for the ratio of work units to heat units, the *mechanical equivalent of heat*.

You will need the following equipment: cardboard tube; lead shot (1–2 kg); stoppers; a thermometer.

Close one end of the tube with a stopper, and put in 1–2 kg of lead shot that has been cooled about 5°C below room temperature. Close the other end of the tube with a stopper in which a hole has been drilled and a thermometer inserted. Carefully roll the shot to this end of the tube and record its temperature. Quickly invert the tube, remove the thermometer, and plug the hole in the stopper. Now invert the tube so the lead falls the full length of the tube and repeat this quickly 100 times. Reinsert the thermometer and measure the temperature. Measure the average distance the shot falls, which is the length of the tube minus the thickness of the layer of shot in the tube.

If the average distance the shot falls is h and the tube is inverted N times, the work you did raising the shot of mass m is

$$\Delta W = N \times ma_g \times h$$

The heat ΔH needed to raise the temperature of the shot by an amount ΔT is

$$\Delta H = cm \; \Delta T$$

where c is the specific heat capacity of lead, 3.1 $\times 10^{-5}$ cal/gC° (or 0.13 J/gC°).

The mechanical equivalent of heat is $\Delta W / \Delta H$. The accepted experimental value is 4.184 Nm/Cal (or Nm/J).

A DIVER IN A BOTTLE

Descartes is a name well known in physics. When you graphed motion in Text Sec. 1.5, you used Cartesian coordinates, which Descartes introduced. Using Snell's law of refraction, Descartes traced 1,000 rays through a sphere and came up with an explanation of the rainbow. He and his astronomer friend Gassendi were a bulwark against Aristotelian physics. Descartes belonged to the generation between Galileo and Newton.

On the lighter side, Descartes is known for a toy called the Cartesian diver that was very popular in the eighteenth century when very elaborate ones were made. To make one, first you will need a column of water. You may find a large cylindrical graduate in the laboratory, the taller the better. If not, you can improvise one out of a large jug or any other tall glass container. Fill the container almost to the top with water. Close the container in a way that permits you to change the pressure in it. For example, take a short piece of glass tubing with fire-polished ends, lubricate the glass tubing and the hole in the stopper with water, and carefully insert the glass tubing. Fit the rubber stopper into the top of the container as shown in Fig. 3-72.

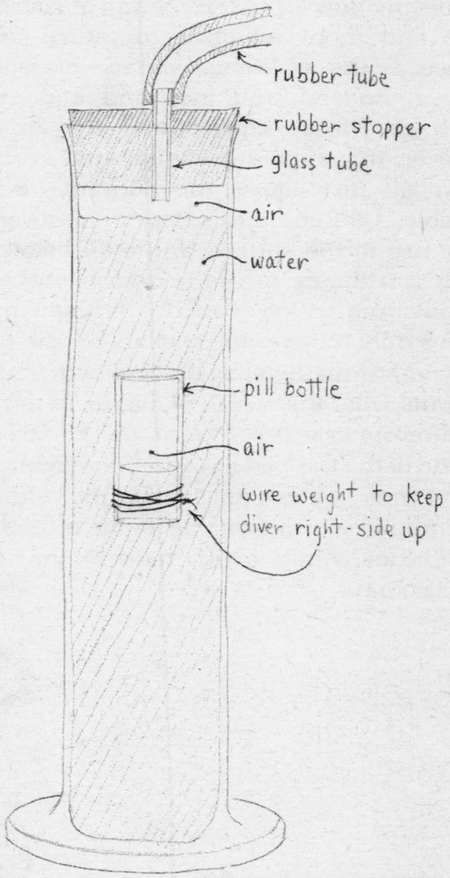

Fig. 3-72

Next construct the diver. You may limit yourself to pure essentials, namely a small pill bottle or vial, which may be weighted with wire.

Partially fill it with water so it just barely floats *upside down* at the top of the water column. If you are so inclined, you can decorate the bottle so it looks like a real underwater swimmer (or creature, if you prefer). The essential things are that you have a diver that just floats and that the volume of water in the diver can be changed.

Now to make the diver perform, blow momentarily on the rubber tube. According to Boyle's law, the increased pressure (transmitted by the water) decreases the volume of trapped air and water enters the diver. The buoyant force decreases, according to Archimedes' principle, and the diver begins to sink. (Archimedes' principle simply says that the buoyant force on an object is equal to the weight of the liquid displaced. This is the reason objects can float.)

If the original pressure is restored, the diver rises again. However, if you are lucky, you will find that as you cautiously make it sink deeper and deeper down into the column of water, it is more and more reluctant to return to the surface as the additional surface pressure is released. Indeed, you may find a depth at which the diver remains almost stationary. However, this apparent equilibrium, at which its weight just equals the buoyant force, is unstable. A bit above this depth, the diver will freely rise to the surface, and a bit below this depth it will sink to the bottom of the water column from which it can be brought to the surface only by vigorous sucking on the tube.

If you are mathematically inclined, you can compute what this depth would be in terms of the atmospheric pressure at the surface, the volume of the trapped air, and the weight of the diver. If not, you can juggle with the volume of the trapped air so that the point of unstable equilibrium comes about halfway down the water column.

The diver raises interesting questions. Suppose you have a well-behaved diver that "floats" at room temperature just halfway down the water column. Where will it "float" if the atmospheric pressure drops? Where will it "float" if the water is cooled or heated? If the ideal gas law is not enough to answer this question, you may have to do a bit of reading about the *vapor pressure* of water.

HOW TO WEIGH A CAR WITH A TIRE PRESSURE GAUGE

Reduce the pressure in all four auto tires so that the pressure is the same in each and somewhat below recommended tire pressure.

Drive the car onto four sheets of graph paper placed so that you can outline the area of the tire in contact with each piece of paper. The car should be on a reasonably flat surface (garage floor or smooth driveway). Then spray water on the graph paper. After the car is moved off the paper, you can measure the dry area. The flattened part of the tire is in equilibrium between the vertical force of the ground upward and the downward force of air pressure within.

Measure the air pressure in the tires, and the area of the flattened areas. If you use centimeter graph paper, you can determine the area in square centimeters by counting squares.

Pressure P (in pascals, Pa) is defined as F/A, where F is the downward force (in newtons) acting perpendicularly on the flattened area A (in square meters). Since the tire pressure gauge indicates the pressure *above* the normal atmospheric pressure of 101 kPa, you must add this value to the gauge reading. Compute the four forces as pressure times area. Their sum gives the weight of the car.

PERPETUAL-MOTION MACHINES?

You must have heard of "perpetual-motion" machines which, once started, will continue running and doing useful work forever. These proposed devices are inconsistent with the laws of thermodynamics. (It is tempting to say that they *violate* the laws of thermodynamics, but this implies that laws are rules by which Nature *must* run, instead of descriptions scientists have thought up.) It is now believed that it is in principle impossible to build such a machine.

But the dream dies hard! New proposals are made almost daily. Thus, S. Raymond Smedile, in *Perpetual Motion and Modern Research for Cheap Power* (Science Publications of Boston, 1962), maintains that this attitude of "it can't be done" negatively influences the search for new sources of cheap power. His book gives 16 examples of proposed machines, of which two are shown here.

Number 5 (Fig. 3-73) represents a wheel composed of 12 chambers marked A. Each chamber contains a lead ball B, which is free to roll. As the wheel turns, each ball rolls to the lowest level possible in its chamber. As the balls roll out to the right edge of the wheel, they create a preponderance of turning effects on the right side as against those balls that roll toward the hub on the left side. Thus, it is claimed that the wheel is driven clockwise perpetually. If you think this machine will not work, explain why not.

Number 7 (Fig. 3-74) represents a water-driven wheel marked A. D represents the buckets on the perimeter of the waterwheel for receiving water draining from the tank marked F. The waterwheel is connected to pump B by a belt and wheel. As the overshot wheel is operated by water dropping in it, it operates the pump that sucks water into C from which it enters into tank F. This operation is supposed to go on perpetually. If you think otherwise, explain why.

If such machines could operate, would the conservation laws necessarily be wrong?

Is the absence of perpetual-motion machines due to "theoretical" or "practical" deficiencies?

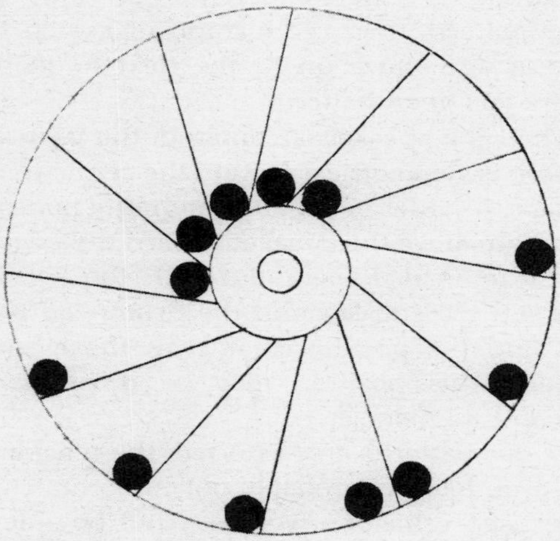

Fig. 3-73 Number 5.

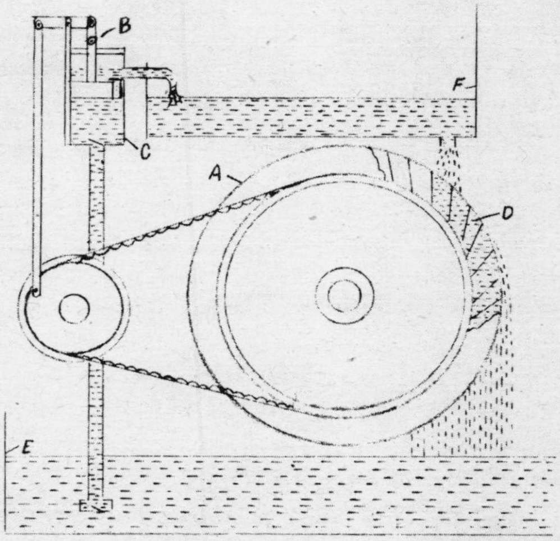

Fig. 3-74 Number 7.

The cartoons on pages 150–151 (and others of the same style that are scattered through the *Handbook*) were drawn in response to some ideas in the *Project Physics* course by a cartoonist who was unfamiliar with physics. On being informed that the drawing on the left did not represent conservation because the candle was not a closed system, he offered the solution above. (Whether a system is "closed" depends, of course, upon what you are trying to conserve.)

STANDING WAVES ON A DRUM AND A VIOLIN

You can demonstrate many different patterns of standing waves on a rubber membrane using a method very similar to that used in Film Loop 42, "Vibrations of a Drum." If you have not yet seen this loop, view it, if possible, before setting up the demonstration in your lab (see Fig. 3-75).

Fig. 3-75

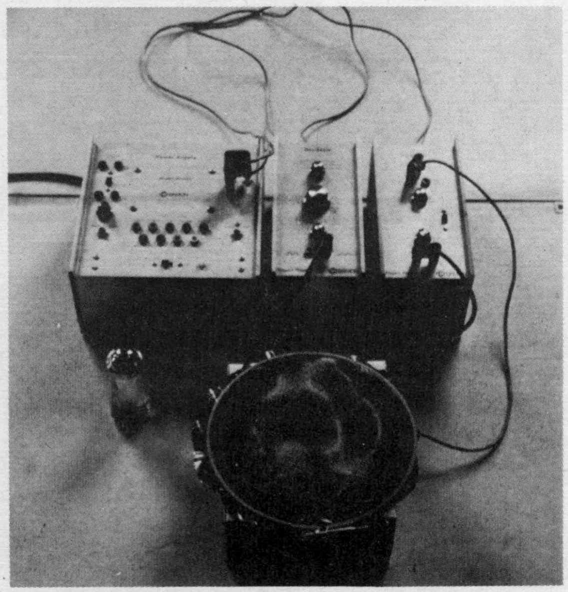

Fig. 3-76

Figure 3-76 shows the apparatus in action, producing one pattern of standing waves. The drumhead in the figure is an ordinary 17.5-cm embroidery hoop with the end of a large balloon stretched over it. If you make your drumhead in this way, use as large and as strong a balloon as possible, and cut its neck off with scissors. A flat piece of sheet rubber (dental dam) gives better results, since even tension over the entire drumhead is much easier to maintain if the rubber is not curved to begin with. Try other sizes and shapes of hoops, as well as other drumhead materials, such as a plastic wrap.

A 10-cm, 45-ohm speaker, lying under the drum and facing upward toward it, drives the vibrations. Connect the speaker to the output of an oscillator. If necessary, amplify the oscillator output.

Turn on the oscillator and sprinkle salt or sand on the drumhead. If the frequency is near one of the resonant frequencies of the surface, standing waves will be produced. The salt will collect along the nodes and be thrown off from the antinodes, thus outlining the pattern of the vibration. Vary the frequency until you get a clear pattern, then photograph or sketch the pattern and move on to the next frequency where you get a pattern.

When the speaker is centered, the vibration pattern is symmetrical around the center of the surface. In order to get antisymmetric nodes of vibration, move the speaker toward the edge of the drumhead. Experiment with the spacing between the speaker and the drumhead until you find the position that gives the clearest pattern; this position may be different for different frequencies.

If your patterns are distorted, the tension of the drumhead is probably not uniform. If you have used a balloon, you may not be able to remedy the distortion, since the curvature of the balloon makes the edges tighter than the center. By pulling gently on the rubber, however, you may at least be able to make the tension even all around the edge.

A similar procedure, used 150 years ago and still used in analyzing the performance of violins, is shown in Fig. 3-77, reprinted from *Scientific American,* "Physics and Music."

REFLECTION

Two-dimensional water surface waves exhibit a fascinating variety of reflection phenomena. If you have never watched closely as water waves are reflected from a fixed barrier, you should do so. Any still pool or water-filled wash basin or tub will do. Watch the circular waves radiate outward, reflect from rocks or walls, run through each other, and finally die out. Dip your fingertip into and out of the water quickly,

Fig. 3-77 Chladni plates indicate the vibration of the body of a violin. These patterns were produced by covering a violin-shaped brass plate with sand and drawing a violin bow across its edge. When the bow caused the plate to vibrate, the sand concentrated along quiet nodes between the vibrating areas. Bowing the plate at various points, indicated by the round white marker, produces different frequencies of vibration and different patterns. Low tones produce a pattern of a few large areas; high tones a pattern of many small areas. Violin bodies have a few such natural modes of vibration that tend to strengthen certain tones sounded by the strings. Poor violin bodies accentuate squeaky top notes. This sand-and-plate method of analysis was devised over 150 years ago by the German acoustical physicist Ernst Chladni.

or let a drop of water fall from your finger into the water. Now watch the circular wave approach and then bounce off of a straight wall or board. The long side of a bathtub is a good straight barrier. (See the illustrations on p. 376, Unit 3.)

MOIRÉ PATTERNS

You will probably notice a disturbing visual effect from the patterns in Figs. 3-78 and 3-79. Some types of art depend on similar effects, many of which are caused by moiré patterns.

If you make a photographic negative of the pattern in Fig. 3-78 or Fig. 3-79 and place it on top of the same figure, you can use it to study the interference pattern produced by two point sources.

There are an increasing number of scientific applications of moiré patterns. Because of the great visual changes caused by very small shifts in two regular overlapping patterns, they can

Fig. 3-78

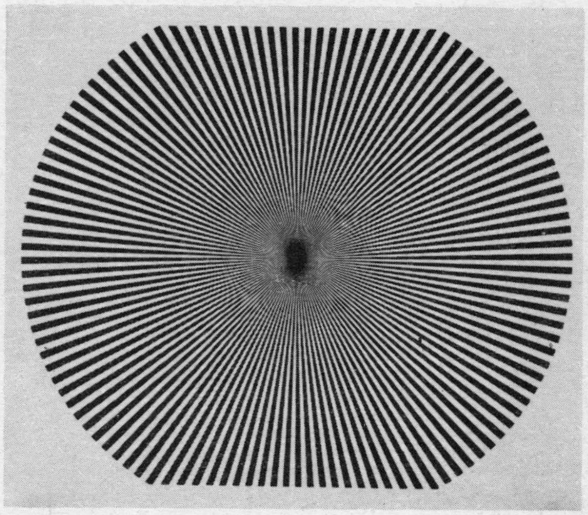

Fig. 3-79

be used to make measurements to an accuracy of +0.0000001%. Some specific examples of the useof moiré patterns are visualization of two- or multiple-source interference patterns, the measurement of small angular shifts, measurements of diffusion rates of solids into liquids, and representations of electric, magnetic, and gravitational fields. Some of the patterns created still cannot be expressed mathematically.

Scientific American (May, 1963) has an excellent article, "Moiré Patterns" by Gerald Oster and Yasunori Nishijima. *The Science of Moiré Patterns,* a book by G. Oster, is available from Edmund Scientific Co., Barrington, N. J. Edmund also has various inexpensive sets of different patterns, which save much drawing time, and that are much more precise than hand-drawn patterns.

MUSIC AND SPEECH ACTIVITIES

(a) Frequency ranges: Set up a microphone and oscilloscope so you can display the pressure variations in sound waves. Play different instruments and see how "high C" differs on them.

(b) Some beautiful oscilloscope patterns result when you display the sound of computer music records, which use sound synthesizers instead of conventional instruments.

(c) For interesting background, see the following articles in *Scientific American:* "Physics and Music," July, 1948; "The Physics of Violins," November, 1962; "The Physics of Wood-winds," October, 1960; and "Computer Music," December , 1959.

(d) The Bell Telephone Company has an interesting educational item, which may be available through your local Bell Telephone office. A 33⅓ LP record, "The Science of Sounds," has 10 bands demonstrating different ideas about sound. For instance, racing cars demonstrate the Doppler shift, and a soprano, a piano, and a factory whistle all sound alike when overtones are filtered out electronically. The record is also available on the Folkways label FX 6136.

(e) "Test records" are available for stereo hi-fi equipment. Many of these records let you check the "frequency response" of a system by giving a series of steady tones at various frequencies. Try playing one of these records and checking for nodes. Be sure to place the speakers far enough apart. (How far apart do the speakers have to be so that you get two nodes between them? Compute this for $f = 100$; 500; 1,000; and 5,000 Hz.) Listen for nodes by moving your head back and forth.

MEASUREMENT OF THE SPEED OF SOUND

For this experiment you need to work outside in the vicinity of a large flat wall that produces a good echo. You also need some source of loud pulses of sound at regular intervals, about one per second or less. A friend beating on a drum or something with a higher pitch will do. The important thing is that the time between one pulse and the next does not vary, so a metronome would help. The sound source should be fairly far away from the wall, about 200 m in front of it.

Stand somewhere between the reflecting wall and the source of pulses (see Fig. 3-80). You will hear both the direct sound and the sound reflected from the wall. The direct sound will reach you first because the reflected sound must travel the additional distance from you to the wall and back again. As you approach the wall, this additional distance decreases, as does the time interval between the direct sound and the echo. Movement away from the wall increases the interval.

Fig. 3-80

If the distance from the source to the wall is great enough, the added time taken by the echo to reach you can amount to more than the time between drum beats. You will be able to find a position at which you hear the *echo* of one pulse at the same time you hear the *direct* sound of the next pulse. Then you know that the sound took a time equal to the interval between pulses to travel from you to the wall and back to you.

Measure your distance from the wall. Find the time interval between pulses by measuring the time for a large number of pulses. Use these two values to calculate the speed of sound.

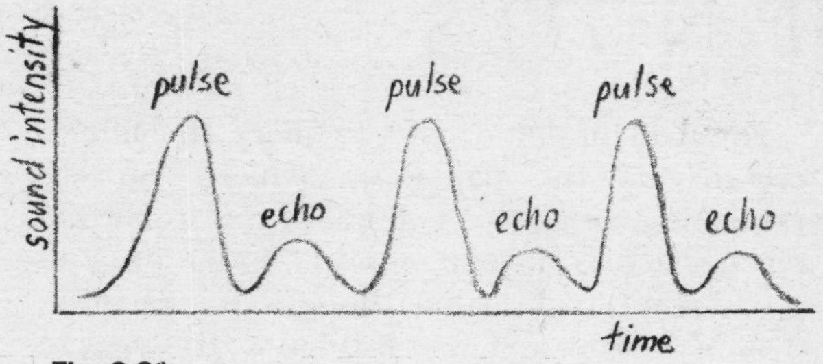

Fig. 3-81

(If you cannot get far enough away from the wall to get this synchronization, shorten the interval between pulses. If this is impossible, you may be able to find a place where you hear the echoes exactly halfway between the pulses as shown in Fig. 3-81. You will hear a pulse, then an echo, then the next pulse. Adjust your position so that these three sounds seem equally spaced in time. At this point you know that the time taken for the return trip from you to the wall and back is equal to *half* the time interval between pulses.)

MECHANICAL WAVE MACHINES

Several types of mechanical wave machines are described below. They help a great deal in understanding the various properties of waves.

Slinky

The spring called a Slinky behaves much better when it is freed of friction from the floor or table. Hang a Slinky horizontally from strings at least 1 m long tied to rings on a wire stretched from two solid supports. Tie strings to the Slinky at every fifth spiral for proper support. (See Fig. 3-82.)

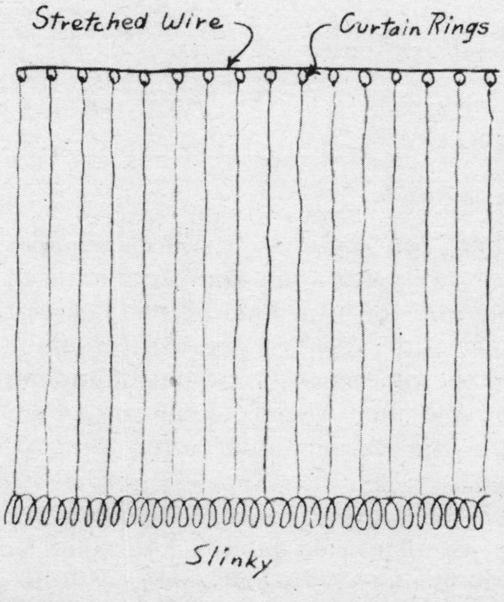

Fig. 3-82

Fasten one end of the Slinky *securely* and then stretch it out to about 5–10 m. By holding onto a 3-m piece of string tied to the end of the Slinky, you can illustrate the "open-ended" reflection of waves.

See Experiment 3-15 for more details on demonstrating the various properties of waves.

Rubber Tubing and Welding Rod

Clamp both ends of a 1-m piece of rubber tubing to a table so it is under slight tension. Punch holes through the tubing every 2.5 cm with a hammer and nail. (Put a block of wood under the tubing to protect the table.)

Obtain enough 30-cm lengths of welding rod for all the holes you punched in the tubing. Unclamp the tubing, and insert one rod in each of the holes. Hang the rubber tubing vertically, as shown in Fig. 3-83, and give its lower end a twist to demonstrate transverse waves. Performance and visibility are improved by adding weights to the ends of the rods or to the lower end of the tubing. (See Fig. 3-83.)

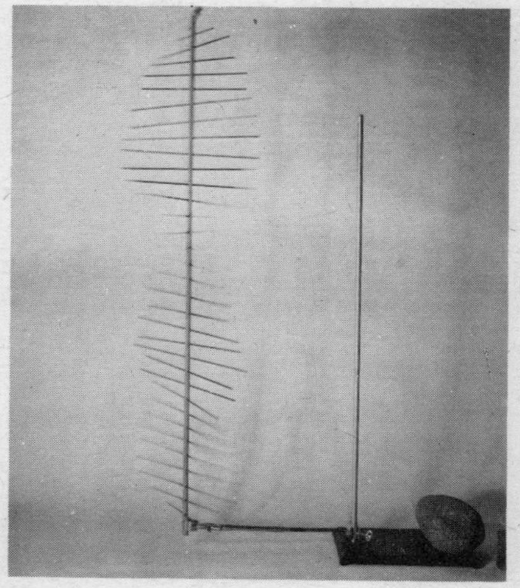

Fig. 3-83

A Better Wave Machine

An inexpensive paperback, *Similarities in Wave Behavior,* by John N. Shive of Bell Telephone Laboratories, has instructions for building a better torsional-wave machine than that described above.

FILM LOOP NOTES

Film Loop 18
ONE-DIMENSIONAL COLLISIONS. I

Two different head-on collisions of a pair of steel balls are shown. The balls hang from long, thin wires that confine each ball's motion to the same circular arc. The radius is large compared with the part of the arc, so the curvature is hardly noticeable. Since the collisions take place along a straight line, they can be called one-dimensional.

In the first example, ball B, weighing 350 g, is initially at rest. In the second example, ball A, with a mass of 532 g, is the one at rest.

With this film, you can make detailed measurements of the total momentum and energy of the balls before and after collision. Momentum is a vector, but in this one-dimensional case you need only worry about its sign. Since momentum is the product of mass and velocity, its sign is determined by the sign of the velocity.

You know the masses of the balls. Velocities can be measured by finding the distance traveled in a known period of time.

After viewing the film, you can decide on what strategy to use for distance and time measurements. One possibility would be to time the motion through a given distance with a stopwatch, perhaps making two lines on the paper. You need the velocity just before and after the collision. Since the balls are hanging from wires, their velocity is not constant. On the other hand, using a small arc increases the chances of distance–time uncertainties. As with most measuring situations, a number of conflicting factors must be considered.

You will find it useful to mark the filmed crosses on the paper on which you are projecting, since this will allow you to correct for projector movement and film jitter. You might want to give some thought to measuring distances. You may use a ruler with marks in millimeters, so you can estimate to one-tenth of a millimeter. Is it wise to try to use the zero end of the ruler, or should you use positions in the middle? Should you use the thicker or the thinner marks on the ruler? Should you rely on one measurement, or should you make a number of measurements and average them?

Estimate the uncertainty in distance and time measurements, and the uncertainty in velocity. What can you learn from this about the uncertainty in momentum?

Fig. 3-84

When you compute the total momentum before and after collision (the sum of the momentum of each ball), remember that you must consider the direction of the momentum.

Are the differences between the momentum before and after collision significant, or are they within the experimental error already estimated?

Save the data you collect so that later you can make similar calculations on total kinetic energy for both balls just before and just after collision.

Film Loop 19
ONE-DIMENSION COLLISIONS. II

Two different head-on collisions of a pair of steel balls are shown, with the same setup as that used in Film Loop 18, "One-Dimensional Collisions. I."

In the first example, ball A with a mass of 1.8 kg collides head on with ball B, with a mass of 532 g. In the second example, ball A catches up with ball B. The instructions for Film Loop 18 may be followed for completing this investigation also.

Film Loop 20
INELASTIC ONE-DIMENSIONAL COLLISIONS

In this film, two steel balls covered with plasticene hang from long supports. Two collisions are shown. The two balls stick together after colliding, so the collision is *inelastic*. In the first example, ball A, weighing 443 g, is at rest when ball B, with a mass of 662 g, hits it. In the second example, the same two balls move toward each other. Two other films, "One-Dimensional Collisions. I" and "One-Dimensional Collisions. II" show collisions in which the two balls bounce off of each other. What different results might you expect from measurements of an inelastic one-dimensional collision?

The instructions for Film Loop 18 may be followed for completing this investigation.

Are the differences between momentum before and after collision significant, or are they within the experimental error already estimated?

Save your data so that later you can make similar calculations on total kinetic energy for both balls just before and just after the collision. Is whatever difference you may have obtained explainable by experimental error? Is there a noticeable difference between elastic and inelastic collisions as far as the conservation of kinetic energy is concerned?

Film Loop 21
TWO-DIMENSIONAL COLLISIONS. I

Two hard steel balls, hanging from long, thin wires, collide. Unlike the collisions in Film Loops 18 and 20, the balls do not move along the *same* straight line before or after the collisions. Although the balls do not all strictly move in one plane because each motion is an arc of a circle, to a good approximation everything occurs in one plane. Therefore, the collisions are two-dimensional. Two collisions are filmed in slow motion, with ball A having a mass of 539 g, and ball B having a mass of 361 g. Two more cases are shown in Film Loop 22.

Using this film, you can find both the momentum and the kinetic energy of each ball before and after the collision, and thus study total momentum and total kinetic energy conservation in this situation. Thus, you should save your momentum data for later use when studying energy.

Both direction and magnitude of momentum should be taken into account, since the balls do not move on the same line. To find momentum you need velocities. Distance measurements accurate to a fraction of a millimeter, and time measurements to about one-tenth of a second are suggested, so choose measuring instruments accordingly.

You can project directly onto a large piece of paper. An initial problem is to determine lines on which the balls move. If you make many marks at the centers of the balls, running the

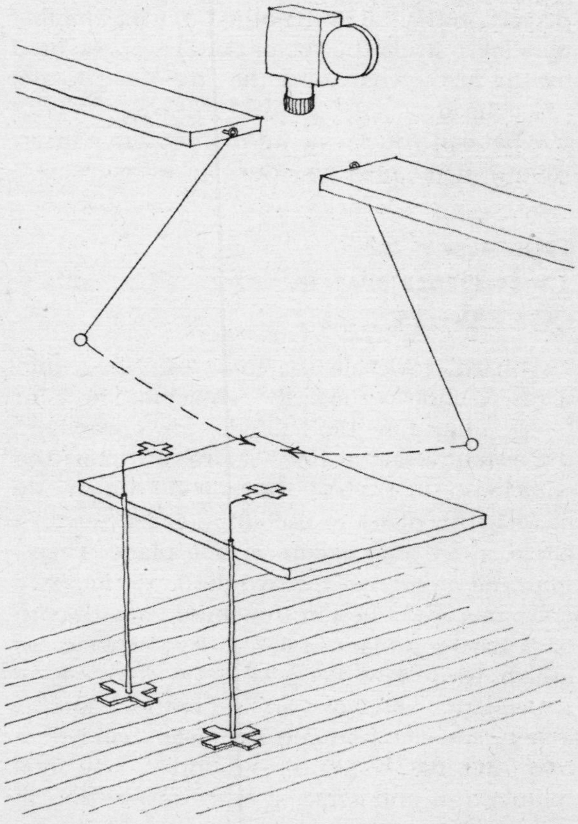

Fig. 3-85

film several times, you may find that these do not form a straight line. This is due both to the inaccuracies in your measurements and to the inherent difficulties of high-speed photography. Cameras photographing at a rate of 2,000 to 3,000 frames a second "jitter," because the film moves so rapidly through the camera that accurate frame registration is not possible. Decide which line is the "best" approximation to determine direction for velocities of the balls before and after collision.

To find the *magnitude* of the velocity, the speed, measure the time it takes the ball to move across two lines marked on the paper. For the sake of accuracy, take a number of different measurements to determine which values to use for the speeds and how much error is present.

Compare the sum of the momentum before collision for both balls with the total momentum after collision. If you do not know how to add vector diagrams, you should consult your teacher or the *Programmed Instruction Booklet*, "Vectors II." The momentum of each object is represented by an arrow whose direction is that of the motion and whose length is proportional to the magnitude of the momentum. Then, if the head of one arrow is placed on the tail of the other, moving the line parallel to itself, the vector sum is represented by the arrow that joins the "free" tail to the "free" head.

What can you say about momentum conservation? Remember to consider measurement errors.

Film Loop 22
TWO-DIMENSIONAL COLLISIONS. II

Two hard steel balls, hanging from long, thin wires, collide. Unlike the collisions in Film Loops 18 and 20, the balls do not move along the *same* straight line before or after the collisions. Although all the balls do not strictly move in one plane, as each motion is an arc of a circle, everything occurs in one plane. Therefore, the collisions are two-dimensional. Two collisions are filmed in slow motion, with both balls having a mass of 367 g. Two other cases are shown in Film Loop 21.

Using this film, you can find both the kinetic energy and the momentum of each ball before and after the collision, and thus study total momentum and total energy conservation in this situation. Follow the instructions given for Film Loop 21 in completing this investigation.

Film Loop 23
INELASTIC TWO-DIMENSIONAL COLLISIONS

Two hard steel balls, hanging from long, thin wires, collide. Unlike the collisions in Film Loops 18 and 20, the balls do not move along the *same* straight line before or after the collision. Although all the balls do not strictly move in one plane, as each motion is an arc of a circle, to a good approximation the motion occurs in one plane. Therefore, the collisions are two-dimensional. Two collisions are filmed in slow motion. Each ball has a mass of 500 g. The plasticene balls stick together after collision, and move as a single mass.

Using this film, you can find both the kinetic energy and the momentum of each ball before and after the collision, and thus study total momentum and total energy conservation in this situation. Follow the instructions given for Film Loop 21 in completing this investigation.

Film Loop 24
SCATTERING OF A CLUSTER OF OBJECTS

This film and also Film Loop 25 each contain one of the Events 8–13 of the series, "Stroboscopic Still Photographs of Two-Dimensional one of the Events 8–13 of the series, "Stroboscopic Still Photographs of Two-Dimensional Collisions," or one of the examples in Film Loops 22 and 23. All these examples involve two-body collisions, whereas the film described here involves seven objects and Film Loop 25, five objects.

In this film, seven balls are suspended from long, thin wires. The camera sees only a small portion of their motion, so the balls all move approximately along straight lines. The slow-motion camera is above the balls. Six balls are initially at rest. A hardened steel ball strikes the cluster of resting balls. The diagram in Fig. 3-86 shows the mass of each of the balls.

Part of the film is photographed in slow motion at 2,000 frames per second. By projecting this section of the film on paper several times and making measurements of distances and times, you can determine the directions and magnitudes of the velocities of each of the balls. Distance and time measurements are needed. Discussions of how to make such measurements are contained in the Film Loop Notes for one-dimensional and two-

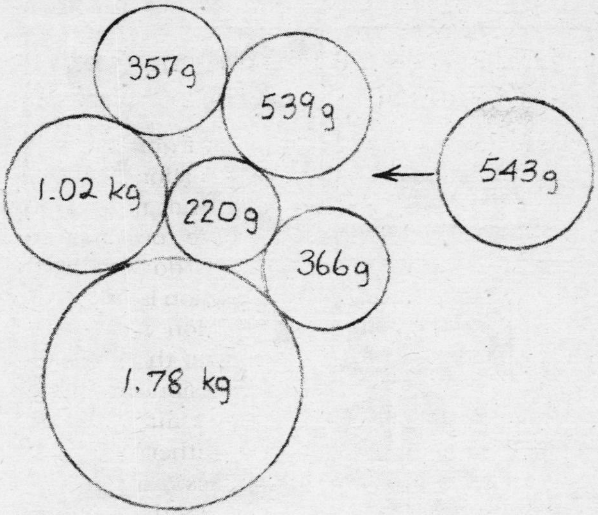

Fig. 3-86

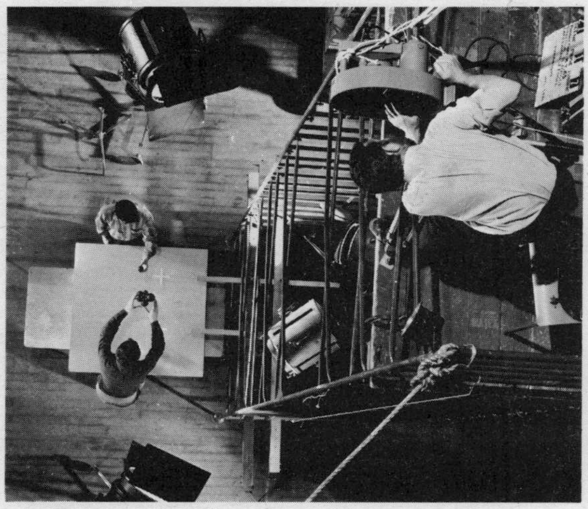

Fig. 3-87

dimensional collisions. (See Film Loops 18 and 21.)

Compare the total momentum of the system both before and after the collision. Remember that momentum has both direction and magnitude. You can add momenta after collision by representing the momentum of each ball by an arrow, and "adding" arrows geometrically. What can you say about the accuracy of your calculations and measurements? Is momentum conserved? You might also wish to consider energy conservation.

Film Loop 25
EXPLOSION OF A CLUSTER OF OBJECTS

Five balls are suspended independently from long, thin wires. The balls are initially at rest, with a small cylinder containing gunpowder in the center of the group of balls. The masses and initial positions of the balls are shown in Fig. 3-88. The charge is exploded and each of the balls moves off in an independent direction. In the slow-motion sequence, the camera is mounted directly above the resting balls. The camera sees only a small part of the motion, so that the paths of the balls are almost straight lines.

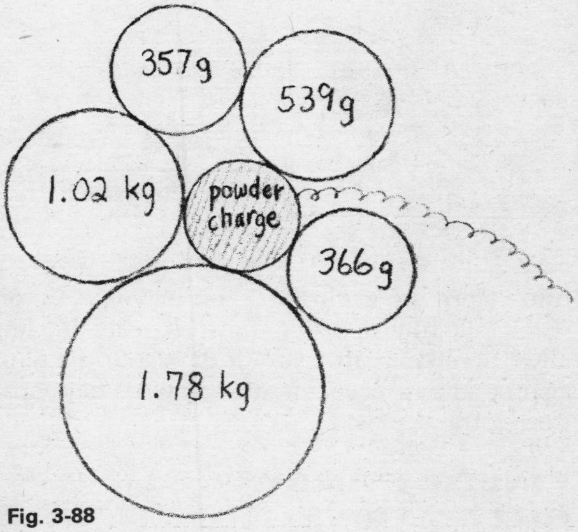

Fig. 3-88

In your first viewing, you might try to predict where the "missing" balls will emerge. Several of the balls are hidden at first by the smoke from the charge of powder. All the balls except one are visible for some time. What information could you use that would help you make a quick decision about where this last ball will appear? What physical quantity is important? How can you use this quantity to make a quick estimate? When you see the ball emerge from the cloud, you can determine whether or not your prediction was correct. The animated elliptical ring toward the end of the film identifies this final ball.

You can also make detailed measurements, similar to the momentum conservation measurements you may have made using other *Project Physics* Film Loops. During the slow-motion sequence, find the magnitude and direction of the velocity of each of the balls after the explosion by projecting the film on paper, and measuring distances and times. The notes on previous films in this series, Film Loops 18 and 21, will provide you with information about how to make such measurements if you need assistance.

Determine the total momentum of all the balls after the explosion. What was the momen-

B.C. by John Hart

tum before the explosion? You may find these results slightly puzzling. Can you account for any discrepancy that you find? Watch the film again and pay close attention to what happens during the explosion.

KINETIC ENERGY CALCULATIONS

You may have used one or more of Film Loops 18 through 25 in your study of momentum. You will find it helpful to view these slow-motion films of one- and two-dimensional collisions again, but this time in the context of the study of energy. The data you collected previously will be sufficient for you to calculate the kinetic energy of each ball before and after the collision. Remember that kinetic energy $\frac{1}{2}mv^2$ is *not* a vector quantity; therefore, you need only use the magnitude of the velocities in your calculations.

On the basis of your analysis you may wish to try to answer such questions as: Is kinetic energy consumed in such interactions? If not, what happened to it? Is the loss in kinetic energy related to such factors as relative speed, angle of impact, or relative masses of the colliding balls? Is there a difference in the kinetic energy lost in elastic and inelastic collisions?

Film Loop 26
FINDING THE SPEED OF A RIFLE BULLET. I

In this film, a rifle bullet of 13.9 g is fired into an 8.44-kg log. The log is initially at rest, and the bullet imbeds itself in the log. The two bodies move together after this violent collision. The height of the log is 15.0 cm. You can use this information to convert distances to centimeters. The setup is illustrated in Fig. 3-89.

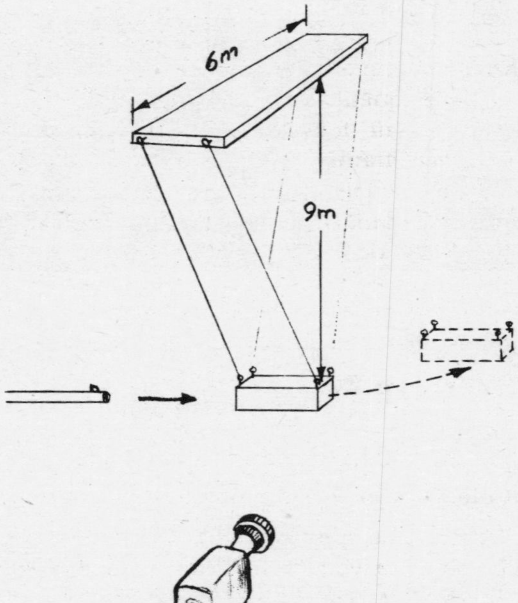

Fig. 3-89 Schematic diagram of ballistic pendulum (not to scale).

You can make measurements in this film using the extreme slow-motion sequence. The high-speed camera used to film this sequence operated at an average rate of 2,850 frames per second; if your projector runs at 18 frames per second, the slow-motion factor is 158. Although there was some variation in the speed of this camera, the average frame rate of 2,850 is quite accurate. For velocity measurements in centimeters per second (a convenient unit to use in considering a rifle bullet) convert the apparent time of the film to seconds. Find the exact duration with a timer or a stopwatch by timing the interval from the yellow circle at the beginning to the one at the end of the film. There are 3,490 frames in the film, so you can determine the precise speed of the projector.

Project the film onto a piece of white paper or graph paper to make your measurements of distance and time. View the film before making decisions about which measuring instruments to use. As suggested above, you can convert your distance and time measurements to centimeters and seconds.

After measuring the speed of the log after impact, calculate the bullet speed at the moment when it entered the log. What physical laws do you need for the calculation? Calculate the kinetic energy given to the bullet, and also calculate the kinetic energy of the log after the bullet enters it. Compare these two energies and discuss any differences that you might find. Is kinetic energy conserved?

A final sequence in the film allows you to find a *lower limit* for the bullet's speed. Three successive frames are shown, so the time between each is 1/2,850 sec. The frames are each printed many times, so each is held on the screen. How does this lower limit compare with your measured velocity?

Film Loop 27
FINDING THE SPEED OF A RIFLE BULLET. II

The problem proposed by this film is to determine the speed of the bullet just before it hits a log. The wooden log with a mass of 4.05 kg is initially at rest. A bullet fired from a rifle enters the log (Fig. 3-90). The mass of the bullet is 7.12 g. The bullet is imbedded in the thick log and the two move together after the impact. The extreme slow-motion sequence is intended for taking measurements.

The log is suspended from thin wires so that it behaves like a pendulum that is free to swing. As the bullet strikes the log, the log starts to rise. When the log reaches its highest point, it momentarily stops, and then begins to swing back down. This point of zero velocity is visible in the slow-motion sequence in the film.

Fig. 3-90

The bullet plus the log *after* impact form a closed system, so you would expect the total amount of mechanical energy of such a system to be conserved. The total mechanical energy is the sum of kinetic energy plus potential energy. If you conveniently take the potential energy as zero at the moment of impact for the lowest position of the log, then the energy at that time is all kinetic energy. As the log begins to move, the potential energy is proportional to the vertical distance above its lowest point, and it increases while the kinetic energy, depending upon the speed, decreases. The kinetic energy becomes zero at the point where the log reverses its direction because the log's speed is zero at that point. All the mechanical energy at the reversal point is potential energy. Because energy is conserved, the initial kinetic energy at the lowest point should equal the potential energy at the top of the swing. On the basis of this result, write an equation that relates the initial log speed to the final height of rise. You might check this result with your teacher or with other students in the class.

If you measure the vertical height of the rise of the log, you can calculate the log's initial speed, using the equation just derived. What is the initial speed that you find for the log? If you wish to convert distance measurements to centimeters, it is useful to know that the vertical dimension of the log is 9.0 cm.

Find the speed of the rifle bullet at the moment it hits the log, using conservation of momentum.

Calculate the kinetic energy of the rifle bullet before it strikes and the kinetic energy of the log plus bullet after impact. Compare the two kinetic energies, and discuss any difference.

Film Loop 28
RECOIL

Conservation laws can be used to determine recoil velocity of a gun, given the experimental information that this film provides.

The preliminary scene shows the recoil of a cannon firing at the fort on St. Hélène Island, near Montreal, Canada (Fig. 3-91). The small brass laboratory "cannon" in the rest of the film is suspended by long wires. It has a mass of 350 g. The projectile has a mass of 3.5 g. When the firing is photographed in slow motion, you can see a time lapse between the time the fuse is lighted and the time when the bullet emerges from the cannon. Why is this delay observed? The camera used here exposes 8,000 frames per second.

Project the film on paper. It is convenient to use a horizontal distance scale in centimeters. Find the bullet's velocity by timing the bullet over a large fraction of its motion. (Only relative values are needed, so it is not necessary to convert this velocity into centimeters per second.)

Use momentum conservation to predict the gun's recoil velocity. The system (gun plus bullet) is one dimensional; all motion is along one straight line. The momentum before the gun is fired is zero in the coordinate system in which the gun is at rest, so the momentum of the cannon after firing should be equal and opposite to the momentum of the bullet.

Test your prediction of the recoil velocity by running the film again and timing the gun to find its recoil velocity experimentally. What margin of error might you expect? Do the predicted and observed values agree? Give reasons for any difference you observe. Is kinetic energy conserved? Explain your answer.

Film Loop 29
COLLIDING FREIGHT CARS

This film shows a test of freight-car coupling. The collisions, in some cases, were violent enough to break the couplings. The "hammer car," coasting down a ramp, reaches a speed of about 10 km/hr. The momentary force between the cars is about 4,400,000 N. The photograph (Fig. 3-92) shows coupling pins that were sheared off by the force of the collision. The slow-motion collision allows you to measure speeds before and after impact, and thus to test conservation of momentum. The collisions are *partially* elastic, as the cars separate to some extent after collision. The masses of the cars are: hammer car: m_1 = 95,000 kg; target car: m_2 = 120,000 kg. To find velocities, measure the film time for the car to move through a given

Fig. 3-91

Fig. 3-92 Broken coupling pins from colliding freight cars.

distance. (You may need to run the film several times.) Use any convenient units for velocities.

Simple timing will give v_1 and v_2. The film was made on a cold winter day and friction was appreciable for the hammer car after collision. One way to allow for friction is to make a velocity time graph, assume a uniform negative acceleration, and extrapolate to the instant after impact.

An example might help. Suppose the hammer car coasts 3 squares on graph paper in 5 sec after collision, and it also coasts 6 squares in 12 sec after collision. The *average* velocity during the first 5 sec was v_1 = (3 squares)/(5 sec) = 0.60 squares/sec. The average velocity during any short interval approximately equals the instantaneous velocity at the midtime of that interval, so the car's velocity was about v_1 = 0.60 squares/sec at t = 2.5 sec. For the interval 0–12 sec, the velocity was v_1 = 0.50 squares/sec at t = 6.0 sec. Now plot a graph like that shown in Fig. 3-93. This graph shows, by extrapolation, that v_1 = 0.67 squares/sec at t = 0, just after the collision.

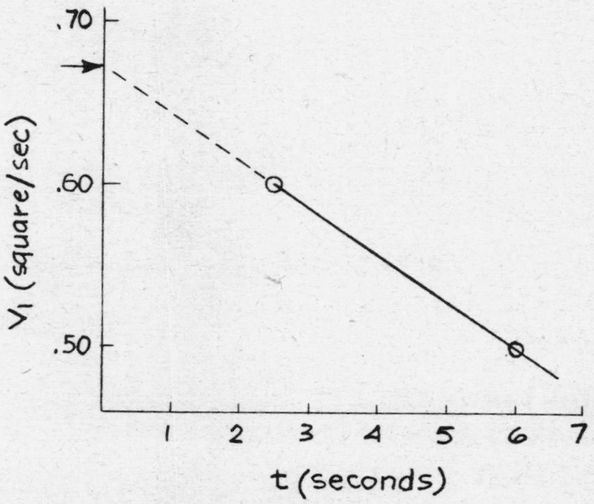

Fig. 3-93 Extrapolation backwards in time allow for friction in estimating the value of v_1 immediately after the collision.

Compare the total momentum of the system before collision with the total momentum after collision. Calculate the kinetic energy of the freight cars before and after collision. What fraction of the hammer car's original kinetic energy has been "lost"? Can you account for this loss?

Film Loop 30
DYNAMICS OF A BILLIARD BALL

The event pictured in this film is one you have probably seen many times, the striking of a ball, in this case a billiard ball, by a second ball. (See Fig. 3-94.) Here, the camera is used to "slow down" time so that you can see details in this event that you probably have never observed. The ability of the camera to alter space and time is important in both science and art. The slow-motion scenes were shot at 3,000 frames per second.

Fig. 3-94 Billiard balls near impact. The two cameras took side views of the collision, which are not shown in this Film Loop.

The "world" of your physics course often has some simplifications in it. Thus, in your textbook, much of the discussion of mechanics of bodies probably assumes that the objects are point objects, with no size. Clearly these massive billiard balls have size, as do all the things you encounter.

For a point particle, you can speak in a simple, meaningful way of its position, its velocity, and so on. But the particles photographed here are billiard balls and not points. What information might be needed to describe their positions and velocities? Looking at the film may suggest possibilities. What motions can you see besides simply the linear forward motion? Watch each ball carefully, just before and just after the collision. Watch not only the overall motion of the ball, but also "internal" motions. Can any of these motions by appropriately described by the word "spin"? Can you distinguish the cases where the ball is rolling along the table, so that there is no slippage between the ball and the table, from the situations where the ball is skidding along the table without rolling? Does the first ball move *immediately* after the collision? You can see that even this simple phenomenon is more complex than you might have expected.

Can you write a careful verbal description of the event? How might you go about giving a more careful mathematical description?

Using the slow-motion sequence, you can make a momentum analysis, at least partially, of this collision. Measure the velocity of the cue ball before impact and the velocity of both balls after impact. Remember that there is friction between the ball and the table, so velocity is *not* constant. Since the balls have the same mass, conservation of momentum predicts that

velocity of cue ball just before collision = sum of velocities of the balls just after collision

How closely do the results of your measurements agree with this principle? What reasons, considering the complexity of the phenomenon, might you suggest to account for any disagreement? What motions are you neglecting in your analysis?

Film Loop 31
A METHOD OF MEASURING ENERGY: NAILS DRIVEN INTO WOOD

Some physical quantities, such as distance, can be measured directly in simple ways. Other concepts can be connected with the world of experience only through a long series of measurements and calculations. In certain situations, simple and reliable methods of determining *energy* are possible. Here, you are concerned with the energy of a moving object.

This film allows you to check the validity of one way of measuring mechanical energy. If a moving object strikes a nail, the object will lose all of its energy. This energy has some effect, in that the nail is driven into the wood. The energy of the object becomes work done on the nail, driving it into the block of wood.

The first scenes in the film show a construction site. A pile driver strikes a pile over and over again, "planting" it in the ground. The laboratory situation duplicates this situation under more controlled circumstances. Each of the blows is the same as any other because the massive object is always raised to the same height above the nail. The nail is hit 10 times. Because the conditions are kept the same, you expect the energy of the impact to be the same for each blow. Therefore, the work from each blow is the same. Use the film to determine if the distance the nail is driven into the wood is proportional to the energy or work. How can you find the energy expended if you know the depth of penetration of the nail?

The simplest way to display the measurements made with this film may be to plot the depth of nail penetration versus the number of blows. Do the experimental points that you obtain lie approximately along a straight line? If the line is a good approximation, then the energy is about proportional to the depth of penetration of the nail. Thus, depth of penetration can be used in the analysis of other films to measure the energy of the striking object.

If the graph is not a straight line, you can still use these results to calibrate your energy-measuring device. By use of penetration versus the number of blows, an observed penetration (in centimeters, as measured on the screen), can be converted into a number of blows, and therefore an amount proportional to the work done on the nail, or the energy transferred to the nail. Thus, in Fig. 3-95, a penetration of 3 cm signifies 5.6 units of energy.

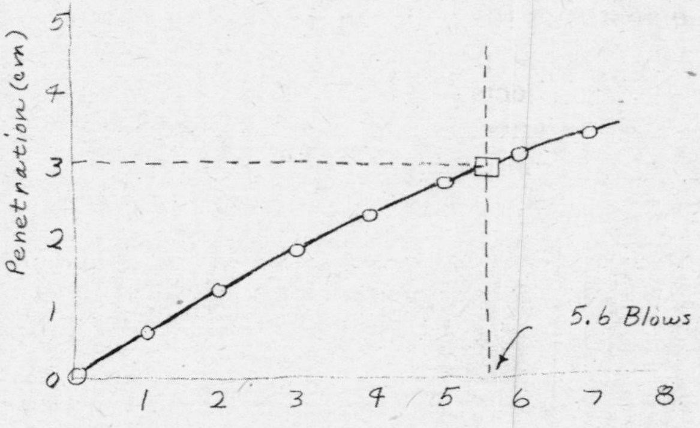

Fig. 3-95

Film Loop 32
GRAVITATIONAL POTENTIAL ENERGY

Introductory physics courses usually do not give a complete definition of potential energy because of the mathematics involved. Only particular kinds of potential energy, such as gravitational potential energy, are considered.

The expression for the *gravitational potential energy* of an object near the earth is the product of the weight of the object and its height. The height is measured from a location chosen arbitrarily as the zero level for potential energy. It is almost impossible to "test" a formula without other physics concepts. Here, a method of measuring energy is needed. The previous Film Loop 31, "A Method of Measuring Energy," demonstrated that the depth of penetration of a nail into wood, due to a blow, is a good measure of the energy at the moment of impact of the object.

Although you are concerned with potential energy, you will calculate it by first finding kinetic energy. Where there is no loss of energy through heat, the sum of the kinetic energy and potential energy is constant. If you measure potential energy from the point at which the weight strikes the nail, at the moment of striking all the energy will be kinetic energy. On the other hand, at the moment an object is released, the kinetic energy is zero, and all the energy is potential energy. These two must, by conservation of energy, be equal.

Since total energy is conserved, you can determine the initial potential energy that the object had from the depth of penetration of the nail by using the results of the measurement connecting energy and nail penetration.

Two types of measurements are possible with this film. The numbered scenes are all photographed from the same position. In the first scenes (Fig. 3-96), you can determine how gravitational potential energy depends upon weight. Objects of different masses fall from the same distance. Project the film on paper and measure the positions of the nailheads before and after the impact of the falling objects. Make a graph relating the penetration depth and the weight ma_g. Use the results of the previous experiment to convert this relation into a relation between gravitational potential energy and weight. What can you learn from this graph? What factors are you holding constant? What conclusions can you reach from your data?

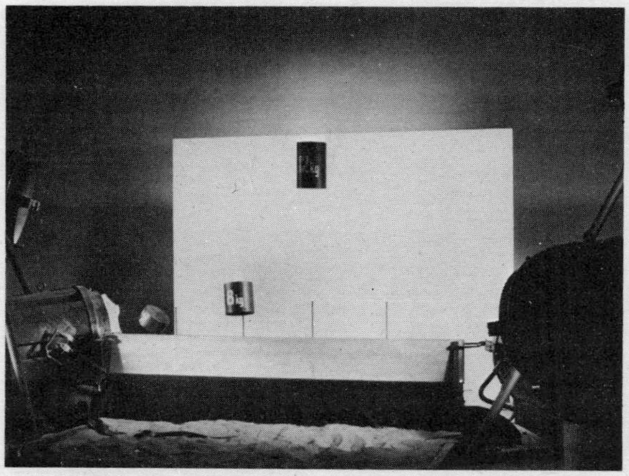

Fig. 3-96

Later scenes (Fig. 3-97) provide information for studying the relationship between gravitational potential energy and position. Bodies of equal mass are raised to different heights and allowed to fall. Study the relationship between the distance of fall and the gravitational potential energy. What graphs might be useful? What conclusion can you reach from your measurements?

Fig. 3-97

Can you relate the results of these measurements to statements in the Text concerning gravitational potential energy?

Film Loop 33
KINETIC ENERGY

In this film, you can test how kinetic energy *KE* depends on speed *v*. You will measure both *KE* and *v*, keeping the mass *m* constant.

The penetration of a nail driven into wood is a good measure of the work done on the nail, and thus is a measure of the energy lost by whatever object strikes the nail. The speed of the moving object can be measured in several ways.

The preliminary scenes show that the object falls on the nail. Only the speed just before the object strikes the nail is important. The scenes intended for measurement were photographed with the camera on its side, so the body appears to move horizontally toward the nail.

The speeds can be measured by timing the motion of the leading edge of the object as it moves from one reference mark to the other. The clock in the film (Fig. 3-98) is a disk that rotates at 3,000 revolutions per minute. Project the film on paper and mark the positions of the clock pointer when the body crosses each reference mark. The time is proportional to the angle through which the pointer turns. The speeds are proportional to the reciprocals of the times since the distance is the same in each case. Since you are testing only the *form* of the kinetic energy dependence on speed, any convenient unit can be used. Measure the speed for each of the five trials.

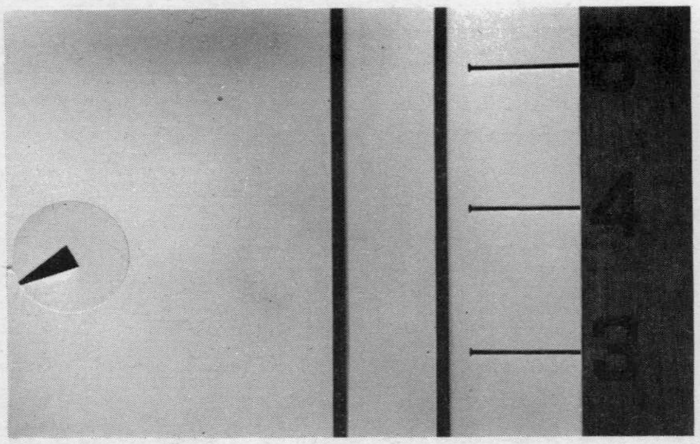

Fig. 3-98

The kinetic energy of the moving object is transformed into the work required to drive the nail into the wood. In Film Loop 31, you related the work to the distance of penetration. Measure the nail penetration for each trial, and use your results from the previous film.

How does *KE* depend on *v*? The conservation law derived from Newton's laws indicates that *KE* is proportional to v^2, the square of the speed, not proportional to *v*. Test this by making two graphs. In one graph, plot *KE* vertically and plot v^2 horizontally. For comparison, plot *KE* versus *v*. What can you conclude? Do you have any assurance that a similar relation will hold if the speeds or masses are very different from those found here? How might you go about determining this?

Film Loop 34
CONSERVATION OF ENERGY: POLE VAULT

This quantitative film can help you study conservation of energy. A pole vaulter (mass 68 kg, height 180 cm) is shown (first at normal speed and then in slow motion) clearing a bar at 3.45 m. You can measure the total energy of the system at two moments in time: (1) just before the jumper starts to rise and (2) part of the way up, when the pole is bent. The total energy of the system is constant, although it is divided differently at different times. Since it takes work to bend the pole, the pole has elastic potential energy when bent. This elastic energy comes from some of the kinetic energy the vaulter has as he runs horizontally before inserting the pole in the socket. Later, the elastic potential energy of the bent pole is transformed into some of the jumper's gravitational potential energy when he is at the top of the jump (Fig. 3-99).

Fig. 3-99

Position 1

The energy is entirely kinetic energy, $\frac{1}{2}mv^2$. To help you measure the runner's speed, successive frames are held as the runner moves past two markers 1 m apart. Each "freeze frame" represents a time interval of 1/250 sec, the camera speed. Find the runner's average speed over this meter, and then find the kinetic energy. If *m* is in kilograms and *v* is in meters per second, *E* will be in joules.

Position 2

The jumper's center of gravity is about 1.02 m above the soles of his feet. Three types of energy are involved at the intermediate positions. Use the stop-frame sequence to obtain the speed of the jumper. The seat of his pants can be used as a reference. Calculate the kinetic energy and gravitational potential energy as already described.

The work done in deforming the pole is stored as elastic potential energy. In the final scene, a chain windlass bends the pole to a shape similar to that which it assumed during the jump in Position 2. When the chain is shortened, work is done on the pole: work = (average force) × (displacement). During the cranking sequence, the force varied. The average force can be approximated by adding the initial and final values (found from the scale) and dividing by two. Convert this force to newtons. The displacement can be estimated from the number of times the crank handle is pulled. A close-up shows how far the chain moves during a single stroke. Calculate the work done to crank the pole into its distorted shape.

You now can add and find the total energy. How does this compare with the original kinetic energy?

Position 3

Gravitational potential energy is the work done to raise the jumper's center of gravity. From the given data, estimate the vertical rise of the center of gravity as the jumper moves from Position 1 to Position 3. (His center of gravity clears the bar by about 0.3 m.) Multiply this height by the jumper's weight to get potential energy. If weight is in newtons and height is in meters, the potential energy will be in joules. A small additional source of energy is in the jumper's muscles; judge for yourself how far he lifts his body by using his arm muscles as he nears the highest point. This is a small correction, so a relatively crude estimate will suffice. Perhaps he pulls with a force equal to his own weight through a vertical distance of 0.7 m.

How does the initial kinetic energy, plus the muscular energy expended in the pull-up, compare with the final gravitational potential energy? (An agreement to within about 10% is about as good as you can expect from a measurement of this type.)

Film Loop 35
CONSERVATION OF ENERGY: AIRCRAFT TAKEOFF

The pilot of a Cessna 150 (Fig. 3-100) holds the plane at constant speed in level flight, just above the surface of the runway. Then, keeping the throttle fixed, the pilot pulls back on the stick, and the plane begins to rise. With the same throttle setting, the plane levels off at about 100 m. At this altitude, the aircraft's speed is less than at ground level. You can use this film to make a crude test of energy conservation. The plane's initial speed was constant, indicating that the net force on it was zero. In terms of an approximation, air resistance remained the same after lift-off. How good is this approximation? What would you expect air resistance to depend on? When the plane rose, its gravitational potential energy increased, at the expense of the initial kinetic energy of the plane. At the upper level, the plane's kinetic energy is less, but its potential energy is greater. According to the principle of conservation of energy, the total energy (*KE* + *PE*) remained constant, assuming that air resistance and any other similar factors are neglected. But are these factors negligible? Here are the data concerning the film and the airplane:

Fig. 3-100

length of plane: 7.5 m
mass of plane: 550 kg
weight of plane:
550 kg × 9.8 m/sec^2 = 5400 N
camera speed: 45 frames/sec

Project the film on paper. Mark the length of the plane to calibrate distances.

Stop-frame photography allows you to measure the speed of 45 frames per second. In printing the measurement section of the film, only every third frame was used. Each of these frames was repeated ("stopped") a number of times, enough to allow time to mark a position on the screen. The effect is one of "holding" time, and then jumping one-fifteenth of a second.

Measure the speeds in all three situations, and also the heights above the ground. You have the data needed for calculating kinetic energy ($\frac{1}{2}mv^2$) and gravitational potential energy (ma_gh) at each of the three levels. Calculate the total energy at each of the three levels.

Can you make any comments concerning air resistance? Make a table showing (for each level) *KE, PE,* and *E* totals. Do your results substantiate the law of conservation of energy within experimental error?

Film Loop 36
REVERSIBILITY OF TIME

It may sound strange to speak of "reversing time." In the world of common experience, we have no control over time direction, in contrast to the many aspects of the world that we can modify. Yet physicists are much concerned with the reversibility of time; perhaps no other issue so clearly illustrates the imaginative and speculative nature of modern physics.

The camera provides a way of manipulating time. If you project movie film backwards, the events pictured happen in reverse time order. This film has sequences in both directions, some shown in their "natural" time order and some in reverse order.

The film concentrates on the motion of objects. Consider each scene from the standpoint of time direction: Is the scene being shown as it was taken, or is it being reversed and shown backward? Many sequences are paired, the same film being used in both time senses. Is it always clear which one is forward in time and which is backward? With what types of events is it difficult to tell the "natural" direction?

The Newtonian laws of motion do *not* depend on time direction. Any filmed motion of particles following strict Newtonian laws should look completely "natural" whether seen forward or backward. Since Newtonian laws are "invariant" under time reversal (changing the direction of time), you could not tell by examining a motion obeying these laws whether the sequence is forward or backward. Any motion that could occur forward in time can also occur, under suitable conditions, with the events in the opposite order.

With more complicated physical systems, with an extremely large number of particles, the situation changes. If ink were dropped into water, you would have no difficulty in determining which sequence was photographed forward in time and which backward. So certain physical phenomena at least *appear* to be irreversible, taking place in only one time direction. Are these processes *fundamentally* irreversible, or is this only a limitation on human powers? This is not an easy question to answer. It could still be considered, in spite of a 50-year history, a frontier problem.

Reversibility of time has been used in many ways in twentieth-century physics. For example, an interesting way of viewing the two kinds of charge in the universe, positive and negative, is to think of some particles as "moving" backward in time. Thus, if the electron is viewed as moving forward in time, the positron can be considered as exactly the same particle moving backward in time. This backward motion is equivalent to the forward-moving particle having the *opposite* charge! This was one of the keys to the development of the space–time view of quantum electrodynamics which R. P. Feynman described in his Nobel Prize lecture.

For a general introduction to time reversibility, see the Martin Gardner article, "Can Time Go Backward?" originally published in *Scientific American* (January, 1967).

Film Loop 37
SUPERPOSITION

Using this film, you will study an important physical idea, superposition. The film was made by photographing patterns displayed on the face of the cathode-ray tube (CRT) of an oscilloscope, similar to a television set. You may have such an oscilloscope in your laboratory.

Still photographs of some of these patterns appearing on the CRT screen are shown in Figs. 3-101 and 3-102. The two patterns at the top of the screen are called *sinusoidal*. They are not just any wavy lines, but lines generated in a precise fashion. If you are familiar with the sine and cosine functions, you will recognize them here. The sine function is the special case where the origin of the coordinate system is located where the function is zero and starting to rise. No origin is shown, so it is arbitrary as to whether one calls these sine curves, cosine curves, or some other sinusoidal curve. What physical situations might lead to curves of this type? (You might want to consult books about simple harmonic oscillators.) Here the curves are produced by electronic circuits that generate an electrical voltage changing in time so as to cause the curve to be displayed on the cathode-ray tube. The oscilloscope operator can adjust the magnitudes and phases of the two top functions.

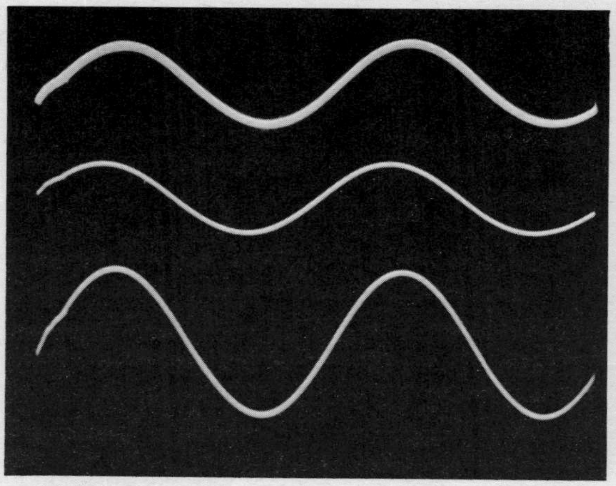
Fig. 3-101

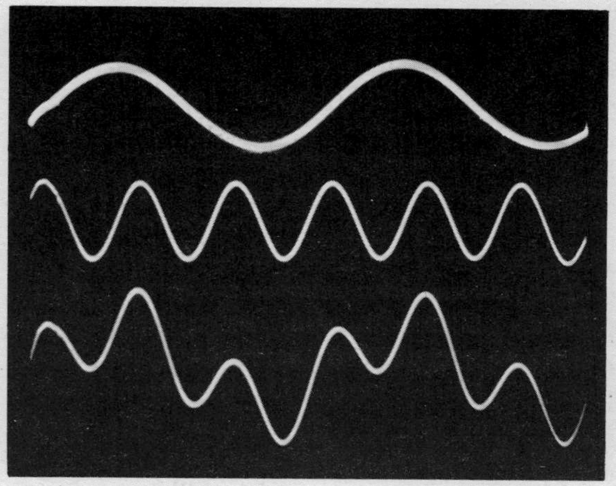
Fig. 3-104

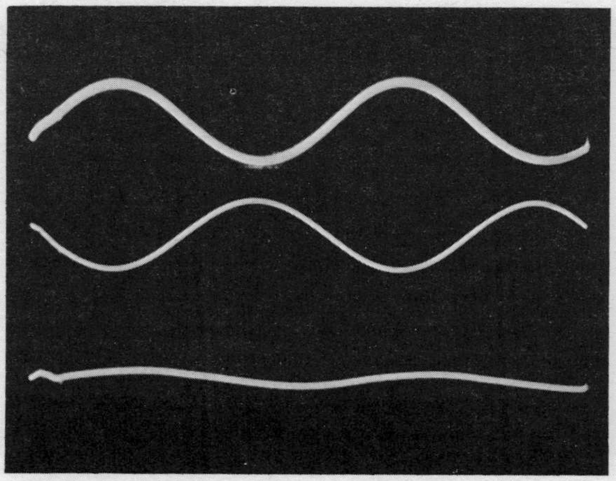
Fig. 3-102

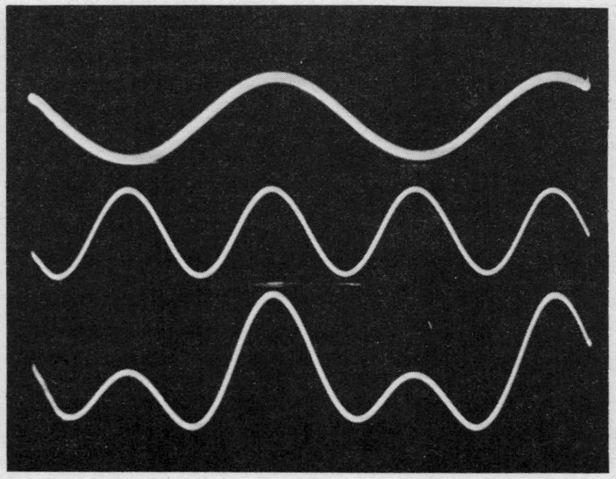
Fig. 3-105

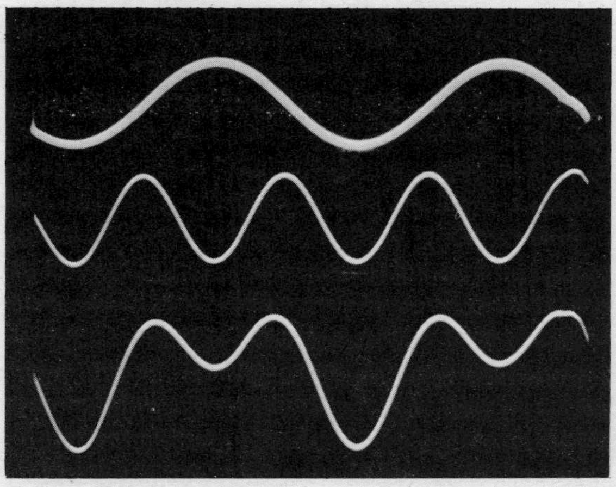
Fig. 3-103

The bottom curve is obtained by a point-by-point adding of the top curves. Imagine a horizontal axis going through each of the two top curves, and positive and negative distances measured vertically from this axis. The bottom curve is at each point the algebraic sum of the two points above it on the top curves, as measured from their respective axes. This point-by-point algebraic addition, when applied to actual waves, is called *superposition*.

Two cautions are necessary. First, you are not seeing waves, but *models* of waves. A wave is a disturbance that propagates in time, but, at least in some of the cases shown, there is no propagation. A model always has some limitations. Second, you should not think that all waves are sinusoidal. The form of whatever is propagating can be any shape. Sinusoidal waves constitute only one important class of waves. Another common wave is the pulse, such as a sound wave produced by a sharp blow on a table. The pulse is *not* a sinusoidal wave.

Several examples of superposition are shown in the film. If, as approximated in Fig. 3-101, two sinusoidal curves of equal period and amplitude are in phase, both having zeroes at the same places, the result is a double-sized function of the same shape. On the other hand, if the curves are combined out of phase, so that one has a positive displacement while the other one has an equal negative displacement, the result is *zero* at each point (Fig. 3-102). If functions of different periods are combined (Figs. 3-103 to 3-105), the result of the superposition is not sinusoidal, but more complex in shape. You are asked to interpret, both verbally and quantitatively, the superpositions shown in the film.

Film Loop 38
STANDING WAVES ON A STRING

Tension determines the speed of a wave traveling down a string. When a wave reaches a fixed end of a string, it is reflected back again. The reflected wave and the original wave are superimposed or added together. If the tension (and therefore the speed) is just right, the resulting wave will be a *standing wave*. Certain nodes will always stand still on the string. Other points on the string will continue to move in accordance with superposition. When the tension in a vibrating string is adjusted, standing waves appear when the tension has one of a set of "right" values.

In the film, one end of a string is attached to a tuning fork with a frequency of 72 vibrations per second. The other end is attached to a cylinder. The tension of the string is adjusted by sliding the cylinder back and forth.

Several standing wave patterns are shown. For example, in the third mode the string vibrates in three segments with two nodes (points of no motion) between the nodes at each end. The nodes are half a wavelength apart. Between the nodes are points of maximum possible vibration called antinodes.

You tune the strings of a violin or guitar by changing the tension on a string of fixed length, higher tension corresponding to higher pitch. Different notes are produced by placing a finger on the string to shorten the vibrating part. In this film, the frequency of vibration of a string is fixed, because the string is always driven at 72 vib/sec. When the frequency remains constant, the wavelength changes as the tension is adjusted because velocity depends on tension.

A high-speed snapshot of the string at any time would show its instantaneous shape. Sections of the string move, except at the nodes. The eye sees a blurred or "time-exposure" superposition of string shapes because of the frequency of the string. In the film, this blurred effect is reproduced by photographing at a slow rate; each frame is exposed for about 1/15 sec.

Some of the vibration modes are photographed by a stroboscopic method. If the string vibrates at 72 vib/sec and frames are exposed in the camera at the rate of 70 times per second, the string seems to go through its complete cycle of vibration at a slower frequency when projected at a normal speed. In this way, a slow-motion effect is shown.

Film Loop 39
STANDING WAVES IN A GAS

Standing waves are set up in air in a large glass tube (Fig. 3-106). The tube is closed at one end by an adjustable piston. A loudspeaker at the other end supplies the sound wave. The speaker is driven by a variable-frequency oscillator and amplifier. About 20 watts of audio-power are used, telling everyone in a large building that filming is in progress! The waves are reflected from the piston.

Fig. 3-106

A standing wave is formed when the frequency of the oscillator is adjusted to one of several discrete values. Most frequencies do *not* give standing waves. Resonance is indicated in each mode of vibration by nodes and antinodes. There is always a node at the fixed end (where air molecules cannot move) and an antinode at the speaker (where air is set into motion). Between the fixed end and the speaker there may be additional nodes and antinodes.

The patterns can be observed in several ways, two of which are used in the film. One method of making visible the presence of a stationary acoustic wave in the gas in the tube is to place cork dust along the tube. At resonance, the dust is blown into a cloud by the movement of air at the antinodes; the dust remains stationary at the nodes where the air is not moving. In the first part of the film, the dust shows standing wave patterns for various frequencies (Table 3-2).

Frequency (vib/sec)	Number of half-wavelengths
230	1.5
370	2.5
530	3.5
670	4.5
1900	12.5

The pattern for $f = 530$ is shown in Fig. 3-107. From node to node is $\frac{1}{2}\lambda$, and the length of the pipe is $3\lambda + \frac{1}{2}\lambda$ (the extra $\frac{1}{2}\lambda$ is from the speaker antinode to the first node). There are, generally, $(n + \frac{1}{2})$ half-wavelengths in the fixed length, so $\lambda \propto 1/(n + \frac{1}{2})$. Since $f \propto 1/\lambda$, $f \propto (n + \frac{1}{2})$. Divide each frequency in the table by $(n + \frac{1}{2})$ to find whether the result is reasonably constant.

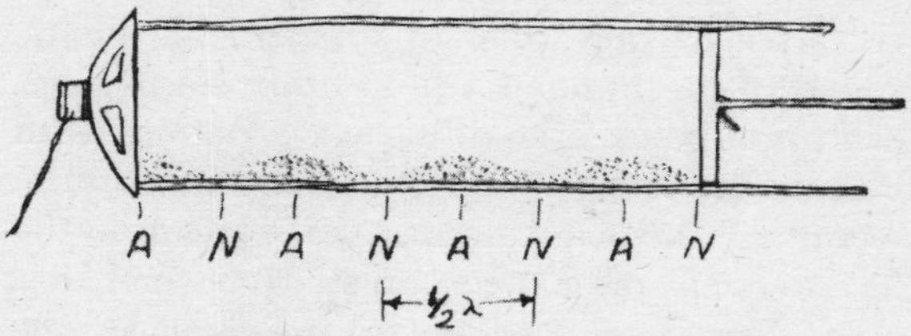

Fig. 3-107

In all modes, the dust remains motionless near the stationary piston that is a node.

In the second part of the film, nodes and antinodes are made visible by a different method. A wire is placed in the tube near the top. This wire is heated electrically to a dull red. When a standing wave is set up, the wire is cooled at the antinodes, because the air carries heat away from the wire when it is in vigorous motion. So the wire is cooled at the antinodes and glows less. The bright regions correspond to nodes where there are no air currents. The oscillator frequency is adjusted to give several standing wave patterns with successively smaller wavelengths. How many nodes and antinodes are there in each case? Can you find the frequency used in each case?

Film Loop 40
VIBRATIONS OF A WIRE

This film shows standing wave patterns in thin but stiff wires. The wave speed is determined by the wire's cross section and by the elastic constants of the metal. There is no external tension. Two shapes of wire, straight and circular, are used.

The wire passes between the poles of a strong magnet. When a switch is closed, a steady electric current from a battery is set up in one direction through the wire. The interaction of this current and the magnetic field leads to a downward force on the wire. When the direction of the current is reversed, the force on the wire is upward. Repeated rapid reversal of the current direction can make the wire vibrate up and down.

The battery is replaced by a source of variable-frequency alternating current whose frequency can be changed. When the frequency is adjusted to match one of the natural frequencies of the wire, a standing wave builds up. Several modes are shown, each excited by a different frequency.

The first scenes show a straight brass wire, 2.4 mm in diameter (Fig. 3-108). The "boundary conditions" for motion require that, in any mode, the fixed end of the wire is a node and the free end is an antinode. (A horizontal plastic rod is used to support the wire at another node.) The wire is photographed in two ways: (1) in a blurred "time exposure," as the eye sees it; and (2) in "slow motion," simulated through stroboscopic photography.

Fig. 3-108

Study the location of the nodes and antinodes in one of the higher modes of vibration. They are not equally spaced along the wire, as for vibrating string (see Film Loop 38). This is because the wire is stiff whereas the string is perfectly flexible.

In the second sequence, the wire is bent into a horizontal loop, supported at one point (Fig. 3-109). The boundary conditions require a node at this point; there can be additional nodes, equally spaced around the ring. Several modes are shown, both in "time exposure" and in "slow motion." To some extent, the vibrating circular wire is a helpful model for the wave behavior of an electron orbit in an atom such as hydrogen; the discrete modes correspond to discrete energy states for the atom.

Fig. 3-109

Film Loop 41
VIBRATIONS OF A RUBBER HOSE

You can generate standing waves in many physical systems. When a wave is set up in a medium, it is usually reflected at the boundaries. Characteristic patterns will be formed, depending on the shape of the medium, the frequency of the wave, and the material. At certain points or lines in these patterns, there is no vibration because all the partial waves passing through these points just manage to cancel each other through superposition (as you saw in the ripple tank).

Standing wave patterns only occur for certain frequencies. The physical process selects a *spectrum* of frequencies from all the possible ones. Often there are an infinite number of such discrete frequencies. Sometimes there are simple mathematical relations between the selected frequencies, but for other bodies the relations are more complex. Several films in this series show vibrating systems with such patterns.

This film uses a rubber hose, clamped at the top. Such a stationary point is called a *node*. The bottom of the stretched hose is attached to a motor whose speed is increased during the film. An eccentric arm attached to the motor shakes the bottom end of the hose. Thus, this end moves slightly, but this motion is so small that the bottom end also is a node. (See Fig. 3-110.)

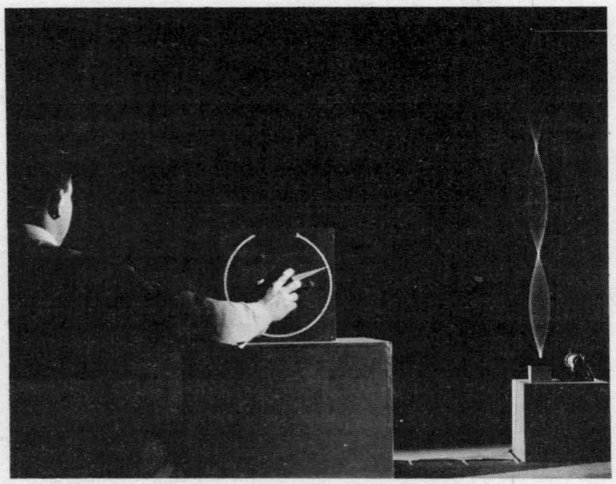

Fig. 3-110

The motor begins at a frequency below that for the first standing wave pattern. As the motor is gradually speeded up, the amplitude of the vibrations increases until a well-defined steady loop is formed between the nodes. This loop has its maximum motion at the center. The pattern is half a wavelength long. Increasing the speed of the motor leads to other harmonics, each one being a standing wave pattern with both nodes and antinodes, points of maximum vibration. These resonances can be seen in the film to occur only at certain sharp frequencies. For other motor frequencies, no such simple pattern is seen. You can count as many as 11 loops with the highest frequency case shown.

It would be interesting to have a sound track for this film. The sound of the motor is by no means constant during the process of increasing the frequency. The stationary resonance pattern corresponds to points where the motor is running much more quietly, because the motor does not need to "fight" against the hose. This sound distinction is particularly noticeable for the higher harmonics.

If you play a violin, cello, or other stringed instrument, you might ask how the harmonics observed in this film are related to musical properties of vibrating strings. What can be done with a violin string to change the frequency of vibration? What musical relation exists between two notes if one of them is twice the frequency of the other?

What would happen if you kept increasing the frequency of the motor? Would you expect to get arbitrarily high resonances, or would something "give"?

Film Loop 42
VIBRATIONS OF A DRUM

The standing wave patterns in this film are formed in a stretched, circular rubber membrane driven by a loudspeaker (see Fig. 3-111). The loudspeaker is fed large amounts of power, about 30 watts — more power than you would want to use with a television set or phonograph. The frequency of the sound can be changed electronically. The lines drawn on the membrane make it easier for you to see the patterns.

Fig. 3-111

The rim of the drum cannot move, so in all cases it must be a nodal circle — a circle that does not move as the waves bounce back and forth on the drum. By operating the camera at a frequency only slightly different from the resonant frequency, a stroboscopic effect enables you to see the rapid vibrations in slow motion.

In the first part of the film, the loudspeaker is directly under the membrane and the vibratory patterns are symmetrical. In the fundamental harmonic, the membrane rises and falls as a whole. At a higher frequency, a second circular node shows up between the center and the rim.

In the second part of the film, the speaker is placed to one side so that a different set of modes is generated in the membrane. You can see an asymmetrical mode where there is a node along the diameter, with a hill on one side and a valley on the other.

Various symmetric and asymmetric vibration modes are shown. Describe each mode, identifying the nodal lines and circles.

In contrast to the one-dimensional hose in Film Loop 41, there is no *simple* relation of the resonant frequencies for this two-dimensional system. The frequencies are *not* integral multiples of any basic frequency. The relation between values in the frequency spectrum is more complex than that for the hose.

Film Loop 43
VIBRATIONS OF A METAL PLATE

The physical system in this film is a square metal plate (see Fig. 3-112). The various vibrational modes are produced by a loudspeaker, as with the vibrating membrane in Film Loop 42. The metal plate is clamped at the center, so that point is always a node for each of the standing wave patterns. Because this is a stiff metal plate, the vibrations are too slight in amplitude to be seen directly. The trick used to make the patterns visible is to sprinkle sand along the plates. This sand moves away from the parts of the plates that are in rapid motion, and tends to fall along the nodal lines, which are not moving. The beautiful patterns of sand are known as Chladni figures. These patterns have often been much admired by artists. These and similar patterns are also formed when a metal plate is caused to vibrate by means of a violin bow, as seen at the end of this

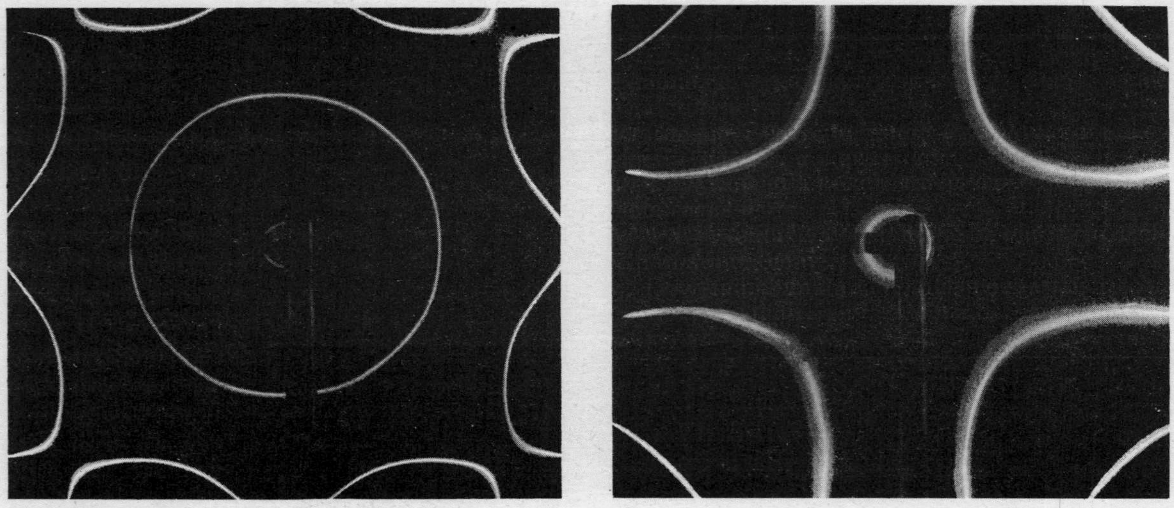

Fig. 3-112

film and in the Activity, "Standing Waves on a Drum and a Violin."

Not all frequencies will lead to stable patterns. As in the case of the drum, these harmonic frequencies for the metal plate obey complex mathematical relations, rather than the simple arithmetic progression seen in a one-dimensional string; but, again, they are discrete events. Only at certain well-defined frequencies are these patterns produced.

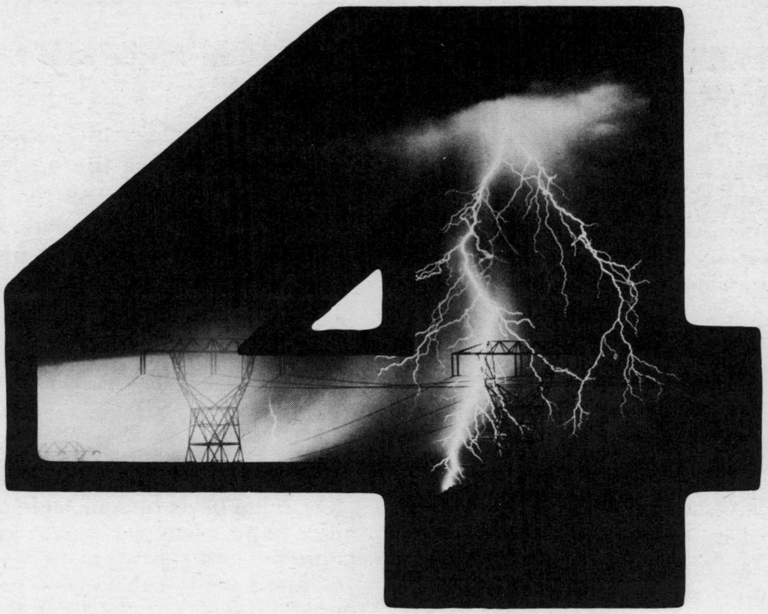

Light and Electromagnetism

EXPERIMENTS

Experiment 4-1
REFRACTION OF A LIGHT BEAM

You can easily demonstrate the behavior of a light beam as it passes from one transparent material to another. All you need is a semicircular plastic dish, a lens, a small light source, and a cardboard tube. The light source from the Millikan apparatus (Unit 5) and the telescope tube with objective lens (Units 1 and 2) will serve nicely.

Making a Beam Projector

To begin with, slide the Millikan apparatus light source over the end of the telescope tube (Fig. 4-1). When you have adjusted the bulb–lens distance to produce a parallel beam of light, the beam will form a spot of constant size on a sheet of paper moved toward and away from it by as much as 50 cm.

Make a thin flat light beam by sticking two pieces of black tape, about 1 mm apart, over the lens end of the tube, creating a slit. Rotate the bulb filament until it is parallel to the slit.

When this beam projector is pointed slightly downward at a flat surface, a thin path of light falls across the surface. By directing the beam into a plastic dish filled with water, you can observe the path of the beam emerging into the air. The beam direction can be measured precisely by placing protractors inside and outside the dish, or by placing the dish on a sheet of polar graph paper (Fig. 4-2).

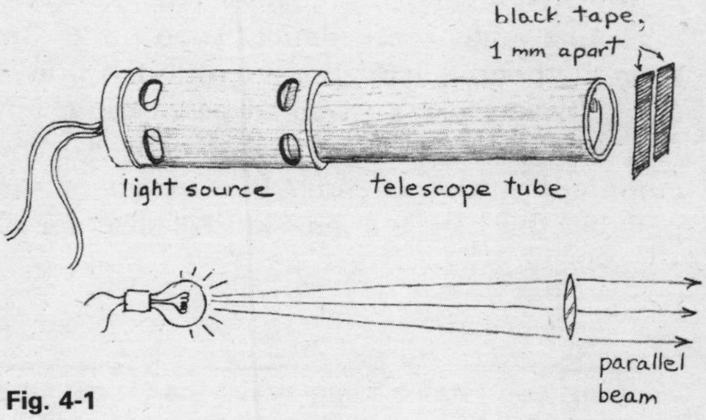

Fig. 4-1

Fig. 4-2

Behavior of a Light Beam at the Boundary between Two Media

Direct the beam at the center of the flat side of the dish, keeping the slit vertical. Tilt the projector until you can see the path of light both before it reaches the dish and after it leaves the other side.

To describe the behavior of the beam, you need a convenient way of referring to the angle the beam makes with the boundary. In physics, the system of measuring angles relative to a surface assigns a value of 0° to the perpendicular, or straight-in, direction. The angle at which a beam strikes a surface is called *angle of incidence* (θ_i); it is the number of degrees away from the straight-in direction. Similarly, the angle at which a refracted beam leaves the boundary is called the *angle of refraction* (θ_r). It is measured as the deviation from the straight-out direction (Fig. 4-3).

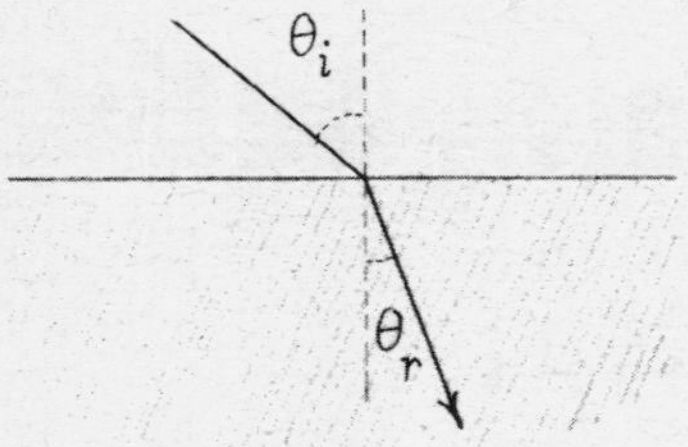

Fig. 4-3

Note the direction of the refracted beam for a particular angle of incidence. Then direct the beam perpendicularly into the rounded side of the dish where the refracted beam came out (Fig. 4-4). At what angle does the beam now come out on the flat side? Does reversing the path like this have the same kind of effect for all angles?

?

1. Can you state a general rule about the passage of light beams through the medium?

2. What happens to the light beam when it reaches the edge of the container along a radius?

Change the angle of incidence and observe how the angles of the reflected and refracted beams change. (It may be easiest to leave the projector supported in one place and to rotate the sheet of paper on which the dish rests.) You will see that the angle of the *reflected* beam is always equal to the angle of the incident beam, but the angle of the *refracted* beam is not related to the angle of incidence in so simple a fashion.

Refraction Angle and Change in Speed

Change the angle of incidence in 5° steps from 0° to 85°, recording the angle of the refracted beam for each step. As the angles in air get larger, the beam in the water begins to spread, so it becomes more difficult to measure its direction precisely. You can avoid this difficulty by directing the beam into the round side of the dish instead of into the flat side. This will give the same result since, as you have seen, the light path is reversible.

?

3. On the basis of your table of values, does the angle in air seem to increase in proportion to the angle in water?

4. Make a plot of the angle in air against the angle in water. How would you describe the relation between the angles?

According to both the simple wave and simple particle models of light, it is not the ratio of angles in two media that will be constant, but the ratio of the *sines* of the angles. Add two columns to your data table and, referring to a table of the sine function, record the sines of the angles you observed. Then plot the sine of the angle in water against the sine of the angle in air.

?

5. Do your results support the prediction made from the models?

6. Write an equation that describes the relationship between the angles.

According to the wave model, the ratio of the sines of the angles in two media is the same as the ratio of the light speeds in the two media.

?

7. According to the wave model, what do your results indicate is the speed of light in water?

Color Differences

You have probably observed in this experiment that different colors of light are not refracted by the same amount. (This effect is called *dispersion*.) This is most noticeable when you direct the beam into the round side of the dish, at an angle such that the refracted beam leaving the flat side lies very close to the flat side. The different colors of light making up the white beam separate quite distinctly.

?

8. What color of light is refracted most?
9. Using the relation between sines and speeds, estimate the difference in the speeds of different colors of light in water.

Other Phenomena

In the course of your observations you probably have observed that for some angles of incidence, no refracted beam appears on the other side of the boundary. This phenomenon is called *total internal reflection*.

?

10. When does total internal reflection occur?

By immersing blocks of glass or plastic in the water, you can observe what heppens to the beam in passing between these media and water. (Liquids other than water can be used, but be sure you do not use one that will dissolve the plastic dish!) If you lower a smaller transparent container upside-down into the water so as to trap air in it, you can observe what happens at another water– air boundary (Fig. 4-5). A round container so placed will show what effect an air bubble in water has on light.

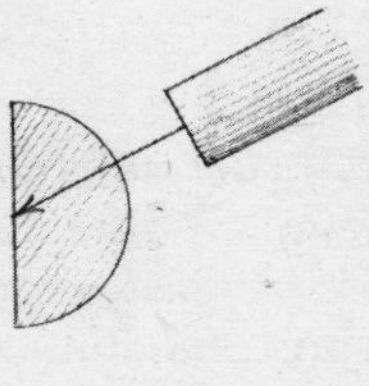

Fig. 4-4

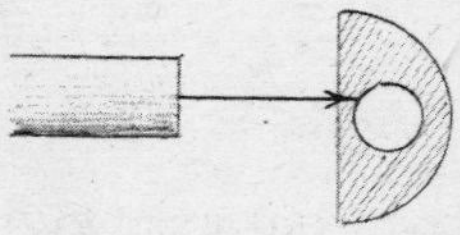

Fig. 4-5

?

11. Before trying this last suggestion, make a sketch of what you believe will happen. If your prediction is wrong, explain what happened.

Experiment 4-2
YOUNG'S EXPERIMENT: THE WAVELENGTH OF LIGHT

You have seen how ripples on a water surface are diffracted, spreading out after having passed through an opening. You have also seen wave interference when ripples, spreading out from two sources, reinforce each other at some places and cancel out at others.

Sound and ultrasound waves behave like water waves. These diffraction and interference effects are characteristic of all wave motions. If light has a wave nature, must it not also show diffraction and interference effects?

You may shake your head when you think about this. If light is diffracted, this must mean that light spreads around corners. But you learned in Unit 4 that "light travels in straight lines." How can light both spread around corners and move in straight lines?

Simple Tests of Light Waves

Have you ever noticed light spreading out after passing through an opening or around an obstacle? Try this simple test: Look at a narrow light source several meters away from you. (A straight-filament lamp is best, but a single fluorescent tube far away will do.) Hold two fingers in front of one eye and parallel to the light source. Look at the light through the gap between your fingers (Fig. 4-6). Slowly squeeze your fingers together to decrease the width of the gap. What do you see? What happens to the light as you reduce the gap between your fingers to a very narrow slit?

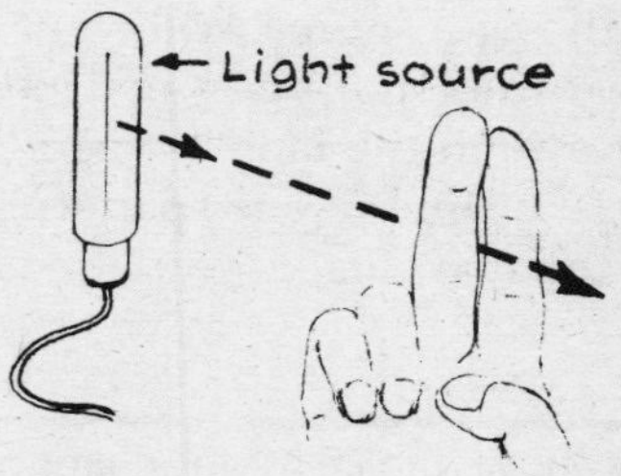

Fig. 4-6

Evidently light *can* spread out in passing through a very narrow opening between your fingers. For the effect to be noticeable, the opening must be small in comparison to the wavelength. The opening must be much smaller than those used in the ripple tank, in the case of light, or with sound waves. This suggests that light is a wave, but that it has a much shorter wavelength than the ripples on water, or sound or ultrasound in the air.

Do light waves show interference? Your immediate answer might be "no." Have you ever seen dark areas formed by the cancellation of light waves from two sources?

As with diffraction, to see interference you must arrange for the light sources to be small and close to each other. A dark photographic negative with two clear lines or slits running across it works very well. Hold up this film in front of one eye with the slits parallel to a narrow light source. What evidence do you see of interference in the light coming from the two slits?

Two-Slit Interference Pattern

To examine this interference pattern of light in more detail, fasten the film with the double slit on the end of a cardboard tube, such as the telescope tube without the lens. Make sure that the end of the tube is "light-tight," except for the two slits. (It helps to cover most of the film with black tape.) Stick a piece of translucent "frosted" tape over the end of a narrower tube that fits snugly inside the first one. Insert this end into the wider tube, as shown in Fig. 4-7.

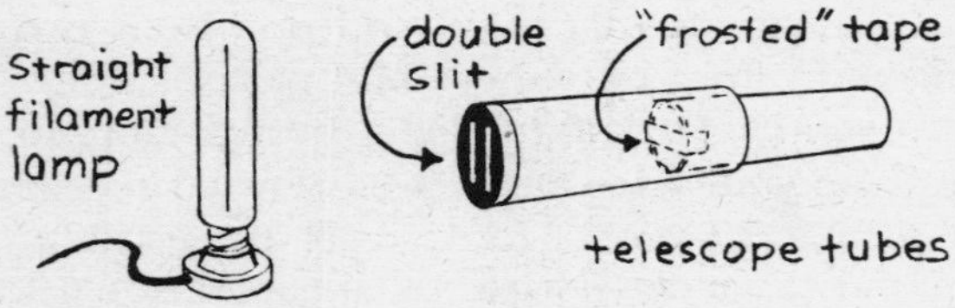

Fig. 4-7

Set up your double tube at least 1.5 m away from the narrow light source with the slits parallel to the light source. With your eye about 30 cm away from the open end of the tube, focus your eye on the tape "screen." There on the screen is the interference pattern formed by light from the two slits.

?

1. Describe how the pattern changes as you move the screen farther away from the slits.
2. Try putting different colored filters in front of the double slits. What are the differences between the pattern formed in blue light and the patterns formed in red or yellow light?

Measurement of Wavelength

Remove the translucent tape screen from the inside end of the narrow tube. Insert a magnifying eyepiece and scale unit in the end toward your eye and look through it at the light. (See Fig. 4-8.) What you see is a magnified view of the interference pattern in the plane of the scale. Try changing the distance between the eyepiece and the double slits.

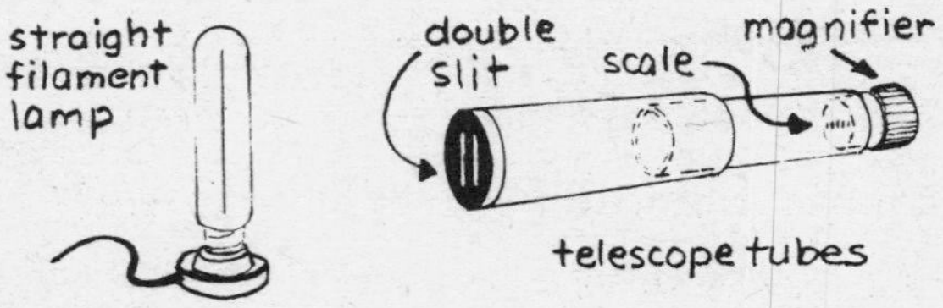

Fig. 4-8

In an earlier experiment, you calculated the wavelength of sound from the relationship

$$\lambda = \frac{x}{l}d$$

The relationship was derived for water waves from two in-phase sources, but the mathematics is the same for any kind of wave. (The use of two closely spaced slits gives a reasonably good approximation of in-phase sources.) See Fig. 4-9.

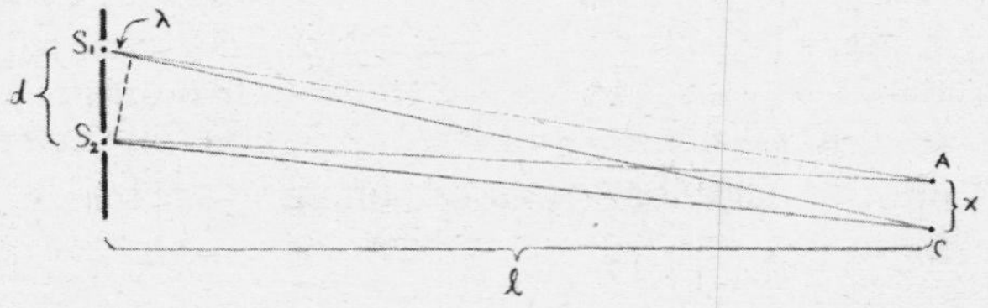

Fig. 4-9

Use the formula to find the wavelength of the light transmitted by the different colored filters. To do so, measure x, the distance between neighboring dark fringes, with the measuring magnifier (Fig. 4-10). (Remember that the smallest divisions on the scale are 0.1 mm.) You can also use the magnifier to measure d, the distance between the two slits. Place the film against the scale and then hold the film up to

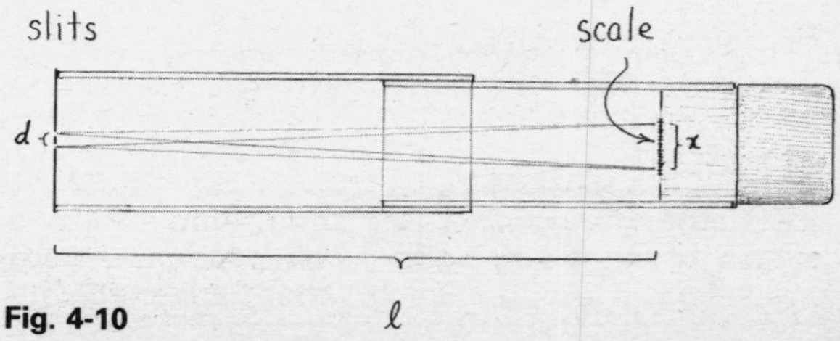

Fig. 4-10

the light. In the drawing, l is the distance from the slits to the plane of the pattern you measure.

The speed of light in air is approximately 3×10^8 m/sec. Use your measured values of wavelength to calculate the approximate light *frequencies* for each of the colors you used.

?

3. Could you use the method of "standing waves" (Experiment 3-18, "Sound") to measure the wavelength of light? Why?
4. Is there a contradiction between the statement, "Light consists of waves" and the statement, "Light travels in straight lines"?
5. Can you think of a common experience in which the wave nature of light is noticeable?

Suggestions for Some More Experiments

1. Examine light diffracted by a circular hole instead of by a narrow slit. The light source should now be a small point, such as a distant flashlight bulb. Look also for the interference effect with light that passes through two small circular sources (pinholes in a card) instead of the two narrow slits. (Thomas Young used circular openings rather than slits in his original experiment in 1802.)
2. Look for the diffraction of light by an obstacle. For example, use straight wires of various diameters, parallel to a narrow light source. Or use circular objects such as tiny spheres, the head of a pin, etc., and a point source of light. You can use either method of observation: the translucent tape screen or the magnifier. You may have to hold the magnifier fairly close to the diffracting obstacle.
3. Try some of the experiments listed under "Two-Slit Interference Pattern," using four or six slits instead of two. Note any differences in results.

Instructions on how to photograph some of these effects are in the activities that follow.

Experiment 4-3
ELECTRIC FORCES. I

If you walk across a carpet on a dry day and then touch a metal doorknob, a spark may jump across between your fingers and the knob. Your hair may crackle as you comb it (see Fig. 4-11). You have probably noticed other examples of the electrical effect of rubbing two objects together. Does your hair ever stand on end after you pull a sweater over your head?

Fig. 4-11

(This effect is particularly strong if the sweater is made of a synthetic fiber.)

Small pieces of paper are attracted to a plastic comb or ruler that has been rubbed on a piece of cloth. Try it. The attractive force is often large enough to lift scraps of paper off the table, showing that the attractive force is stronger than the gravitational force between the paper and the entire earth!

The force between the rubbed plastic and the paper is an electrical force, one of the four basic forces of nature.

In this experiment, you will make some observations of the nature of the electrical force. If you do the next experiment, "Electric Forces. II," you will make quantitative measurements of the force.

Forces between Electrified Objects

Stick a 20-cm length of transparent tape to the tabletop. Press the tape down well with your finger leaving about 2.5 cm loose as a "handle." Carefully remove the tape from the table by pulling on this loose end, preventing the tape from curling up around your fingers.

To test whether or not the tape became electrically charged when you stripped it from the table, see if the nonsticky side will pick up a scrap of paper. Even better, will the paper jump up from the table to the tape? Is the tape charged? Is the paper charged?

So far, you have considered only the effect of a charged object (the tape) on an uncharged object (the scrap of paper). What effect does a charged object have on another charged object? Here is one way to test it.

Charge a piece of tape by sticking it to the table and peeling it off as you did before. Suspend the tape from a horizontal wooden rod, or over the edge of the table. (Do not let the lower end curl around the table legs.)

Now charge a second strip of tape in the same way and bring it close to the first one. Have the two nonsticky sides facing (see Fig. 4-12).

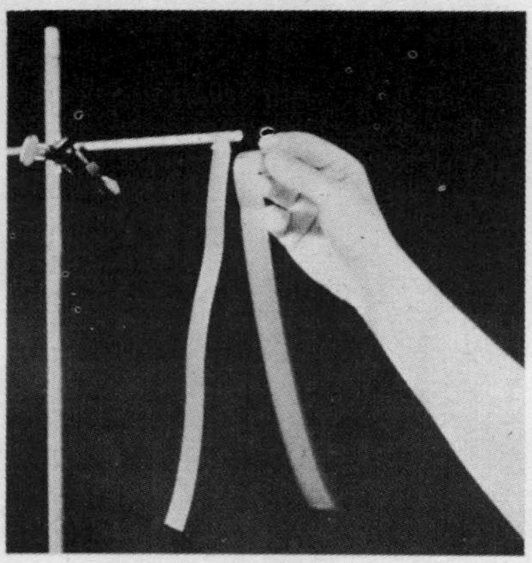

Fig. 4-12

Do the two tapes affect each other? Is the force between the tapes attractive or repulsive?

Hang the second tape about 10 cm away from the first one. Proceed as before and electrify a third piece of tape. Observe the reaction between this tape and your first two tapes. Record all your observations. Leave only the first tape hanging from its support; you will need it again shortly. Discard the other two tapes.

Stick a new piece of tape (A) on the table and stick a second tape (B) over it. Press them down well. Peel the stuck-together tapes from the table. To remove the net charge the pair will have picked up, run the nonsticky side of the pair over a water pipe or your lips. Check the pair with the original test strip to be sure the pair is electrically neutral. Now carefully pull the two tapes apart.

?

1. As you separated the tapes did you notice any interaction between them (other than that due to the adhesive)?
2. Hold one of these tapes in each hand and bring them slowly towards each other (nonsticky sides facing). What do you observe?
3. Bring first one, then the other of the tapes near the original test strip. What happens?

Mount A and B on the rod or table edge to serve as test strips. If you have rods of plastic, glass, or rubber available, or a plastic comb, ruler, etc., charge each one in turn by rubbing on cloth or fur. Bring the rod or comb close to A and then B.

Although you cannot prove it from the results of a limited number of experiments, there seem to be only two classes of electrified objects. No one has ever produced an electrified object that either attracts or repels *both A and B* (where A and B are themselves electrified objects). The two classes are called *positive* (+) and *negative* (−). Write a general statement summarizing how all members of the same class behave with each other (attract, repel, or remain unaffected by) and with all members of the other class.

A Puzzle

Your system of two classes of electrified objects was based on observations of the way charged objects interact. How can you account for the fact that a charged object (like a rubbed comb) will attract an *un*charged object (like a scrap of paper)?

?

4. Is the force between a charged body (either + or −) and an uncharged body always attractive, always repulsive, or is it sometimes one, sometimes the other?
5. Can you explain how a force arises between charged and uncharged bodies and why it is always the way it is? The clue here is the fact that the negative charges can move about slightly, even in materials called nonconductors, like plastic and paper.

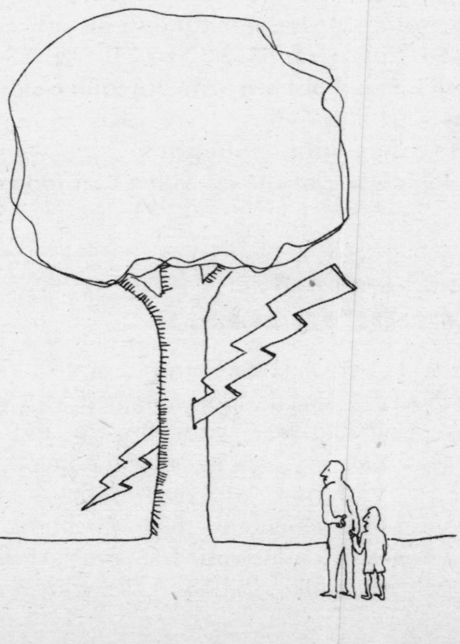

Fig. 4-13

Experiment 4-4
ELECTRIC FORCES. II: COULOMB'S LAW

You have seen that electrically charged objects exert forces on each other, but so far your observations have been qualitative; you have observed but not measured. In this experiment, you will find out how the amount of electrical force between two charged bodies depends on the amounts of electrical charges and on the separation of the bodies. In addition, you will experience some of the difficulties in using sensitive equipment.

The electric forces between charges that you can conveniently produce in a laboratory are small. To measure them at all requires a sensitive balance, good technique, and a day that is not too humid (otherwise the charges leak off too rapidly).

Constructing the Balance

(If your balance is already assembled, you need not read this section; go on to "Using the Balance.") A satisfactory balance is shown in Fig. 4-13.

Coat a small Styrofoam ball with a conducting paint and fix it to the end of a plastic sliver or toothpick by sticking the pointed end of the sliver into the ball. Since it is very important that the plastic be clean and dry (to reduce leakage of charge along the surface), *handle the plastic slivers as little as possible, and then only with clean, dry fingers.* Push the sliver into one end of a soda straw leaving at least 2.5 cm of plastic exposed, as shown in Fig. 4-14.

Next, fill the plastic support for the balance with glycerin, oil, or some other liquid. Cut a *shallow* notch in the top of the straw about 2 cm from the axle on the side away from the sphere (see Fig. 4-14).

Locate the balance point of the straw, ball, and sliver unit. Push a pin through the straw at this point to form an axle. Push a second pin through the straw directly in front of the axle and perpendicular to it. (As the straw rocks back and forth, this pin moves through the fluid in the support tube. Friction with the fluid reduces the swings of the balance.) Place the straw on the support, the pin hanging down into the liquid. Now, adjust the balance by sliding the plastic sliver slightly in or out of the straw, until the straw rests horizontally. If necessary, stick small bits of tape to the straw to make it balance. Make sure the balance can swing freely while making this adjustment.

Finally, cut five or six small, equal lengths of thin, bare wire (such as #30 copper). Each should be about 2 cm long, and they *must all be as close to the same length as you can make them.* Bend them into small hooks (Fig. 4-14) that can be hung over the notch in the straw or hung from each other. These are your "weights."

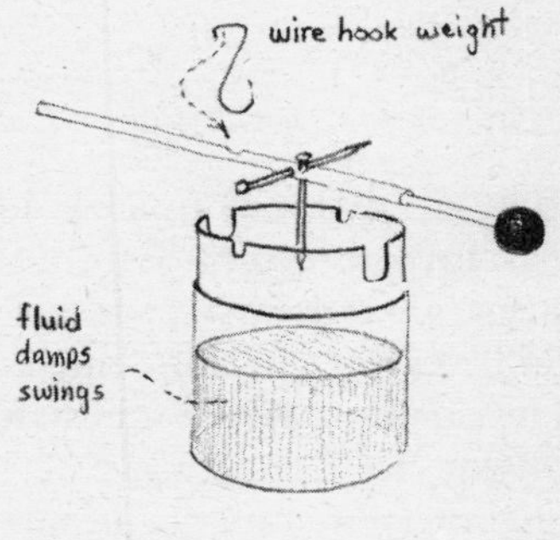

Fig. 4-14

Mount another coated ball on a pointed, plastic sliver and fix it in a clamp on a ring stand, as shown in Fig. 4-13.

Using the Balance

Charge both balls by wiping them with a rubbed plastic comb or ruler. Then bring the ring-stand ball down toward the balance ball.

?

1. What evidence have you that there is a force between the two balls?
2. Is the force due to the charges?
3. Can you compare the size of the electrical force between the two balls with the size of gravitational force between them?

Your balance is now ready, but in order to do the experiment, you need to solve two technical problems. During the experiment you will adjust the position of the ring-stand ball so that the force between the charged balls is balanced by the wire weights. The straw will then be horizontal. First, therefore, you must check quickly to be sure that the straw *is* balanced horizontally each time. Second, *measure* the distance between the centers of the two balls. You cannot put a ruler near the charged balls, or its presence will affect your results; however, if the ruler is not close to the balls, it is very difficult to make the measurement accurately.

Here is a way to make the measurement. With the balance in its horizontal position, you can record its balanced position with a mark on a folded card placed near the end of the straw (at least 5 cm away from the charges). (See Fig. 4-13.)

How can you avoid the parallax problem? Try to devise a method for measuring the distance between the centers of the balls. Ask your instructor if you cannot think of one.

You are now ready to make measurements to see how the force between the two balls depends on their separation and on their charges.

Doing the Experiment

From now on, work as quickly as possible but move carefully to avoid disturbing the balance or creating air currents. It is not necessary to wait for the straw to stop moving before you record its position. When it is swinging slightly, but *equally*, to either side of the balanced position, you can consider it balanced.

Charge both balls, touch them together briefly, and move the ring-stand ball until the straw is returned to the balanced position. The weight of one hook now balances the electric force between the charged spheres at this separation. Record the distance between the balls.

Without recharging the balls, add a second hook and readjust the system until balance is again restored. Record this new separation. Repeat until you have used all the hooks; do not reduce the air space between the balls to less than 0.5 cm. Then quickly retrace your steps by removing one hook (or more) at a time and raising the ring-stand ball each time to restore balance.

?

4. The separations recorded on the "return trip" may not agree with your previous measurements with this same number of hooks. If they do not, can you suggest a reason why?
5. Why must you not recharge the balls between one reading and the next?

Interpreting Your Results

Make a graph of your measurements of force F against separation d between centers. Clearly F and d are inversely related; that is, F increases as d decreases. You can go further to find the relationship between F and d. For example, it might be $F \propto 1/d$, $F \propto 1/d^2$, or $F \propto 1/d^3$, etc.

?

6. How would you test which of these relationships best represents your results?
7. What relationship do you find between F and d?

Further Investigation

In this experiment, you can determine how the force F varies with the charges on the spheres, when d is kept constant.

Charge both balls and then touch them together briefly. Since they are nearly identical, it is assumed that when touched, they will share the total charge almost equally.

Hang four hooks on the balance and move the ring-stand ball until the straw is in the balanced position. Note this position.

Touch the upper ball with your finger to discharge the ball. If the two balls are again brought into contact, the charge left on the balance ball will be shared equally between the two balls.

?

8. What is the charge on each ball now (as a fraction of the original charge)?

Return the ring-stand ball to its previous position; how many hooks must you remove to restore the balance?

?

9. Can you state this result as a mathematical relationship between quantity of charge and magnitude of force?

10. Consider why you had to follow two precautions in doing the experiment:

(a) Why can a ruler placed too close to the charge affect results?

(b) Why should you get the spheres no closer than about 0.5 cm?

11. How might you modify this experiment to see if Newton's third law applies to these electric forces?

Experiment 4-5
FORCES ON CURRENTS

If you did Experiment 4-4, you used a simple but sensitive balance to investigate how the electric force between two charged bodies depends on the distance between them and on the amount of charge. In this and the next experiment you will examine a related effect: the force between *moving* charges, that is, between electric currents. You will investigate the effect of the magnitudes and the directions of the currents. Before starting the experiment, you should have read the description of Oersted's and Ampère's work (Text Sections 14.11 and 14.12).

The apparatus for these experiments (like that in Fig. 4-15) is similar in principle to the

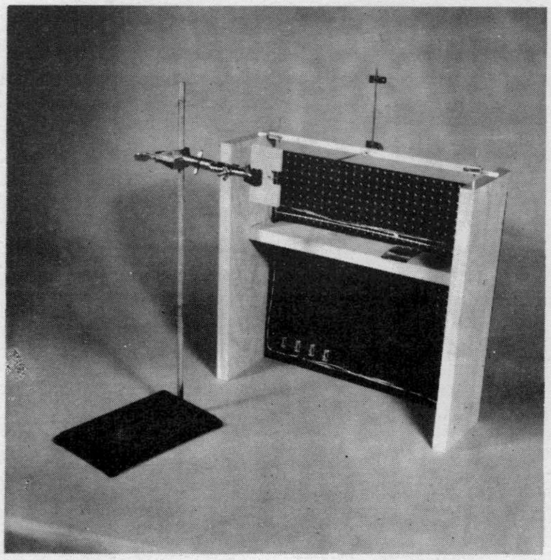

Fig. 4-15

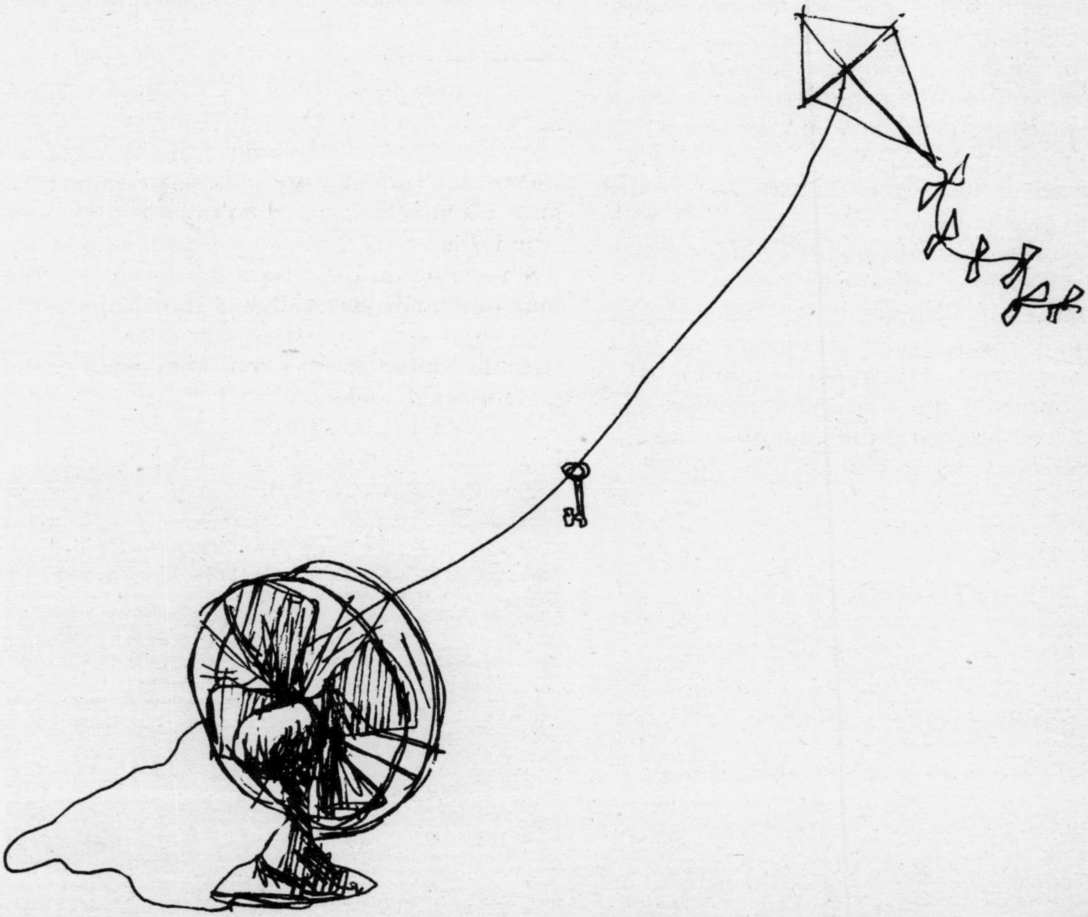

balance apparatus you used to measure electric forces. The current balance measures the force on a horizontal rod suspended so that it is free to move sideways at right angles to its length. You can study the forces exerted by a magnetic field on a current by bringing a magnet up to this rod while there is a current in it. A force on the current-carrying rod causes it to swing away from its original position.

You can also pass a current through a fixed wire parallel to the pivoted rod. Any force exerted on the rod by the current in the fixed wire will again cause the pivoted rod to move. You can measure these forces simply by measuring the counterforce needed to return the rod to its original position.

Adjusting the Current Balance

This instrument is more complicated than most of those you have worked with so far. Therefore, it is worthwhile spending a little time getting to know how the instrument operates before you start taking readings.

1. You have three or four light metal rods bent into └─┘ or ⌐└┘¬ shapes. These are the movable "loops." Set up the balance with the longest loop clipped to the pivoted horizontal bar. Adjust the loop so that the horizontal part of the loop hangs level with the bundle of wires (the fixed coil) on the pegboard frame. Adjust the balance on the frame so that the loop and coil are parallel as you look down at them. They should be at least 5 cm apart. Make sure the loop swings freely.

2. Adjust the "counterweight" cylinder to balance the system so that the long pointer arm is approximately horizontal. Mount the ⊟-shaped plate (zero-mark indicator) in a clamp and position the plate so that the zero line is opposite the horizontal pointer (Fig. 4-16). (If you are using the equipment for the first time, draw the zero-index line yourself.)

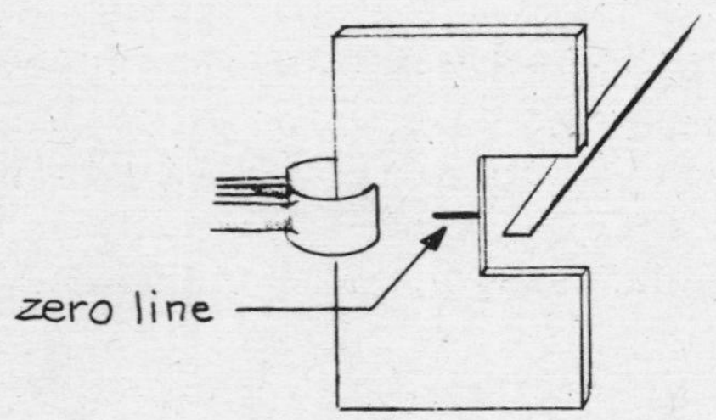

Fig. 4-16 Set the zero mark level with the pointer when there is current in the balance loop and no current in the fixed coil.

3. Now set the balance for maximum sensitivity. To do this, move the sensitivity clip up the vertical rod (Fig. 4-17) until the loop slowly swings back and forth. These oscillations may take as much as 4 or 5 sec per swing. If the clip is raised too far, the balance may become unstable and flop to either side without "righting" itself.

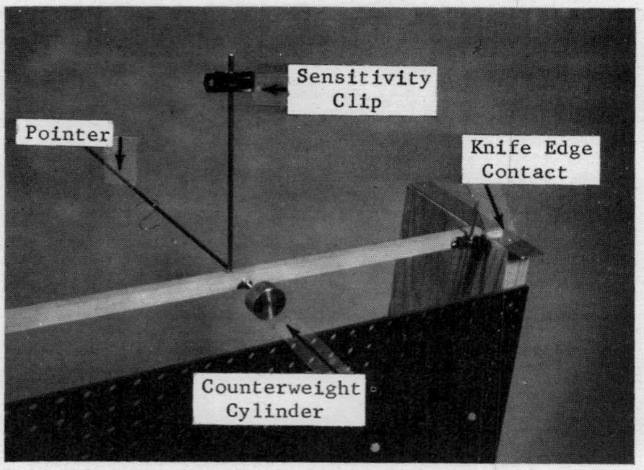

Fig. 4-17

4. Make sure that the pivots (knife-edge contacts) are clean and shiny (use fine-grade abrasive paper), and remain clean throughout the experiment; otherwise, they will not let the current pass reliably. Now connect a 6 V/5 A max power supply that can supply up to 5 A through an ammeter to one of the flat horizontal plates on which the pivots rest. Connect the other plate to the second terminal of the power supply (Fig. 4-18). To limit the current and keep it from tripping the circuit breaker, it may be necessary to put one or two 1-Ω resistors in the circuit. (If your power supply does not have variable control, it should be connected to the plate through a rheostat.)

Fig. 4-18

5. Set the variable control for minimum current, and turn on the power supply. If the ammeter deflects the wrong way, interchange the leads to it. Slowly increase the current to about 4.5 A.

6. Now bring a small magnet close to the pivoted conductor.

?

1. How must the magnet be placed to produce the greatest effect on the rod? What determines the direction in which the rod swings?

You will make quantitative measurements of the forces between magnet and current in the next experiment, "Currents, Magnets, and Forces." The rest of this experiment is concerned with the interaction between two currents.

7. Connect a similar circuit consisting of a power supply, ammeter, and rheostat (if there is no variable control on the power supply) to the fixed coil on the vertical pegboard (the bundle of ten wires, not the single wire). The two circuits (fixed-coil and movable-hook) must be independent. Your setup should now look like the one shown in Fig. 4-19. Only one meter is actually *required*, because you can move it from one circuit to the other as needed. It is, however, more convenient to work with two meters.

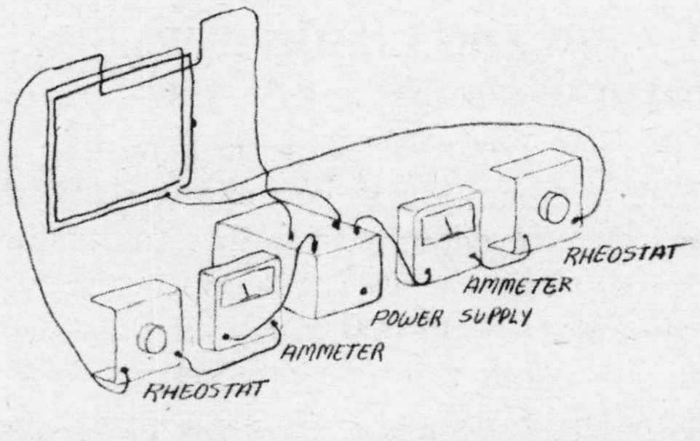

Fig. 4-19 Current balance connections using rheostats when variable power supply is not available.

8. Turn on the currents in both circuits and check to see which way the pointer rod on the balance swings. It should move *up*. If it does not, see if you can make the pointer swing *up* by changing something in your setup.

?

2. Do currents flowing in the same direction attract or repel each other? What about currents flowing in opposite directions?

9. Prepare some "weights" from the thin wire given to you. You will need a set that contains wire lengths of 1 cm, 2 cm, 5 cm, and 10 cm. You may want more than one of each; you can make more as needed during the experiment. Bend the wires into small S-shaped hooks so that they can hang from the notch on the pointer or from each other. This notch is the same distance from the axis of the balance as the bottom of the loop; therefore, when there is a force on the horizontal section of the loop, the total weight F hung at the notch will equal the magnetic force acting horizontally on the loop. (See Fig. 4-20).

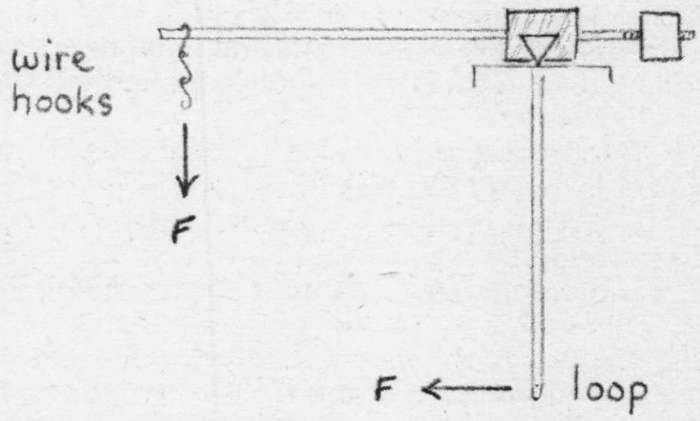

Fig. 4-20 Side view of a balanced loop. The distance from the pivot to the wire hook is the same as the distance to the horizontal section of the loop, so the weight of the additional wire hooks is equal in magnitude to the horizontal magnetic force on the loop.

These preliminary adjustments are common to all the investigations. But from here on, there are separate instructions for three different experiments. Different members of the class will investigate how the force depends upon:

(a) the *current* in the wires,
(b) the *distance* between the wires, or
(c) the *length* of one of the wires.

When you have finished your experiment, read the section "For Class Discussion."

(a) How Force Depends on Current in the Wires

By keeping a constant separation between the loop and the coil, you can investigate the effects of varying the currents. Set the balance on the frame so that, as you look down at them, the loop and the coil are parallel and about 1 cm apart.

Set the current in the balance loop to about 3 A. Do not change this current throughout the experiment. With this current in the balance loop and no current in the fixed coil, set the zero mark in line with the pointer rod.

Starting with a relatively small current in the fixed coil (about 1 A), determine how many centimeters of wire you must hang on the pointer notch until the pointer rod returns to the zero mark.

Record the current I_f in the fixed coil and the length of wire added to the pointer arm. The weight of wire is the balancing force F.

Increase I_f step by step, checking the current in the balance loop as you do so, until you reach a current of about 5 A in the fixed coil.

?

3. What is the relationship between the current in the fixed coil and the force on the balance loop? One way to discover this is to plot force F against current I_f. Another way is to find what happens to the balancing force when you double, then triple, the current I_f.

4. Suppose you had held I_f constant and measured F as you varied the current in the balance loop I_b. What relationship do you think you would have found between F and I_b? Check your answer experimentally (for example, by doubling I_b) if you have time.

5. Can you write a symbolic expression for how F depends on *both* I_f and I_b? Check your answer experimentally (by doubling both I_f and I_b) if you have time.

6. How do you convert the force, as measured in centimeters of wire hung on the pointer arm, into the conventional unit for force (newtons)?

(b) How Force Varies with the Distance between Wires

To measure the distance between the two wires, you have to look down. Put a scale on the wooden shelf below the loop. Because there is a gap between the wires and the scale, the number you read on the scale changes as you move your head back and forth. This effect is called *parallax*, and it must be reduced if you are to get good measurements. If you look down into a mirror set on the shelf, you can tell when you are looking straight down because the wire and its image will line up. Try it (Fig. 4-21).

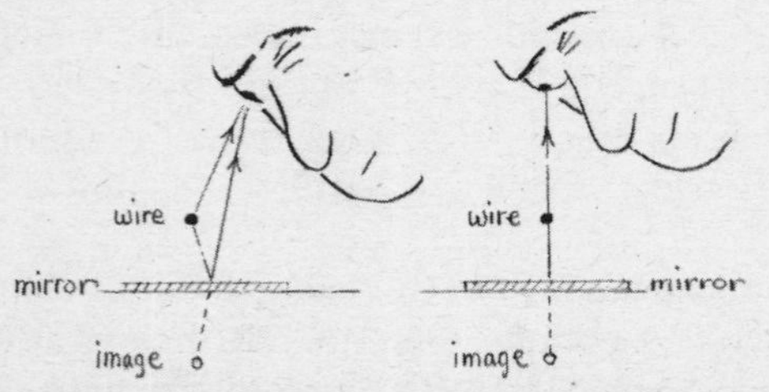

Fig. 4-21 Only when your eye is perpendicularly above the moving wire will it line up with its reflection in the mirror.

Stick a length of centimeter tape along the side of the mirror so that you can sight down and read off the distance between one edge of the fixed wire and the corresponding edge of the balance loop. Set the zero mark with a current I_b of about 4.5 A in the balance loop and no current I_f in the fixed coil. Then adjust the distance to about 0.5 cm.

Begin the experiment by adjusting the current passing through the fixed coil to 4.5 A. Hang weights on the notch in the pointer arm until the pointer is again at the zero position. Record the weight and distance carefully.

Repeat your measurements for four or five greater separations. Between each set of measurements make sure the loop and coil are still parallel; check the zero position, and see that the currents are still 4.5 A.

?

7. What is the relationship between the force F on the balance loop and the distance d between the loop and the fixed coil? One way to discover this is to find some function of d (such as $1/d^2$, $1/d$, d^3, etc.) that gives a straight line when plotted against F. Another way is to find what happens to the balancing force F when you double, then triple, the distance d.

8. If the force on the balance loop is F, what is the force on the fixed coil?

9. Can you convert the force, as measured in centimeters of wire hung on the pointer arm, into force in newtons?

(c) How Force Varies with the Length of One of the Wires

By keeping constant currents I_f and I_b and a constant separation d, you can investigate the effects on the *length* of the wires. In the current balance setup, it is the bottom, horizontal section of the loop that interacts most strongly with the coil. Loops with several different lengths of horizontal segment are provided.

To measure the distances between the two wires, you have to look down at them. Put a scale on the wooden shelf below the loop. Because there is a gap between the wires and the scale, what you read on the scale changes as you move your head back and forth. This effect is called *parallax*; parallax must be reduced if you are to get good measurements. If you look down into a mirror set on the shelf, you can tell when you are looking straight down because the wire and its image will line up. Try it (Fig. 4-21).

Stick a length of centimeter tape along the side of the mirror. Then you can sight down and read off the distance between one edge of the fixed wire and the corresponding edge of the balance loop. Adjust the distance to about 0.5 cm. With a current I_b of about 4.5 A in the balance loop and no current I_f in the fixed coil, set the pointer at the zero mark.

Begin the experiment by passing 4.5 A through both the balance loop and the fixed coil. Hang weights on the notch in the pointer arm until the pointer is again at the zero position.

Record the value of the currents, the distance between the two wires, and the weights added.

Turn off the currents, and carefully remove the balance loop by sliding it out of the holding clips (Fig. 4-22). Measure the length l of the horizontal segment of the loop.

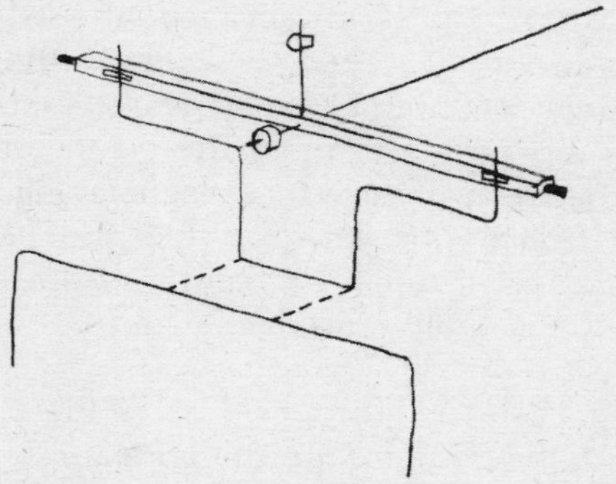

Fig. 4-22

Insert another loop. Adjust it so that it is level with the fixed coil and so that the distance between loop and coil is just the same as you had before. This is important. The loop must also be parallel to the fixed coil, both as you look down at the wires from above and as you look at them from the side. Also reset the clip on the balance for maximum sensitivity. Check the zero position, and see that the currents are still 4.5 A.

Repeat your measurements for each balance loop.

?

10. What is the relationship between the length l of the loop and the force F on it? One way to discover this is to find some function of l (such as l, l^2, $1/l$, etc.) that gives a straight line when plotted against F. Another way is to find what happens to F when you double l.
11. Can you convert the force, as measured in centimeters of wire hung on the pointer arm, into force in newtons?
12. If the force on the balance loop is F, what is the force on the fixed coil?

For Class Discussion

Be prepared to report the results of your particular investigation to the rest of the class. As a class, you will be able to combine the individual experiments into a single statement about how the force varies with current, with distance, and with length. In each part of this experiment, one factor was varied while the other two were kept constant. In combining the three separate findings into a single expression for force, you are assuming that the effects of the three factors are *independent*. For example, you are assuming that doubling one current will *always* double the force, *regardless* of what constant values d and l have.

?

13. What reasons can you give for assuming such a simple independence of effects? What could you do experimentally to support the assumption?
14. To make this statement into an equation, what other facts do you need; that is, what must you know to be able to predict the force (in newtons) existing between the currents in two wires of given length and separation?

Experiment 4-6
CURRENTS, MAGNETS, AND FORCES

If you did Experiment 4-5, "Forces on Currents," you discovered how the force between two wires depends on the current in them, their length, and the distance between them. You also know that a nearby magnet exerts a force on a current-carrying wire. In this experiment, you will use the current balance to study further the interaction between a magnet and a current-carrying wire. You may need to refer

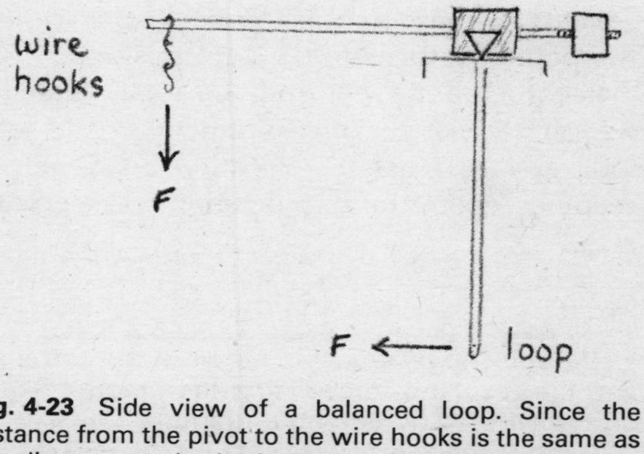

Fig. 4-23 Side view of a balanced loop. Since the distance from the pivot to the wire hooks is the same as the distance to the horizontal section of the loop, the weight of the additional wire hooks is equal to the horizontal magnetic force on the loop.

back to the notes on Experiment 4-5 for details on the equipment.

In this experiment, you will *not* use the fixed coil. The frame on which the coil is wound will serve merely as a convenient support for the balance and the magnets.

Attach the longest of the balance loops to the pivotal horizontal bar, and connect it through an ammeter to a variable source of current. Hang weights on the pointer notch until the pointer rod returns to the zero mark (see Fig. 4-23).

(a) How the Force between Current and Magnet Depends on the Current

1. Place two small ceramic magnets on the inside of the iron yoke. Their orientation is important; they must be turned so that the two near faces attract each other when they are moved close together. (Caution: Ceramic magnets are brittle. They break if you drop them.) Place the yoke and magnet unit on the platform so that the balance loop passes through the center of the region between the ceramic magnets (Fig. 4-24).

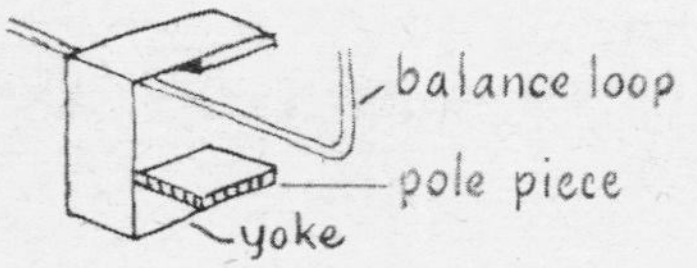

Fig. 4-24 Each magnet consists of a yoke and a pair of removable ceramic-magnet pole pieces.

2. Check whether the horizontal pointer moves *up* when you turn on the current. If it moves down, change something (the current? the magnets?) so that the pointer swings up.

3. With the current off, mark the zero position of the pointer arm with the indicator. Adjust the current in the coil to about 1 A. Hang wire weights in the notch of the balance arm until the pointer returns to the zero position.

Record the current and the total balancing weight. Repeat the measurements for at least four greater currents. Between each pair of readings, check the zero position of the pointer arm.

?

1. What is the relationship between the current I_b and the resulting force F that the magnet exerts on the wire? (Try plotting a graph.)

2. If the magnet exerts a force on the current, do you think the current exerts a force on the magnet? How would you test this?

3. How would a stronger or a weaker magnet affect the force on the current? If you have time, try the experiment with different magnets or by doubling the number of pole pieces. Then plot F against I_b on the same graph as in Question 1 above. How do the plots compare?

(b) How the Force between a Magnet and a Current Depends on the Length of the Region of Interaction

1. Place two small ceramic magnets on the inside of the iron yoke to act as pole pieces (Fig. 4-24). (Caution: Ceramic magnets are brittle. They break if you drop them.) Their orientation is important; they must be turned so that the two near faces attract each other when they are moved close together. Place the yoke and magnet unit on the platform so that the balance loop passes through the center of the region between the ceramic magnets (Fig. 4-24).

Place the yoke so that the balance loop passes through the center of the magnet and the pointer moves *up* when you turn on the current.

With the current off, mark the zero position of the pointer with the indicator.

2. Hang 10 cm or 15 cm of wire on the notch in the balance rod, and adjust the current to return the pointer rod to its zero position. Record the current and the total length of wire, and set aside the magnet for later use.

3. Put a second yoke and pair of pole pieces in position and see if the balance is restored. You have changed neither the current nor the length of wire hanging on the pointer. Therefore, if balance is restored, this magnet must be of the same strength as the first one. If it is not, try other combinations of pole pieces until you have two magnets of the same strength. You can produce small variations in strength by moving the pole pieces: a) into the yoke to make it stronger; b) to the ends of the yoke to make it weaker. If possible, try to get three matched magnets.

4. Now you are ready for the important test. Place two of the magnets on the platform at the same time (Fig. 4-25). To keep the magnets from affecting each other's field appreciably, they should be at least 10 cm apart. Of course, each magnet must be positioned so that the pointer is deflected upward. With the current just what it was before, hang wire weights in the notch until the balance is restored.

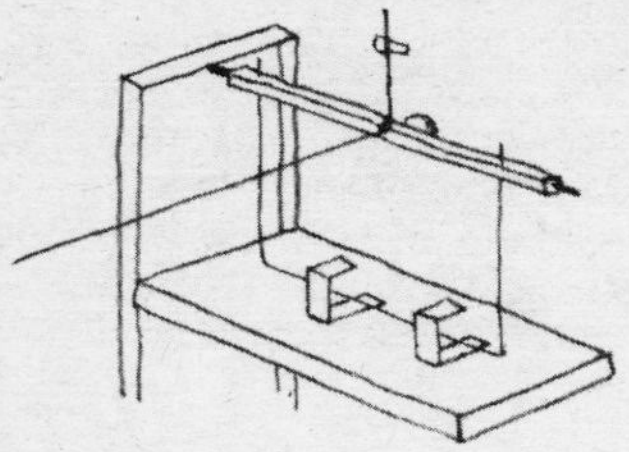

Fig. 4-25

If you have three magnet units, repeat the process using three units at a time. Again, keep the units well apart.

Interpreting Your Data

Your problem is to find a relationship between the length l of the region of interaction and the force F on the wire.

You may not know the exact length of the region of interaction between magnet and wire for a single unit. It certainly extends beyond the region between the two pole pieces. But the force decreases rapidly with distance from the magnets. As long as the separate units are far from each other, neither will be influenced by the prescence of the other. You can then assume that the total length of interaction with two units is double that for one unit.

?

4. How does F depend on l?

(c) A Study of the Interaction Between the Earth and an Electric Current

The magnetic field of the earth is much weaker than the field near one of the ceramic magnets and the balance must be adjusted to its maximum sensitivity. The following sequence of detailed steps will make it easier for you to detect and measure the small forces on the loop.

1. Set the balance, with the longest loop, to maximum sensitivity by sliding the sensitivity clip to the top of the vertical rod. The sensitivity can be increased further by adding a second clip; be careful not to make the balance top-heavy so that it falls over.

2. With no current in the balance loop, align the zero mark with the end of the pointer arm.

3. Turn on the current and adjust it to about 5 A. Turn off the current and let the balance come to rest.

4. Turn on the current, and observe carefully: Does the balance move when you turn the current on? Since there is no current in the fixed coil, and there are no magnets nearby, any force acting on the current in the loop must be due to an interaction between it and the earth's magnetic field.

5. To make measurements of the force on the loop, you must set up the experiment so that the pointer swings up when you turn on the current. If the pointer moves down, try to find a way to make it go up. (If you have trouble, consult your instructor). Turn off the current, and bring the balance to rest. Mark the zero position with the indicator.

6. Turn on the current. Hang weights on the notch, and adjust the current to restore balance. Record the current and the length of wire on the notch. Repeat the measurement of the force needed to restore balance for several different values of current.

If you have time, repeat your measurements of force and current for a shorter loop.

Interpreting Your Data

Try to find the relationship between the current I_b in the balance loop and the force F on it. Make a plot of F against I_b.

?

5. How can you convert your weight unit (centimeters of wire) into newtons of force?
6. What force (in newtons) does the earth's magnetic field in your laboratory exert on a current I_b in the loop?

For Class Discussion

Different members of the class have investigated how the force F between a current and a magnet varies with current I and with the length of the region of interaction with the current l. It should also be clear that in any statement that describes the force on a current due to a magnet, you must include another term that takes into account the "strength" of the magnet.

Be prepared to report to the class the results of your own investigations and to help formulate an expression that includes all the relevant factors investigated by members of the class.

?

7. The strength of a magnetic field can be expressed in terms of the force exerted on a wire carrying 1 A of current when the length of the wire interacting with the field is 1 m. Try to express the strength of the magnetic field of your magnet yoke or of the earth's magnetic field in these units, newtons per ampere-meter. (That is, what force would the fields exert on a horizontal wire 1 m long carrying a current of 1 A?)

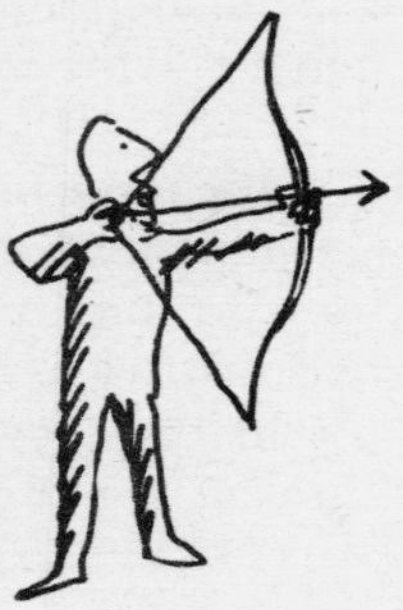

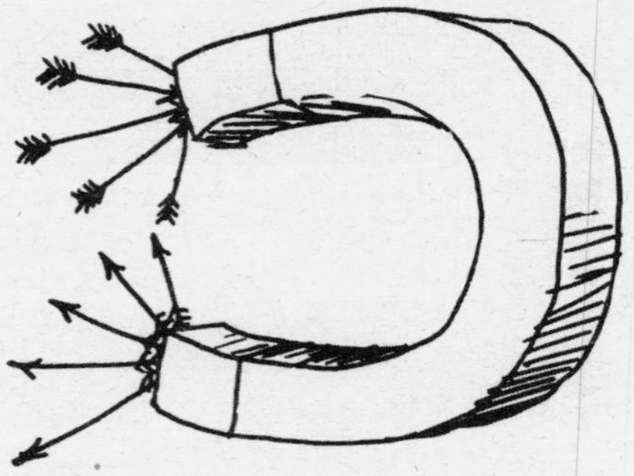

In using the current balance in this experiment, all measurements were made in the zero position when the loop was at the very bottom of the swing. In this position a vertical force will not affect the balance. Therefore, you have measured only *horizontal* forces on the bottom of the loop.

However, since the force exerted on a current by a magnetic field is always at right angles to the field, you have therefore measured only the *vertical* component of the magnetic fields. From the symmetry of the magnet yoke, you might guess that the field is entirely vertical in the region directly between the pole pieces. But the earth's magnetic field is exactly vertical only at the magnetic poles. (See the drawing on page 457 of the Text.)

?

8. How would you have to change the experiment to measure the horizontal component of the earth's magnetic field?

Experiment 4-7
ELECTRON BEAM TUBE. I

If you did the experiment "Electric Forces. II: Coulomb's Law," you found that the force on a test charge, in the vicinity of a second charged body, decreases rapidly as the distance between the two charged bodies is increased. In other words, the *electric field* strength due to a single small charged body decreases with distance from the body. In many experiments it is useful to have a region where the field is uniform, that is, a region where the force on a test charge is the same at all points. The field everywhere between two closely spaced parallel, flat, oppositely charged plates is very nearly uniform (Fig. 4-26).

The nearly uniform magnitude E depends upon the potential difference between the plates and upon their separation d:

$$E = \frac{V}{d}$$

In addition to electric forces on charged bodies, you found (if you did either of the previous two experiments with the current balance) that there is a force on a current-carrying wire in a magnetic field.

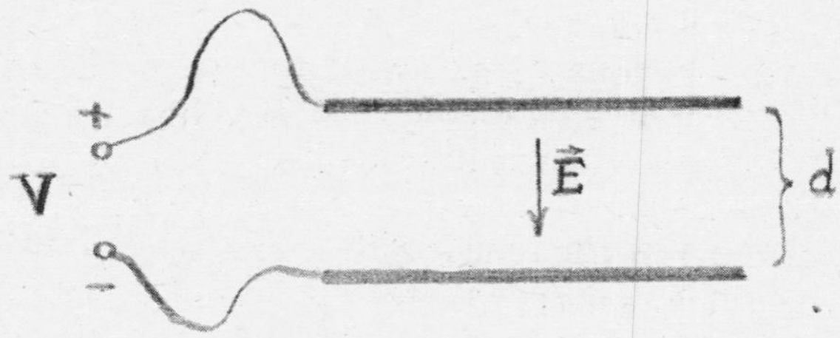

Fig. 4-26 The field between two parallel flat plates is uniform $E = V/d$ where V is the potential difference (volts) between the two plates.

Free Charges

In this experiment, the charges will not be confined to a Styrofoam ball or to a metallic conductor. Instead they will be free charges, free to move through the field on their own in air at low pressure.

You will build a special tube for this experiment. The tube will contain a filament wire and a metal can with a small hole in one end. Electrons emitted from the heated filament are accelerated toward the positively charged can and some of them pass through the hole into the space beyond, forming a beam of electrons. If the tube is carefully constructed, the air pumped out to the right pressure, and the tube well sealed, it is possible to observe how the beam is affected by electric and magnetic fields.

When one of the air molecules remaining in the partially evacuated tube is struck by an electron, the molecule emits some light. Molecules of different gases emit light of different colors. (Neon gas, for example, glows red.) The bluish glow of the air left in the tube shows the path of the electron beam.

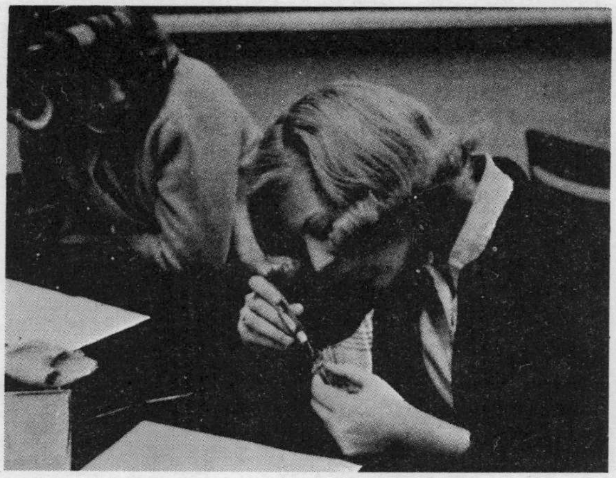

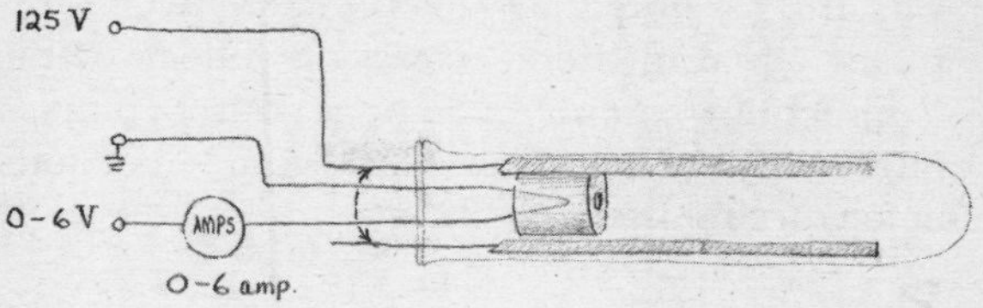

Fig. 4-28 The pins to the two plates are connected so that they will be at the same potential and there will be no electric field between them.

Building Your Electron Beam Tube

Full instructions on how to build the tube are included with the parts. Note that one of the plates is connected to the can. The other plate must not touch the can.

After you have assembled the filament and plates on the pins of the glass tube base, you can see how good the alignment is if you look in through the narrow glass tube. You should be able to see the filament across the center of the hole in the can. Do not seal the header in the tube until you have checked this alignment. Then leave the tube undisturbed overnight while the sealant hardens. It is not unusual that some of the tubes in a given class do not work well. In that case, try to share the tubes that function successfully.

Operating the Tube

With the power supply turned off, connect the tube as shown in Figs. 4-27 and 4-28. The low-voltage connection provides a current to heat the filament and make it emit electrons.

Fig. 4-27

The ammeter in this circuit allows you to keep a close check on the current and avoid burning out the filament. Be sure the 0– 6 V control is turned down to 0.

The high-voltage connection provides the field that accelerates these electrons toward the can. Let the instructor check the circuits before you proceed further.

Turn on the vacuum pump and let it run for several minutes. If you have done a good job of putting the tube together, and if the vacuum pump is in good condition, you should not have much difficulty getting a glow in the area where the electron beam comes through the hole in the can.

You should work with the faintest glow that you can see clearly. Even then, it is important to keep a close watch on the brightness of the glow. There is an appreciable current from the filament wire to the can. As the residual gas gets hotter, it becomes a better conductor, thus increasing the current. The increased current will cause further heating, and the process can build up; the back end of the tube will glow intensely blue-white and the can will become red hot. You must immediately reduce the current to prevent the tube from being destroyed. *If the glow in the back end of the tube begins to increase noticeably, turn down the filament current very quickly, or turn off the power supply altogether.*

Deflection by an Electric Field

When you get an electron beam, try to deflect it in an electric field by connecting the deflecting plate to the ground terminal (see Fig. 4-29). You

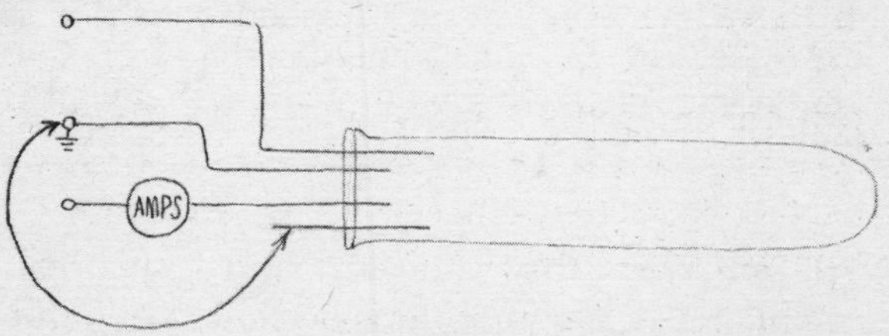

Fig. 4-29 Connecting one deflecting plate to ground will put a potential difference of 125 V between the plates.

will put a potential difference between the plates equal to the accelerating voltage. Other connections can be made to get other voltages, but check your ideas with your instructor before trying them.

?

1. Make a sketch showing the direction of the electric field and of the force on the charged beam. Does the deflection in the electric field confirm that the beam consists of negatively charged particles?

Deflection by a Magnetic Field

Now try to deflect the beam in a magnetic field, using the yoke and magnets from the current balance experiments.

?

2. Make a vector sketch showing the direction of the magnetic field, the velocity of the electrons, and the force on them.

Balancing the Electric and Magnetic Effects

Try to orient the magnets so as to cancel the effect of an electric field between the two plates, permitting the charges to travel straight through the tube.

?

3. Make a sketch showing the orientation of the magnetic yoke relative to the plates.

The Speed of the Charges

As explained in Chapter 14 of the Text, the magnitude of the magnetic force is qvB, where q is the electron charge, v is the speed, and B is the magnetic field strength. The magnitude of the electric force is qE, where E is the strength of the electric field. If you adjust the voltage on the plates until the electric force just balances the magnetic force, then $qvB = qE$ and, therefore, $v = B/E$.

?

4. Show that B/E will be in speed units if B is expressed in newtons/ampere-meter and E is expressed in newtons/coulomb. *Hint:* Remember that 1 A = 1 C/sec.

If you know the value of B and E, you can calculate the speed of the electrons. The value of E is easy to find since, in a uniform field between parallel plates, $E = V/d$, where V is the potential difference between the plate (in volts) and d is the separation of the plates (in meters). (The unit volts/meter is equivalent to newtons/coulomb.)

A rough value for the strength of the magnetic field between the poles of the magnet yoke can be obtained as described in Experiment 4-6, "Currents, Magnets, and Forces."

?

5. What value do you get for E (in volts per meter)?
6. What value did you get for B (in newtons per ampere-meter)?
7. What value do you calculate for the speed of the electrons in the beam?

An Important Question

One of the problems facing physicists at the end of the nineteenth century was to decide on the nature of these "cathode rays" (so called because they are emitted from the negative electrode or cathode). One group of scientists (mostly German) thought that cathode rays were a form of radiation, like light, while others (mostly English) thought they were streams of particles. J.J. Thomson at the Cavendish Laboratory in Cambridge, England, did experiments much like the one described here which showed that the cathode rays behaved like particles; these particles are now called *electrons*.

These experiments were of great importance in the early development of atomic physics. In Unit 5, you will do an experiment to determine the ratio of the charge of an electron to its mass.

Experiment 4-8 ELECTRON BEAM TUBES. II*

1. Focusing the Electron Beam

A current in a wire coiled around the electron tube will produce inside the coil a magnetic field parallel to the axis of the tube. (Ring-shaped magnets slipped over the tube will produce the same kind of field.) An electron moving directly along the axis will experience no force since its velocity is parallel to the magnetic field. An electron moving perpendicular to the axis, however, will experience a force ($F = qvB$) at right angles to both velocity and field. If the curved path of the electron remains

**Note:* This experiment is more complex than usual.

in the uniform field, it will be a circle. The centripetal force $F = mv^2/R$ that keeps it in the circle is just the magnetic force qvB, so

$$qvB = \frac{mv^2}{R}$$

where R is the radius of the orbit. In this simple case, therefore,

$$R = \frac{mv}{qB}$$

Suppose the electron is moving down the tube only slightly off axis, in the presence of a field parallel to the axis [Fig. 4-30(a)]. The electron's velocity can be thought of as made up of two components: an axial portion of v_a and a transverse portion (perpendicular to the axis) v_t [Fig. 4-30(b)]. Consider these two components of the electron's velocity independently. You know that the axial component will be unaffected; the electron will continue to move down the tube with speed v_a [Fig. 4-30(c)]. The transverse component, however, is perpendicular to the field, so the electron will also move in a circle [Fig. 4-30(d)]. In this case,

$$R = \frac{mv_t}{qB}$$

The resultant motion (uniform speed down the axis plus circular motion perpendicular to the axis) is a helix, like the thread on a bolt [Fig. 4-30(e)].

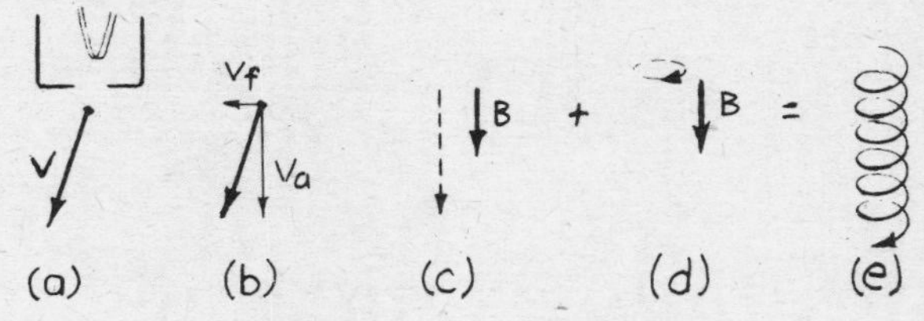

Fig. 4-30

In the absence of any field, electrons traveling off-axis would continue toward the edge of the tube. In the presence of an axial magnetic field, however, the electrons move down the tube in helixes; that is, they have been focused into a beam. The radius of this beam depends on the field strength B and the transverse velocity v_t.

Wrap heavy-gauge copper wire, such as #18, around the electron beam tube (about two turns per centimeter) and connect the tube to a low-voltage (3 – 6 V), high-current source to give a noticeable focusing effect. Observe the shape of the glow, using different coils and currents. (Alternatively, you can vary the number and spacing of ring magnets slipped over the tube to produce the axial field.)

2. Reflecting the Electron Beam

If the pole of a very strong magnet is brought near the tube (with great care being taken that it does not pull the iron mountings of the tube toward it), the beam glow will be seen to spiral more and more tightly as it enters stronger field regions. If the field lines diverge enough, the path of the beam may start to spiral back. The reason for this is suggested in SG 39 in Chapter 14.

This kind of reflection operates on particles in the radiation belt around the earth as they approach the earth's magnetic poles. (See drawing at the end of Text Chapter 14.) Such reflection is what makes it possible to hold tremendously energetic, charged particles in magnetic "bottles." One kind of coil used to produce a "bottle" field is called a magnetic torus.

3. Diode and Triode Characteristics

This experiment gives suggestions for how you can explore some characteristics of electronic vacuum tubes with your electron beam tube materials.

These experiments are performed at accelerating voltages below those that cause ionization (a visible glow) in the electron beam tube.

Rectification

Connect an ammeter between the can and high-voltage supply to show the direction of the current, and to show that there is a current only when the can is at a higher potential than the filament (Fig. 4-31).

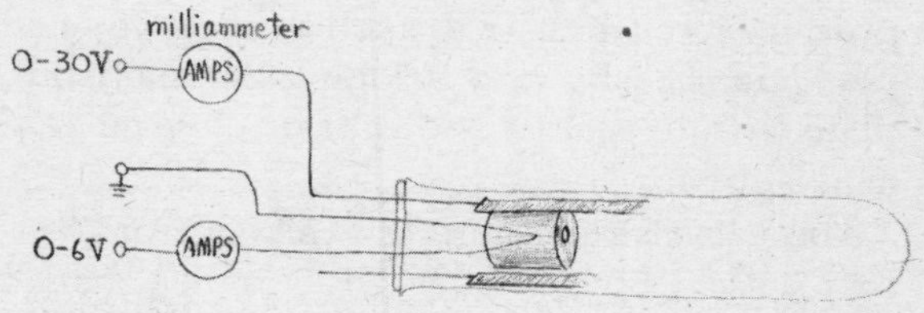

Fig. 4-31

Note that there is a measurable current at voltages far below those needed to give a visible glow in the tube. Then apply an alternating potential difference between the can and filament (for example, from a Variac). Use an oscilloscope to show that the can is alternately above and below filament potential. Then connect the oscilloscope across a resistor in the plate circuit to show that the current is only in one direction. (See Fig. 4-32.)

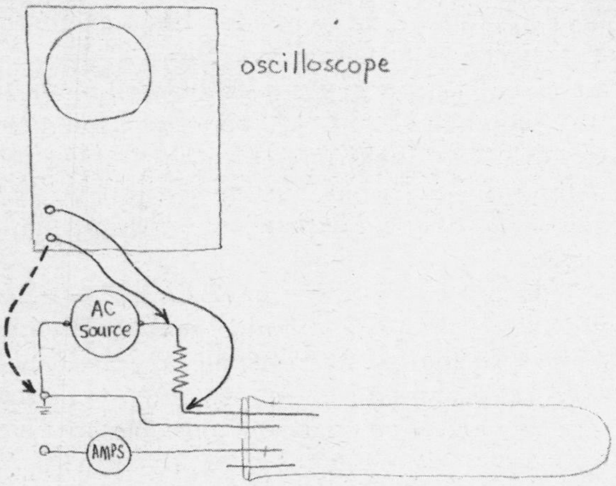

Fig. 4-32 The one-way valve (rectification) action of a diode can be shown by substituting an ac voltage source for the dc accelerating voltage, and connecting a resistor (about 1,000 ohms) in series with it. When an oscilloscope is connected, as shown by the solid lines above, it will indicate the current in the can circuit. When the one wire is changed to the connection shown by the dashed arrow, the oscilloscope will indicate the voltage on the can.

Triode

The "triode" in Fig. 4-33 was made with a thin aluminum sheet for the plate and nichrome wire for the grid. The filament is the original one from the electron beam tube kit, and thin aluminum tubing from a hobby shop was used for the connections to plate and grid. (For reasons lost in the history of vacuum tubes, the can is usually called the *plate*.) It is interesting to plot graphs of plate current versus filament heating current, and plate current versus voltage. Note that these characteristic curves apply only to voltages too low to produce ionization. With such a triode, you can plot curves showing triode characteristics: plate current against grid voltage; plate current against plate voltage.

You can also measure the voltage amplification factor, which describes how large a change in plate voltage is produced by a change in grid voltage. More precisely, the amplification factor is

$$\mu = \frac{-\Delta V_{plate}}{\Delta V_{grid}}$$

when the plate current is kept constant.

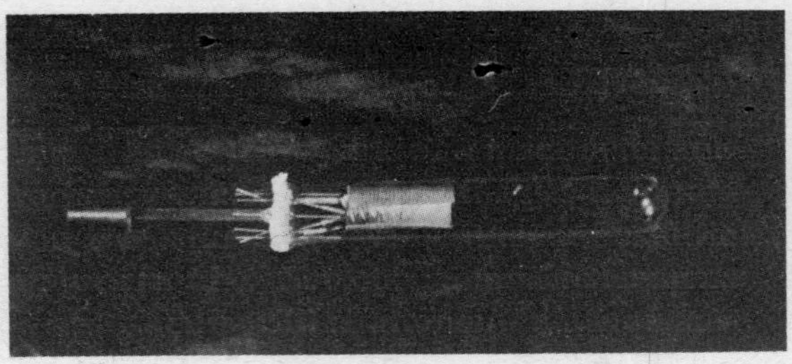

Fig. 4-33

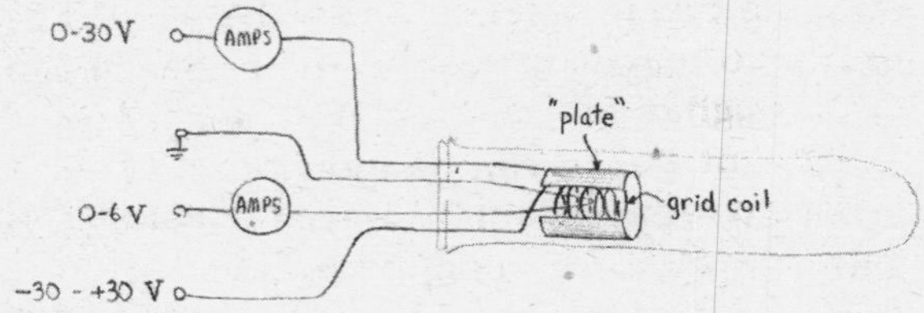

Fig. 4-34

Change the grid voltage by a small amount, then adjust the plate voltage until you have regained the original plate current. The magnitude of the ratio of these two voltage changes is the amplification factor. (Commercial vacuum tubes commonly have amplification factors as high as 500.) The tube gave noticeable amplification in the circuit shown in Fig. 4-35 and Fig. 4-36.

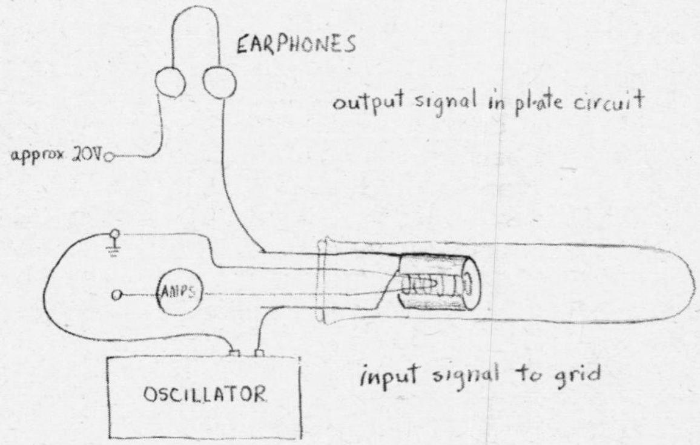

Fig. 4-35 An amplifying circuit.

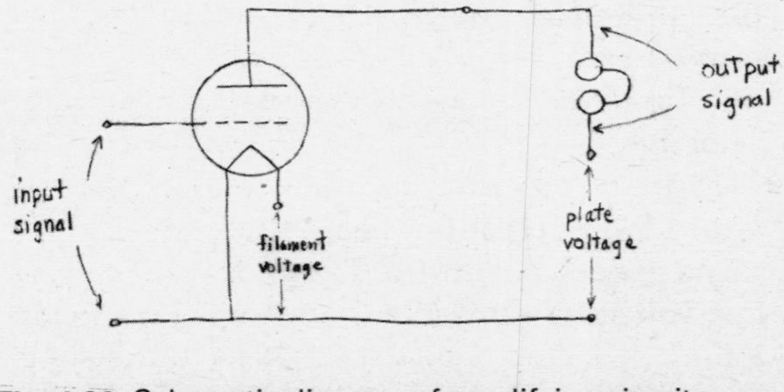

Fig. 4-36 Schematic diagram of amplifying circuit.

Experiment 4-9
WAVES AND COMMUNICATION

Having studied many kinds and characteristics of waves in Units 3 and 4 of the Text, you are now in a position to see how they are used in communications. Here are some suggestions for investigations with equipment that you have probably already seen demonstrated. The following notes assume that you understand how to use the equipment. If you do not, then do not go on until you ask for instructions. Although different groups of students may use different equipment, all the investigations are related to the same phenomenon: communication by means of waves.

A. Turntable Oscillators

Turn on the oscillator with the pen attached to it. (See Unit 5, page 231.) Turn on the chart recorder, but do not turn on the oscillator on which the recorder is mounted. The pen will trace out a sine curve as it goes back and forth over the moving paper. When you have recorded about 7 cm – 8 cm, turn off the oscillator and bring the pen to rest in the middle of the paper. Now turn on the second oscillator at the same rate that the first one was going. The pen will trace out a similar sine curve as the moving paper goes back and forth under it. The wavelengths of the two curves are probably very nearly, but not exactly, equal.

?

1. What do you predict will happen if you turn on *both* oscillators? Try it. Look carefully at the pattern that is traced out with both oscillators on and compare it to the curves previously drawn by the two oscillators running alone.

Change the wavelength of one of the components slightly by putting weights on one of the platforms to slow it down a bit. Then make more traces from other pairs of sine curves. Each trace should consist of three parts, as in Fig. 4-37: the sine curve from one oscillator; the sine curve from the other oscillator; and the composite curve from both oscillators.

?

2. According to a mathematical analysis of the addition of sine waves, the wavelength of the envelope (λ_e in Fig. 4-37) will *increase* as the wavelengths of the two components (λ_1, λ_2) become more nearly equal. Do your results confirm this?
3. If the two wavelengths λ_1 and λ_2 were exactly equal, what pattern would you get when both turntables were turned on; that is, when the two sine curves were superimposed? What else would the pattern depend on, in addition to λ_1 and λ_2?

As the difference between λ_1 and λ_2 gets smaller, λ_e gets bigger. You can thus detect a very small difference in the two wavelengths by examining the resultant wave for changes in amplitude that take place over a relatively long distance. This method, called the *method of beats*, provides a sensitive way of comparing two oscillators, and of adjusting one until it has the same frequency as the other.

This method of beats is also used for tuning musical instruments. If you play the same note on two instruments that are not quite in tune, you can hear the beats. The more nearly in tune the two instruments are, the lower the frequency of the beats. You might like to try this with two guitars or other musical instruments (or two strings on the same instrument).

In radio communication, a signal can be transmitted by using it to modulate a *carrier wave* of much higher frequency. (See Part *E* for further explanation of modulation.) A snapshot of the modulated wave looks similar to the beats you have been producing, but it results from one wave being used to control the amplitude of the other, not from simply adding the waves together.

B. Resonant Circuits

You have probably seen a demonstration of how a signal can be transmitted from one

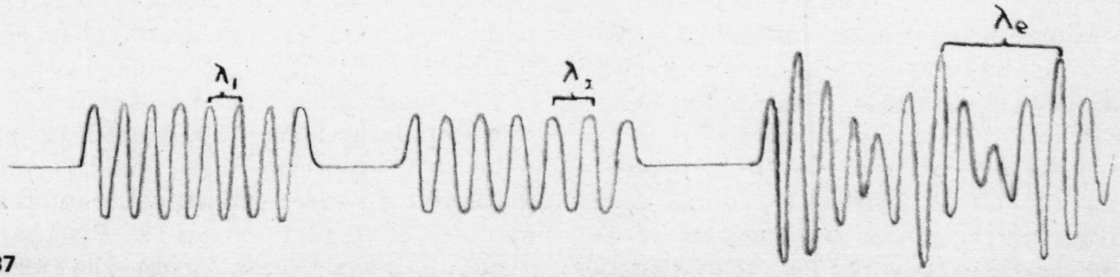

Fig. 4-37

circuit to another that is tuned to it. (If you have not seen the demonstration, you should set it up for yourself using the apparatus shown in Fig. 4-38.)

Fig. 4-38 Two resonant circuit units. Each includes a wire coil and a variable capacitor. The unit on the right has an electric cell and ratchet to produce pulses of oscillation in its circuit.

This setup is represented by the schematic drawing in Fig. 4-39.

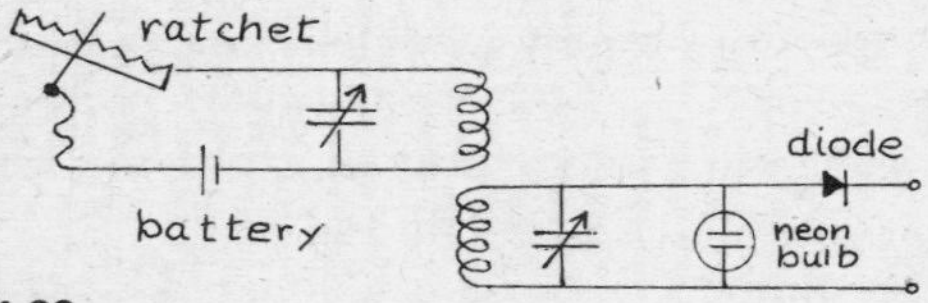

Fig. 4-39

The two coils have to be quite close to each other for the receiver circuit to pick up the signal from the transmitter. The effect is due to the fluctuation of the magnetic field from one coil inducing a fluctuating current in the other coil. It works on the same principle as a transformer.

Investigate the effect of changing the position of one of the coils. Try turning one of them around, moving it farther away, etc.

?

4. What happens when you put a sheet of metal, plastic, wood, cardboard, wet paper, or glass between the two coils?
5. Why does an automobile always have an outside antenna, while a home radio does not?

You have probably learned that to transmit a signal from one circuit to another, the two circuits must be tuned to the same frequency. To investigate the range of frequencies obtainable with your resonant circuit, connect an antenna (length of wire) to the resonant receiving circuit, in order to increase its sensitivity, and replace the speaker by an amplifier and oscilloscope (Fig. 4-40). Set the oscilloscope to "Internal Sync" and the sweep rate to about 100 kHz.

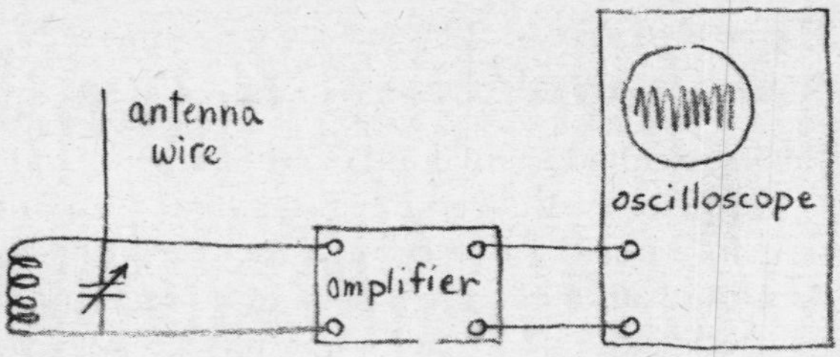

Fig. 4-40

?

6. Change the setting of the variable capacitor ( ) and see how the trace on the oscilloscope changes. Which setting of the capacitor gives the highest frequency? which setting the lowest? By how much can you change the frequency by adjusting the capacitor setting?

When you tune a radio, you are usually, in the same way, changing the setting of a variable capacitor to tune the circuit to a different frequency.

The coil also plays a part in determining the resonant frequency of the circuit. If the coil has a different number of turns, a different setting of the capacitor would be needed to get the same frequency.

C. Elementary Properties of Microwaves

With a microwave generator, you can investigate some of the characteristics of short waves in the radio part of the electromagnetic spectrum. In Experiment 3-17 "Measuring Wavelength" and Experiment 3-18 "Sound," you explored the behavior of several different kinds of waves. These earlier experiments contained a number of ideas that will help you conclude that the energy emitted by your microwave generator is in the form of waves.

Refer to your notes on these experiments. Then, using the arrangements suggested there or ideas of your own, explore the transmission of microwaves through various materials as well as microwave reflection and refraction. Try to detect their diffraction around obstacles and through narrow openings in some material that is opaque to them. Finally, if you have two transmitters available or a metal horn attachment with two openings, see if you can measure the wavelength using the interference method of Experiment 3-18. Discuss your results with students doing the following

experiment (*D*) on the interference of reflected microwaves.

D. Interference of Reflected Microwaves

With microwaves it is easy to demonstrate interference between direct radiation from a source and radiation reflected from a flat surface, such as a metal sheet. At points where the direct and reflected waves arrive in phase, there will be maxima; at points where they arrive one-half cycle out of phase, there will be minima. The maxima and minima are readily found by moving the detector along a line perpendicular to the reflector. (Fig. 4-41)

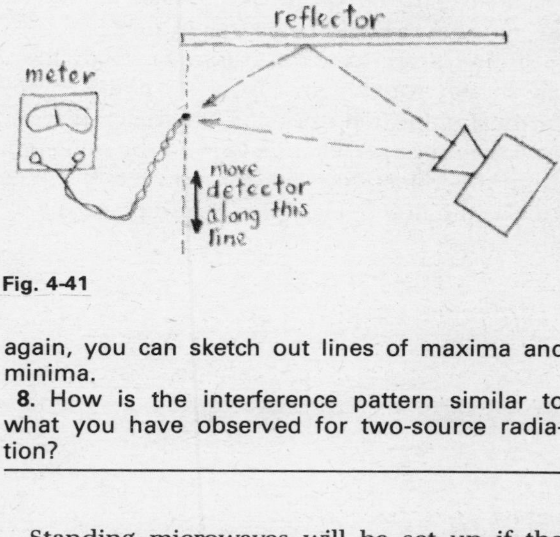

Fig. 4-41

?

7. Can you state a rule with which you could predict the positions of maxima and minima?

By moving the detector back and scanning again, you can sketch out lines of maxima and minima.

8. How is the interference pattern similar to what you have observed for two-source radiation?

Standing microwaves will be set up if the reflector is placed exactly perpendicular to the source (Fig. 4-42). (As with other standing

Fig. 4-42

waves, the nodes are one-half wavelength apart.) Locate several nodes by moving the detector along a line between the source and reflector, and from the node separation calculate the wavelength of microwaves.

?

9. What is the wavelength of your microwaves?
10. Microwaves, like light, propagate at 3×10^8 m/sec. What is the frequency of your microwaves? Check your answer against the chart of the electromagnetic spectrum given on page 511 of Text Chapter 16.

The interference between direct and reflected radio waves has important practical consequences. There are layers of partly ionized (and therefore electrically conducting) air, collectively called the *ionosphere*, that surround the earth roughly 30 km – 300 km above its surface. One of the layers at about 300 km is a good reflector for radio waves; it is used to bounce radio messages to points that, because of the curvature of the earth, are too far away to be reached in a straight line.

If the transmitting tower is 100 m high, then, as shown roughly in Fig. 4-43, point A, the farthest point that the signal can reach directly in flat country, is 35 km away. But by reflection from the ionosphere, a signal can reach points not in the line of sight like B and beyond.

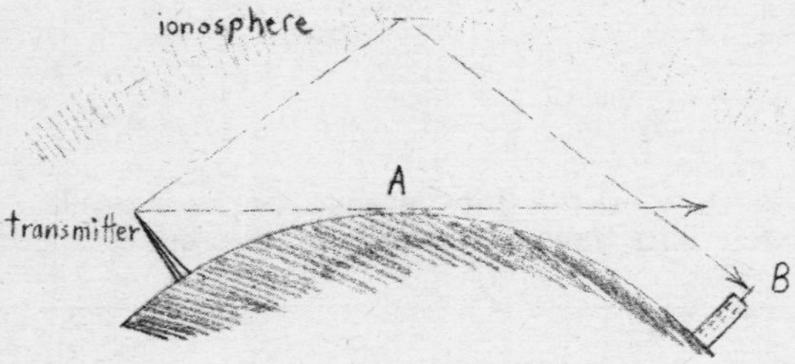

Fig. 4-43

Sometimes both a direct and a reflected signal will arrive at the same place and interference occurs; if the two signals are out of phase and have identical amplitudes, the receiver will pick up nothing. Such destructive interference is responsible for radio fading. It is complicated by the fact that the height of the ionosphere and the intensity of reflection from it vary with the amount of sunlight and with time.

The setup in Fig. 4-44 is a model of this situation. Move the reflector (the "ionosphere") back and forth. What happens to the signal strength?

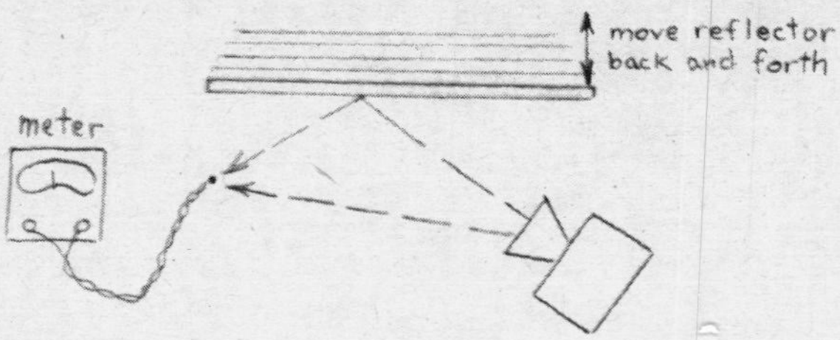

Fig. 4-44

There can also be multiple reflections; the radiation can bounce back and forth between earth and ionosphere several times on its way from, say, New York to Calcutta, India. Perhaps you can simulate this situation, too, with your microwave equipment.

E. Signals and Microwaves

Thus far, you have been learning about the behavior of microwaves of a single frequency and constant amplitude. A *signal* can be added to these waves by changing their amplitude at the transmitter. The most obvious way to change the amplitude of the waves would be just to turn them on and off as represented in Fig. 4-45. Coded messages (dots and dashes) can be transmitted in this primitive fashion. But the wave amplitude can be varied in a more elaborate way to carry music or voice signals. For example, a 1,000-Hz sine wave fed into part of the microwave transmitter will cause the amplitude of the microwave to vary smoothly at 1,000 Hz.

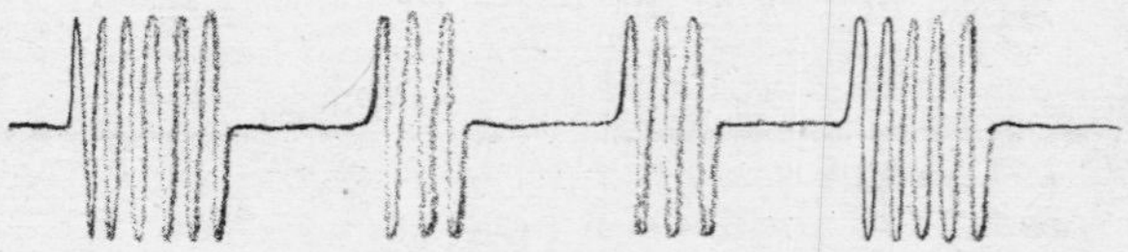

Fig. 4-45

Controlling the amplitude of the transmitted wave like this is called *amplitude modulation*; in Fig. 4-46, A represents the unmodulated microwave, B represents a modulating signal, and C represents the modulated microwave. The Damon microwave oscillator has an input for a modulating signal. You can modulate the microwave output with a variety of signals, for example, with an audio frequency oscillator or with a microphone and amplifier.

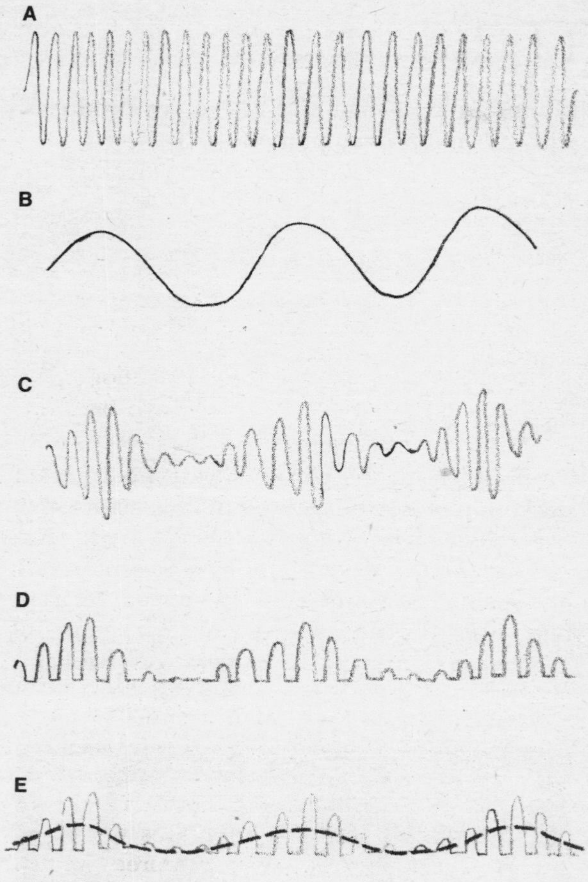

Fig. 4-46

The microwave detector probe is a one-way device; it passes current in only one direction. If the microwave reaching the probe is represented in C, then the electric signal from the probe will be as shown in D.

You can see this signal on the oscilloscope by connecting the oscilloscope to the microwave probe (through an amplifier if necessary).

The detected modulated signal from the probe can be turned into sound by connecting an amplifier and loudspeaker to the probe. The speaker will not be able to respond to the 10^9 individual pulses per second of the "carrier" wave, but only to their averaged effect, represented by the broken line in E. Consequently, the sound output of the speaker will correspond very nearly to the modulating signal.

?

11. Why must the carrier frequency be much greater than the signal frequency?

12. Why is a higher frequency needed to transmit television signals than radio signals? (The highest frequency necessary to convey radio sound information is about 12,000 Hz. The electron beam in a television tube completes one picture of 525 lines in 0.03 sec, and the intensity of the beam should be able to vary several hundred times during a single line scan.)

ACTIVITIES

THIN FILM INTERFERENCE

Press two *clean* microscope slides together. Look at the light they reflect from a source (like a mercury lamp or sodium flame) that emits light at only a few definite wavelengths. What you see is the result of interference between light waves reflected at the two inside surfaces that are almost, but not quite, touching. (The thin film is the layer of air between the slides.)

The phenomenon can also be used to determine the flatness of surfaces. If the two inside surfaces are planes, the interference fringes are parallel bands. Bumps or depressions as small as a fraction of a wavelength of light can be detected as wiggles in the fringes. This method is used to measure very small distances in terms of the known wavelength of light of a particular color. If two very flat slides are placed at a slight angle to each other, an interference band appears for every wavelength of separation. (See Fig. 4-47.)

How could this phenomenon be used to measure the thickness of a very fine hair or very thin plastic?

Fig. 4-47

HANDKERCHIEF DIFFRACTION GRATING

Stretch a linen or cotton handkerchief of good quality and look through it at a distant light source, such as a street light about one block away. You will see an interesting diffraction pattern. (A window screen or cloth umbrella will also work.)

PHOTOGRAPHING DIFFRACTION PATTERNS

Diffraction patterns like those pictured here can be produced in your lab or at home. The photos in Figs. 4-49 and 4-50 were produced with the setup diagrammed in Fig. 4-48.

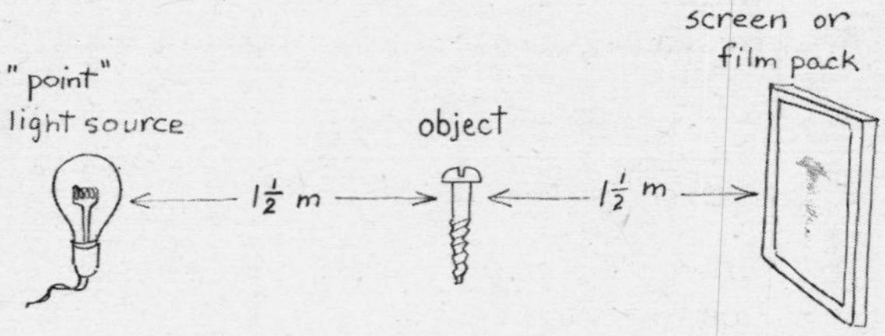

Fig. 4-48

To photograph the patterns, you must have a darkroom or a large, light-tight box. Figure 4-49 was taken using a Polaroid 4 × 5 back on a Graphic press camera. The lens was removed, and a single sheet of 3,000-ASA-speed Polaroid film was exposed for 10 sec; a piece of cardboard in front of the camera was used as a shutter.

Fig. 4-49

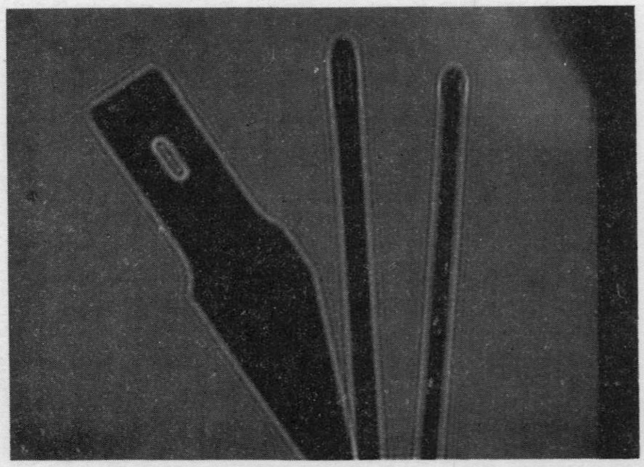

Fig. 4-50

As a light source, use a 1.5-volt flashlight bulb and AA cell. Turn the bulb so the end of the filament acts as a point source. A red (or blue) filter makes the fringes sharper. You can see the fringes by examining the shadow on the screen with the 10× magnifier. Razor blades, needles, or wire screens make good objects.

POISSON'S SPOT

A bright spot can be observed in a photograph of the center of some shadows, like that shown in Fig. 4-51. To see this, set up a light source, obstacle, and screen as shown in Fig. 4-52. Satisfactory results require complete darkness. Try a 2-sec exposure with Polaroid 3,000-ASA film.

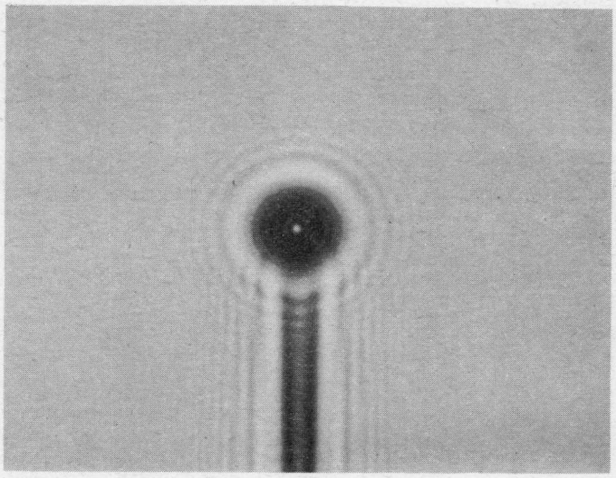

Fig. 4-51

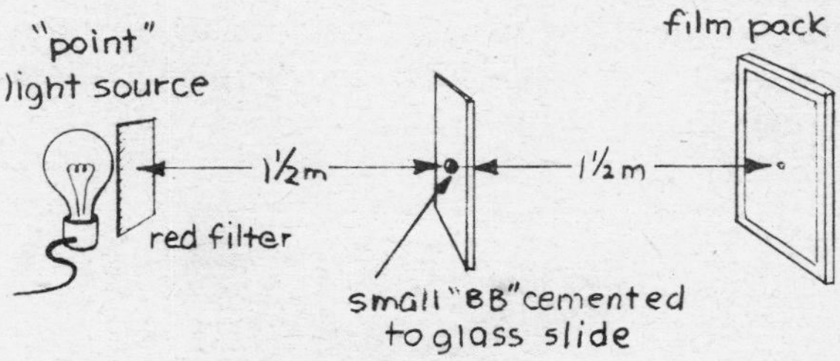

Fig. 4-52

PHOTOGRAPHIC ACTIVITIES

The number of photography activities is limitless, so we shall not try to describe many in detail. Rather, this is a collection of suggestions to give you a "jumping-off" point for classroom displays, demonstrations, and creative work.

History of Photography

Life magazine, December 23, 1966, had an excellent special issue on photography. How the world's first trichromatic color photograph was made by James Clerk Maxwell in 1861 is described in the Science Study Series paperback, *Latent Image*, by Beaumont Newhall. Much of the early history of photography in the United States is discussed in *Mathew Brady*, by James D. Horan (Crown Publishers).

Schlieren Photography

For a description and instructions for equipment, see *Scientific American*, February, 1964, pp. 132 – 133.

Infrared Photography

Try to make some photos like the one shown on page 513 of Unit 4 in the Text. Kodak infrared film is no more expensive than normal black and white film, and can be developed with normal developers. If you have a 4 × 5 camera with a Polaroid back, you can use 4 × 5 Polaroid infrared film sheets. You may find the Kodak Data Book M-3, "Infrared and Ultraviolet Photography," very helpful.

COLOR

You can easily carry out many intriguing experiments and activities related to the physical, physiological, and psychological aspects of color. Some of these are suggested here.

Scattered Light

Add about one-quarter teaspoon of milk to a drinking glass full of water. Set a flashlight about 60 cm away so that it shines into the glass. When you look through the milky water toward the light, it has a pale orange color. As you move around the glass, the milky water appears to change color. Describe the change and explain what causes it.

The Rainbow Effect

The way in which rainbows are produced can be demonstrated by using a glass of water as a large cylindrical raindrop. Place the glass on a piece of white paper in the early morning or late afternoon sunlight. To make the rainbow more visible, place two books upright, leaving a space a little wider than the glass between them, so that the sun shines on the glass but the white paper is shaded (Fig. 4-53). The rainbow will be seen on the backs of the books. What is the relationship between the arrange-

Fig. 4-53

ment of colors of the rainbow and the side of the glass that the light entered? This and other interesting optical effects are described in *Science from Your Airplane Window*, by Elizabeth A. Wood (Dover, 1975, paperback).

Color Vision by Contrast (Land Effect)

Hook up two small lamps as shown in Fig. 4-54. Place an obstacle in front of the screen so that adjacent shadows are formed on the screen. Do the shadows have any tinge of color? Now cover one bulb with a red filter and notice that the other shadow appears green by contrast. Try this with different colored filters and vary the light intensity by moving the lamps to various distances.

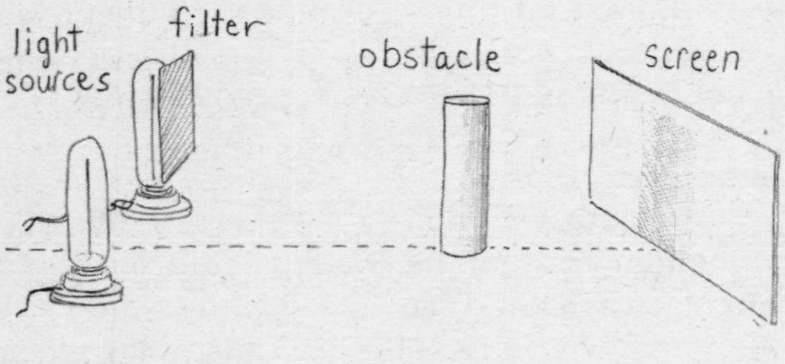

Fig. 4-54

Land Two-Color Demonstrations

A different and interesting activity is to demonstrate that a full-color picture can be created by simultaneously projecting two black-and-white transparencies taken through a red and a green filter. For more information see *Scientific American*, May 1959; September 1959; and January 1960.

The eye itself responds to color in a way that is neither obvious nor completely understood, even today. It does *not* work like a camera in this respect. For a discussion of a contemporary theory, see *Scientific American*, December, 1977, p. 108.

POLARIZED LIGHT

The use of polarized light in basic research is spreading rapidly in many fields of science. The laser, the most intense laboratory source of polarized light, was invented by researchers in electronics and microwaves. Botanists have discovered that the direction of growth of certain plants can be determined by controlling the polarization form of illumination, and zoologists have found that bees, ants, and various other creatures routinely use the polarization of sky light as a navigational "compass." High-energy physicists have found that the most modern particle accelerator, the synchrotron, is a superb source of polarized X rays. Astronomers find that the polarization of radio waves from planets and from stars offers important clues to the dynamics of those bodies. Chemists and mechanical engineers are finding new uses for polarized light as an analytical tool. Theoreticians have discovered shortcut methods of dealing with polarized light algebraically. From all sides, the onrush of new ideas is imparting new vigor to this classical subject.

A discussion of many of these aspects of the nature and application of polarized light, including activities such as those discussed below, can be found in *Polarized Light*, by W.A. Shurcliff and S.S. Ballard (Van Nostrand Momentum Book #7, 1964).

Detection

Polarized light can be detected directly by the unaided human eye, provided you know what to look for. To develop this ability, begin by staring through a sheet of Polaroid film at the sky for about 10 sec. Then quickly turn the polarizer 90° and look for a pale yellow brush-shaped pattern similar to Fig. 4-55.

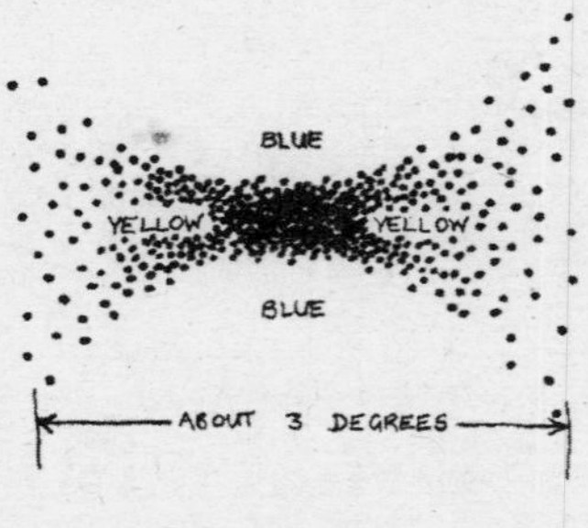

Fig. 4-55

The color will fade in a few seconds, but another pattern will appear when the Polaroid is again rotated 90°. A light-blue filter behind the Polaroid may help.

B.C. By John Hart

How is the axis of the brush related to the direction of polarization of light transmitted by the Polaroid? (To determine the polarization direction of the filter, look at light reflected from a horizontal nonmetallic surface, such as a tabletop. Turn the Polaroid until the reflected light is brightest. Put tape on one edge of the Polaroid parallel to the floor to show the direction of polarization.) Does the axis of the yellow pattern always make the same angle with the axis of polarization?

Some people see the brush most clearly when viewed with circularly polarized light. To make a circular polarizer, place a piece of Polaroid in contact with a piece of cellophane with its axis of polarization at a 45° angle to the fine stretch lines of the cellophane.

Picket-Fence Analogy

At some time, you may have heard the polarization of light explained in terms of a rope tied to a fixed object at one end, and being shaken at the other end. In between, the rope passes through two picket fences (as in Fig. 4-56), or through two slotted pieces of cardboard. This analogy suggests that when the slots are parallel the waves pass through, but when the slots are perpendicular the waves are stopped. (You may want to use a rope and slotted boards to see if this really happens.)

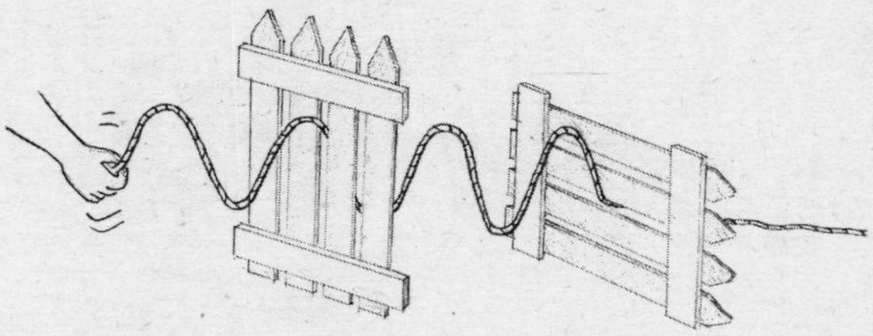

Fig. 4-56

Place two Polaroid filters parallel to each other and turn one so that it blacks out the light completely. Then place a third filter between the first two, and rotate it about the axis of all three. What happens? Does the picket-fence analogy still hold?

A similar experiment can be done with microwaves using parallel strips of tinfoil on cardboard instead of Polaroid filters. The electric field in the microwaves is "shorted out," however, when the pickets are *parallel* to the field. This is just the opposite of the rope-and-fence analogy. Prove this for yourself.

MAKE AN ICE LENS

Dr. Clawbonny, in Jules Verne's *The Adventures of Captain Hatteras*, was able to light a fire in −48° weather (thereby saving stranded travelers) by shaping a piece of ice into a lens and focusing it on some tinder. If ice is clear, the sun's rays pass through with little scattering. You can make an ice lens by freezing water in a round-bottomed bowl. Use boiled, distilled water, if possible, to minimize problems due to gas bubbles in the ice. Measure the focal length of the lens and relate this length to the radius of the bowl. (Adapted from *Physics for Entertainment,* Y. Perelman, Foreign Languages Publishing House, Moscow, 1936.)

DETECTING ELECTRIC FIELDS

Many methods can be used to explore the shape of electric fields. Two very simple ones are described here.

Gilbert's Versorium

A sensitive electric "compass" is easily constructed from a toothpick, a needle, and a cork. An external electric field induces surface charges on the toothpick. The forces on these induced charges cause the toothpick to line up along the direction of the field.

To construct the versorium, first bend a flat toothpick into a slight arc. When it is mounted horizontally, the downward curve at the ends will give the toothpick stability by lowering its center of gravity below the pivot point of the toothpick. With a small nail, drill a hole at the balance point almost all the way through the pick. Balance the pick horizontally on the needle, being sure it is free to swing like a compass needle. Try bringing charged objects near it.

For details of Gilbert's and other experiments, see Holton and Roller, *Foundations of Modern Physical Science,* Chapter 26.

Charged Ball

A charged pithball (or conductor-coated Styrofoam ball) suspended from a stick on a thin insulating thread can be used as a rough indicator of fields around charged spheres, plates, and wires.

Use a point source of light to project a shadow of the thread and ball. The angle between the thread and the vertical gives a rough measure of the forces. Use the charged pithball to explore the nearly uniform field near a large, charged plate suspended by tape strips, and the $1/r$ drop-off of the field near a long charged wire.

Plastic strips rubbed with cloth are adequate for charging well-insulated spheres, plates, or wires. (To prevent leakage of the charge from the pointed ends of a charged wire, fit the ends with small metal spheres. Even a smooth small blob of solder at the ends should help.)

AN 11¢ BATTERY

Using a penny (95% copper) and a silver dime (90% silver) you can make an 11¢ battery. Cut a 2.5-cm square of filter paper or paper towel, dip it in salt solution, and place it between the penny and the dime. Connect the penny and the dime to the terminals of a galvanometer with two lengths of copper wire. Does your meter indicate a current? Will the battery also produce a current with the penny and dime in direct dry contact?

VOLTAIC PILE

Cut 20 or more disks each of two different metals. Copper and zinc make a good combination. (The round metal "slugs" from electrical outlet-box installations can be used for zinc disks because of their heavy zinc coating.) Pennies and nickels or dimes will work, but not as well. Cut pieces of filter paper or paper towel to fit in between each pair of two metals in contact. Make a pile of the metal disks and the salt-water soaked paper, as Volta did. Keep the pile in order; for example, copper-paper-zinc, copper-paper-zinc, etc. Connect copper wires to the top and bottom ends of the pile. Touch the free ends of the wires with two fingers of one hand. What is the effect? Can you increase the effect by moistening your fingers? In what other ways can you increase the effect? How many disks do you need in order to light a flashlight bulb?

If you have metal fillings in your teeth, try biting a piece of aluminum foil. Can you explain the sensation?

MEASURING MAGNETIC FIELD INTENSITY

Many important devices used in physics experiments make use of a uniform magnetic field of known intensity. Cyclotrons, bubble chambers, and mass spectrometers are examples. Use the current balance described in Experiments 4-4 and 4-6. Measure the magnetic field intensity in the space between the pole faces of two ceramic disk magnets placed close together. Then when you are learning about radioactivity you can observe the deflection of beta particles as they pass through this space, and determine the average energy of the particles.

Bend two strips of thin sheet aluminum or copper (not iron), and tape them to two disk magnets as shown in Fig. 4-57.

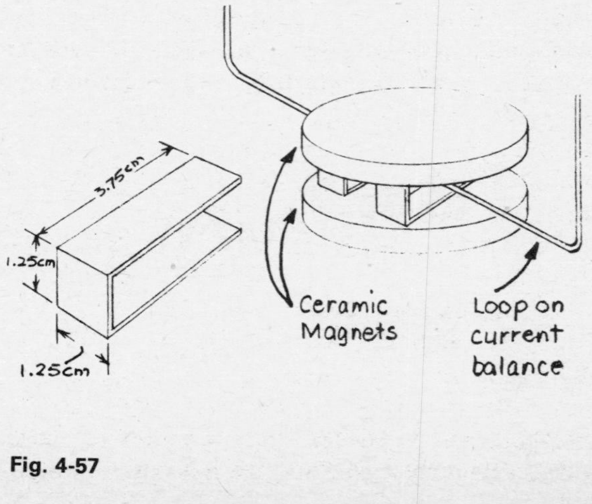

Fig. 4-57

Be sure that the pole faces of the magnets are parallel and are attracting each other (unlike poles facing each other). Suspend the movable loop of the current balance midway between the pole faces. Determine the force needed to restore the balance to its initial position when a measured current is passed through the loop. You learned in Experiment 4-5 that there is a simple relationship between the magnetic field intensity, the length of the part of the loop that is in the field, and the current in the loop. It is $F = BIl$, where F is the force of the loop (in newtons), B is the magnetic field intensity (in newtons per ampere-meter), I is the current (in amperes), and l is the length (in meters) of that part of the current-carrying loop that is actually in the field. With your current balance, you can measure F, I, and l, and thus compute B.

For this activity, make two simplifying assumptions that are not strictly true but which enable you to obtain reasonably good values for B: (a) the field is fairly uniform throughout the space between the poles, and (b) the field drops to zero outside this space. With these approximations you can use the diameter of the magnets as the quantity l in the above expression.

Try the same experiment with two disk magnets above and two below the loop. How does B change? Bend metal strips of different shapes so you can vary the distance between pole faces. How does this affect B?

An older unit of magnetic field intensity still often used is the *gauss* (G). To convert from one unit to the other, use the conversion factor, 1 N/A (1 tesla, T) = 10^4 G.

MORE PERPETUAL MOTION MACHINES

The diagrams in Figs. 4-58 and 4-59 show two more of the perpetual motion machines discussed by R. Raymond Smedile in his book, *Perpetual Motion and Modern Research for Cheap Power*. (See also page 150 of Unit 3, *Handbook*.) What is the weakness of the argument for each of them? (Also see "Perpetual Motion Machines," Stanley W. Angrist, *Scientific American*, January, 1968.)

In Fig. 4-58, A represents a stationary wheel around which is a larger, movable wheel, E. Three magnets marked B are placed on stationary wheel A in the position shown in the drawing. On rotary wheel E are placed eight magnets marked D. The magnets are attached

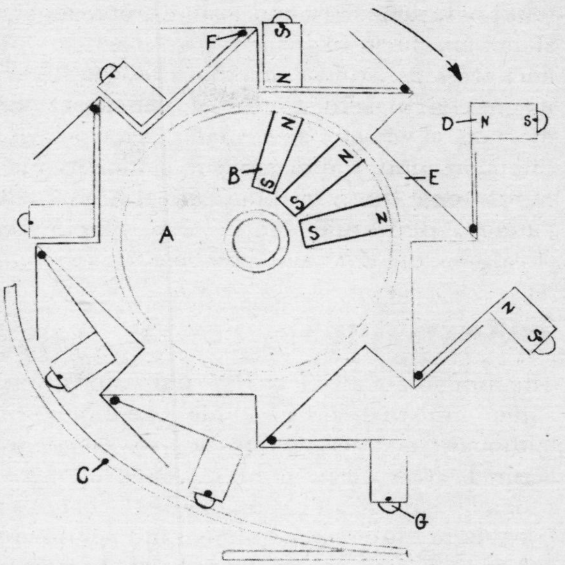

Fig. 4-58

to eight levers and are securely hinged to wheel E at the point marked F. Each magnet is also provided with a roller wheel, G, to prevent friction as it rolls on the guide marked C.

Guide C is supposed to push each magnet toward the hub of this mechanism as it is being carried upward on the left-hand side of the mechanism. As each magnet rolls over the top, the fixed magnets facing it cause the magnet on the wheel to fall over. This creates an overbalance of weight on the right of wheel E and thus perpetually rotates the wheel in a clockwise direction.

In Fig. 4-59, A represents a wheel in which eight hollow tubes marked E are placed. In each of the tubes a magnet, B, is inserted so

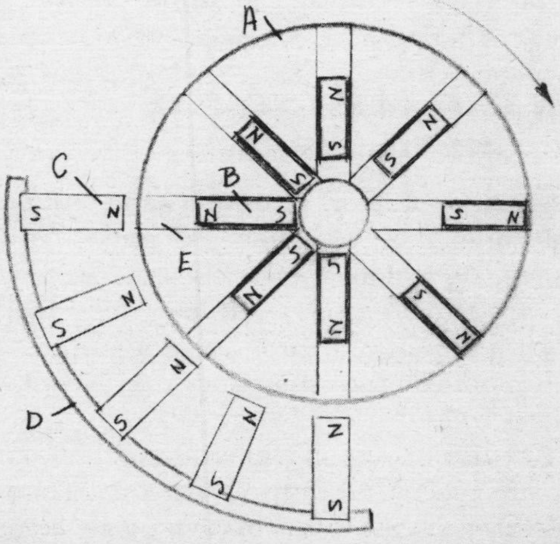

Fig. 4-59

that it will slide back and forth. D represents a stationary rack in which five magnets are anchored as shown in the drawing. Each magnet is placed so that it will repel the magnets in wheel A as it rotates in a clockwise direction. Since the magnets in stationary rack D will repel those in rotary wheel A, this will cause a perpetual overbalance of magnet weight on the right side of wheel A.

TRANSISTOR AMPLIFIER

The function of a PNP or NPN transistor is very similar to that of a triode vacuum tube (although its operation is not so easily described). The diagram in Fig. 4-60 shows a schematic transistor circuit that is analogous to the vacuum tube circuit shown in Experiment 4-8. In both cases, a small input signal controls a large output current.

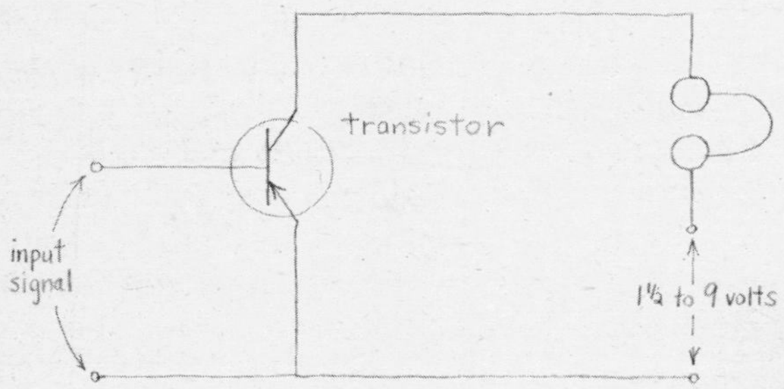

Fig. 4-60

Some inexpensive transistors can be bought at almost any radio supply store, and almost any PNP or NPN will do. Such stores also usually carry a variety of paperback books that give simplified explanations of how transistors work and how you can use cheap components to build some simple electronic equipment.

AN ISOLATED NORTH MAGNETIC POLE?

Magnets made of a rather soft rubber-like substance are available in some hardware stores. Typical magnets are flat pieces 20 mm × 25 mm and about 5 mm thick, with a magnetic north pole on one 20 × 25 mm surface and a south pole on the other. They may be cut with a sharp knife.

Cut six of these magnets so that you have six square pieces, 20 mm on an edge. Then level the edges on the S side of each piece so that the pieces can be fitted together to form a hollow cube with all the N sides facing outward. The pieces repel each other strongly and may be either glued (with rubber cement) or tied together with thread.

Do you now have an isolated north pole, that is, a north pole all over the outside (and south pole on the inside)?

Is there a magnetic field directed outward from all surfaces of the cube?

FARADAY DISK DYNAMO

You can easily build a disk dynamo similar to the one shown in Fig. 4-61. Cut a 20-cm diameter disk of sheet copper. Drill a hole in the center of the disk and put a bolt through the hole. Run a nut tight up against the disk so the disk will not slip on the bolt. Insert the bolt in a hand drill and clamp the drill in a ring stand so that the disk passes through the region between the poles of a large magnet. Connect one wire of a 100-microamp (μA) dc meter to the metal part of the drill that does not turn. Tape the other wire to the magnet so it brushes lightly against the copper disk as the disk is spun between the magnet poles.

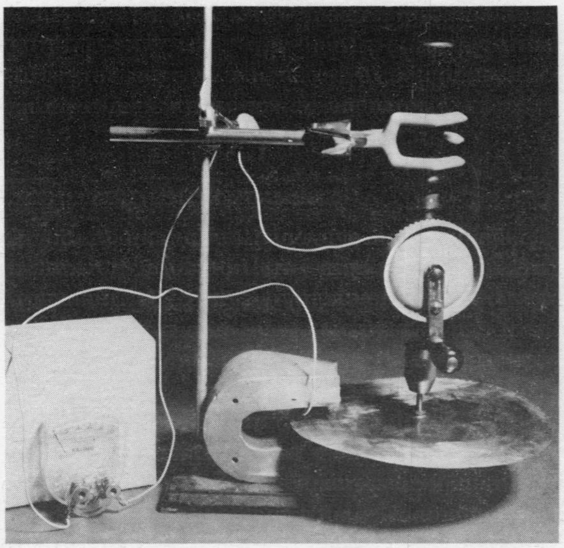

Fig. 4-61

Frantic cranking can create a 10-μA current with the magnetron magnet shown in Fig. 4-61. If you use one of the metal yokes from the current balance, with three ceramic magnets on each side of the yoke, you may be able to get the needle to move from the zero position just noticeably.

The braking effect of currents induced in the disk can also be noticed. Remove the meter, wires, and magnet. Have one person crank while another holds the magnet so that the disk is spinning between the magnet poles. Compare the effort needed to turn the disk with and without the magnet over the disk.

If the disk will coast, compare the coasting times with and without the magnet in place. (If there is too much friction in the hand drill for the disk to coast, loosen the nut and spin the disk by hand on the bolt.)

GENERATOR JUMP ROPE

With a piece of wire about twice the length of a room, and a sensitive galvanometer, you can generate an electric current using only the earth's magnetic field. Connect the ends of the wire to the meter. Pull the wire out into a long loop and twirl half the loop like a jump rope. As the wire cuts across the earth's magnetic field, a voltage is generated. If you do not have a sensitive meter, connect the input of one of the amplifiers, and connect the amplifier to a less sensitive meter.

How does the current generated when the axis of rotation is along a north–south line compare with the current generated with the same motion along an east–west line? What does this tell you about the earth's magnetic field? Is there any effect if the people stand on two landings and hang the wire (while swinging it) down a stairwell?

SIMPLE METERS AND MOTORS

You can make workable current meters and motors from very simple parts:

2 ceramic magnets, 1 steel yoke, 1 #7 cork (from current balance kit)
1 metal rod, about 2 mm in diameter and 12 cm long (a piece of bicycle spoke, coat-hanger wire, or a large finishing nail will do)
1 block of wood, about 10 cm × 5 cm × 1 cm
about 2.7 m of insulated #30 copper magnet wire
2 thumbtacks
2 safety pins
2 carpet tacks or small nails, 1 white card (10 cm × 12.5 cm), stiff black paper, for pointer (for meter only)
electrical insulating tape (for motor only)

Meter

To build a meter, follow the steps below paying close attention to the diagrams. Push the rod through the cork. Make the rotating coil, or *armature*, by winding about 20 turns of wire around the cork, keeping the turns parallel to the rod. Leave about 30 cm of wire at both ends (Fig. 4-62).

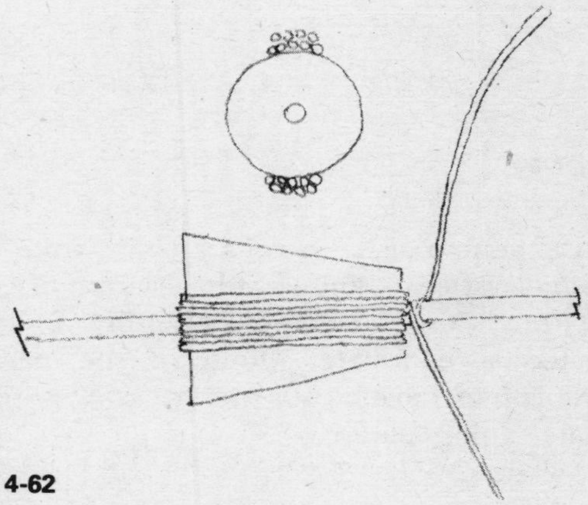

Fig. 4-62

Use nails or carpet tacks to fix two safety pins firmly to the ends of the wooden-base block (Fig. 4-63).

Fig. 4-63

Make a pointer out of the black paper and push it onto the metal rod. Pin a piece of the white card to one end of the base. Then suspend the armature between the two safety pins from the free ends of wire into two loose coils, and attach them to the base with thumbtacks. Put the two ceramic magnets on the yoke (unlike poles facing), and place the yoke around the armature (Fig. 4-64). Clean the insulation off the ends of the leads, and you are ready to connect your meter to a low-voltage dc source.

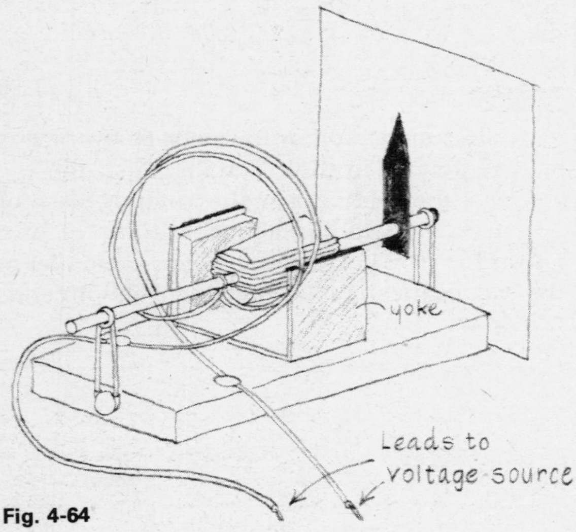

Fig. 4-64

Calibrate a scale in volts on the white card using a variety of known voltages from dry cells or from a low-voltage power supply, and your meter is complete. Minimize the parallax problem by having your pointer as close to the scale as possible.

Motor

To make a motor, wind an armature as you did for the meter. Leave about 6 cm of wire at each end; carefully scrape the insulation from the wire. Bend each into a loop and then twist into a tight pigtail. Tape the two pigtails along opposite sides of the metal rod (Fig. 4-65).

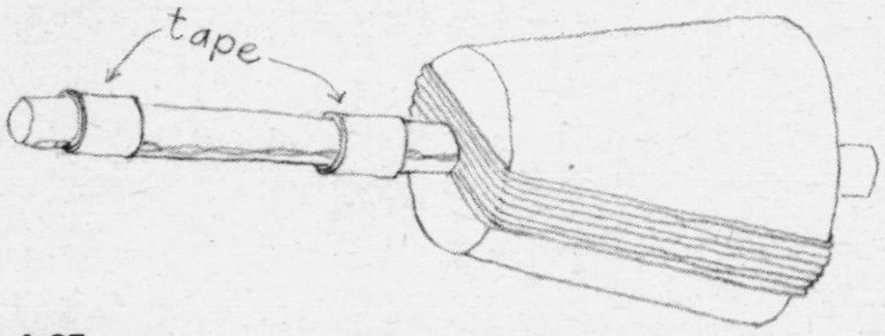

Fig. 4-65

Fix the two safety pins to the base as for the meter, and mount the coil between the safety pins.

The leads into the motor are made from two pieces of wire attached to the baseboard with thumbtacks at points X (Fig. 4-66).

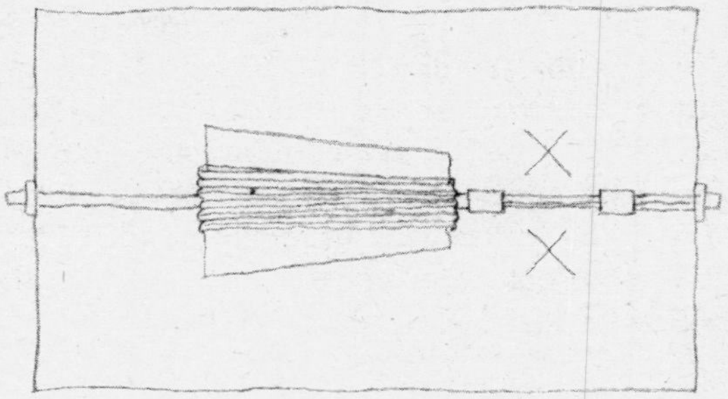

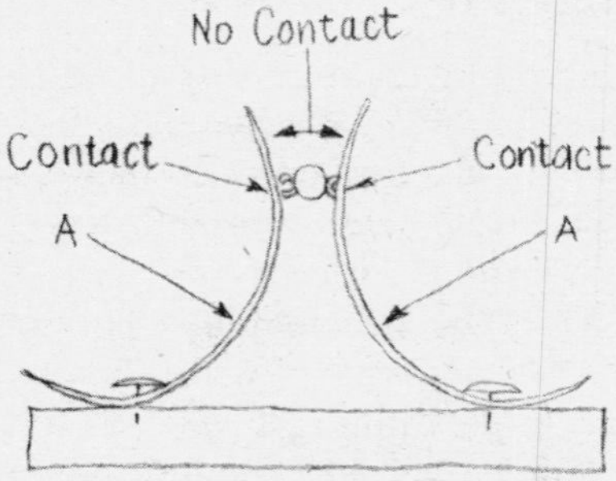

Fig. 4-66

Place the magnet yoke around the coil. The coil should spin freely (Fig. 4-67).

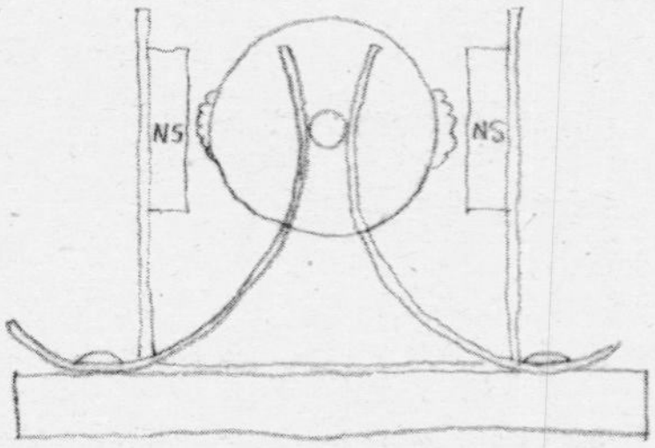

Fig. 4-67

Connect a 1.5-volt battery to the leads. Start the motor by spinning it with your finger. If it does not start, check the contacts between leads and the contact wires on the rod. You may not have removed all the enamel from the wires. Try pressing lightly at points A (Fig. 4-66) to improve the contact. Also check to see that the two contacts touch the armature wires at the same time (Fig. 4-67).

SIMPLE MOTOR – GENERATOR DEMONSTRATION

With two fairly strong U-magents and two coils, which you wind yourself, you can prepare a simple demonstration showing the principles of a motor and a generator. Wind two flat coils of magnet wire 100 turns each. The cardboard tube from a roll of paper towels makes a good form. Leave about 0.5 m of wire free at each end of the coil. Tape the coil so it does not wind when you remove it from the cardboard tube.

Hang the coils from two supports as shown (Fig. 4-68) so the coils pass over the poles of two U-magnets set on the table about 1 m apart. Connect the coils. Pull one coil to one side and release it. What happens to the other coil? Why? Does the same thing happen if the coils are not connected to each other? What if the magnets are reversed?

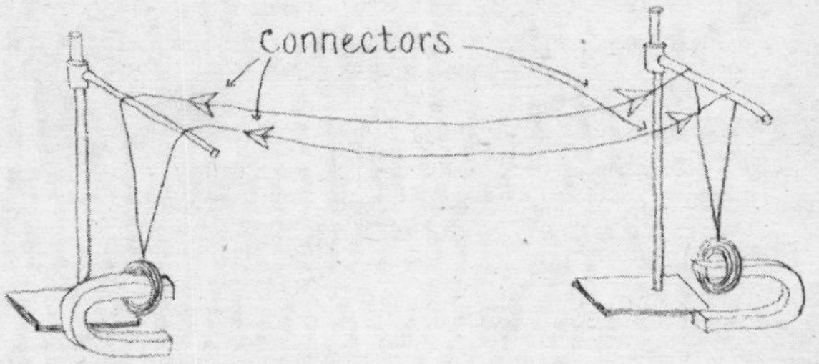

Fig. 4-68

Try various other changes, such as turning one of the magnets over while both coils are swinging, or starting with both coils at rest and then sliding one of the magnets back and forth.

If you have a sensitive galvanometer, it is interesting to connect it between the two coils.

PHYSICS COLLAGE

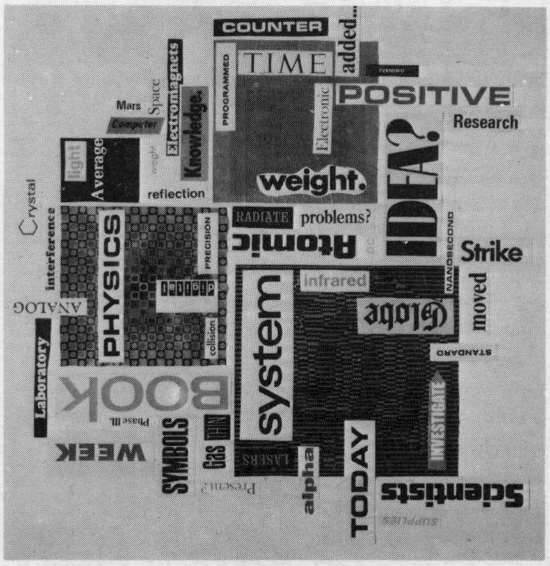

Many of the words used in physics class enjoy wide usage in everyday language. Cut "physics words" out of magazines, newspapers, etc., and make your own collage. You may wish to take on a more challenging art problem by trying to give a visual representation of a physical concept, such as speed, light, or waves.

BICYCLE GENERATOR

The generator on a bicycle (Fig. 4-69) operates on the same basic principle as that described in the Text, but with a different, and extremely simple, design. Take apart such a generator and see if you can explain how it works. *Note:* You may not be able to reassemble it.

Fig. 4-69

LAPIS POLARIS, MAGNES

The etching in Fig. 4-70 shows a philosopher in his study surrounded by the scientific equipment of his time. In the left foreground, in a basin of water, a natural magnet or lodestone floating on a piece of wood orients itself north and south. Traders from the great Mediterranean port of Amalfi probably introduced the floating compass, having learned of it from Arab mariners. An Amalfi historian, Flavius Blondus, writing about A.D. 1450, indicates the uncertain origin of the compass, but later historians in repeating this early reference warped it and gave credit for the discovery of the compass to Flavius.

2.
LAPIS POLARIS, MAGNES.
Lapis recluſit iſte Flauio abditum Poli ſuum hunc amorem, at ipse nauitæ.

Fig. 4-70

Can you identify the various devices lying around the study? When do you think the etching was made? (If you have some background in art, you might consider whether your estimate on the basis of scientific clues is consistent with the style of the etching.)

MICROWAVE TRANSMISSION SYSTEMS

Microwaves of about 6-cm wavelength are used to transmit telephone conversations over long distances. Because microwave radiation has a limited range, a series of relay stations has been erected about 50 km apart. At each station the signal is detected and amplified before being retransmitted to the next one. If you have several microwave generators that can be amplitude modulated, see if you can put together a demonstration of how this system works. You will need an audio frequency oscillator (or microphone), amplifier, microwave generator and power supply, detector; another amplifier and a loudspeaker; another microwave generator; and another detector, a third amplifier, and a loudspeaker.

GOOD READING

Several good paperbacks in the Science Study Series (Anchor Books, Doubleday and Co.) are appropriate for Unit 4, including *The Physics of Television,* by Donald G. Fink and David M. Lutyens; *Waves and Messages,* by John R. Pierce; *Quantum Electronics,* by John R. Pierce; *Electrons and Waves,* by John R. Pierce; *Computers and the Human Mind,* by Donald G. Fink. Throughout this course, you should make a point of checking your library for books or articles on topics that interest you.

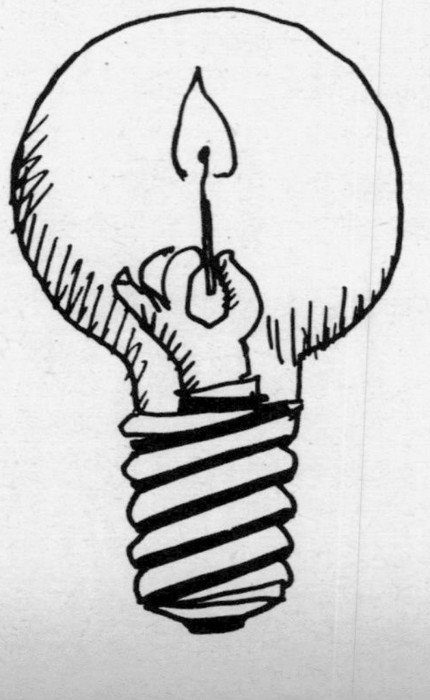

FILM LOOP NOTES

Film Loop 44
STANDING ELECTROMAGNETIC WAVES

Standing waves are not confined to mechanical waves in strings or in gas. It is only necessary to reflect the wave at the proper distance from a source so that two oppositely moving waves superpose in just the right way. In this film, standing electromagnetic waves are generated by a radio transmitter.

The transmitter produces electromagnetic radiation at a frequency of 435×10^6 Hz. Since all electromagnetic waves travel at the speed of light, the wavelength is $\lambda = c/f = 0.69$ m. The output of the transmitter oscillator (Fig. 4-71) passes through a power-indicating meter, then to an antenna of two rods each one-quarter wavelength ($\frac{1}{4}\lambda$), or a total antenna of $\frac{1}{2}\lambda$, long.

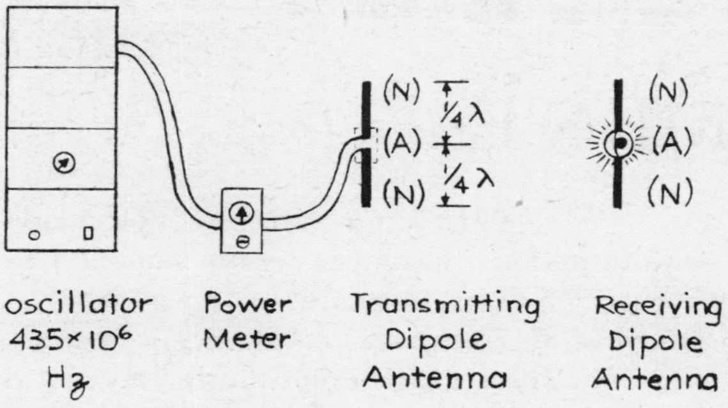

Fig. 4-71

The receiving antenna (Fig. 4-72) is also $\frac{1}{2}\lambda$ long. The receiver is a flashlight bulb connected between two stiff wires each $\frac{1}{4}\lambda$ long. If the electric field of the incoming wave is parallel to the receiving antenna, the force on the electrons in the wire drives them back and forth through the bulb. The brightness of the bulb indicates the intensity of the electromagnetic radiation at the antenna. A rectangular aluminum cavity, open toward the camera, confines the waves to provide sufficient intensity.

Initial scenes show how the intensity depends on the distance of the receiving antenna from the transmitting antenna. The radiated

Fig. 4-72

power is about 20 watts. Does the received intensity decrease as the distance increases? The radiation has vertical polarization, so the response falls to zero when the receiving antenna is rotated to the horizontal position.

Standing waves are set up when a metal reflector is placed at the right end of the cavity. The reflector must be placed at a node. The distance from source to reflector must be an integral number of half-wavelengths plus one-quarter wavelength. The cavity length must be "tuned" to the wavelength. Nodes and antinodes are identified by moving a receiving antenna back and forth. Then a row of vertical receiving antennas is placed in the cavity, and the nodes and antinodes are shown by the pattern of brilliance of the lamp bulbs. How many nodes and antinodes can be seen in each trial?

Standing waves of different types can all have the same wavelength. In each case, a source is required (tuning fork, loudspeaker, or dipole antenna). A reflector is also necessary (support for string, wooden piston, or sheet aluminum mirror). If the frequencies are 72 Hz for the string, 505 Hz for the gas, and 435×10^6 Hz for the electromagnetic waves and they all have the same wavelength, what can you conclude about the speeds of these three kinds of waves? Discuss the similarities and differences among the three cases. What can you say about the "medium" in which the electromagnetic waves travel?

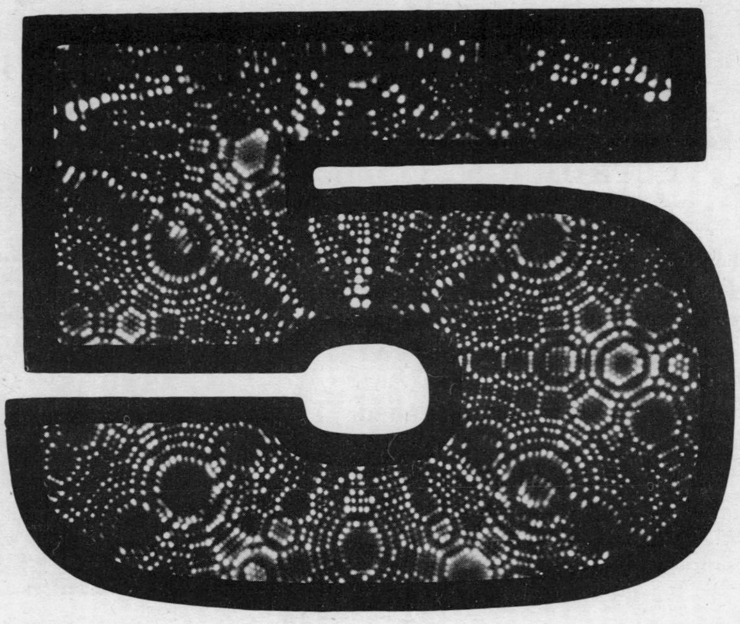

Models of the Atom

EXPERIMENTS

Experiment 5-1
ELECTROLYSIS

Volta and Davy discovered that electric currents create chemical changes never observed before. As you have already learned, these scientists were the first to use electricity to break down apparently stable compounds and to isolate certain chemical elements.

Later, Faraday and other experimenters compared the *amount* of electric charge used with the *amount* of chemical products formed in such electrochemical reactions. Their measurements fell into a regular pattern that hinted at some underlying link between electricity and matter.

In this experiment, you will use an electric current just as Volta and Davy did to decompose a compound. By comparing the charge used with the mass of one of the products, you can compute the mass and volume of a single atom of the product.

Theory Behind the Experiment

A beaker of copper sulfate ($CuSO_4$) solution in water is supported under one arm of a balance (Fig. 5-1). A negatively charged copper electrode is supported in the solution by the balance arm so that you can measure its mass without removing it from the solution. A second, positively charged copper electrode fits around the inside wall of the beaker. The beaker, its solution, and the positive electrode are *not* supported by the balance arm.

If you have studied chemistry, you probably know that in solution copper sulfate breaks down into separate, charged particles, called *ions*, of copper (Cu^{++}) and sulfate ($SO_4^{=}$), which move about freely in the solution.

When a voltage is applied across the copper electrodes, the electric field causes the $SO_4^{=}$ ions to drift to the positive electrode (or anode) and the Cu^{++} ions to drift to the negative electrode (or cathode). At the cathode, the Cu^{++} particles acquire enough negative charge to form neutral copper atoms that deposit on the cathode and add to its weight. The motion of charged particles toward the electrodes is a continuation of the electric current in the wires and the rate of transfer of charge (coulombs per second) is equal to it in magnitude. The electric current is provided by a power supply that

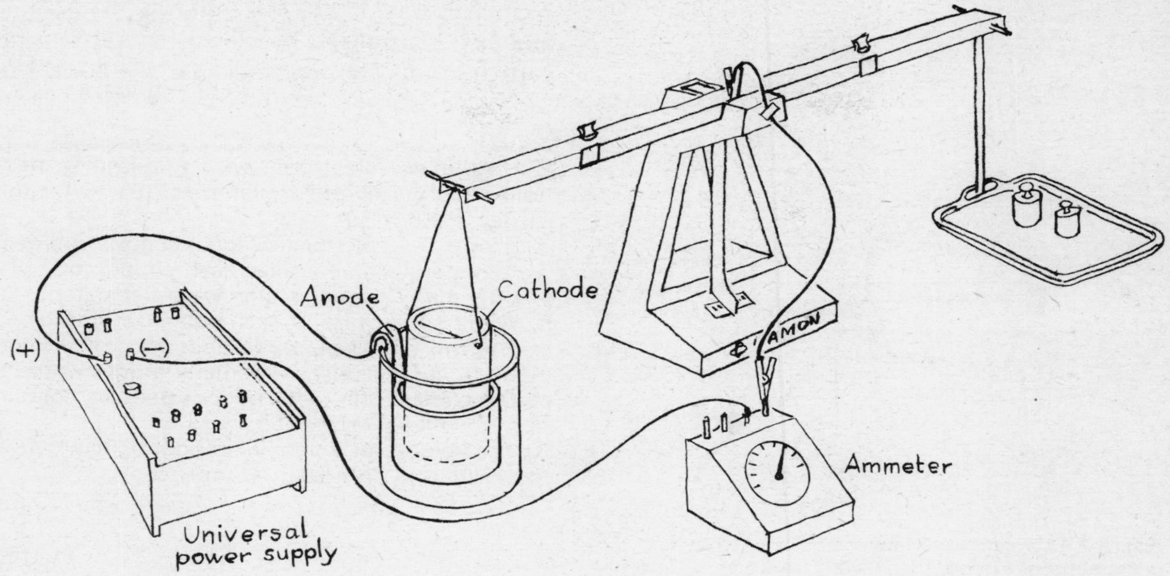

Fig. 5-1

converts a 100-V alternating current into a low-voltage direct current. The current is set by a variable control on the power supply (or by an external rheostat) and measured by an ammeter in series with the electrolytic cell as shown in Fig. 5-1.

With a watch to measure the time the current flows, you can compute the electric charge that passed through the cell. By definition, the current I is the rate of transfer of charge, $I = \Delta Q / \Delta t$. It follows that the charge transferred is the product of the current and the time.

$$\Delta Q = I \times \Delta t$$

$$\left(\text{coulombs} = \frac{\text{coulombs}}{\text{sec}} \times \text{sec}\right)$$

Since the amount of charge carried by a single electron is known ($q_e = 1.6 \times 10^{-19}$ C), the number of electrons transferred, N_e, is

$$N_e = \frac{\Delta Q}{q_e}$$

If n electrons are needed to neutralize each copper ion, then the number of copper atoms deposited, N_{Cu}, is

$$N_{Cu} = \frac{N_e}{n}$$

If the mass of each copper atom is m_{Cu}, then the total mass of copper deposited, M_{Cu}, is

$$M_{Cu} = N_{Cu} m_{Cu}$$

Thus, if you measure I, Δt, and M_{Cu} and you know q_e and n, you can calculate a value for m_{Cu}, the mass of a single copper atom!

Combining the above equations algebraically gives

$$m_{Cu} = \frac{M_{Cu} n q_e}{I \, \Delta t}$$

Setup and Procedure

Either an equal-arm or a triple-beam balance can be used for this experiment. First arrange the cell and the balance as shown in Fig. 5-1. The cathode cylinder must be supported far enough above the bottom of the beaker so that the balance arm can move up and down freely when the cell is full of the copper sulfate solution.

Next connect the circuit as illustrated in the figure. Note that the electrical connection from the negative terminal of the power supply to the cathode is made through the balance beam. The knife-edge and its seat *must* be bypassed by a short piece of thin flexible wire, as shown in Fig. 5-1 for equal-arm balances, or in Fig. 5-2 for triple-beam balances. The positive terminal of the power supply is connected directly to the anode in any convenient manner.

Before any measurements are made, operate the cell long enough (10 – 15 min) to form a preliminary deposit on the cathode unless this has already been done. In any case, run the current long enough to set it at the value recommended by your instructor, probably about 5 A.

When all is ready, adjust the balance and record its reading. Pass the current for the length of time recommended by your instruc-

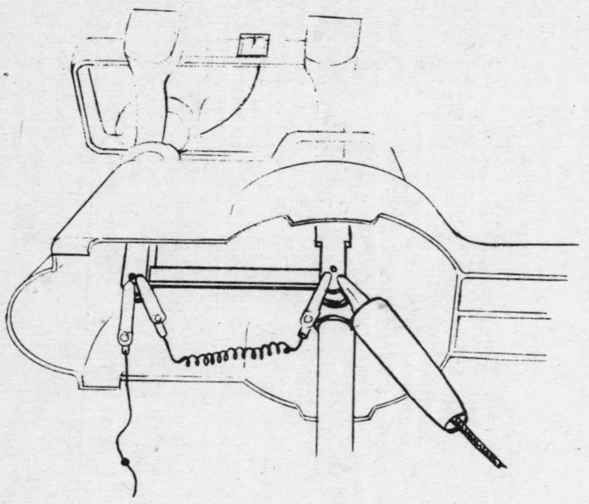

Fig. 5-2 This cutaway view shows how to bypass the knife-edge of a typical balance. The structure of other balances may differ.

tor. Measure and record the current I and the time interval Δt during which the current passes. Check the ammeter occasionally and, if necessary, adjust the control in order to keep the current set at its original value.

At the end of the run, record the new reading of the balance and find, by subtraction, the increase in mass of the cathode.

Calculating Mass and Volume of an Atom

Since the cathode is buoyed by a liquid, the masses you have measured are not the true masses. Because of the buoyant force exerted by the liquid, the mass of the cathode and its increase in mass will both appear to be less than they would be in air. To find the true mass increase, you must divide the observed mass increase by the factor $(1 - D_s/D_c)$, where D_s is the density of the solution and D_c is the density of the copper.

Your instructor will give you the values of these two densities if you cannot find values for them yourself, and also explain how the correction factor is derived. The important thing for you to understand here is why a correction factor is necessary.

?

1. How much positive or negative charge was transferred to the cathode?

In the solution, this positive charge is carried from anode to cathode by doubly charged copper ions, Cu^{++}. At the cathode the copper ions are neutralized by electrons and neutral copper atoms are deposited: $Cu^{++} + 2e^- = Cu$.

?

2. How many electrons were required to neutralize the total charge transferred? (Each electron carries -1.6×10^{-19} C.)
3. How many electrons (single negative charges) were required to neutralize each copper ion?
4. How many copper atoms were deposited?
5. What is the mass of each copper atom?
6. The mass of a penny is about 3 g. If it were made of copper only, how many atoms would a penny contain? (In fact, modern pennies contain zinc as well as copper.)
7. The volume of a penny is about 0.3 cm³. How much volume does each atom occupy?

Experiment 5-2
THE CHARGE-TO-MASS RATIO FOR AN ELECTRON

In this experiment, you will make measurements on cathode rays. A set of similar experiments by J. J. Thomson convinced physicists that these rays are not waves but streams of identical, charged particles, each with the same ratio of charge to mass. If you did the experiment on the "Electron Beam Tube," you have already worked with cathode rays and have seen how they can be deflected by electric and magnetic fields.

Thomson's use of this deflection is described on page 543 of Unit 5, Text. Read that section of the Text before beginning this experiment.

Theory of the Experiment

The basic plan of the experiment is to measure the bending of the electron beam by a known magnetic field. From these measurements and a knowledge of the voltage accelerating the electrons, you can calculate the electron charge-to-mass ratio. The reasoning behind the calculation is illustrated in Fig. 5-3. The algebraic steps are described below.

When the beam of electrons (each of mass m and charge q_e) is bent into a circular arc of radius R by a uniform magnetic field B, the centripetal force mv^2/R on each electron is supplied by the magnetic force Bq_ev. Therefore,

$$\frac{mv^2}{R} = Bq_ev$$

or, rearranging to get v by itself,

$$v = \frac{Bq_eR}{m}$$

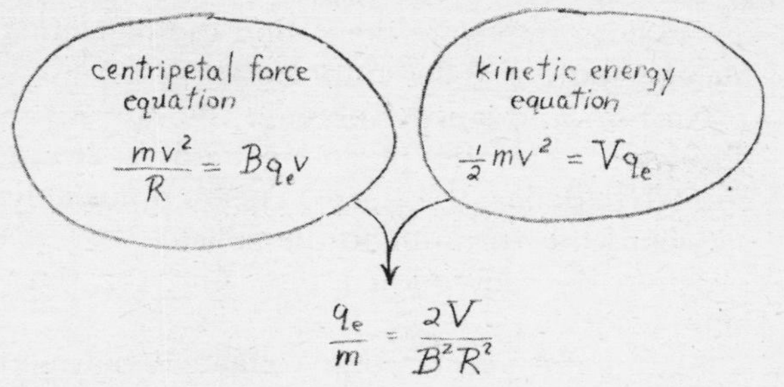

Fig. 5-3 The combination of two relationships, for centripetal and kinetic energy, with algebraic steps that eliminate velocity, *v*, lead to an equation for the charge-to-mass ratio of an electron.

The electrons in the beam are accelerated by a voltage *V*, which gives them a kinetic energy

$$\frac{mv^2}{2} = Vq_e$$

If you replace *v* in this equation by the expression for *v* in the preceding equation, you get

$$\frac{m}{2}\left(\frac{Bq_eR}{m}\right)^2 = Vq_e$$

or, after simplifying,

$$\frac{q_e}{m} = \frac{2V}{B^2R^2}$$

You can measure with your apparatus all the quantities on the right-hand side of this expression, so you can use it to calculate the charge-to-mass ratio for an electron.

Preparing the Apparatus

You will need a tube that gives a beam at least 5 cm long. If you kept the tube you made in Experiment 4-7, you may be able to use that. If your class did not have success with that experiment, you will have to use another method.

In this experiment, you need to be able to adjust the strength of the magnetic field until the magnetic force on the charges just balances the force due to the electric field. To enable you to change the magnetic field, you will use a pair of coils instead of permanent magnets. A current in a pair of coils, which are separated by a distance equal to the coil radius, produces a nearly uniform magnetic field in the central region between the coils. You can vary the magnetic field by changing the current in the coils.

Into a cardboard tube about 7.5 cm in diameter and 7.5 cm long, cut a slot 3 cm wide (Fig. 5-4). Your electron beam tube should fit into this slot as shown in the photograph of the completed setup (Fig. 5-5). A current in the pair of coils will create a magnetic field at right angles to the axis of the cathode rays.

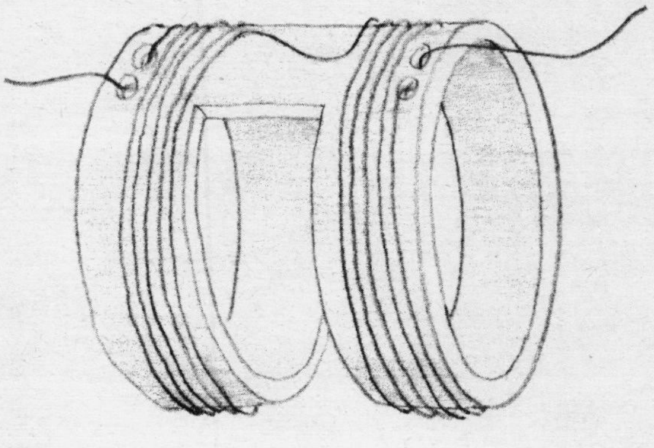

Fig. 5-4

Fig. 5-5 The magnetic field is parallel to the axis of the coils; the electric and magnetic fields are perpendicular to each other and to the electron beam.

Now wind the coils, one on each side of the slot, using a single length of insulated copper wire (magnet wire). Wind about 20 turns of wire for each of the two coils, one coil on each side of the slot, leaving 25 cm of wire free at both ends of the coil. Do not cut the wire off the reel until you know how much you will need. Make the coils as neat as you can and keep them close to the slot. Wind both coils in the same direction (for example, make both clockwise).

When you have made your set of coils, you must *calibrate* it; that is, you must find out what magnetic field strength *B* corresponds to what value of current *I* in the coils. To do this, you can use the current balance, as you did in Experiment 4-5. Use the shortest of the balance "loops" so that it will fit inside the coils as shown in Fig. 5-6.

Fig. 5-6

Connect the two leads from your coils to a power supply capable of giving up to 5 A of direct current. There must be a variable control on the power supply (or a rheostat in the circuit) to control the current; and an ammeter to measure it.

Measure the force F for a current I in the loop. To calculate the magnetic field due to the current in the coils, use the relationship $F = BIl$ where l is the length of short section of the loop. Do this for several different values of current in the coil and plot a calibration graph of magnetic field B against coil current I.

Set up your electron beam tube as in Experiment 4-7. Reread the instructions for operating the tube.

Connect a shorting wire between the pins for the deflecting plates. This will insure that the two plates are at the same electric potential, so the electric field between them will be zero. Pump the tube out and adjust the filament current until you have an easily visible beam. Since there is no field between the plates, the electron beam should go straight up the center of the tube between the two plates. (If it does not, it is probably because the filament and the hole in the anode are not properly aligned.)

Turn down the filament current and switch off the power supply. Now, without releasing the vacuum, mount the coils around the tube as shown in Fig. 5-6.

Connect the coils as before to the power supply. Connect a voltmeter across the power supply terminals that provide the accelerating voltage V.

Your apparatus is now complete.

Performing the Experiment

Turn on the beam and make sure it is travelling in a straight line. The electric field remains off throughout the experiment, and the deflecting plates should still be connected.

Turn on and slowly increase the current in the coils until the magnetic field is strong enough to deflect the electron beam noticeably.

Record the current I in the coils.

Using the calibration graph, find the magnetic field B.

Record the accelerating voltage V between the filament and the anode plate.

Finally, you need to measure R, the radius of the arc into which the beam is bent by the magnetic field. The deflected beam is slightly fan-shaped because some electrons are slowed by collisions with air molecules and are bent into a curve of smaller radius R. You need to know the largest value of R (the "outside" edge of the curved beam), which is the path of electrons that have made no collisions. You will not be able to measure R directly, but you can find it from measurements that are easy to make (Fig. 5-7).

You can measure x and d. It follows from Pythagoras' theorem that $R^2 = d^2 + (R - x)^2$, so

$$R = \frac{d^2 + x^2}{2x}$$

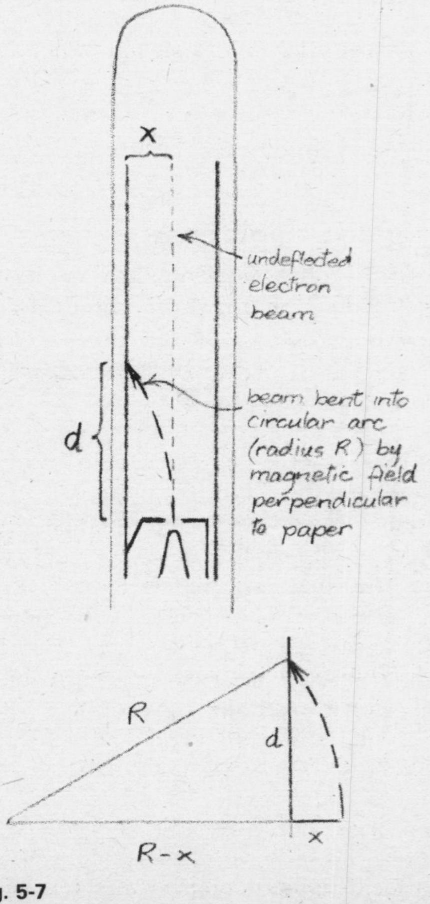

Fig. 5-7

?

1. What is your calculation of R on the basis of your measurements?

Now that you have values for V, B, and R, you can use the formula $q_e/m = 2V/B^2R^2$ to calculate your value for the charge-to-mass ratio for an electron.

?

2. What is your value for q_e/m, the charge-to-mass ratio for an electron?

Experiment 5-3 THE MEASUREMENT OF ELEMENTARY CHARGE

In this experiment, you will investigate the charge of the electron, which is a fundamental physical constant in electricity, electromagnetism, and nuclear physics. This experiment is substantially the same as Millikan's famous oil-drop experiment, described on page 547 of Unit 5, Text. The following instructions assume that you have read that description. Like Millikan, you are going to measure very small electric charges to see if there is a limit to how small an electric charge can be. Try to answer the following three questions before you begin to do the experiment in the lab.

?

1. What is the electric field between two parallel plates separated by a distance d meters if the potential difference between them is V volts?
2. What is the electric force on a particle carrying a charge of q coulombs in an electric field of E volts/meter?
3. What is the gravitational force on a particle of mass m in the earth's gravitational field?

Background

Electric charges are measured through the forces they experience and produce. The extremely small charges that you are seeking require that you measure extremely small forces. Objects on which such small forces can have a visible effect must also, in turn, be very small.

Millikan used the electrically charged droplets produced in a fine spray of oil. The varying size of the droplets complicated his measurements. Fortunately, you can now use suitable objects whose sizes are accurately known. You will use tiny latex spheres (about 10^{-4} cm diameter), which are almost identical in size in any given sample. In fact, these spheres, shown magnified (about 5,000 ×) in Fig. 5-8, are used as a convenient way to find the magnifying power of electron microscopes. The spheres can be bought in a water suspension, with their diameter recorded on the bottle. When the suspension is sprayed into the air, the water quickly evaporates and leaves a cloud of these particles, which have become charged by friction during the spraying. In the space between the plates of the Millikan apparatus, they appear through the 50-power microscope as bright points of light against a dark background.

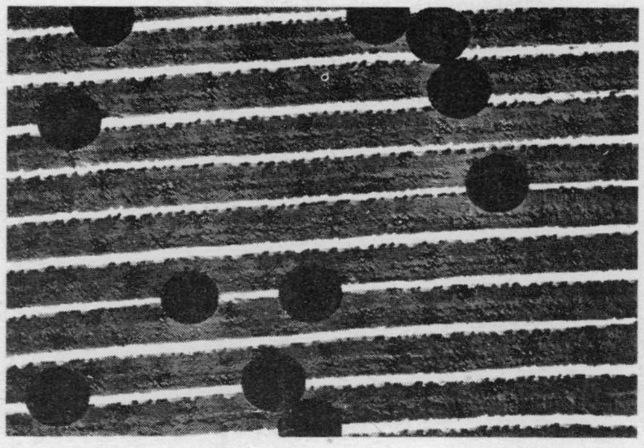

Fig. 5-8 Electron micrograph of latex spheres 1.1×10^{-4} cm, silhouetted against diffraction grating of 11,340 lines/cm. What magnification does this represent?

You will find that an electric field between the plates can pull some of the particles upward against the force of gravity, so you will know that they are charged electrically.

In your experiment, you will adjust the voltage producing the electric field until a particle hangs motionless. On a balanced particle carrying a charge q, the upward electric force Eq and the downward gravitational force ma_g are equal; therefore,

$$ma_g = Eq$$

The field $E = V/d$, where V is the voltage between the plates (the voltmeter reading) and d is the separation of the plates. It follows that

$$q = \frac{ma_g d}{V}$$

Notice that $ma_g d$ is a constant for all measurements and need be found only once.

Each value of q will be this constant ma_gd times $1/V$ as the equation above shows. That is, the value of q for a particle is proportional to $1/V$; the greater the voltage required to balance the weight of the particle, the smaller the charge of the particle must be.

Using the Apparatus

If the apparatus is not already in operating condition, consult your teacher. Study Figs. 5-9 and 5-10 until you can identify the various parts. Then switch on the light source and look through the microscope. You should see a series of lines in clear focus against a uniform gray background.

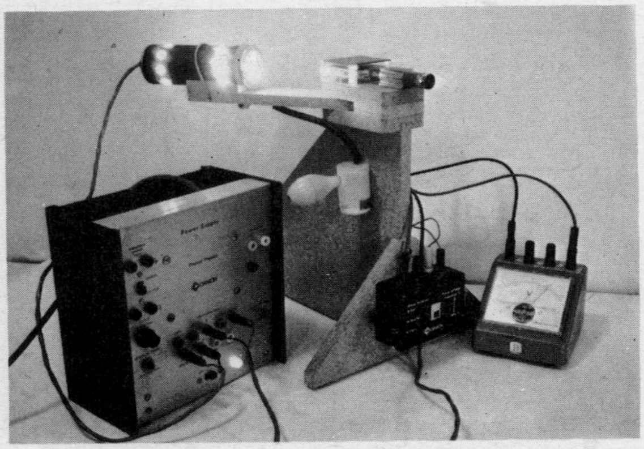

Fig. 5-9 A typical set of apparatus. Details may vary considerably.

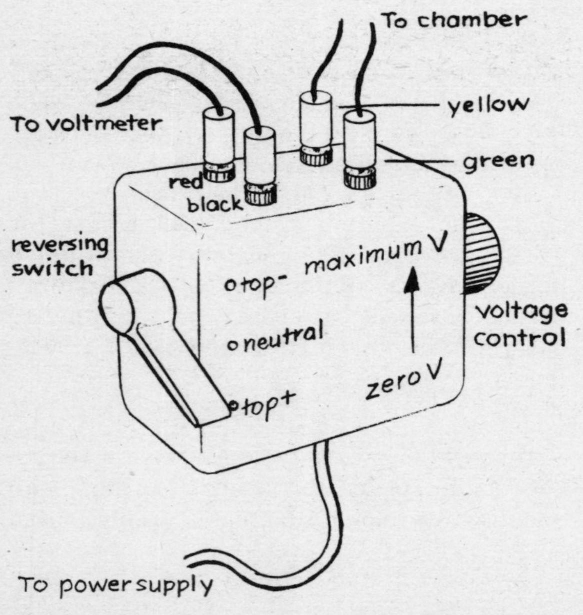

Fig. 5-10 A typical arrangement of connections to the high-voltage reversing switch.

The lens of the light source may fog up as the heat from the lamp drives moisture out of the light-source tube. If this happens, remove the lens and wipe it on a clean tissue. Wait for the tube to warm up thoroughly before replacing the lens.

Squeeze the bottle of latex suspension two or three times until five or ten particles drift into view. You will see them as tiny bright spots of light. You may have to adjust the focus slightly to see a specific particle clearly. Notice how the particle appears to move upward. The view is inverted by the microscope; the particles are actually falling in the earth's gravitational field.

Now switch on the high voltage across the plates by turning the switch up or down. Notice the effect on the particles of varying the electric field by means of the voltage-control knob.

Notice the effect when you reverse the electric field by reversing the switch position. (When the switch is in its mid-position, there is zero field between the plates.)

?

4. Do all the particles move in the same direction when the field is on?

5. How do you explain this?

6. Some particles move much more rapidly in the field than others. Do the rapidly moving particles have larger or smaller charges than the slowly moving particles?

Sometimes a few particles cling together, making a clump that is easy to see. The clump falls more rapidly than single particles when the electric field is off. Do not try to use these clumps for measuring q.

Try to balance a particle by adjusting the field until the particle hangs motionless. Observe it carefully to make sure it is not slowly drifting up or down. The smaller the charge, the greater the electric field must be to hold up the particle.

Taking Data

It is not worth working at voltages much below 50 V. Only highly charged particles can be balanced in these small fields, and you are interested in obtaining the least charge possible.

Set the potential difference between the plates to about 75 V. Reverse the field a few times so that the more quickly moving particles (those with the greater charge) are swept out of the field of view. Any particles that remain have low charges. If *no* particles remain, squeeze in some more and look again for some with small charges.

When you have isolated one of these particles carrying a low charge, adjust the voltage carefully until the particle hangs motionless. Observe it for some time to make sure that it is not moving up or down very slowly, and that the adjustment of voltage is as precise as possible. (Because of uneven bombardment by air molecules, there will be some slight, uneven drift of the particles.)

Read the voltmeter. Then estimate the precision of the voltage setting by seeing how little the voltage needs to be changed to cause the particle to start moving just perceptibly. This small change in voltage is the greatest amount by which your *setting* of the balancing voltage can be uncertain.

When you have balanced a particle, make sure that the voltage setting is as precise as you can make it before you go on to another particle. The most useful range to work within is 75– 150 V, but try to find particles that can be brought to rest in the 200– 250 V range too, if the meter can be used in that range. Remember that the higher the balancing field the smaller the charge on the particle.

In this kind of an experiment, it is helpful to have large amounts of data. This usually makes it easier to spot trends and to distinguish main effects from the background scattering of data. Thus, you may wish to contribute your findings to a class data pool. Before doing that, however, arrange your values of V in a vertical column of increasing magnitude.

?

7. Do the numbers seem to clump together in groups, or do they spread out more or less evenly from the lowest to the highest values?

Now combine your data with that collected by your classmates. This can conveniently be done by placing your values of V on a class histogram. When the histogram is complete, the results can easily be transferred to a transparent sheet for use on an overhead projector. Alternatively, you may wish to take a Polaroid photograph of the completed histogram for inclusion in your laboratory notebook.

?

8. Does your histogram suggest that all values of q are possible and that electric charge is, therefore, endlessly divisible, or the converse?

If you would like to make a more complete quantitative analysis of the class results, calculate an average value for each of the highest three or four clumps of V values in the class histogram. Next change those values to values of $1/V$ and list them in order. Since q is proportional to $1/V$, these values represent the magnitude of the charges on the particles.

To obtain actual values for the charges, the $1/V$'s must be multiplied by ma_gd. The separation d of the two plates, typically about 5 mm, or 5×10^{-3} m, is given in the specification sheets provided by the manufacturer. You should check this.

The mass m of the spheres is worked out from a knowledge of their volume and the density D of the material they are made from. Mass = volume × density, or $m = \frac{4}{3}r^3 \times D$. The sphere diameter ($2r$) has been previously measured and is given on the supply bottle. The density D is 1,077 kg/m^3 (found by measuring a large batch of latex before it is made into little spheres).

?

9. What is the spacing between the observed average values of $1/V$, and what is the difference in charge that corresponds to this difference in $1/V$?

10. What is the smallest value of $1/V$ that you obtained? What is the corresponding value of q?

11. Do your experimental results support the idea that electric charge is quantized? If so, what is *your value* for the quantum of charge?

12. If you have already measured q_e/m in Experiment 4-9, compute the mass of an electron. Even if your value differs from the accepted value by a factor of 10, perhaps you will agree that its measurement is a considerable intellectual triumph.

Experiment 5-4
THE PHOTOELECTRIC EFFECT

In this experiment, you will make observations of the effect of light on a metal surface; then you will compare the appropriateness of the wave model and the particle model of light for explaining what you observe.

Before doing the experiment, read Text Sec. 18.4 (Unit 5) on the photoelectric effect.

How the Apparatus Works

Light that you shine through the window of the phototube falls on a half-cylinder of metal

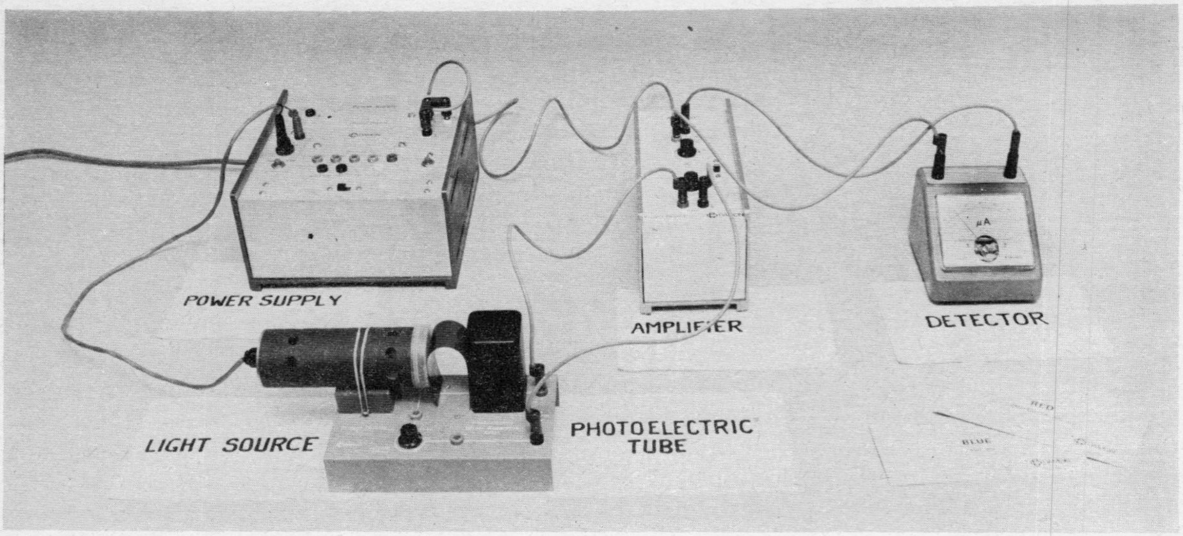

Fig. 5-11

called the *emitter*. The light drives electrons from the emitter surface.

Along the axis of the emitter (the center of the tube) is a wire called the *collector*. When the collector is made a few volts positive with respect to the emitter, practically all the emitted electrons are drawn to it, and will return to the emitter through an external wire. Even if the collector is made slightly negative, some electrons will reach it and there will be a measurable current in the external circuit.

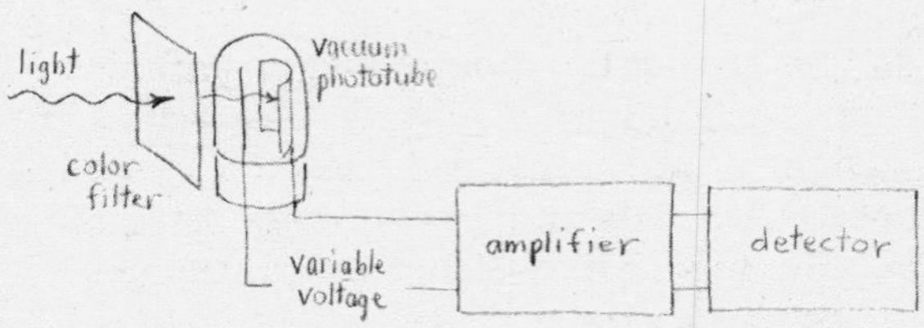

Fig. 5-12 However much the details may differ, any equipment for the photoelectric effect experiment will consist of these basic parts.

The small current cam be amplified several thousand times and detected in any of several different ways. One way is to use a small loudspeaker in which the amplified photoelectric current causes an audible hum; another is to use a cathode-ray oscilloscope. The following description assumes that the output current is read on a microammeter (Fig. 5-11).

The voltage control knob on the phototube unit allows you to vary the voltage between emitter and collector. In its full counterclockwise position, the voltage is zero. As you turn the knob clockwise the *photocurrent* decreases. You are making the collector more and more negative and fewer and fewer electrons get to it. Finally the photocurrent ceases altogether; all the electrons are turned back before reaching the collector. The voltage between emitter and collector that just stops all the electrons is called the *stopping voltage*. The value of this voltage indicates the maximum kinetic energy with which the electrons leave the emitter. To find the value of the stopping voltage precisely, you will have to be able to determine precisely when the photocurrent is reduced to zero. Because there is some drift of the amplifier output, the current indicated on the meter will drift around the zero point even when the actual current remains exactly zero. Therefore you will have to adjust the amplifier *offset* occasionally to be sure the zero level is really zero. An alternative is to ignore the precise reading of the current meter and adjust the collector voltage until *turning the light off and on causes no detectable change in the current.* Turn up the negative collector voltage until blocking the light from the tube (with black paper) has no effect on the meter reading. The exact location of the meter pointer is not important.

The position of the voltage-control knob at the current cutoff gives you a rough measure of stopping voltage. To measure it more precisely, connect a voltmeter as shown in Fig. 5-13.

In the experiment, you will measure the stopping voltages as light of different frequencies falls on the phototube. Good colored filters will allow light of only a certain range of

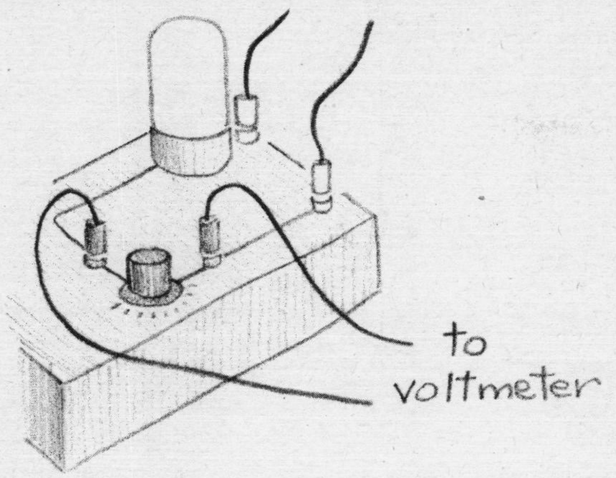

Fig. 5-13

frequencies to pass through. You can use a hand spectroscope to find the highest frequency line passed by each filter. The filters select frequencies from the mercury spectrum emitted by an intense mercury lamp. Useful frequencies of the mercury spectrum are:

yellow	5.2×10^{14}/sec
green	5.5×10^{14}/sec
blue	6.9×10^{14}/sec
violet	7.3×10^{14}/sec
(ultraviolet)	8.2×10^{14}/sec

Doing the Experiment

PART I

The first part of the experiment is qualitative. To see if there is a *time delay* between light falling on the emitter and the emission of photoelectrons, cover the phototube and then quickly remove the cover. Adjust the light source and filters to give the smallest photocurrent that you can conveniently notice on the meter.

?

1. Can you detect any time delay between the moment that light hits the phototube and the moment that the motion of the microammeter pointer (or a hum in the loudspeaker or deflection of the oscilloscope trace) signals the passage of photoelectrons through the phototube?

To see if the *current* in the phototube depends on the intensity of incident light, vary the distance of the light source.

?

2. Does the *number* of photoelectrons emitted from the sensitive surface vary with light intensity, that is, does the output current of the amplifier vary with the intensity of the light?

To find out whether the *kinetic energy* of the photoelectrons depends on the *intensity* of the incident light, measure the stopping voltage with different intensities of light falling on the phototube.

?

3. Does the kinetic energy of the photoelectrons depend on intensity, that is, does the stopping voltage change with light intensity?

Finally, determine how the kinetic energy of photoelectrons depends on the *frequency* of incident light. You will remember (Text Sec. 18.5) that the maximum kinetic energy of the photoelectrons is $V_{stop}q_e$, where V_{stop} is the stopping voltage, and $q_e = 1.60 \times 10^{-19}$ C, the charge on an electron. Measure the stopping voltage with various filters over the window.

?

4. How does the stopping voltage and, thus, the kinetic energy change as the light is changed from red through blue or ultraviolet (no filters)?

PART II

In the second part of the experiment you will make more precise measurements of stopping voltage. To do this, adjust the voltage-control knob to the cutoff (stopping voltage) position and then measure V with a voltmeter (Fig. 5-13). Connect the voltmeter only after the cutoff adjustment is made so that the voltmeter leads will not pick up any ac voltage induced from other conducting wires in the room.

Measure the stopping voltage V_{stop} for three or four different light frequencies, and plot the data on a graph. Along the vertical axis, plot electron energy $V_{stop}q_e$. When the stopping voltage V is in volts, and q_e is in coulombs, Vq_e will be energy, in joules.

Along the horizontal axis, plot the frequency of light f.

Interpretation of Results

As suggested in the opening paragraph, you can compare the wave model of light and the particle model in this experiment. Consider, then, how these models explain your observations.

?

5. If the light striking your phototube acts as *waves*,
(a) can you explain why the stopping voltage should depend on the *frequency* of light?

(b) would you expect the stopping voltage to depend on the *intensity* of the light? Why?
(c) would you expect a delay between the time that light first strikes the emitter and the emission of photoelectrons? Why?
6. If the light is acting as a stream of *particles*, what would be the answer to questions a, b, and c above?

If you drew the graph suggested in Part II of the experiment, you should now be prepared to interpret the graph. It is interesting to recall that Einstein predicted its form in 1905, and by experiments similar to yours, Millikan verified Einstein's prediction in 1916.

Einstein's photoelectric equation (Text Sec. 18.5) describes the energy of the most energetic photoelectrons (the last ones to be stopped as the voltage is increased), as

$$\tfrac{1}{2}mv^2_{max} = V_{stop}q_e$$
$$= hf - W$$

This equation has the form

$$y = kx - c$$

In this equation c is a constant, the value of y at the point where the straight line cuts the vertical axis; and k is another constant, namely the slope of the line. (See Fig. 5-14.) Therefore, the slope of a graph of $V_{stop}q_e$ against f should be h.

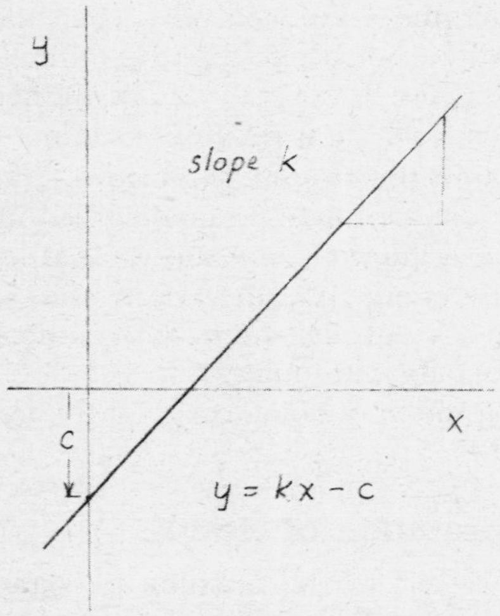

Fig. 5-14

?
7. What is the value of the slope of *your* graph? How well does this value compare with the value of Planck's constant, $h = 6.6 \times 10^{-34}$ J/sec? (See Fig. 5-15.)

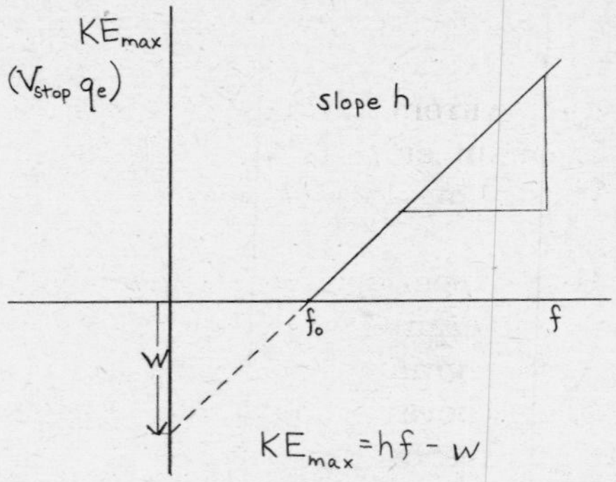

Fig. 5-15

With the equipment you used, the slope is unlikely to agree with the accepted value of h (6.6×10^{-34} J/sec) more closely than an order of magnitude. Perhaps you can give a few reasons why your agreement cannot be closer.

The lowest frequency at which any electrons are emitted from the cathode surface is called the *threshold frequency*, f_0. At this frequency $\tfrac{1}{2}mv_{max} = 0$ and $hf_0 = W$, where W is the *work function*. Your experimentally obtained value of W is not likely to be the same as that found for very clean cathode surfaces, more carefully filtered light, etc. The important thing to notice here is that there *is* a value of W, indicating that there is a minimum energy needed to release photoelectrons from the emitter.

?
8. Einstein's equation was derived from the assumption of a particle (photon) model of light. If your results do not fully agree with Einstein's equation, does this mean that your experiment supports the wave theory?

Experiment 5-5 SPECTROSCOPY

In Text Chapter 19 you learned of the immense importance of spectra to an understanding of nature. You are about to observe the spectra of a variety of light sources to see for yourself how spectra differ from each other and to learn how to measure the wavelengths of spectrum lines. In particular, you will measure the wavelengths of the hydrogen spectrum and relate them to the structure of the hydrogen atom.

Before you begin, review carefully Sec. 19.1 of Text Chapter 19.

Creating Spectra

Materials can be made to give off light (or be "excited") in several different ways: by heating in a flame, by an electric spark between electrodes made of the material, or by an electric current through a gas at low pressure.

The light emitted can be dispersed into a spectrum by either a prism or a diffraction grating.

In this experiment, you will use a diffraction grating to examine light from various sources. A *diffraction grating* consists of many very fine parallel grooves on a piece of glass or plastic. The grooves can be seen under a 400-power microscope.

In Experiment 4-2 (Young's Experiment), you saw how two narrow slits spread light of different wavelengths through different angles, and you used the double slit to make approximate measurements of the wavelengths of light of different colors. The distance between the two slits was about 0.2 mm. The distance between the lines in a diffraction grating is about 0.002 mm. A diffraction grating may have about 10,000 grooves instead of just two. Because there are more lines and they are closer together, a grating diffracts more light and separates the different wavelengths more than a double slit, and can be used to make very accurate measurements of wavelength.

Observing Spectra

You can observe diffraction when you look at light that is reflected from a phonograph record. Hold the record so that light from a distant source is almost parallel to the record's surface, as in Fig. 5-16. Like a diffraction grating, the grooved surface disperses light into a spectrum.

Fig. 5-16

Use a real diffraction grating to see spectra simply by holding the grating close to your eye with the lines of the grating parallel to a distant light source. Better yet, arrange a slit about 25 cm in front of the grating, as shown in Fig. 5-17, or use a pocket spectroscope.

Look through the pocket spectroscope at a fluorescent light, at an ordinary (incandescent) light bulb, at mercury-vapor and sodium-vapor street lamps, at neon signs, at light from the sky (but *do not* look directly at the sun), and at a flame into which various compounds are

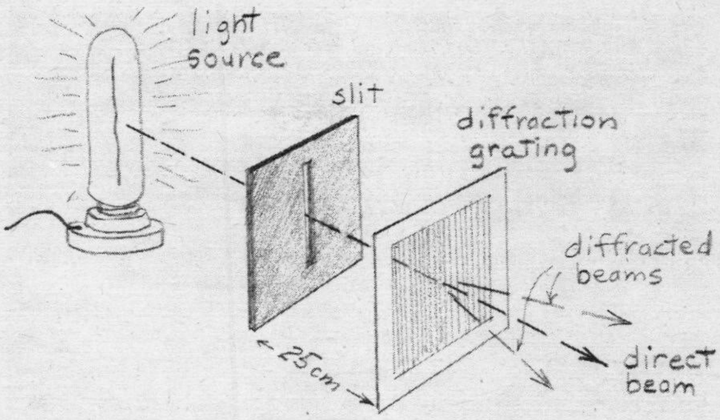

Fig. 5-17

introduced (such as salts of sodium, potassium, strontium, barium, and calcium).

?

1. Which color does the grating diffract into the widest angle and which into the narrowest? Are the long wavelengths diffracted at a wider angle than the short wavelengths, or vice versa?

2. The spectra discussed in the Text are (a) either emission or absorption, and (b) either line or continuous. What different *kinds* of spectra have you observed? Make a table showing the type of spectra produced by each of the light sources you observed. Do you detect any relationship between the nature of the source and the kind of spectrum it produces?

Photographing the Spectrum

A photograph of a spectrum has several advantages over visual observation. A photograph reveals a greater range of wavelengths; also, it allows greater convenience for your measurement of wavelengths.

When you hold the grating up to your eye, the lens of your eye focuses the diffracted rays to form a series of colored images on the retina. If you put the grating in front of the camera lens (focused on the source), the lens will produce sharp images on the film.

The spectrum of hydrogen is particularly interesting to measure because hydrogen is the simplest atom and its spectrum is fairly easily related to a model of its structure. In this experiment, hydrogen gas in a glass tube is excited by an electric current. The electric discharge separates most of the H_2 molecules into single hydrogen atoms.

Set up a meter stick just behind the tube (Fig. 5-18). This is a scale against which to observe and measure the position of the spectrum lines. The tube should be placed at about the 70-cm mark since the spectrum viewed through the grating will appear nearly 70 cm long.

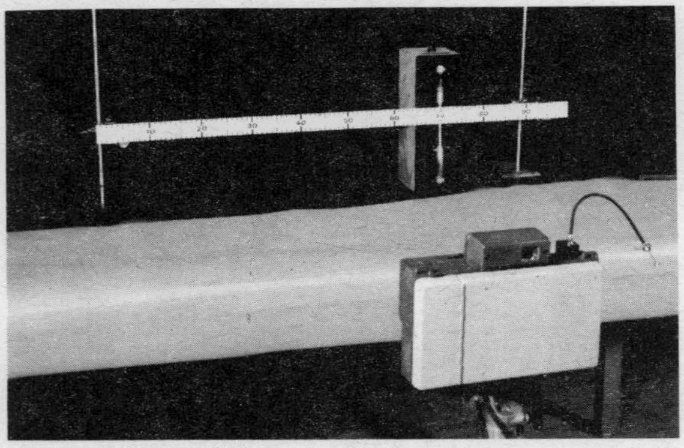

Fig. 5-18

From the camera position, look through the grating at the glowing tube to locate the positions of the visible spectral lines against the meter stick. Then, with the grating fastened over the camera lens, set up the camera with its lens in the same position your eye was. The lens should be aimed perpendicularly at the 50-cm mark, and the grating lines must be parallel to the source.

Now take a photograph that shows both the scale on the meter stick and the spectral lines. You may be able to take a single exposure for both, or you may have to make a double exposure: first the spectrum, and then, with more light in the room, the scale. It depends on the amount of light in the room. Consult your instructor.

Analyzing the Spectrum

Count the number of spectral lines on the photograph, using a magnifier to help pick out the faint ones.

?

3. Are there more lines than you can see when you hold the grating up to your eye? If you do see additional lines, are they located in the visible part of the spectrum (between red and violet) or in the infrared or ultraviolet part?

The angle θ through which light is diffracted by a grating depends on the wavelength λ of the light and the distance d between lines on the grating. The formula is a simple one: $\lambda = d \sin \theta$.

To find θ, you need to find $\tan \theta = x/l$ as shown in Fig. 5-19. Here x is the distance of the spectral line along the meter stick from the source, and l is the distance from the source to the grating. Use a magnifier to read x from your photograph. Calculate $\tan \theta$, and then look up the corresponding values of θ and $\sin \theta$ in trigonometric tables.

To find d, remember that the grating space is probably given as lines per centimeter. You must convert this to the distance between lines in meters. One centimeter is 1.00×10^{-2} m, so if there are 5,275 lines per centimeter, then d is $(1.00 \times 10^{-2})/(0.52 \times 10^{4}) = 1.92 \times 10^{-6}$ m.

Calculate the values of λ for the various spectral lines you have measured.

?

4. How many of these lines are visible to the eye?
5. What would you say is the shortest wavelength to which your eye is sensitive?
6. What is the shortest wavelength that you can measure on the photograph?

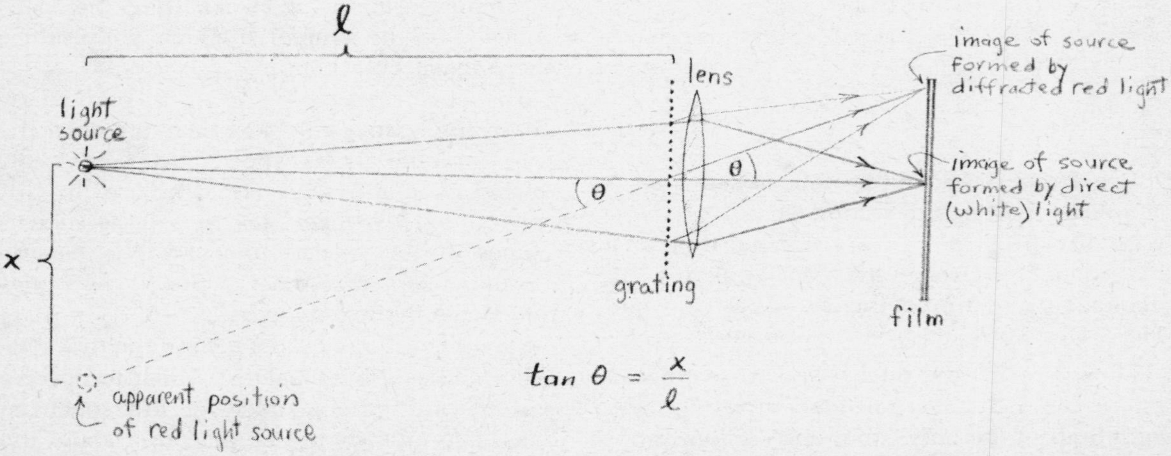

Fig. 5-19 Different images of the source are formed on the film by different colors of diffracted light. The angle of diffraction is equal to the apparent angular displacement angle of the source in the photograph, so $\tan \theta = \frac{x}{l}$.

Compare your values for the wavelengths with those given in the Text, or in a more complete list (for instance, in the *Handbook of Chemistry and Physics*). The differences between your values and the published ones should be less than the experimental uncertainty of your measurement. Are they?

This is not all that you can do with the results of this experiment. You could, for example, work out a value for the Rydberg constant for hydrogen (mentioned in Text Sec. 19.2).

More interesting perhaps is to calculate some of the energy levels for the excited hydrogen atom. Using Planck's constant (h = 6.6×10^{-34}), the speed of light in vacuum (c = 3.0×10^{8} m/sec), and your measured value of the wavelength λ of the separate lines, you can calculate the energy of various wavelengths of photons ($E = hf = hc/\lambda$) emitted when hydrogen atoms change from one state to another. The energy of the emitted photon is the difference in energy between the initial and final states of the atom.

Make the assumption (which is correct) that for all lines of the series you have observed, the final energy state is the same. The energies that you have calculated represent the energy of various excited states above this final level.

Draw an energy-level diagram something like the one shown in Fig. 5-20. Show on it the energy of the photon emitted in transition from each of the excited states to the final state.

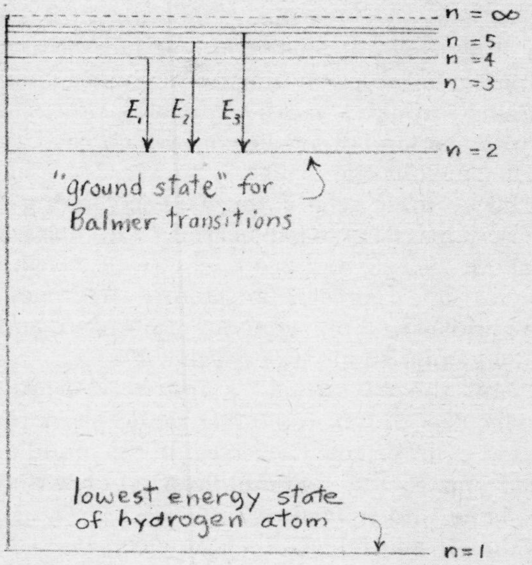

Fig. 5-20

?

7. How much energy does an excited hydrogen atom lose when it emits red light?

ACTIVITIES

DALTON'S PUZZLE

Once Dalton had his theory to work with, the job of figuring out relative atomic masses and empirical formulas became nothing more than working through a series of puzzles. Here is a very similar kind of puzzle with which you can challenge your classmates.

Choose three sets of objects, each having a different mass. Large ball bearings with masses of about 70, 160, and 200 g work well. Let the smallest one represent an atom of hydrogen, the middle-sized one an atom of nitrogen, and the large one an atom of oxygen.

From these "atoms" construct various "molecules." For example, NH_3 could be represented by three small objects and one middle-sized one; N_2O by two middle-sized ones and one large, and so forth.

Conceal one molecule of your collection in each one of a series of covered Styrofoam cups (or other lightweight, opaque containers). Mark on each container the symbols (but not the formula!) of the elements contained in the compound. Dalton would have obtained this information by qualitative analysis.

Give the covered cups to other students. Instruct them to measure the "molecular" mass of each compound and to deduce the relative atomic masses and empirical formulas from the set of masses, making Dalton's assumption of simplicity. If the objects you have used for "atoms" are so light that the mass of the Styrofoam cups must be taken into account, you can either supply this information as part of the data or leave it as a complication in the problem.

If the assumption of simplicity is relaxed, what other atomic masses and molecular formulas would be consistent with the data?

ELECTROLYSIS OF WATER

The fact that electricity can decompose water was an amazing and exciting discovery, yet the process is one that you can easily demonstrate with materials at your disposal. Figure 5-21 provides all the necessary information. Set up an electrolysis apparatus and demonstrate the process for your classmates.

In the sketch it looks as if about twice as many bubbles are coming from one electrode as from the other. Which electrode is it? Does this happen in your apparatus? Would you expect it to happen?

How would you collect the two gases that bubble off the electrodes? How could you prove their identity?

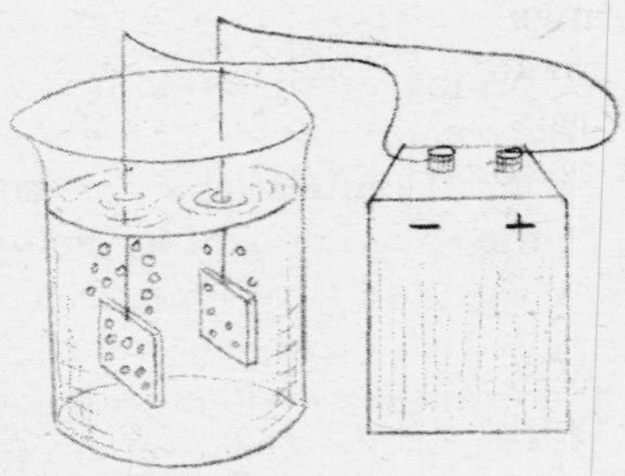

Fig. 5-21

If water is really just two gases "put together" chemically, you should be able to put the gases together again and get back the water with which you started. Using your knowledge of physics, predict what must then happen to all the electrical energy you sent flowing through the water to separate it.

SINGLE-ELECTRODE PLATING

A student asked if copper would "plate out" from a solution of copper sulfate if only a negative electrode were placed in the solution. It was tried and no copper was observed even when the electrode was connected to the negative terminal of a high-voltage source for 5 min. Another student suggested that only a very small (invisible) amount of copper was deposited since copper ions should be attracted to a negative electrode.

A more precise test was devised. A nickel sulfate solution was made containing several microcuries of radioactive nickel (no radiocopper was available). A single carbon electrode was immersed in the solution, and connected to the negative terminal of the high-voltage source again for 5 min. The electrode was removed, dried, and tested with a Geiger counter. The rod was slightly radioactive. A control test was run using identical test conditions, except that *no* electrical connection was made to the electrode. The control showed *more* radioactivity.

Repeat these experiments and see if the effect is true generally. What explanation would you give for these effects? (Adapted from *Ideas for Science Investigations*, N.S.T.A., 1966).

ACTIVITIES FROM *SCIENTIFIC AMERICAN*

Articles from the "Amateur Scientist" section of *Scientific American* often relate to Unit 5. These articles range widely in difficulty. In your library, scan the index of *Scientific American* for the past few years for such topics as accelerators, cloud chambers, spectroscopy, etc.

WRITINGS BY OR ABOUT EINSTEIN

In addition to his scientific works, Einstein wrote many perceptive essays on other areas of life that are easy to read and are still very timely. The chapter titles from *Out of My Later Years* (Citadel, 1973) indicate the scope of these essays: Convictions and Beliefs; Science; Public Affairs; Science and Life; Personalities; My People. This book includes his writings from 1934 to 1950. The best (and most inexpensive) book of Einstein's writings is *Albert Einstein, Ideas and Opinions* (Dell Publishing Co., 1973). The most scholarly biography is Philip Frank, *Einstein, His Life and Times* (Knopf, 1947).

Albert Einstein: Philosopher-Scientist, edited by P. Schilpp (Library of Living Philosophers, Vol. 7, 1973) contains Einstein's autobiographical notes, left-hand pages in German and right-hand pages in English, and essays by 12 physicist contemporaries of Einstein's about various aspects of his work. An informative and very readable book on Einstein is *Albert Einstein: Creator and Rebel* by Banesh Hoffmann in collaboration with Helen Dukas, Einstein's secretary for nearly 30 years (Viking Press, 1972). Also see *Einstein* (Penguin, 1976, paperback) by Jeremy Bernstein.

MEASURING q/m FOR THE ELECTRON

With the help of a "tuning eye" tube such as you may have seen in radio sets, you can measure the charge-to-mass ratio of the electron in a way that is very close to J.J. Thomson's original method.

Complete instructions appear in the PSSC *Physics Laboratory Guide*, Second Edition, D.C. Heath Company, Experiment IV-12, "The Mass of the Electron," pp. 79–81.

CATHODE RAYS IN A CROOKES TUBE

A Crookes tube having a metal barrier inside it for demonstrating that cathode rays travel in straight lines may be available in your classroom. In use, the tube is excited by a Tesla coil or induction coil.

Use a Crookes tube to demonstrate to the class the deflection of cathode rays in magnetic fields. To show how a magnet focuses cathode rays, bring one pole of a strong bar magnet toward the shadow of the cross-shaped obstacle near the end of the tube. Watch what happens to the shadow as the magnet gets closer and closer to it. What happens when you switch the poles of the magnet? What do you think would happen if you had a stronger magnet?

Can you demonstrate deflection by an electric field? Try using static charges as in Experiment 4-3, "Electric Forces. I," to create a deflecting field. Then, if you have an electrostatic generator, such as a small Van de Graaff or a Wimshurst machine, try deflecting the rays using parallel plates connected to the generator.

LIGHTING AN ELECTRIC LAMP WITH A MATCH

Here is a trick with which you can challenge your friends. It illustrates one of the many amusing and useful applications of the photoelectric effect in real life. You will need the phototube from Experiment 5-4 "The Photoelectric Effect," together with the *Project Physics* Amplifier and Power Supply. You will also need a 1.5-V dry cell or power supply and a 6-V light source such as the one used in the Millikan apparatus. (If you use this light source, remove the lens and cardboard tube and use only the 6-V lamp.) Mount the lamp on the photoelectric-effect apparatus and connect it to the 0.5-V, 5 A variable output on the power supply. Adjust the output to maximum. Set the *transistor switch input* switch to *switch*.

Connect the photoelectric-effect apparatus to the amplifier as shown in Fig. 5-22. Notice that the polarity of the 1.5-V cell is reversed and

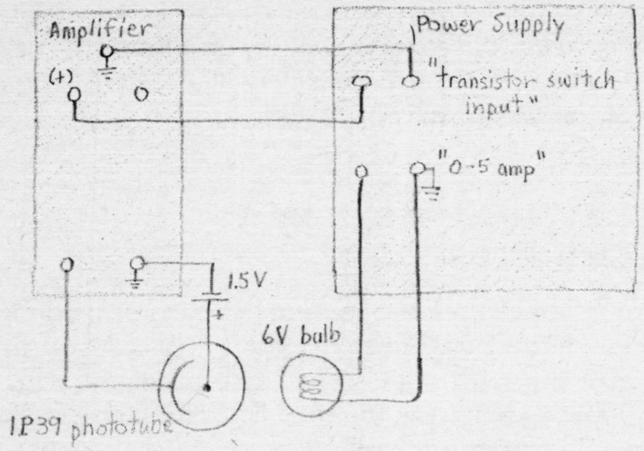

Fig. 5-22

that the output of the amplifier is connected to the *transistor switch input*.

Advance the gain control of the amplifier to maximum, then adjust the offset control in a positive direction until the filament of the 6-V lamp ceases to glow. Ignite a match near the apparatus (the wooden type works the best) and bring it quickly to the window of the phototube while the phosphor of the match is still glowing brightly (Fig. 5-23). The phosphor flare of the match head will be bright enough to cause sufficient photocurrent to operate the transistor switch that turns on the bulb. Once the bulb is lit, it keeps the photocell activated by its own light; you can remove the match and the bulb will stay lit.

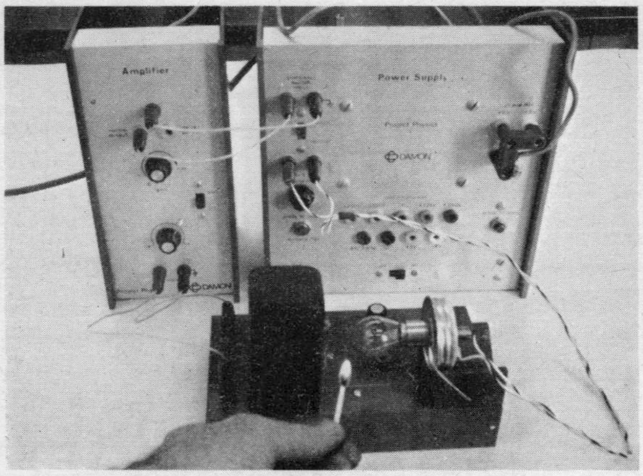

Fig. 5-23

When you are demonstrating this effect, tell your audience that the bulb is really a candle and that it should not surprise them that you can light it with a match. Of course, one way to put out a candle is to moisten your fingers and pinch out the wick. When your fingers pass between the bulb and the photocell, the bulb turns off, although the filament may glow a little, just as the wick of a freshly snuffed candle does. You can also make a "candle snuffer" from a little cone of any reasonably opaque material and use this instead of your fingers. Or you can "blow out" the bulb. It will go out obediently if you take care to remove it from in front of the photocell as you blow it out.

X RAYS FROM A CROOKES TUBE

To demonstrate that X rays penetrate materials that stop visible light, place a sheet of 10 cm × 12.5 cm 3,000-ASA-speed Polaroid Land film, still in its protective paper jacket, in contact with the end of the Crookes tube. (A film pack cannot be used, but any other photographic film in a light-tight paper envelope could be substituted.) Support the film on books or the table so that it does not move during the exposure. The photo in Fig. 5-24 is from a 1-min exposure using a hand-held Tesla coil to excite the Crookes tube.

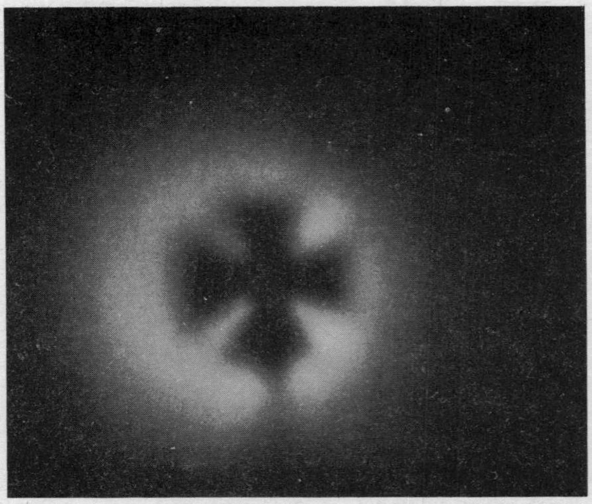

Fig. 5-24

SCIENTISTS ON STAMPS

Scientists are often pictured on the stamps of many countries, often being honored by countries other than their homelands. You may want to visit a stamp shop and assemble a display for your classroom.

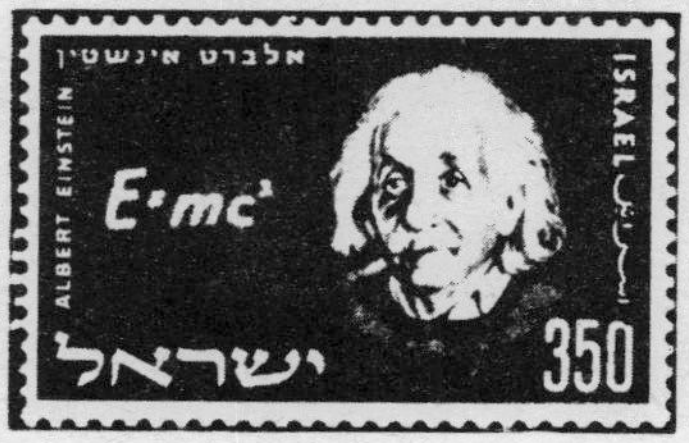

MEASURING IONIZATION: A QUANTUM EFFECT

With an inexpensive thyratron 884 tube, you can demonstrate an effect that is closely related to the famous Franck–Hertz effect.

Theory

According to the Rutherford–Bohr model, an atom can absorb and emit energy only in certain amounts that correspond to permitted "jumps" between states.

If you keep adding energy in larger and larger "packages," you will finally reach an amount large enough to separate an electron entirely from its atom, that is, to ionize the atom. The energy needed to do this is called the *ionization energy*.

Now imagine a beam of electrons being accelerated by an electric field through a region of space filled with argon atoms. This is the situation in a thyratron 884 tube with its grid and anode both connected to a source of variable voltage, as shown schematically in Fig. 5-25.

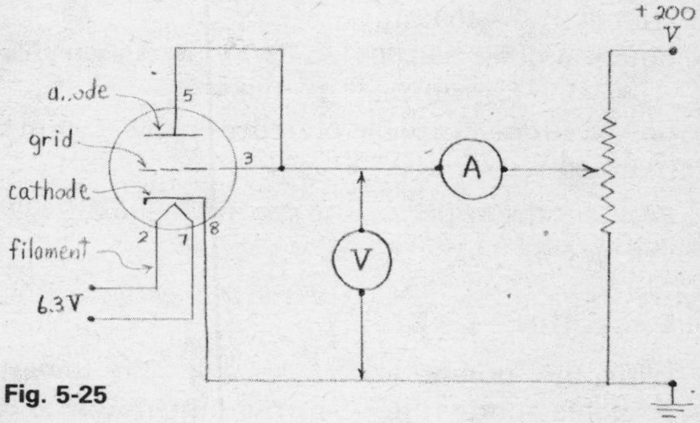

Fig. 5-25

In the form of its kinetic energy, each electron in the beam carries energy in a single "package." The electrons in the beam collide with argon atoms. As you increase the accelerating voltage, the electrons eventually become energetic enough to excite the atoms, as in the Franck–Hertz effect. However, your equipment is not sensitive enough to detect the resulting small energy absorptions, so nothing seems to happen. The electron current from cathode to anode appears to increase quite linearly with the voltage, as you would expect, until the electrons get up to the ionization energy of argon. This happens at the *ionization potential* V_i, which is related to the ionization energy E_i and to the charge q_e on the electron as follows:

$$E_i = q_e V_i$$

As soon as electrons begin to ionize argon atoms, the current increases sharply. The argon is now in a different state, called an *ionized state*, in which it conducts electric current much more easily than before. Because of this sudden decrease in electrical resistance, the thyratron tube can be used as an "electronic switch" in such devices as stroboscopes. (A similar process ionizes the air so that it can conduct lightning.) As argon ions recapture electrons, they emit photons of ultraviolet and of visible violet light. When you see this violet glow, the argon gas is being ionized.

For theoretical purposes, the important point is that ionization takes place in any gas at a particular energy that is characteristic of that gas. This is easily observed evidence of one special case of Bohr's postulated discrete energy states.

Equipment

thyratron 884 tube
octal socket to hold the tube (not essential but convenient)
voltmeter (0– 30 V dc)
ammeter (0– 100 mA)
potentiometer (10,000 Ω, 2 W or larger) or variable transformer, 0– 120 V ac
power supply, capable of delivering 50– 60 mA at 200 V dc

Connect the apparatus as shown schematically in Fig. 5-25.

Procedure

With the potentiometer set for the lowest available anode voltage, turn on the power and wait a few seconds for the filament to heat. Now increase the voltage by small steps. At each new voltage, call out to your partner the voltmeter reading. Pause only long enough to permit your partner to read the ammeter and to note both readings in your data table. Take data as rapidly as accuracy permits: Your potentiometer will heat up quickly, especially at high currents. If it gets too hot to touch, turn the power off and wait for it to cool before beginning again.

Watch for the onset of the violet glow. Note in your data table the voltage at which you first observe the glow, and then note what happens to the glow at higher voltages.

Plot current versus voltage, and mark the point on your graph where the glow first appeared. From your graph, determine the first ionization potential of argon. Compare your experimental value with published values, such as the one in the *Handbook of Chemistry and Physics*.

What is the amount of energy an electron must have in order to ionize an argon atom?

MODELING ATOMS WITH MAGNETS

Here is one easy way to demonstrate some of the important differences between the Thomson "raisin-pudding" atom model and the Rutherford nuclear model.

To show how alpha particles would be expected to behave in collisions with a Thomson atom, represent the spread-out "pudding" of positive charge by a roughly circular arrangement of small disk magnets, spaced 10– 12.5 cm apart, under the center of a smooth tray, as shown in Fig. 5-26. Use tape or putty to fasten the magnets to the underside of the tray. Put the large magnet (representing the alpha particle) down on top of the tray in such a way that the large magnet is repelled by the small magnets. Sprinkle onto the tray enough tiny plastic beads to make the large magnet slide freely. Now push the "alpha particle" from the edge of the tray toward the "atom." As long as the "alpha particle" has enough momentum to reach the other side, its deflection by the small magnets under the tray will be quite small: never more than a few degrees.

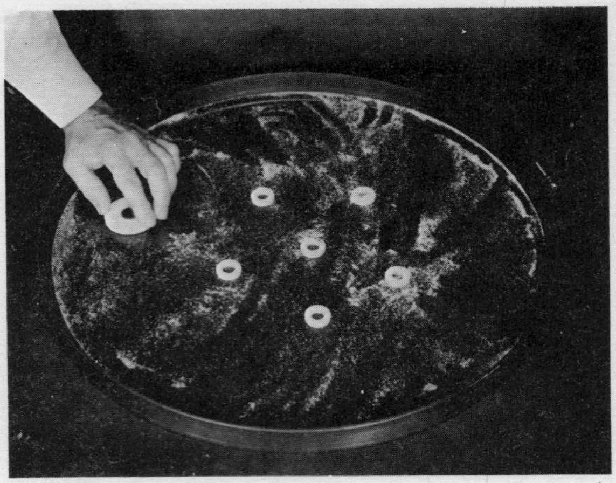

Fig. 5-26 The arrangement of the magnets for a "Thomson atom."

For the Rutherford model, on the other hand, gather all the small magnets into a vertical stack under the center of the tray, as shown in Fig. 5-27. Turn the stack so that it repels "alpha particles" as before. This "nucleus of positive charge" now has a much greater effect on the path of the "alpha particle."

Fig. 5-27 The arrangement of the magnets for a "Rutherford atom."

Have a partner tape an unknown array of magnets to the bottom of the tray; can you determine the arrangement just by scattering the large magnet?

With this magnet analogue you can do some quantitative work with the scattering relationships that Rutherford investigated. (See Text Sec. 19.3 and the notes on Film Loop 47, "Rutherford Scattering" in this *Handbook*.) Try again with different sizes of magnets. Devise a launcher so that you can control the velocity of your projectile magnets and the distance of closest approach.

(1) Keep the initial projectile velocity v constant and vary the distance b (see Fig. 5-28); then plot the scattering angle ϕ versus b.

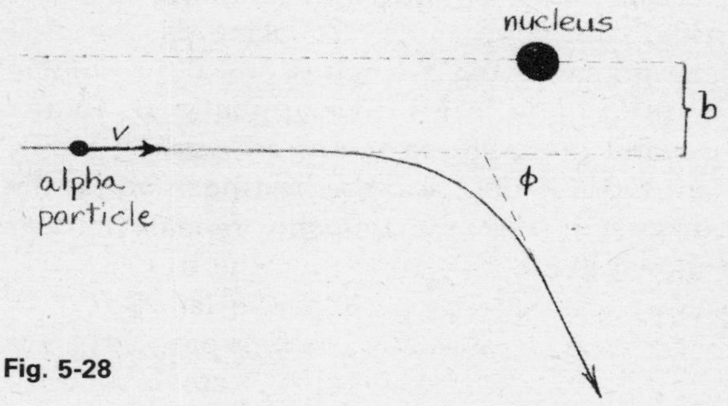

Fig. 5-28

(2) Hold b constant and vary the speed of the projectile, then plot ϕ versus v.

(3) Try scattering hard, nonmagnetized disks off of each other. Plot ϕ versus b and ϕ versus v as before. Contrast the two kinds of scattering-angle distributions.

"BLACK-BOX" ATOMS

Place two or three different objects, such as a battery, a small block of wood, a bar magnet, or a ball bearing, in a small box. Seal the box, and have one of your fellow students try to tell you as much about the contents as possible, without opening the box. For example, sizes might be determined by tilting the box, relative masses by balancing the box on a support, or whether or not the contents are magnetic by checking with a compass.

The object of all this is to get a feeling for what you can or cannot infer about the structure of an atom purely on the basis of secondary evidence. It may help you to write a report on your investigation in the form you may have used for writing a proof in plane geometry, with the property of the box in one column and your reason for asserting that the property is present in the other column.

The analogy can be made even better if you are exceptionally imaginative: Do not let the guesser open the box, ever, to find out what is really inside.

STANDING WAVES ON A BAND-SAW BLADE

Standing waves on a ring can be shown by shaking a band-saw blade with your hand. Wrap tape around the blade for about 15 cm to protect your hand. Then gently shake the blade up and down until you have a feeling for the lowest vibration rate that produces reinforcement of the vibration. Double the rate of shaking, and continue to increase the rate of shaking, watching for standing waves. You should be able to maintain five or six nodes.

TURNTABLE OSCILLATOR PATTERNS RESEMBLING DE BROGLIE WAVES

If you set up two turntable oscillators and a Variac as shown in Fig. 5-29, you can draw pictures resembling de Broglie waves, like those shown in Chapter 20 of the Text.

Fig. 5-29

Place a paper disk on the turntable. Set both turntables at their lowest speeds. Before starting to draw, check the back-and-forth motion of the second turntable to be sure the pen stays on the paper. Turn both turntables on and use the Variac as a precise speed control on the second turntable. Your goal is to get the pen to follow exactly the same path each time the paper disk goes around. Try higher frequencies of back-and-forth motion to get more wavelengths around the circle. For each stationary pattern that you get, check whether the back-and-forth frequency is an integral multiple of the circular frequency.

STANDING WAVES IN A WIRE RING

With the apparatus described here, you can set up circular waves that somewhat resemble the de Broglie wave models of certain electron orbits. You will need a strong magnet, a fairly stiff wire loop, a frequency oscillator, and a power supply with a transistor chopping switch.

The output current of the oscillator is much too small to interact with the magnetic field enough to set up visible standing waves in the wire ring. However, the oscillator current can operate the transistor switch to control ("chop") a much larger current from the power supply (see Fig. 5-30).

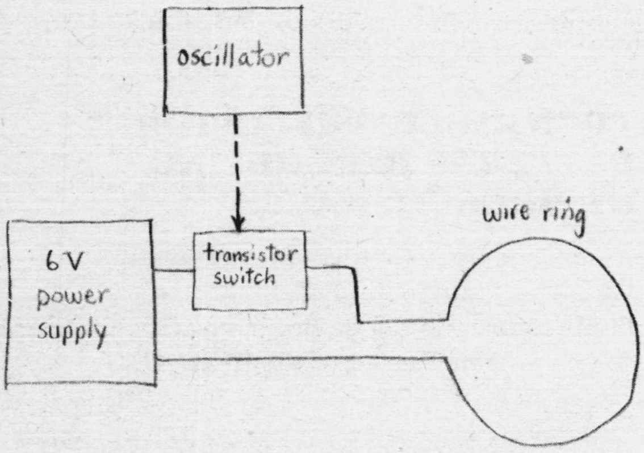

Fig. 5-30 The signal from the oscillator controls the transistor switch, causing it to turn the current from the power supply on and off. The "chopped" current in the wire ring interacts with the magnetic field to produce a pulsating force on the wire.

The wire ring must be of nonmagnetic metal. Insulated copper magnet wire works well. Twist the ends together and support the ring at the twisted portion by means of a binding post, Fahnestock clip, thumbtack, or ringstand clamp. Remove a little insulation from each end for electrical connections.

A ring 10 – 15 cm in diameter made of 22-guage enameled copper wire has its lowest rate of vibration at about 20 Hz. Stiffer wire or a smaller ring will have higher characteristic vibrations that are more difficult to see.

Position the ring as shown, with a section of the wire passing between the poles of the magnet. When the pulsed current passes through the ring, the current interacts with the magnetic field, producing alternating forces that cause the wire to vibrate. In Fig. 5-30, the magnetic field is vertical, and the vibrations are in the plane of the ring. You can turn the magnet so that the vibrations ar eperpendicular to the ring.

Because the ring is clamped at one point, it can support standing waves that have any integral number of half-wavelengths. In this respect they are different from waves on a *free* wire ring, which are restricted to integral numbers of *whole* wavelengths. Such waves are more appropriate for comparison to an atom.

When you are looking for a certain mode of vibration, position the magnet between expected nodes (at antinodes). The first "characteristic state," or "mode of vibration," that the ring can support in its plane is the first harmonic, having two nodes: one at the point of support and the other opposite it. In the second mode, three nodes are spaced evenly around the loop, and the best position for the magnet is directly opposite the support, as shown in Fig. 5-31.

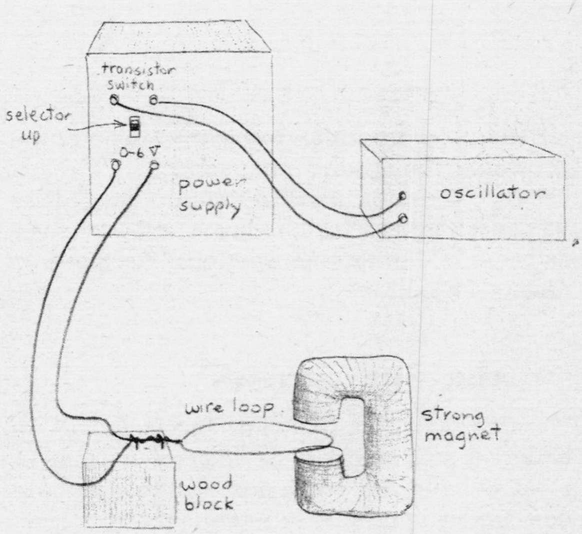

Fig. 5-31

You can demonstrate the various modes of vibration to the class by setting up the magnet, ring, and support on the platform of an overhead projector. Be careful not to break the glass with the magnet, especially if the frame of the projector happens to be made of a magnetic material.

The *Project Physics* Film Loop "Vibrations of a Wire" also shows this type of vibration.

FILM LOOP NOTES

Film Loop 45
PRODUCTION OF SODIUM BY ELECTROLYSIS

In 1807, Humphry Davy produced metallic sodium by the electrolysis of molten lye (sodium hydroxide).

In the film, sodium hydroxide (NaOH) is placed in an iron crucible and heated until it melts at a temperature of 318°C. A rectifier connected to a power transformer supplies a steady current through the liquid NaOH through iron rods inserted in the liquid. Sodium ions are positive and are therefore attracted to the negative electrode; there they pick up electrons and become metallic sodium, as indicated symbolically in this reaction:

$$Na^{+} + e^{-} = Na$$

The sodium accumulates in a thin, shiny layer floating on the surface of the molten sodium hydroxide.

Sodium is a dangerous material that combines explosively with water. The experimenter in the film scoops out a little of the metal and places it in water (Fig. 5-32). Energy is released rapidly, as you can see from the violence of the reaction. Some of the sodium is vaporized and the hot vapor emits the yellow light characteristic of the spectrum of sodium. The same yellow emission is easily seen if common salt, sodium chloride, or some other sodium compound is sprinkled into an open flame.

Film Loop 46
THOMSON MODEL OF THE ATOM

Before the development of the Bohr theory, a popular model for atomic structure was the "raisin-pudding" model of J.J. Thomson. According to this model, the atom was supposed

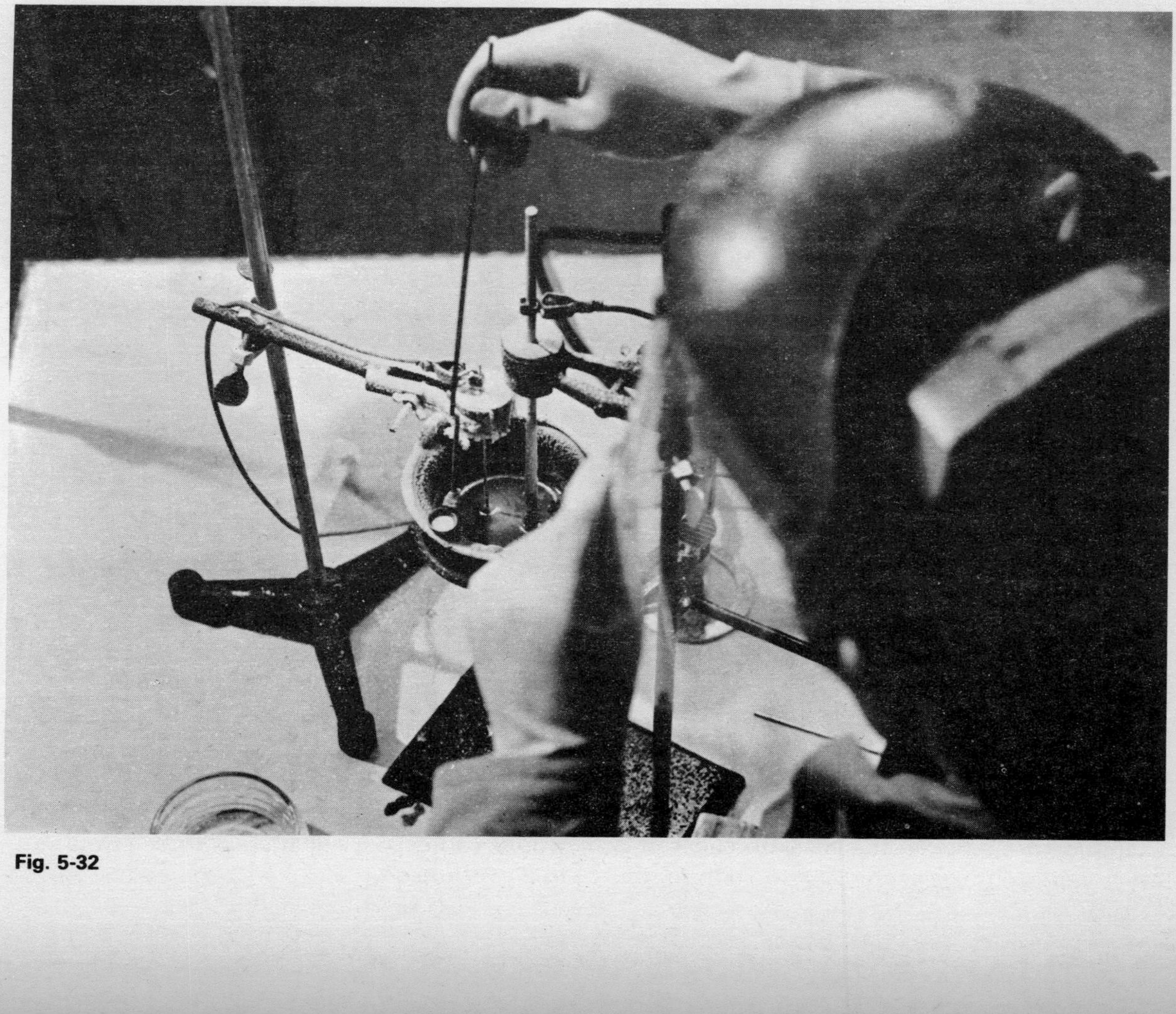

Fig. 5-32

to be a uniform sphere of positive charge in which were embedded small negative "corpuscles" (electrons). Under certain conditions, the electrons could be detached and observed separately, as in Thomson's historic experiment to measure the charge/mass ratio.

The Thomson model did not satisfactorily explain the stability of the electrons and especially their arrangement in "rings," as suggested by the periodic table of the elements. In 1904, Thomson performed experiments which to him showed the *possibility* of a ring structure within the broad outline of the raisin-pudding model. Thomson also made mathematical calculations of the various arrangements of electrons in his model.

In the Thomson model of the atom, the cloud of positive charge created an electric field directed along radii, strongest at the surface of the sphere of charge and decreasing to zero at the center. You are familiar with a gravitational example of such a field. The earth's downward gravitational field is strongest at the surface and it decreases toward the center of the earth.

For his model-of-a-model, Thomson used still another type of field: a magnetic field caused by a strong electromagnet above a tub of water. Along the water surface the field is "radial," as shown by the pattern of iron filings sprinkled on the glass bottom of the tub. Thomson used vertical magnetized steel needles to represent the electrons; these were stuck through corks and floated on the surface of the water. The needles were oriented with like poles pointing upward; their mutual repulsion tended to cause the magnets to spread apart. The outward repulsion was counteracted by the radial magnetic field directed inward toward the center. When the floating magnets were placed in the tub of water, they came to equilibrium configurations under the combined action of all the forces. Thomson saw in this experiment a partial verification of his calculation of how electrons (raisins) might come to equilibrium in a spherical blob of positive fluid.

In the film the floating magnets are 3.8 cm long, supported by ping-pong balls (Fig. 5-33). Equilibrium configurations are shown for various numbers of balls, from 1 to 12. Perhaps you can interpret the patterns in terms of rings, as did Thomson.

Thomson was unable to make an exact correlation with the facts of chemistry. For example, he knew that the eleventh electron is easily removed (corresponding to sodium, the eleventh atom of the periodic table), yet his floating magnet model failed to show this. Instead, the patterns for 10, 11, and 12 floating magnets are rather similar.

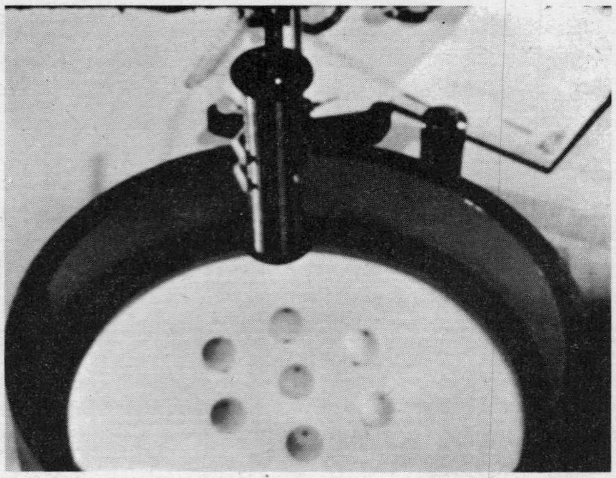

Fig. 5-33

Thomson's work with this apparatus illustrates how physical theories may be tested with the aid of analogies. He was disappointed by the failure of the model to account for the details of atomic structure. A few years later, the Rutherford model of a nuclear atom made the Thomson model obsolete, but in its day the Thomson model did receive some support from experiments such as those shown in the film.

Film Loop 47
RUTHERFORD SCATTERING

This film simulates the scattering of alpha particles by a heavy nucleus, such as gold, as in Ernest Rutherford's famous experiment. The film was made with a digital computer.

The computer program was a slight modification of that used in Film Loops 13 and 14, on program orbits, concerned with planetary orbits. The only difference is that the operator selected an inverse-square law of *repulsion* instead of a law of attraction such as that of gravity. The results of the computer calculation were displayed on a cathode-ray tube and then photographed (Fig. 5-34). Points are shown at equal time intervals. Verify the law of areas for the motion of the alpha particles by projecting the film for measurements. Why would you expect equal areas to be swept out in equal times?

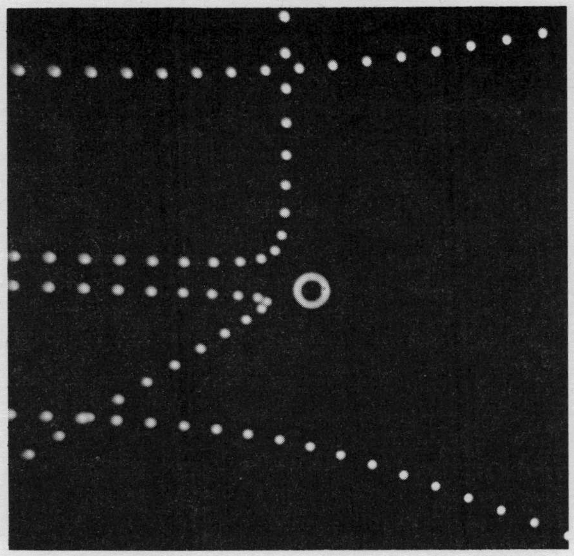

Fig. 5-34

All the scattering particles shown are near a nucleus. If the image from your projector is 30 cm high, the nearest adjacent nucleus would be about 150 m above the nucleus shown. Any alpha particles moving through this large area between nuclei would show no appreciable deflection.

The computer and a mathematical model are used to tell what the result will be if particles are shot at a nucleus. The computer does not "know" about Rutherford scattering. What it does is determined by a program placed in the computer's memory, written, in this particular instance, in a language called FORTRAN. The programmer has used Newton's laws of motion and has assumed an inverse-square repulsive force. It would be easy to change the program to test another force law, for example $F = K/r^3$. The scattering would be computed and displayed; the angle of deflection for the same distance of closest approach would be different than for inverse-square force.

Working backward from the observed scattering data, Rutherford deduced that the inverse-square Coulomb force law is correct for all motions taking place at distances greater than about 10^{-14} m from the scattering center, but he found deviations from Coulomb's law for closer distances. This suggested a new type of force, called *nuclear force*. Rutherford's scattering experiment showed the size of the nucleus (supposedly the same as the range of the nuclear forces) to be about 10^{-14} m, which is about 1/10,000 the distance between the nuclei in solid bodies.

The Nucleus

EXPERIMENTS

Experiment 6-1
RANDOM EVENTS

In Unit 6, after having explored the random behavior of gas molecules in Unit 3, you are learning that some atomic and nuclear events occur in a random manner. The purpose of this experiment is to give you some firsthand experience with random events.

What Is a Random Event?

Dice are useful for studying random behavior. You cannot predict with certainty how many spots will show on a single throw. But you are about to discover that you *can* make useful predictions about a large number of throws. If the behavior of the dice is truly random, you can use probability theory to make predictions. When, for example, you shake a box of 100 dice, you can predict with some confidence how many will fall with one spot up, how many with two spots up, and so on. Probability theory has many applications. For example, it is used in the study of automobile traffic flow, the interpretation of faint radar echoes from the planets, the prediction of birth, death, and accident rates, and the study of the breakup of nuclei.

The theory of probability provides ways of determining whether a set of events is random. An important characteristic of all truly *random* events is that each event is independent of the others. For example, if you throw a legitimate die four times in a row and find that a single spot turns up each time, your chance of observing a single spot on the fifth throw is no greater or smaller than it was on the first throw.

If events are to be independent, the circumstances under which the observations are made must never favor one outcome over another. This condition is met in each of the following parts of this experiment. You are expected to do only one of these parts. The section "Recording Your Data" that follows the descriptions applies to all parts of the experiment. Read this section in preparing to do any part of the experiment.

(a) Random Event Disks

You will use 100 random event disks and a large piece of graph paper. When the disks are spread around on the graph paper, even into the corners, what is the chance that a cross of the heavy grid marks of the graph paper will be covered by a disk (a "hit")?

On your graph paper, spread the disks around fairly evenly and count the number of "hits" for that trial. Record your results in a table like that on p. 238. Then spread the disks again and make another count. The counting will go faster if you divide the graph paper into sections and have helpers count in each area. Repeat the process until you have counted 100 trials.

As described below, calculate the mean number of hits per trial, then divide by 100 to obtain the fraction of one grid area covered by a disk.

From your distribution table, estimate the spread of values around the mean that include two-thirds of the values. That number, called the *standard deviation* (s.d.), is a characteristic of the distribution and, like the mean, should be nearly the same value for each set of trials.

An estimate of the uncertainty of the mean, called the *standard error* (s.e.), is found from s.d./$\sqrt{N}$ where N is the number of trials used to obtain the mean. As N increases, the s.e. decreases.

?

1. What is the s.e. for your set of 100 trials?
2. If you combine several sets of trials, what is the new mean and its s.e.?

(b) Twenty-Sided Dice

A tray containing 120 dice is used for this experiment. Each die has 20 identical faces (the name for a solid with this shape is *icosahedron*). One of the 20 faces on each die should be marked; if it is not, mark one face on each die with a felt-tip pen.

?

3. What is the probability that the marked face will appear at the top for any one throw of one die? To put it another way, *on the average*, how many marked faces would you expect to see face up if you roll all 120 dice?

Now try it, and see how well your prediction holds. Record as many trials as you can in the time available, shaking the dice, pouring them out onto the floor or a large tabletop, and counting the number of marked faces showing face up. (See Fig. 6-1.)

Fig. 6-1 Icosahedral dice in use.

The counting will go faster if the floor area or tabletop is divided into three or four sections, with a different person counting each section and another person recording the total count. Work rapidly, taking turns with others in your group if you get tired, so that you can count at least 100 trials.

(c) Diffusion Cloud Chamber

A *cloud chamber* is a device that makes visible the trail left by the particles emitted by radioactive atoms. One version is a transparent box filled with supercooled alcohol vapor. When an α partical passes through, it leaves a trail of ionized air molecules. The alcohol molecules are attracted to these ions and they condense into tiny droplets which mark the trail.

Your purpose in this experiment is not to learn about the operation of the chamber, but simply to study the randomness with which the α particles are emitted. A barrier with a narrow opening is placed in the chamber near a radioactive source that emits α particles. Count the number of tracks you observe coming through the opening in a convenient time interval, such as 10 sec. Continue counting for as many intervals as you can during the class period. (See Fig. 6-2.)

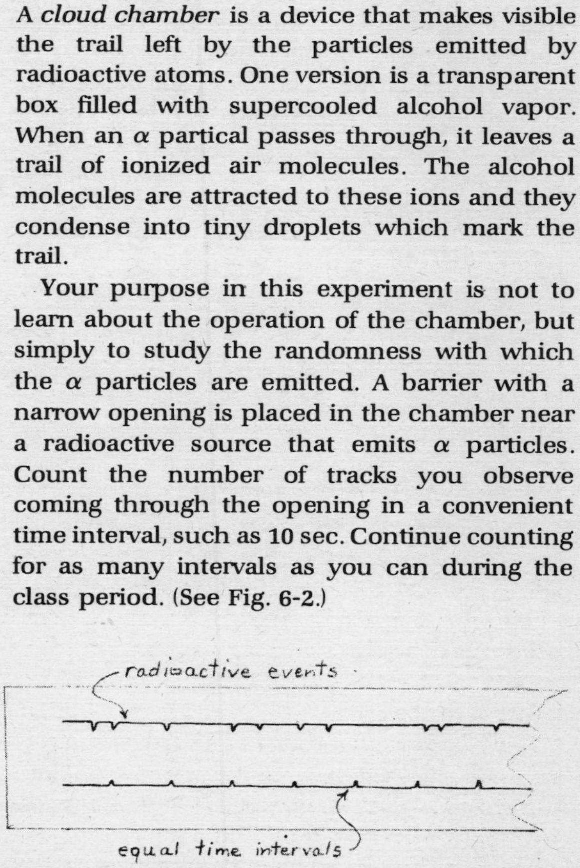

Fig. 6-2 A convenient method of counting events in successive time intervals is to mark them in one slot of the "dragstrip" recorder, while marking seconds (or 10-sec intervals) in the other slot.

(d) Geiger Counter

A *Geiger counter* is another device that detects the passage of invisible particles. A potential difference of several hundred volts is maintained between the two electrodes of the Geiger tube. When a β particle or a γ ray ionizes the gas in the tube, a short pulse of electricity passes through the tube. The pulse may be heard as an audible click in an earphone, seen as a "blip" on an oscilloscope screen, or read as a change in a number on an electronic scaling device. When a radioactive source is brought near the tube, the pulse rate goes up rapidly. But even without the source, an occasional pulse still occurs. These pulses are called *background* and are caused by cosmic radiation and by a slight amount of radiactivity always present in objects around the tube.

Use the Geiger counter to determine the rate of background radiation, counting over and over again the number of pulses in a convenient time interval, such as 10 sec.

Recording Your Data

Whichever of the experiments you do, prepare your data record in the following way:

Down the left-hand edge of your paper write a column of numbers from 0 to the highest number you ever expect to observe in one count. For example, if your Geiger counts seem to range from 3 to 20 counts in each time interval, record numbers from 0 to 20 or 25.

Number of events observed in one time interval (n)	(frequency) (f)	Total number of events observed (n x f)
0	|	0
1		0
2	|	2
3	卌 卌	30
4	卌 卌 |||	52
5	卌 卌 ||	65
6	卌 卌 卌 卌 |	126
7	卌 卌 卌 |	112
8	卌 卌 ||	96
9	||||	36
10	||||	40
11	||||	44
12	|	12
13	|	13
		623

Fig. 6-3 A typical data page.

To record your data, put a tally mark opposite each number in the column for each time this number occurs. Continue making tally marks for as many trial observations as you can make during the time you have. When you are through, add another column in which you multiply each number in the first column by the number of tallies opposite it. Whichever experiment you do, your data sheet will look something like the sample in Fig. 6-3. The third column shows that a total of 623 marked faces (or pulses or tracks) were observed in the 100 trials. The *average* is 623 divided by 100, or about 6.23. You can see that most of the counts cluster around the mean.

This arrangement of data is called a *distribution table*. The distribution shown was obtained by shaking the tray of 20-sided dice 100 times. Its shape is also typical of any random events such as random disks, Geiger-counter, and cloud-chamber results.

A Graph of Random Data

The pattern of your results is easier to visualize if you display your data in the form of a bar graph, or *histogram*, as in Fig. 6-4.

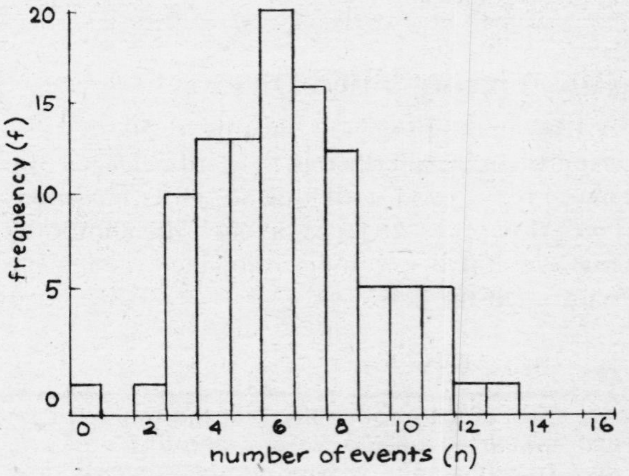

Fig. 6-4 The results obtained when a tray of 20-sided dice (one side marked) was shaken 100 times.

If you were to shake the dice another set of 100 times, your distribution would not be exactly the same as the first one. However, if sets of 100 trials were repeated several times, the *combined* results would begin to form a smoother histogram. Figure 6-5 shows the kind of result you could expect if you did 1,000 trials.

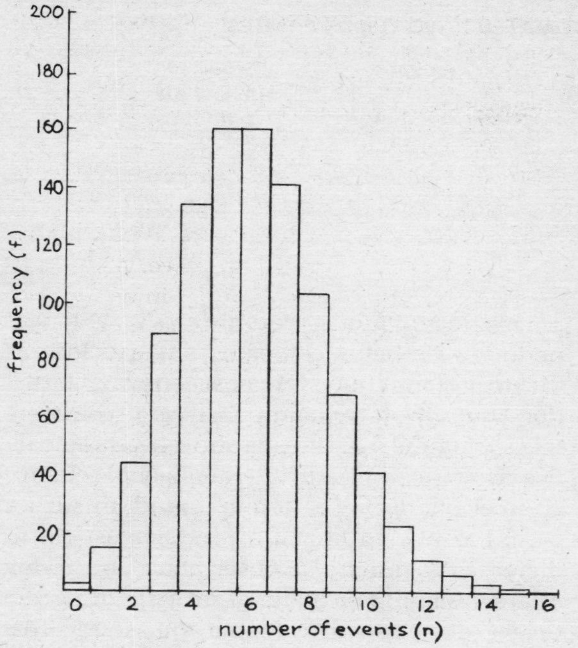

Fig. 6-5 The predicted results of shaking the dice 1,000 times. Notice that the vertical scale is different from that in Fig. 6-4. Do you see why?

Compare this with the results for only 10 trials shown in Fig. 6-6. As the number of trials increases, the distribution generally becomes smoother and more like the distribution in Fig. 6-5.

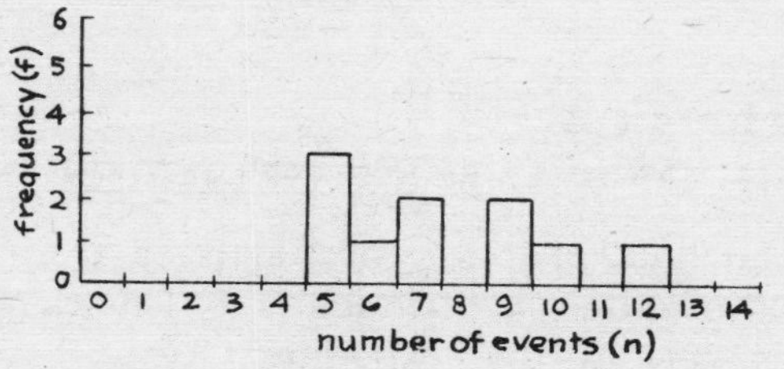

Fig. 6-6 Results of shaking the dice 10 times.

Predicting Random Events

How can data like these be used to make predictions?

On the basis of Fig. 6-5, the best prediction of the number of marked faces turning up would be 5 or 6 out of 120 rolls. Apparently the chance of a die having its marked face up is about 1 in 20, that is, the probability is $^1/_{20}$.

But not all trials had 5 or 6 marked faces showing. In addition to the average of a distribution, you also need to know something about how the data spread out around the average. Examine the histogram and answer the following questions.

?

4. How many of the trials in Fig. 6-5 had from 5 to 7 counts?

5. What fraction is this of the total number of observations?

6. How far, going equally to the left and right of the average, must you go to include one-half of all the observations? to include two-thirds?

For a theoretical distribution like this (which your own results will closely approximate as you increase the number of trials), it turns out that there is a simple rule for expressing the spread: If the average count is A, then two-thirds of the counts will be between $A - \sqrt{A}$ and $A + \sqrt{A}$. Putting it another way, about two-thirds of the values will be in the range of $A \pm \sqrt{A}$.

Another example may help make this clear. For example, suppose you have been counting cloud-chamber tracks and find that the average of a large number of 1-min counts is 100 tracks. Since the square root of 100 is 10, you would find that about two-thirds of your counts would lie between 90 and 110.

Check this prediction in Fig. 6-5. The average is 6. The square root of 6 is about 2.4. The points along the base of the histogram corresponding to 6 ± 2.4 are between 3.6 and 8.4. (Of course, it does not really make sense to talk about a fraction of a marked side. You would need to round off to the nearest whole numbers, 4 and 8.) Therefore, the chances are about two out of three that the number of marked sides showing after any shake of the tray will be in the range of 4 to 8 out of 120 throws.

?

7. How many of the trials did give results in the range 4 to 8? What fraction is this of the total number of trials?

8. Whether you used the disks, rolled dice, counted tracks, or used the Geiger counter, inspect your results to see if two-thirds of your counts do lie in the range $A \pm \sqrt{A}$.

If you counted for only a *single* 1-min trial, the chances are about two out of three that your single count C will be in the range $A \pm \sqrt{A}$, where A is the true average count (which you would find over many trials). This implies that you can predict the true average value fairly well even if you have made only a single 1-min count. The chances are about two out of three that the single count C will be within $\sqrt{A}$ of the true average A. If you assume C is a fairly good estimate of A, you can use $\sqrt{C}$ as an estimate of $\sqrt{A}$ and conclude that the

TABLE 6-1. SAMPLE RESULTS AND ESTIMATED "TWO-THIRDS RANGES"

Time min	Total Count	Expected Uncertainty	Average Count per min	Expected Uncertainty per min
1	100	±10	100	±10
10	1000	±32	100	±3.2
100	10000	±100	100	±1.0

chances are two out of three that *the value obtained for C is within* $\pm\sqrt{C}$ *of the true average.*

You can decrease the uncertainty in predicting a true average like this by counting for a longer period. Suppose you continued the count for 10 min. If you counted 1,000 tracks, the expected "two-thirds range" would be about $1000 \pm \sqrt{1000}$ or 1000 ± 32. The result is 1000 ± 32 counts in *10* min, which gives an average of 100 ± 3.2 counts per minute. If you counted for still longer, say 100 min, the range would be $10{,}000 \pm \sqrt{10{,}000}$ or $10{,}000 \pm 100$ counts in *100* min. Your estimate of the average count rate would be 100 ± 1 counts per minute. Table 6-1 lists these sample results.

Notice that although the expected uncertainty in the *total* count increases as the count goes up, it becomes a smaller *fraction* of the total count. Therefore, the uncertainty in the *average* count rate (number of counts per minute) decreases.

(The percentage of uncertainty can be expressed as $\frac{\sqrt{C}}{C}$, which is equal to $\frac{1}{\sqrt{C}}$. In this expression, you can see clearly that the percentage of uncertainty goes down as C increases.)

You can see from these examples that the higher the total count (the longer you count or more trials you do), the more precisely you can estimate the true average. This becomes important in the measurement of the activity of radioactive samples and many other kinds of random events. To get a precise measure of the activity (the average count rate), you must work with large numbers of counts.

?

9. If you have time, take more data to increase the precision of your estimate of the mean.

10. If you count 10 cosmic-ray tracks in a cloud chamber during 1 min, for how long would you expect to have to go on counting to get an estimate of the average with a "two-thirds range" that is only 1% of the average value?

This technique of counting over a longer period to get better estimates is fine as long as the true count rate remains constant. But it does not always remain constant. If you were measuring the half-life of a short-lived radioactive isotope, the activity rate would change appreciably during a 10-min period. In such a case, the way to increase precision is still to increase the number of observations, by having a larger sample of material or by putting the Geiger tube closer to it, so that you can record a large number of counts during a short time.

?

11. In a small town it is impossible to predict whether there will be a fire next week. But in a large metropolitan area, firefighters know with remarkable accuracy how many fires there will be. How is this possible? What assumption must the firefighters make?

Experiment 6-2
RANGE OF α AND β PARTICLES

An important property of particles from radioactive sources is their ability to penetrate solid matter. In this experiment, you will determine the distances α and β particles can travel in various materials.

Alpha (α) particles are most easily studied in a cloud chamber, a transparent box containing super-cooled alcohol vapor. Since the α particles are relatively massive and have a double positive charge, they leave a thick trail of ionized air molecules behind them as they move along. The ions then serve as centers about which alcohol condenses to form tracks of visible droplets.

Beta (β) particles also ionize air molecules as they move. But because of their smaller mass and smaller charge, they form relatively few ions, and they are farther apart than those formed by α particles. As a result, the trail of droplets in the chamber from β particles is much harder to see.

A Geiger counter, on the other hand, detects β particles better than α particles. This is because α particles, in forming a heavy trail, lose all their energy long before they get through even the thin window of an ordinary Geiger tube. Beta particles encounter the atoms in the tube window also, but they give up relatively less energy so that their chances of getting through the wall are fairly good.

For these reasons, you count α particles using a cloud chamber and β particles with a Geiger counter.

Observing α Particles

Mark off a distance scale on the bottom of the cloud chamber so that you will be able to estimate, at least to the nearest 0.5 cm, the lengths of the tracks formed (Fig. 6-7). Insert a source of α radiation and a barrier (as in the preceding experiment on random events) with a small slot opening at such a height that the tracks form a fairly narrow beam moving parallel to the bottom of the chamber. Put the cloud chamber into operation according to the instructions supplied with it.

Practice watching the tracks until you can report the length of any of the tracks you see.

Fig. 6-7

When you are ready to take data, count and record the number of α's that come through the opening in the barrier in 1 min. Measure the opening and calculate its area. Measure and record the distance from the source to the barrier.

Actually, you have probably not seen all the particles coming through the opening since the sensitive region in which tracks are visible is rather shallow and close to the chamber floor. You will probably miss the α's above this layer.

The Range and Energy of α Particles

The maximum range of radioactive particles as they travel through an absorbing material depends on several factors, including the density and the atomic number of the absorber. The graph (Fig. 6-8) summarizes the results of many measurements of the range of α particles travelling through air. The range– energy curve for particles in air saturated with alcohol vapor, as the air is in your chamber, does not differ significantly from the curve shown. You are, therefore, justified in using Fig. 6-8 to get a fair estimate of the kinetic energy of the α particles you observed.

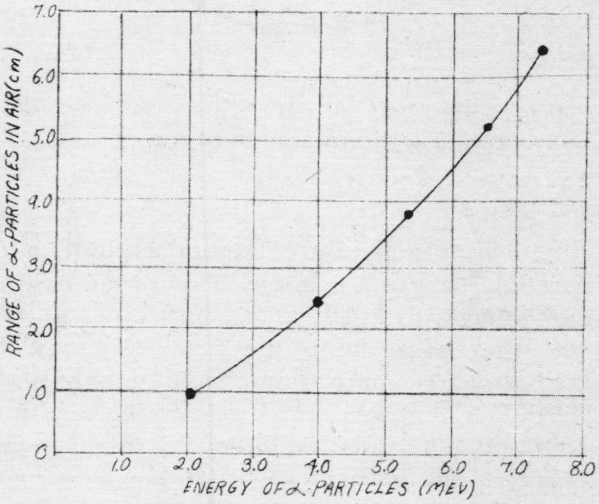

Fig. 6-8 Range of α particles in air as a function of their energy.

?

1. Was there a wide variation in α-particle energies, or did most of the particles appear to have about the same energy? What was the energy of the α particle that caused the longest track you observed?

Now calculate the rate at which energy is being carried away from the radioactive source. Assume that the source is a point. From the number of α particles per minute passing through an opening of known area at a known distance from the source, estimate the number of α particles per minute leaving the source in *all* directions.

For this estimate, imagine a sphere with the source at its center and a radius r equal to the distance from the source to the barrier (Fig. 6-9). From geometry, the surface area of the entire sphere is known to be $4\pi r^2$. You know the approximate rate c at which particles are emerging through the small opening, whose area a you have calculated. By proportion, you

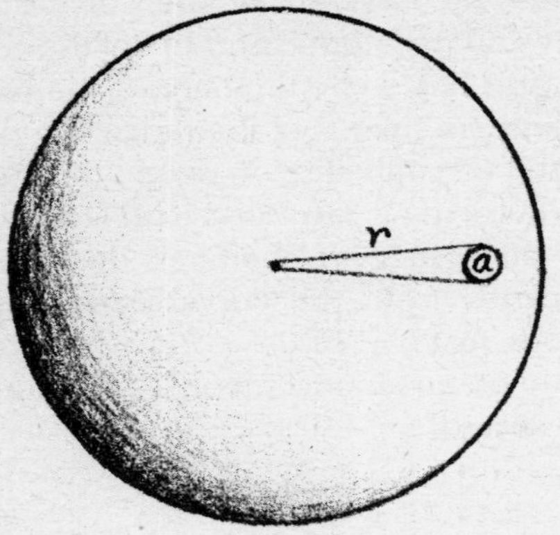

Fig. 6-9

can find the rate C at which the particles must be penetrating the total area of the sphere:

$$\frac{C}{c} = \frac{4\pi r^2}{a}$$

(The α-particle source is not a point, but probably part of a cylinder. This discrepancy, combined with a failure to count those particles that pass above the active layer, will introduce an error of as much as a factor of 10.)

The total number of particles leaving the source per minute, multiplied by the average energy of the particles, is the total energy lost per minute.

To answer the following questions, use the relationships

$$1 \text{ MeV} = 1.60 \times 10^{-13} \text{ J}$$
$$1 \text{ cal} = 4.18 \text{ J}$$

?

2. How many joules of energy are leaving the source per minute?
3. How many calories per minute does this equal?
4. If the source were placed in 1 g of water in a perfectly insulated container, how long would it take to heat the water from 0°C to 100°C?
5. How many joules per *second* are leaving the source? What is the power output in watts?

Observing β Particles

After removing all radioactive sources from near the Geiger tube, count the number of pulses caused by background radiation in several minutes. Calculate the average background radiation in counts per minute. Then place a source of β radiation near the Geiger tube, and determine the new count rate. (Make sure that the source and Geiger tube are not moved during the rest of the experiment.) Since you are concerned only with the particles from the source, subtract the average background count rate.

Next, place a piece of absorbing material (such as a sheet of cardboard or thin sheet metal) between the source and the tube and count again. Place a second, equally thick sheet of the same material in front of the first and count a third time. Keep adding absorbers and recording counts until the count rate ha dropped nearly to the level of background radiation.

Plot a graph on which the horizontal scale is the total thickness (number) of absorbers and the vertical scale is the number of β's getting through the absorber per minute.

In addition to plotting single points, show the uncertainty in your estimate of the count rate for each point plotted. You know that because of the random nature of radioactivity, the count rate actually fluctuates around some average value. You do not know what the *true* average value is; it would ideally take an infinite number of 1-min counts to determine the "true" average. But you know that the distribution of a great number of 1-min counts will have the property that two-thirds of them will differ from the average by less than the square root of the average. (See Experiment 6-1.)

For example, suppose you have observed 100 counts in a given 1-min interval. The chances are two out of three that, if you counted for a very long time, the mean count rate would be between 90 and 110 counts (between $100 - \sqrt{100}$ and $100 + \sqrt{100}$ counts). For this reason, you would mark a vertical line on your graph extending from 90 counts up to 110. In this way you avoid the pitfall of making a single measurement and assuming you know the "correct" value. (For an example of this kind of graph, see notes for Film Loop 9 in Unit 1.)

If other kinds of absorbing material are available, repeat the experiment with the same source and another set of absorbers. For sources that emit very low-energy β rays, it may be necessary to use very thin materials, such as paper or household aluminum foil.

Range and Absorption of β Particles

Examine your graph of the absorption of particles.

?

6. Is it a straight line?
7. What would the graph look like if (as is the case for α particles) all β particles from the source were able to penetrate the same thickness of a given absorbing material before giving up all their energy?

8. If you were able to use different absorbing materials, how did the absorption curves compare?
9. What might you conclude about the kinetic energy of β particles?

Experiment 6-3
HALF-LIFE. I

The more people there are in the world, the more people die each day. The less water there is in a tank, the more slowly water leaks out of a hole in the bottom.

In this experiment, you will observe three other examples of quantities that change at a rate that depends on the total amount of the quantity present. The objective is to find a common principle of change. Your conclusions will apply to many familiar growth and decay processes in nature.

If you experimented earlier with random disks, rolling dice, and radioactive decay (Experiment 6-1), you were studying random events you could observe one at a time. You found that the fluctuations in such small numbers of random events were relatively large. But this time you will deal with a large number of events, and you will find that the outcome of your experiments is therefore more precisely predictable.

A. Random Disks

Use the random disks and graph paper as in Experiment 6-1, Part (a), as an analogue of radioactive decay. Spread the 100 disks, which represent radioactive nuclei, over the graph paper. When a disk covers a heavy cross on the graph paper, consider it a radioactive decay. For each trial, remove the disks that "decayed" and record their number. When removed, these disks can be arranged like a bar graph in a series of piles. Make up to 20 trials, or until you have less than three decays per trial. Graph the decay curve, which shows for each trial the number of nuclei that decayed.

?

1. What was the initial decay rate for your first trial? After how many trials had the decay rate decreased to half the original rate? to one-fourth the original rate?
2. If other groups are making similar experiments, how well do the results agree? Are the differences within the range to be expected from sampling? Consider your results for Experiment 6-1. Does a combination of results from several groups give a more uniform decay curve and a more reliable value for the half-life?
3. Use your best decay curve to estimate the number of trials needed to reduce the decay activity to one-eighth the initial value. How many trials would be needed before you were sure that there was no more radioactivity in the sample?

A variation that is more representative of actual radioactive disintegration series can be made by using a small disk representing a daughter nucleus to replace each of the larger disks that decays. Then the number of nuclei remains constant; none has vanished. Because the daughter nuclei are represented by smaller disks, their decay rate will be slower. Make 20 to 25 trials using both size disks. For each trial, tally the number of decays of large and of small disks. Also record their sum, the total activity of the sample. A plot of the total activity will show the pattern of activity for a sample that is a mixture of two radioactive nuclei.

?

4. What is the half-life of the daughter nucleus?

You could even add a granddaughter nucleus by using pennies or cardboard pieces as a third size disk.

B. Twenty-Sided Dice

Mark any two sides of each 20-sided die with a (washable) marking pen. The chances will therefore be 1 in 10 that a marked surface will be face up on any one die when you shake and roll the dice. When you have rolled the 120 dice, *remove* all the dice that have a marked surface face up. Record the number of dice you removed (or line them up in a column). With the remaining dice, continue this process by shaking, rolling, and removing the marked dice at least 20 times. Record the number you remove each time (or line them up in a series of columns).

Plot a graph in which each roll is represented by one unit on the horizontal axis, and the number of dice *removed* after each roll is plotted on the vertical axis. (If you have lined up columns of removed dice, you already have a graph.)

Plot a second graph with the same horizontal scale, but with the vertical scale representing the number of dice *remaining* in the tray after each roll.

You may find that the numbers you have recorded are too erratic to produce smooth curves. Modify the procedure as follows: Roll the dice and count the dice with marked surfaces face up. Record this number but do not remove the dice. Shake and count again. Do this five times. Now find the mean of the five numbers, and remove that number of dice. The effect will be the same as if you had actually started with 120 × 5 or 600 dice. Continue this procedure as before, and you will find that it is easier to draw smooth curves that pass very nearly through all the points on your two graphs.

?

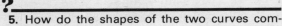

5. How do the shapes of the two curves compare?
6. What is the ratio of the number of dice removed after each shake to the number of dice shaken in the tray?
7. How many shakes were required to reduce the number of dice in the tray from 120 to 60? from 60 to 30? from 100 to 50?

C. Electric Circuit

A *capacitor* is a device that stores electric charge. It consists of two conducting surfaces placed very close together, but separated by a thin sheet of insulating material. When the two surfaces are connected to a battery, negative charge is removed from one plate and added to the other so that a potential difference is established between the two surfaces. (See Sec. 14.6 of Unit 4, Text.) If the conductors are disconnected from the battery and connected together through a resistor, the charge will begin to flow back from one side to the other. The charge will continue to flow as long as there is a potential difference between the sides of the capacitor. As you learned in Unit 4, the rate of flow of charge (the current) through a conducting path depends both on the resistance of the path and the potential difference across it.

To picture this situation, think of two partly filled tanks of water connected by a pipe running from the bottom of one tank to the bottom of the other (Fig. 6-10). When water is transferred from one tank to the other, the additional potential energy of the water is given by the difference in height, just as the potential difference between the sides of a charged capacitor is proportional to the potential energy stored in the capacitor. Water flows through the pipe at the bottom until the water

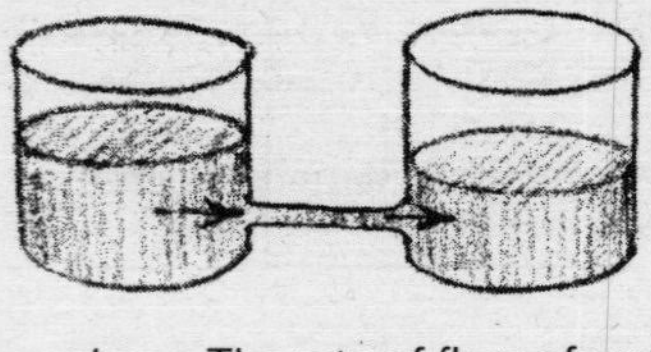

Fig. 6-10 An analogy: The rate of flow of water depends upon the difference in height of the water in the two tanks and upon the resistance the pipe offers to the flow of water.

levels are the same in the two tanks. Similarly, charge flows through the conducting path connecting the sides of the capacitor until there is no potential difference between the two plates.

Connect the circuit as in Fig. 6-11, close the switch, and record the reading on the voltmeter. Now open the switch and take a series of voltmeter readings at regular intervals. Plot a graph using time intervals for the horizontal axis and voltmeter readings for the vertical.

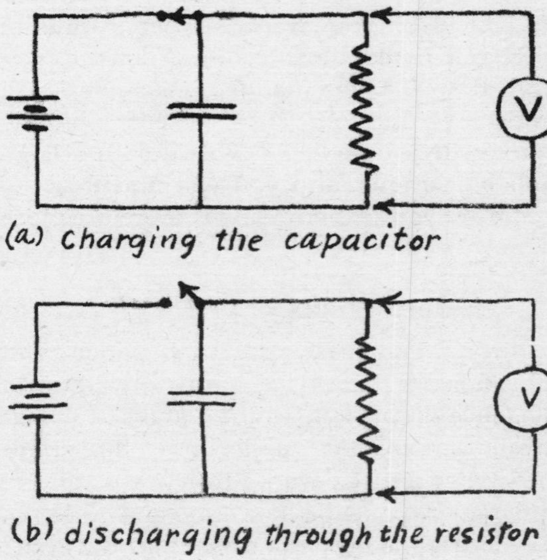

Fig. 6-11

?

8. How long does it take for the voltage to drop to one-half of its initial value? from one-half to one-fourth? from one-third to one-sixth?

Repeat the experiment with a different resistor in the circuit. Find the time required for the voltage to drop to one-half its initial value. Do this for several resistors.

?

9. How does the time required for the voltage to drop to half its initial value change as the resistance in the circuit is changed?

D. Short-Lived Radioisotope

Whenever you measure the radioactivity of a sample with a Geiger counter, you must first determine the level of background radiation. With no radioactive material near the Geiger tube, take a count for several minutes and calculate the average number of counts per minute caused by background radiation. This number must be subtracted from any count rates you observe with a sample near the tube, to obtain what is called the *net count rate* of the sample.

The measurement of background rate can be carried on by one member of your group while another prepares the sample according to the directions given below. Use this measurement of background rate to become familiar with the operation of the counting equipment. You will have to work quite quickly when you begin counting radiation from the sample itself.

First, a sample of a short-lived radioisotope must be isolated from its radioactive parent material and prepared for the measurmeent of its radioactivity.

Although the amount of radioactive material in this experiment is too small to be considered at all dangerous (unless you drink large quantities of it), it is a very good idea to practice caution in dealing with the material. Respect for radioactivity is an important attitude in our increasingly complicated world.

The basic plan is to (1) prepare a solution containing several radioactive substances, (2) add a chemical that absorbs only one of the radioisotopes, (3) wash most of the solution away leaving the absorbing chemical on a piece of filter paper, (4) mount the filter paper close to the end of the Geiger counter.

1. Prepare a funnel — filter assembly by placing a small filter paper in the funnel and wetting it with water.

Pour 12 mL of thorium nitrate solution into one graduated cylinder, and 15 mL of dilute nitric acid into another cylinder.

2. Take these materials to the filter flask set up in your laboratory. Your teacher will connect your funnel to the filter flask and pour in a quantity of ammonium phosphomolybdate precipitate, $(NH_4)_3PMo_{12}O_{40}$. The phosphomolybdate precipitate adsorbs the radioisotope's radioactive elements present in the thorium nitrate solution.

3. Wash the precipitate by sprinkling several milliliters of distilled water over it, and then *slowly* pour the thorium nitrate solution onto the precipitate (Fig. 6-12). Distribute the solution over the whole surface of the precipitate. Wash the precipitate again with 15 mL of dilute nitric acid and wait a few moments while the pump attached to the filter flask dries the sample. By the time the sample is dry, the nitric acid should have carried all the thorium nitrate solution through the filter. Left behind on the phosphomolybdate precipitate should be the short-lived daughter product whose radioactivity you wish to measure.

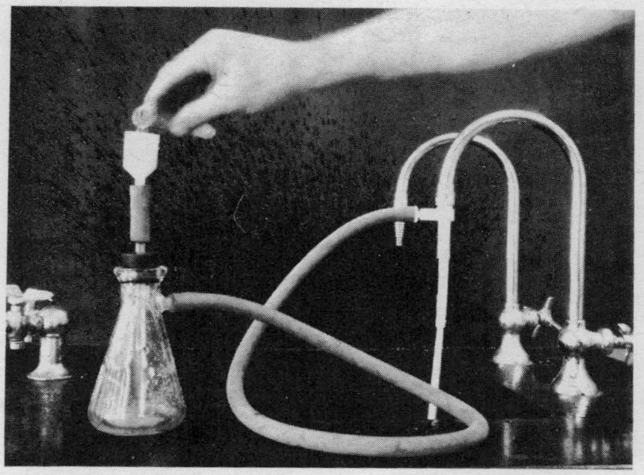

Fig. 6-12

4. As soon as the sample is dry, remove the upper part of the funnel from the filter flask and take it to the Geiger counter. Make sure that the Geiger tube is protected with a layer of thin plastic food wrapping. Then lower it into the funnel carefully until the end of the tube almost touches the precipitate (Fig. 6-13).

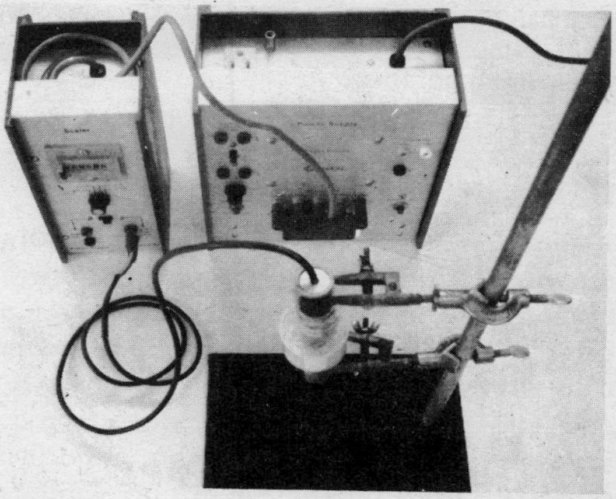

Fig. 6-13

You will probably find it convenient to count for one period of 30 sec in each minute. This will give you 30 sec to record the count, reset the counter, and so on, before beginning the next count. Record your results in a table like Fig. 6-14. Try to make about 10 trials.

Background = 12 counts per minute
= 6 counts per ½ minute

time (mins)	count	net count rate (counts per ½ min)
0 - ½	803	797
1 - 1½	627	621
2 - 2½	'	'
3 - 3½	'	'
4 - 4½	'	'

Fig. 6-14

Plot a graph of *net* count rate as a function of time. Draw the best curve you can through all the points. From the curve, find the time required for the net count rate to decrease to half its initial value.

?

10. How long does it take for the net count rate to decrease from one-half to one-fourth its initial value? one-third to one-sixth? one-fourth to one-eighth?
11. The half-life of a radioisotope is one of the important characteristics used to identify it. Using the *Handbook of Chemistry and Physics*, or another reference source, identify which of the decay products of thorium is present in your sample.
12. Can you tell from the curve you drew whether your sample contains only one radioisotope or a mixture of isotopes?

Discussion

It should be clear from your graphs and those of your classmates that the three kinds of quantities you observed all have a common property: It takes the same time (or number of rolls of the dice) to reduce the quantity to one-half its initial value as it does to reduce from one-half to one-fourth, from one-third to one-sixth, from one-fourth to one-eighth, etc. This quantity is the *half-life*.

In the experiments on the "decay" of 20-sided dice with two marked faces, you knew beforehand that the "decay rate" was one-tenth. That is, over a large number of throws, an average of one-tenth of the dice would be removed for each shake of the tray.

The relationship between the half-life of a process and the decay constant λ is discussed in the special page "Mathematics of Decay" in Chapter 21 of the Text. There you learned that for a large number of truly random events, the half-life $T_{1/2}$ is related to the decay constant λ by the equation:

$$T_{1/2} = \frac{0.693}{\lambda}$$

?

13. From the known decay constant of the dice, calculate the half-life of the dice and compare it with the experimental value found by you or your classmates.
14. If you measured the half-life of capacitor discharge or of radioactive decay, calculate the decay constant for that process.

Experiment 6-4
HALF-LIFE. II

Look at the thorium decay series in Table 6-2. One of the members of the series, radon-220, is a gas. In a sealed bottle containing thorium or one of its salts, some radon gas always gathers in the air space above the thorium. Radon-220 has a very short half-life (51.5 sec). The subsequent members of the series (polonium-214, lead-210, etc.) are solids. Therefore, as the radon-220 decays, it forms a solid deposit of

TABLE 6-2. THE THORIUM DECAY SERIES

Name	Symbol	Mode of Decay	Half-Life
Thorium-232	${}_{90}Th^{232}$	α	1.39×10^{10} yr
Radium-228	${}_{88}Ra^{228}$	β	6.7 yr
Actinium-228	${}_{89}Ac^{228}$	β	6.13 hr
Thorium-228	${}_{90}Th^{228}$	α	1.91 yr
Radium-224	${}_{88}Ra^{224}$	α	3.64 d
Radon-220	${}_{86}Rn^{220}$	α	51.5 sec
Polonium-216	${}_{84}Po^{216}$	α	0.16 sec
Lead-212	${}_{82}Pb^{212}$	β	10.6 hr
Bismuth-212	${}_{83}Bi^{212}$	α or β*	60.5 min
Polonium-212	${}_{84}Po^{212}$	α	3.0×10^{-7} sec
Thallium-208	${}_{81}Tl^{208}$	β	3.10 min
Lead-208	${}_{82}Pb^{208}$	Stable	3.10 min

*Bismuth-212 can decay in two ways: 34% decays by α emission to thallium-208; 66% decays by β emission to polonium-212. Both thallium-208 and polonium-212 decay to lead-208.

radioactive material in the bottle. In this experiment, you will measure the half-life of this radioactive deposit.

Although the amount of radioactive material in this experiment is too small to be considered at all dangerous (unless you drink large quantities of it), it is a very good idea to practice caution in dealing with the material. Respect for radioactivity is an important attitude in our increasingly complicated world.

The setup is illustrated in Fig. 6-15. The thorium nitrate is spread on the bottom of a sealed container. (The air inside should be kept damp by moistening the sponge with water.) Radon gas escapes into the air of the container, and some of its decay products are deposited on the upper foil.

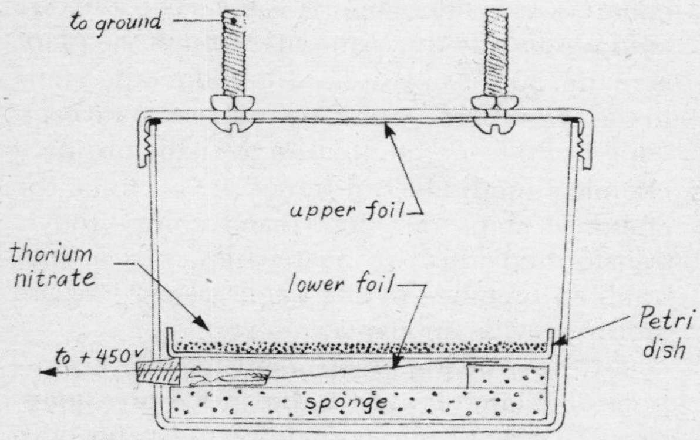

Fig. 6-15

When radon disintegrates in the nuclear reaction

$$_{86}Rn^{220} \rightarrow {}_{84}Po^{216} + {}_{2}He^{4},$$

the polonium atoms formed are ionized, apparently because they recoil fast enough to lose an electron by inelastic collision with air molecules.

Because the atoms of the first daughter element of radon are ionized (positively charged), you can increase the amount of deposit collected on the upper foil by charging it negatively to several hundred volts. Although the electric field helps, it is not essential; you will get some deposit on the upper foil even if you do not set up an electric field in the container.

After two days, so much deposit has accumulated that it is decaying nearly as rapidly as the constant rate at which it is being formed. Therefore, to collect a sample of maximum activity, your apparatus should stand for about two days.

Before beginning to count the activity of the sample, you should take a count of the background rate. Do this far away from the vessel containing the thorium. Remove the cover, place your Geiger counter about 1 mm above the foil, and begin to count. Make sure, by adjusting the distance between the sample and the window of the Geiger tube, that the initial count rate is high (several hundred per minute). Fix both the counter and the foil in position so that the distance will not change. To get fairly high precision, take a count over a period of at least 10 min (see Experiment 6-1). Because the deposit decays rather slowly, you can afford to wait several hours between counts, but you will need to continue taking counts for several days. Make sure that the distance between the sample and the Geiger tube stays constant.

Record the net count rate and its uncertainty (the "two-thirds range" discussed in Experiment 6-1). Plot the net count rate against time.

Remember that the deposit contains several radioactive isotopes and each is decaying. The net count rate that you measure is the sum of the contributions of all the active isotopes. The situation is not as simple as it was in Experiment 6-2, in which the single radioactive isotope decayed into a *stable* isotope.

?

1. Does your graph show a constant half-life or a changing half-life?

Look again at the thorium series and, in particular, at the half-lives of the decay products of radon. Try to interpret your observations of the variation of count rate with time.

?

2. Which isotope is present in the greatest amount in your sample? Can you explain why this is so? Make a sketch to show approximately how the relative amounts of the different isotopes in your sample vary with time. Ignore the isotopes with half-lives of less than 1 min.

You can use your measurement of count rate and half-life to get an estimate of the amount of deposit on the foil. The activity, $\frac{\Delta N}{\Delta t}$, depends on the number of atoms present N:

$$\frac{\Delta N}{\Delta t} = \lambda N.$$

The decay constant λ is related to the half-life $T_{1/2}$ by

$$\lambda = \frac{0.693}{T_{1/2}}$$

Use your values of counting rate and half-life to estimate N, the number of atoms present in the deposit. What mass does this represent? (1 amu = 1.7×10^{-27} kg.) The smallest amount of material that can be detected with a chemical balance is on the order of 10^{-6} g.

Discussion

It is not too difficult to calculate the speed and therefore the kinetic energy of the polonium atom. In the disintegration

$$_{86}Rn^{220} \rightarrow {}_{84}Po^{216} + {}_{2}He^{4},$$

the α particle is emitted with kinetic energy 6.8 MeV. Combining this with the value of its mass, you can calculate v^2 and, therefore, v. What is the momentum of the α particle? Momentum is of course conserved in the disintegration. So what is the momentum of the polonium atom? What is its speed? What is its kinetic energy?

The ionization energy (the energy required to remove an outer electron from the atom) is typically a few electron volts. How does your value for the polonium atom's kinetic energy compare with the ionization energy? Does it seem likely that most of the recoiling polonium atoms would ionize?

Experiment 6-5
RADIOACTIVE TRACERS

In this group of experiments, you have the opportunity to invent your procedures yourself and to draw your own conclusions. Most of the experiments will take more than one class period and will require careful planning in advance.

Caution

All these experiments take cooperation from the biology or the chemistry department, and require that safety precautions be observed very carefully so that neither you nor other students will be exposed to radiation.

For example, handle radioisotopes as you would a strong acid; wear disposable plastic gloves and goggles, and work with all containers in a tray lined with paper to soak up any spills. Never draw radioactive liquids into a pipette by mouth; use a mechanical pipette or a rubber bulb. Your teacher will discuss other safety precautions with you before you begin.

None of these activities is suggested just for the sake of doing tricks with isotopes. You should have a question clearly in mind before you start, and should plan carefully so that you can complete your experiment in the time you have available.

Tagged Atoms

Radioactive isotopes have been called tagged atoms because even when they are mixed with stable atoms of the same element, they can still be detected. To see how tagged atoms are used, consider the following example.

A green plant absorbs carbon dioxide (CO_2) from the air, and by a series of complex chemical reactions, builds the carbon dioxide (and water) into the material of which the plant is made. Suppose you tried to follow the steps in the series of reactions. You can separate each compound from the mixture by using ordinary chemical methods. But how can you trace the chemical steps by which each compound is transformed into the next when they are all jumbled together in the same place? Tagged atoms provide an answer.

Put the growing green plant into an atmosphere containing carbon dioxide. A tiny quantity of CO_2 molecules containing the radioactive isotope carbon-14 in place of normal carbon-12 should be added to this atmosphere. Less than 1 min later, the radioactivity can be detected within some, but not all, of the molecules of complex sugars and amino acids being synthesized in the leaves. As time goes on, the radioactive carbon enters step by step into each of the carbon compounds in the leaves.

With a Geiger counter, one can, in effect, watch each compound in turn to detect the moment when radioactive molecules begin to be added to it. In this way, the mixture of compounds in a plant can be arranged in the order of their formation, which is obviously a useful clue to chemists studying the reactions. Photosynthesis, long a mystery, has been studied in detail in this way.

Radioactive isotopes used in this manner are called *tracers*. The quantity of tracer material needed to do an experiment is astonishingly small. For example, compare the amount of carbon that can be detected by an analytical balance with the amount needed to do a tracer experiment. Your Geiger counter may, typically, need 100 net counts per minute to distinguish the signal from background radiation. If only 1% of the particles emitted by the sample are detected, then, in the smallest detectable sample, 10,000 (or 10^4) atoms are decaying each minute. This is the number of atoms that decay each minute in a sample of only 4×10^{-4} μg of carbon-14. Under ideal conditions, a chemical balance might detect 1 μg.

Thus, in this particular case, measurement by radioactivity is over 10,000 times more sensitive than by the balance.

In addition, tracers give you the ability to find the precise location of a tagged substance *inside* an undisturbed plant or animal. Radiation from thin sections of a sample placed on photographic film produces a visible spot (Fig. 6-16). This method can be made so precise that scientists can tell not only which *cells* of an organism have absorbed the tracer, but also which *parts* of the cell (nucleus, mitochondria, etc.).

Choice of Isotope

The choice of which radioactive isotope to use in an experiment depends on many factors, only a few of which are suggested here.

Carbon-14, for example, has several properties that make it a useful tracer. Carbon compounds are a major constituent of all living organisms. It is usually impossible to follow the fate of any one carbon compound that you inject into an organism, since the added molecules and their products are immediately lost in the sea of identical molecules of which the organism is made. Carbon-14 atoms, however, can be used to tag the carbon compounds, which can then be followed step by step through complex chains of chemical processes in plants and animals. On the other hand, the carbon-14 atom emits only β particles of rather low energy. This low energy makes it impractical to use carbon-14 inside a large liquid or solid sample since all the emitted particles would be stopped inside the sample.

The half-life of carbon-14 is about 6,000 years, which means that the activity of a sample will remain practically constant for the duration of an experiment. But sometimes the experimenter prefers to use a short-lived isotope so that it will rapidly drop to negligibly low activity in the sample or on the laboratory table if it gets spilled.

Some isotopes have chemical properties that make them especially useful for a specific kind of experiment. Phosphorus-32 (half-life: 14.3 days) is especially good for studying the growth of plants because phosphorus is used by the plant in many steps of the growth process. Practically all the iodine in the human body is used for just one specific process, the manufacture of a hormone in the thyroid gland that regulates metabolic rate. Radioactive iodine-131 (half-life: 8.1 days) has been immensely useful as a tracer in unravelling the steps in that complex process.

The amount of tracer to be used is determined by its activity, by how much it will be diluted during the experiment, and by how much radiation can be safely allowed in the laboratory. Since even very small amounts of radiation are potentially harmful to people, safety precautions and regulations must be carefully followed. The Department of Energy has established licensing procedures and regulations governing the use of radioisotopes. As a student you are permitted to use only limited quantities of certain isotopes under carefully controlled conditions. However, the variety of experiments you can do is still so great that these regulations need not discourage you from using radioactive isotopes as tracers.

One unit used to measure the radioactivity of a source is called the *curie*. When 3.7×10^{10} atoms within a source disintegrate or decay in 1 sec, its activity is said to be 1 curie (c). (This number was chosen because it is the approximate average activity of 1 g of pure radium-226.) A more practical unit for tracer experiments is the microcurie (μc), which is 3.7×10^4 disintegrations per second or 2.2×10^6 per minute. The quantity of radioisotope that students may safely use in experiments, without special license, varies from 0.1 μc to 50 μc depending on the type and energy of radiation.

Notice that even when you are restricted to 0.1 μc for your experiments, you may still expect 3,700 disintegrations per second, which would cause 37 counts a second in a Geiger counter that recorded only 1% of them.

?

1. What would be the "two-thirds range" in the activity (disintegrations per minute) of a 1 μc source?

2. What would be the "two-thirds range" in counts per minute for such a source measured with a Geiger counter that detects only 1% of the disintegrations?

3. Why does a Geiger tube detect such a small percentage of the β particles that leave the sample? (Review that part of Experiment 6-2 on the range of β particles.)

A. Autoradiography

One rather simple experiment you can almost certainly do is to reenact Becquerel's original discovery of radioactivity. Place a radioactive object (lump of uranium ore, luminous watch dial with the glass removed, etc.) on a Polaroid film packet or on a sheet of X-ray film in a light-tight envelope. A strong source of radiation will produce a visible image on the film within an hour, even through the paper wrapping. If the source is not so strong, leave it in place overnight. To get a very sharp picture, you must use unwrapped film in a completely dark room and expose it with the radioactive source pressed firmly against the film.

(Most Polaroid film can be developed by placing the packet on a flat surface and passing a metal or hard-rubber roller firmly over the pod of chemicals and across the film. Other kinds of film are processed in a darkroom according to the directions on the developer package.)

This photographic process has grown into an important experimental technique called *autoradiography*. The materials needed are relatively inexpensive and easy to use, and there are many interesting applications of the method. For example, you can grow plants in soil treated with phosphorus-32, or in water to which some phosphorus-32 has been added, and make an autoradiograph of the roots, stem, and leaves (Fig. 6-16). Or each day take a leaf from a fast-growing young plant and show how the phosphorus moves from the roots to the growing tips of the leaves.

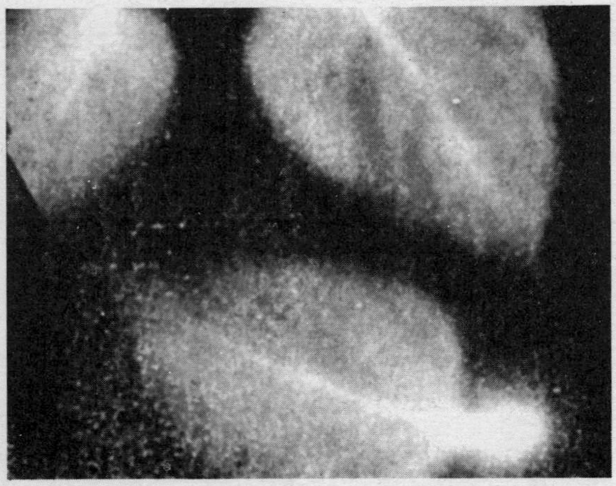

Fig. 6-16 Autoradiograph made by a student to show uptake of phosphorus-32 in Coleus leaves.

B. Chemical Reactions and Separations

Tracers are used as sensitive indicators in chemical reactions. You may want to try a tracer experiment using iodine-131 to study the reaction between lead acetate and potassium iodide solutions. Does the radioactivity remain in the solute or is it carried down with the precipitate? How complete is the reaction?

When you do experiments like this one with liquids containing β sources, transfer them carefully (with a special mechanical pipette or a disposable plastic syringe) to a small, disposable container called a *planchet*, and evaporate them so that you count the dry sample. This is important when you are using β sources since, otherwise, much of the radiation would be absorbed in the liquid before it reached the Geiger tube.

You may want to try more elaborate experiments involving the movement of tracers through chemical or biological systems. Students have grown plants under bell jars in an atmosphere containing radioactive carbon dioxide, fed radioactive phosphorus to earthworms and goldfish, and studied the metabolism of rats with iodine-131. Be sure to review safety and humane guidelines for the use of animals in research before attempting any of these experiments.

Experiment 6-6
MEASURING THE ENERGY OF β RADIATION

With a device called a *β-ray spectrometer*, you can sort out the β particles emitted by a radioactive source according to their energy just as a grating or prism spectroscope spreads out the colors of the visible spectrum. You can make a simple β-ray spectrometer with two disk magnets and a packet of Polaroid film. WIth it you can make a fairly good estimate of the average energy of the β particles emitted from various sources by observing how much they are deflected by a magnetic field of known intensity.

Mount two disk magnets as shown in Fig. 6-17. Be sure the faces of the magnets are parallel and opposite poles are facing each other.

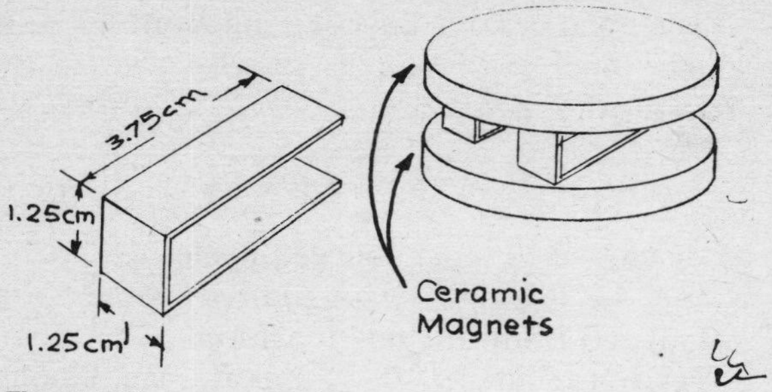

Fig. 6-17

Bend a piece of sheet metal into a curve so that it will hold a Polaroid film packet snugly around the magnets. Place a β source behind a barrier made of thin sheet lead containing two narrow slits that will allow a beam of β particles to enter the magnetic field as shown in Fig. 6-18. Expose the film to the β radiation for two days. Then carefully remove the magnets without changing the relative positions of the film and β source. Expose the film for two more days. The long exposure is necessary because the collimated beam contains only a small fraction of the β radiation given off by the source, and because Polaroid film is not very sensitive to β radiation. (You can shorten the exposure time to a few hours if you use X-ray film.)

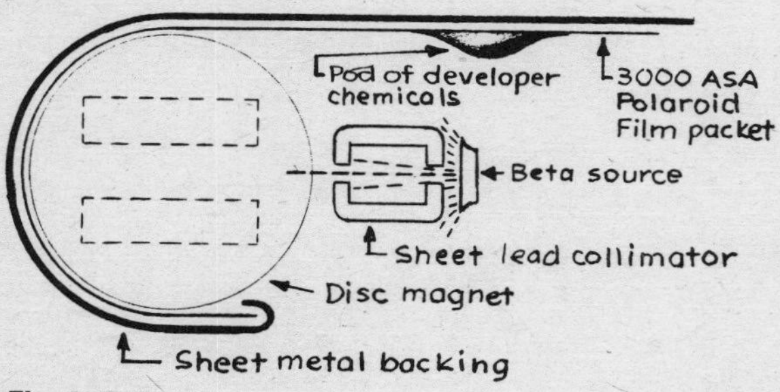

Fig. 6-18

When developed, your film will have two blurred spots on it; the distance between their centers will be the arc length *a* in Fig. 6-19.

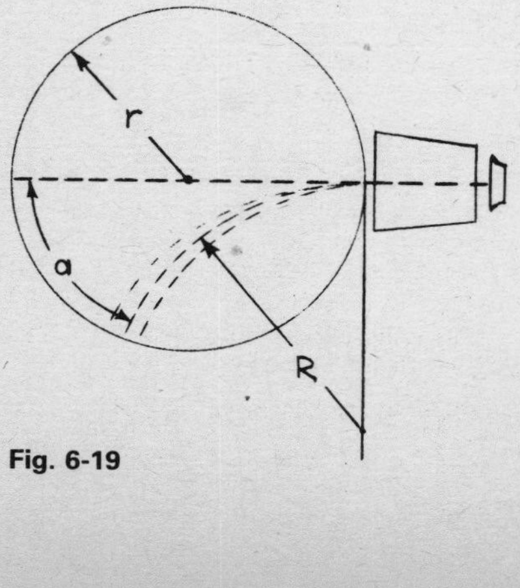

Fig. 6-19

An interesting mathematical problem is to find a relationship between the angle of deflection, as indicated by *a*, and the average energy of the particles. You can calculate the *momentum of* the particle fairly easily. Unfortunately, since the β particles from radioactive sources are travelling at nearly the speed of light, the simple relationships between momentum, velocity, and kinetic energy (which you learned in Unit 3) cannot be used. Instead, you need to use equations derived from the special theory of relativity which, although not at all mysterious, are a little beyond the scope of this course. (The necessary relations are developed in the supplemental unit, "Elementary Particles.") A graph (Fig. 6-20) that gives the values of kinetic energy for various values of momentum is provided.

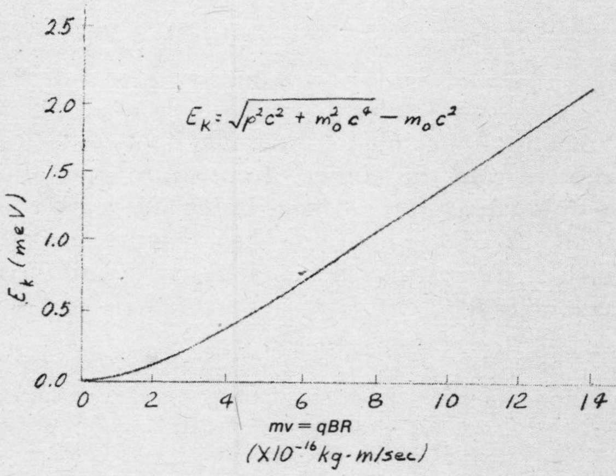

Fig. 6-20 Kinetic energy versus momentum for electrons (m_0c^2 = 0.511 MeV).

First, you need an expression that will relate the deflection to the momentum of the particle. The relationship between the force on a charged particle in a magnetic field and the radius of the circular path is derived on page 542 of Unit 5, Text. Setting the magnetic force equal to the centripetal force gives

$$Bqv = \frac{mv^2}{R}$$

which simplifies to

$$mv = BqR$$

If you know the magnetic field intensity *B* (measured with the current balance as described in the Unit 4 *Handbook*), the charge on the electron, and can find *R*, you can compute the momentum. A little geometry will enable you to calculate *R* from the arc length *a* and the radius *r* of the magnets. A detailed solution will not be given here, but a hint is shown in Fig. 6-21.

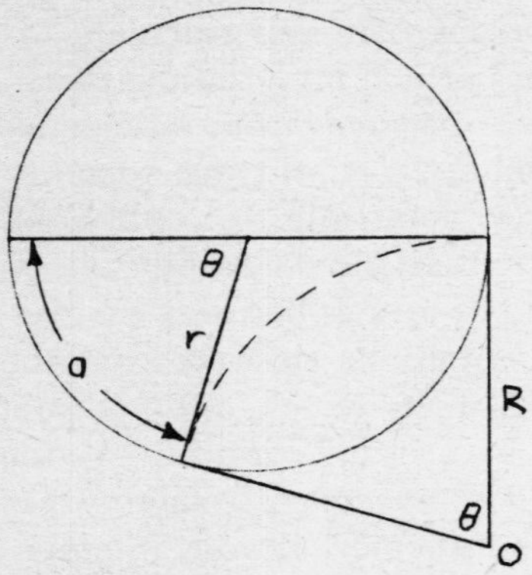

Fig. 6-21

The angle θ is

$$\theta = \frac{a}{2\pi r} \times 360°$$

You should be able to prove that if tangents are drawn from the center of curvature O to the points where the particles enter and leave the field, the angle between the tangents at O is also θ. With this as a start, see if you can calculate R.

The relationship between momentum and kinetic energy for objects travelling at nearly the speed of light

$$E = \sqrt{p^2c^2 + m_0^2c^4}$$

is discussed in most college physics texts. The graph in Fig. 6-21 was plotted using data calculated from this relationship.

From the graph, find the average kinetic energy of the β particles whose momentum you have measured. Compare this with values given in the *Handbook of Chemistry and Physics*, or another reference book, for the particles emitted by the source you used.

You will probably find a value listed which is two to three times higher than the value you found. The value in the reference book is the *maximum* energy that any one β particle from the source can have, whereas the value you found was the *average* of all the β particles reaching the film. This discrepancy between the maximum energy (which all the β's should theoretically have) and the average energy puzzled physicists for a long time. The explanation, suggested by Enrico Fermi in the mid-1930s, led to the discovery of a strange new particle called the *neutrino*, which you will want to find out about.

FILM LOOP NOTES

Film Loop 48
COLLISIONS WITH AN OBJECT OF UNKNOWN MASS

In 1932, Chadwick discovered the neutron by analyzing collision experiments. This film allows a measurement similar to Chadwick's, using the laws of motion to deduce the mass of an unknown object. The film uses balls rather than elementary particles and nuclei, but the analysis, based on conservation laws, is remarkably similar.

The first scene shows collisions of a small ball with stationary target balls, one of similar mass and one of larger mass (Fig. 6-22). The incoming ball always has the same velocity, as you can see.

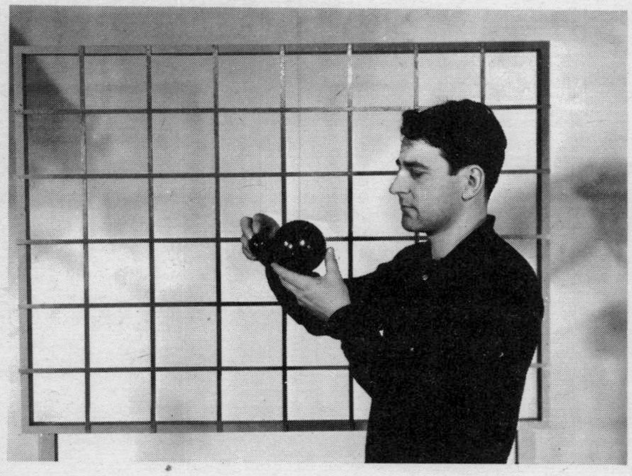

Fig. 6-22

The slow-motion scenes allow you to measure the velocity acquired by the targets. The problem is to find the mass and velocity of the incoming ball without measuring them directly. The masses of the targets are $M_1 = 352$ g, $M_2 = 4{,}260$ g.

Chadwick used hydrogen and nitrogen nuclei as targets and measured their recoil velocities. The target balls in the film do not have the same mass ratio, but the idea is the same.

The analysis is shown in detail on a special page in Chapter 23 of the Text. For each of the two collisions, equations can be written expressing conservation of energy and conservation of momentum. These four equations contain three quantities that Chadwick could not measure: the initial neutron velocity and the two final neutron velocities. Some algebraic manipulation results in the elimination of these quantities, leaving a single equation that can be solved for the neutron mass. If v_1' and v_2' are the speeds of targets 1 and 2 after collision, and M_1 and M_2 the masses, the neutron mass m can be found from

$$m(v_1' - v_2') = M_2 v_2' - M_1 v_1'$$

or

$$m = \frac{M_2 v_2' - M_1 v_1'}{v_1' - v_2'}$$

Make measurements only on the targets, as the incoming ball (representing the neutron) is supposed to be unobservable both before and after the collisions. Measure v_1' and v_2' in any convenient unit, such as divisions per second. (Why is the choice of units not important here?) Calculate the mass m of the invisible, unknown particle. In what ways might your results differ from Chadwick's?

Answers to End-of-Section Questions

Chapter 1

1. We have no way of knowing the lengths of time involved in going the observed distances.
2. No. The time between stroboscope flashes is constant and the distance intervals shown are not equal.
3. An object has a uniform speed if it travels equal distances in equal time intervals; or, if the *distance traveled* divided by time taken = constant, regardless of the particular distances and times chosen.
4.
$$v_{av} = \frac{\Delta d}{\Delta t} = \frac{720\ \text{m}}{540\ \text{sec}} = 1.3\ \text{m/sec}$$
5.
$$\Delta d = v_{av}\ \Delta t = (20\ \text{m/sec})(15\ \text{sec}) = 300\ \text{m}$$
6.

Δt	$\Delta d/\Delta t$
(5.0)	(1.0)
(6.0)	(0.8)
(4.5)	1.1
(5.5)	0.9
7.5	0.67
8.0	0.62
8.6	0.58

(Entries in parentheses are those already given in the text.)

7. Hint: To determine the location of the left edge of the puck relative to readings on the meter stick, line up a straight edge with the edge of the puck and *both* marks on the meter stick corresponding to a given reading.

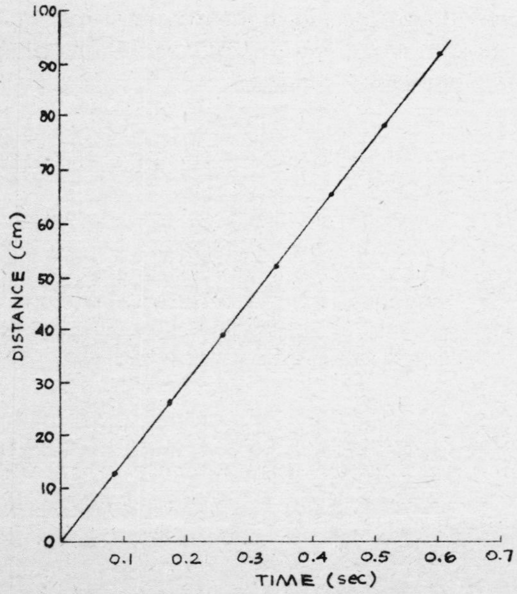

d(cm)	t(sec)
0	0
13	0.1
26	0.2
39	0.3
52	0.4
65	0.5
78	0.6
92	0.7

8. The one on the left has the larger slope mathematically; it corresponds to 160 km/hr, whereas the one on the right corresponds to 80 km/hr.
9. most rapidly at the beginning when the slope is steepest; most slowly toward the end where the slope is most shallow
10.
$$\frac{\Delta d}{\Delta t} = \frac{2\ \text{m}}{4\ \text{sec}}$$
$$= 0.5\ \text{m/sec (from table)}$$
$$= \frac{4.5\ \text{m}}{8.5\ \text{sec}}$$
$$= 0.5\ \text{m (from graph)}$$
The two results are the same.
11. (a) 1.7 cm at 42 min (interpolation)
(b) 3 cm at 72 min (extrapolation)

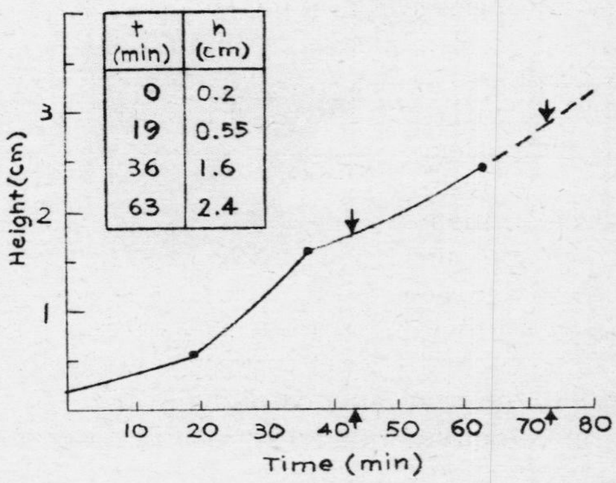

12. an estimate for an additional lap (extrapolation)
13. $v_{inst} = \Delta d/\Delta t$ as the time interval Δt becomes very small. Instantaneous speed is the limit of the average speeds computed for very small time intervals at a given point.
14. (a) about 5 m/sec
(b) about 25 m/sec
(c) zero
Average speed is equal to the distance traveled divided by the elapsed time. Instantaneous speed is the speed at a particular moment of time.
15.
$$a_{av} = \frac{\text{final speed} - \text{initial speed}}{\text{elapsed time}} = \frac{100\ \text{km/hr} - 0\ \text{km/hr}}{5\ \text{sec}} = 20\ \text{km/hr/sec}$$

16. $$a_{av} = \frac{2\text{ km/hr} - 5\text{ km/hr}}{15\text{ min}} = \frac{-3\text{ km/hr}}{15\text{ min}} = -0.2\text{ km/hr/min}$$
No, since *average* acceleration is specified.

17. $$a_{av} = \frac{\Delta v}{\Delta t}$$
$$\Delta t = \frac{\Delta v}{a_{av}} = \frac{12\text{ m/sec}}{4\text{ m/sec}_2} = 3\text{ sec}$$

18. $$\Delta v = a_{av}\,\Delta t = (4\text{ m/sec}_2)(6\text{ sec}) = 24\text{ m/sec}$$

Chapter 2

1. shape, size, and weight
   Drop several different objects, being sure to change only one quality with each object, and observe the results.
2. Composition: Terrestrial objects are composed of combinations of earth, water, air, and fire; celestial objects of nothing but a unique fifth element.
   Motion: Terrestrial objects seek their natural positions of rest depending on their relative contents of earth (heaviest), water, air, and fire (lightest); celestial objects moved endlessly in circles.
3. (a), (b), and (c)
4. Aristotle: The nail is heavier than the toothpick so it falls faster.
   Galileo: Air resistance slows down the toothpick more than the nail.
5. (1) The bag, since it is lighter than the rock, slows the rock down. Therefore, the rock and bag together fall *slower* than the rock alone.
   (2) The bag and the rock together are heavier than the rock alone. Therefore, the bag and rock together fall *faster* than the rock alone.
6. See question 3 of Chapter 1.
7. An object is uniformly accelerated if its speed increases by equal amounts during equal time intervals; $\Delta v/\Delta t$ = constant.
8. $$a_{av} = \frac{\Delta v}{\Delta t} = \frac{32\text{ m/sec} - 22\text{ m/sec}}{5\text{ sec}} = 2\text{ m/sec}_2$$
   $$\Delta t = \frac{40\text{ m/sec} - 32\text{ m/sec}}{2\text{ m/sec}_2} = 4\text{ sec}$$
9. The definition should (1) be mathematically simple and (2) correspond to actual free-fall motion.
10. (b)
11. Distances are relatively easy to measure as compared with speeds; measuring short time intervals remained a problem, however.
12. The expression $d = vt$ can only be used if v is constant. The second equation refers to accelerated motion in which v is *not* constant. Therefore, the two equations cannot be applied to the same event.
13. (c) and (e)
14. (d)
15. (a), (c), and (d)
16. (a)

Chapter 3

1. kinematic: (a), (b), (d) dynamic: (c), (e)
2. a continuously applied force.
3. The air pushed aside by the puck moves around to fill the space left behind the puck as it moves along and so provides the propelling force needed.
4. the force of gravity downward and an upward force of equal size exerted by the table
   The sum of the forces must be zero, because the vase is not accelerating.
5. the first three
6. 6; zero; 6; 2 up; 6 in
7. no
   When an object is in equilibrium, the net force acting on it equals zero.
   no
8. (a) vector (d) vector
   (b) scalar (e) scalar
   (c) scalar (f) vector
9. Vector quantities (1) have magnitude and direction, (2) can be represented graphically by arrows, and (3) can be combined to form a single resultant vector by using either the head-to-tail or the parallelogram method. (Note: Only vectors of the same kind are combined in this way; that is, we add force vectors to force vectors, *not* force vectors to velocity vectors, for example.)
10. Direction is now taken into account. (We must now consider a change of direction to be as valid a case of acceleration as speeding up or slowing down.)
11. Aristotle: A continuous force is required to sustain motion.
    Newton: Since a force is not needed to sustain motion, the pedaling force must exactly oppose air resistance, and the net force is zero.
    Aristotle's ideas represent a "common sense" interpretation of natural events, which we all tend to accept before studying the events carefully.
12. W downward, 0,0,0
13. Galileo's "straight line forever" motion may have meant motion at a constant height above the earth, whereas Newton's meant moving in a straight line through empty space.
14. 1 newton = 1 kg · m/sec^2
15. $$m = \frac{F}{a} = \frac{10\text{ N}}{4\text{m/sec}^2} = 2.5\text{ kg}$$

16. false (Frictional forces must be taken into account in determining the actual net force exerted.)

17. $a = \frac{\Delta v}{\Delta t}$
$= \frac{10 \text{ m/sec}}{5 \text{ sec}}$
$= 2 \text{ m/sec}^2$

$F = ma$
$= (2 \text{ kg})(2 \text{ m/sec}^2)$
$= 4 \text{ N}$

18. the magnitude and direction of the applied force, and the mass of the object

19.

Mass (kg)	Acceleration (m/sec^2)
1	30
2	15
3	10
1/5	150
0.5	60
45	0.67
10	3
0.4	75

20. Weight ($m \times a_g$) is the force acting on a mass at the earth's surface, caused by the earth's gravitational attraction. Weight is a vector quantity.
Mass (m) is the unchanging quality of a body that describes its resistance to any acceleration. Mass is a scalar quantity.

21. Because the hammer is 20 times more massive than the nail, it also requires 20 times more force to accelerate the hammer. Thus, the accelerations are nearly equal.

22.
$F = ma$
$a = \frac{F}{m}$
$= \frac{30 \text{ N}}{3 \text{ kg}}$
$= 10 \text{ m/sec}^2$

The acceleration on the moon or in deep space would be the same since the mass of the object remains the same.

23. (c) and (f)

24. (1) The second force is equal in magnitude (3 N).
(2) The second force acts in the opposite direction (to the left).
(3) The second force acts on the first object.

25. The horse pushes against the earth; the earth pushes against the horse causing the horse to accelerate forward. (The earth accelerates also, but can you measure it?) The swimmer pushes backward against the water; the water, according to the third law, pushes forward against the swimmer; however, there is also a backward frictional force of drag exerted by the water on the swimmer. The two forces acting *on* the swimmer add up to zero, since he is not accelerating.

26. No. The force "pulling the string apart" is still only 300 N; the 500 N would have to be exerted at both ends to break the line.

27. See *Text* p. 68

28.
$F = ma$
$= (70 \text{ kg})(3 \text{ m/sec}^2)$
$= 210 \text{ N}$

29.
$a = \frac{F}{m}$
$= \frac{29.4 \text{ N}}{3 \text{ kg}}$
$= 9.8 \text{ m/sec}^2$

The weight of the object is equal to its mass (3 kg) times the acceleration of gravity (9.8 m/sec^2); that is, 3 kg × 9.8 m/sec^2 = 29.4 N.

Chapter 4

1. It agrees with the experiment. Also, the horizontal motion agrees with Newton's first law, while the vertical motion agrees with Newton's second law.

2. a_g, because the two motions are independent

3. $\Delta d = v_i \Delta t + \frac{1}{2} a (\Delta t)^2$

vertical ($v_i = 0$)	horizontal ($a = 0$)
$\Delta y = 0 + \frac{1}{2} a (\Delta t)^2$	$\Delta x = v_x \Delta t$
$\Delta y = \frac{1}{2} a (\Delta x / \Delta v_x)^2$	$\Delta t = \Delta x / \Delta v_x$

4. (a), (c), and (e)

5. $\Delta x = v_x \Delta t$
$= (4 \text{ m/sec})(10 \text{ sec})$
$= 40 \text{ m}$
$\Delta y = \frac{1}{2} a_g (\Delta t)^2$
$= \frac{1}{2} (10 \text{ m/sec}^2)(10 \text{ sec})^2$
$= 500 \text{ m}$

6. They must be moving with a uniform speed relative to each other.

7. A person in the stands sees a parabolic path. Two different observers see different motions. The moving observer is in the plane of the motion and does not observe a sideways motion.

8. (a) $T = 1/f = 1/45 = 2.2 \times 10^{-2}$ min
(b) 2.2×10^{-2} min × 60 sec/min = 1.32 sec
(c) f = 45 rpm × 1/60 min/sec = 0.75 rps

9. $v = \frac{2\pi R}{T} = \frac{2 \times 3.14 \times 3}{60} = 0.3\dot{1}$ cm/min

10. f = 80 vibrations/min = 1.3 vib/sec
$T = 1/f = 1/1.3 = 0.75$ sec

11. (a) and (b)

12. $\frac{mv^2}{R}$

13.
$F = m\left(\frac{v^2}{R}\right)$
$= \frac{m(4\pi^2 R)}{T^2}$
$= \frac{4\pi^2(1)(0.5)}{1}$
$= 19.7$ N (to center)

It would move in a straight path.
$= 19.7$ N (from center)

14. the value of the gravitational acceleration and the radius of the moon (to which 110 km is added to determine R)

15. The force we calculate to be holding it in orbit is the earth's gravity at that height.

16.
$$F = \frac{4\pi^2 m (R_{earth} + h)}{T^2}$$
$$= 1{,}660 \text{ N}$$

Chapter 5

1. The sun would set 4 min later each day.
2. Calendars were needed to schedule agricultural activities and religious rites.
3. The sun has a westward motion each day, an eastward motion with respect to the fixed stars, and a north–south variation.
4.

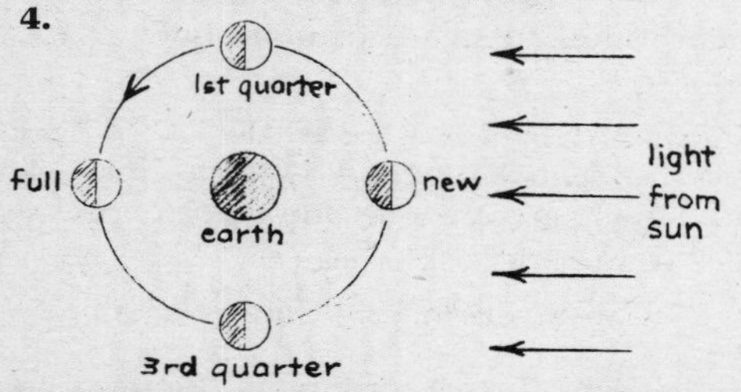

5. Eclipses do not occur each month because the moon and the earth do not have the same planes of orbit.
6. (a) The moon continuously moves in an easterly direction through the sky, exhibits changing phases, and rises and sets each day.
   (b) Throughout one year, the motion is similar month by month.
   (c) The words "moon" and "month" have the same origin.
7. Mercury and Venus are always fairly near the sun: either east of the sun (evening sky) or west of the sun (morning sky).
8. When in opposition, a planet is opposite the sun; therefore, the planet would rise at sunset and be on the north–south line at midnight.
9. after they have been farthest east of the sun and are visible in the evening sky
10. when they are near opposition
11. No. They are always close to the ecliptic.
12. Mercury and Venus remain near the sun. Mars, Jupiter, and Saturn (as well as Uranus, Neptune, and Pluto) can be found in different locations along the zodiac. When they are nearly opposite the sun, Mars, Jupiter, and Saturn cease their eastward motion and move westward (in retrograde motion) for several months. They then return to their eastward motion.
13. How may the irregular motions of the planets be accounted for by combinations of constant speeds along circles?
14. Many of their written records have been destroyed by fire, weathering, and decay.
15. Only perfect circles and uniform speeds were suitable for the perfect and changeless heavenly bodies.
16. A geocentric system is an earth-centered system. The yearly motion of the sun is accounted for by assuming that it is attached to a separate sphere that moves contrary to the motion of the stars.
17. The first solution, as proposed by Eudoxus, consisted of a system of transparent crystalline spheres that turned at various rates around various axes.
18. Aristarchus assumed that the earth rotated daily, which accounted for all the daily motions observed in the sky. He also assumed that the earth revolved around the sun, which accounted for the many annual changes observed in the sky.
19. When the earth moved between one of these planets and the sun (with the planet being observed in opposition), the earth would be moving faster than the planet. So the planet would appear to us to be moving westward.
20. The direction to the stars should show an annual shift: the annual parallax. (This involves a very small angle and so could not be observed with instruments available to the Greeks. It was first observed in A.D. 1836.)
21. Aristarchus was considered to be impious because he suggested that the earth, the abode of human life, might not be at the center of the universe.

    His system was neglected for a number of reasons:
    (1) "Religious": It displaced humanity from the center of the universe.
    (2) Scientific: Stellar parallax was not observed.
    (3) Practical: It predicted celestial events no better than other, less offensive, theories.

Chapter 6

1. The lack of uniform velocity associated with equants was (1) not sufficiently absolute and (2) not sufficiently pleasing to the mind.
2. (a) P, C (d) C
   (b) P, C (e) P, C
   (c) P (f) C
3. the relative size of the planetary orbits as compared with the distance between the earth and the sun; these were related to the calculated periods of revolution about the sun.
4. (b) and (d)
5. 2° in both cases
6. No. Precise computations required more small motions than in the system of Ptolemy.
7. Both systems were about equally successful in explaining observed phenomena.
8. The position of humanity and its abode, the earth, were important in interpreting the divine plan of the universe.
9. They are equally valid; for practical purposes we prefer the Copernican because of its simplicity.
10. He challenged the earth-centered world outlook of his time and opened the way for later modifications and improvements by Kepler, Galileo, and Newton.
11. the appearance in 1572 of a "new star" of varying brightness
12. It included expensive equipment and facilities

and involved the coordinated work of a staff of people.

13. They showed that comets were distant astronomical objects, not local phenomena as had been believed.

14. He made them larger and sturdier and devised scales with which angle measurements could be read more precisely.

15. He analyzed the probable errors inherent in each piece of his equipment; also, he made corrections for the effects of atmospheric refraction.

16. He kept the earth fixed, as did Ptolemy, and he had the planets going around the sun, as did Copernicus.

Chapter 7

1. finding out the correct motion of Mars through the heavens

2. Epicycles were only a convenient computational device for describing some of the cyclical motions. Kepler began to seek objects and forces that caused the motions.

3. By means of circular motion, Kepler could not make the position of Mars agree with Tycho Brahe's observations. (There was a discrepancy of 8 min of arc in latitude.)

4. By means of triangulation, based on observations of the directions of Mars and the sun 687 days apart, he was able to plot the orbit of the earth.

5. A line drawn from the sun to a planet sweeps out equal areas during equal time intervals.

6. where it is closest to the sun

7. You would probably guess that he went to the store every day at 8 A.M. An empirical law is a general statement based on observations. Empirical laws can result in over-generalizations.

8. Mars has the largest eccentricity of the planets Kepler could study.

9. (a) Law of elliptical orbits
(b) Law of areas
(c) Both (plus date of passage of perihelion, for example)

10. The square of the period of any planet is proportional to the cube of its average distance to the sun.

11. Jupiter: $\frac{T^2}{R^3} = \frac{(11.86)^2}{(5.2)^3} = 1$

Saturn: $\frac{T^2}{R^3} = \frac{(29.46)^2}{(9.54)^3} = 1$

All objects orbiting the sun have $T^2/R^3 = 1$ in units of years and astronomical units.

12. $R^3 = (5)^2$
$R^3 = 25$
$R = 2.9$ AU

13. Kepler's periods described motion around the sun. The complex motions of an earth-centered system did not include such periods.

14. Kepler based his laws upon observations, and expressed them in a mathematical form.

15. popular language, concise mathematical expression

16. The common beliefs contradicted by Galileo's observations were:
(a) Changes do not occur in the starry heavens.
(b) Stars are meant to provide light at night; therefore, no "invisible" stars can exist.
(c) The moon is perfectly spherical.
(d) Only the earth is the center of revolutions.
(e) The sun, like other heavenly bodies, is perfect.
(f) Venus is always between the sun and the earth, and therefore cannot show a "full" phase.

17. both the heliocentric and Tychonic theories

18. The sunspots and the mountains on the moon refuted the Ptolemaic assertion that all heavenly bodies were perfect spheres.

19. No. They only supported a belief that he already held.

20. Some believed that distortions in the telescope (which were plentiful) could have caused the peculiar observations. Others believed that established physics, religion, and philosophy far outweighed a few odd observations.

21. b, c (d is not an unreasonable answer, since it was by writing in Italian that he stirred up many people.)

Chapter 8

1. (1) Scholars published journals and formed societies where they presented reports.
(2) Time and money were available.
(3) Interest in science was growing.
(4) Capable scientists and artisans were available.
(5) Experimental and mathematical tools were improving.
(6) Interesting problems were clearly stated.

2. "From the phenomena of motions, to investigate (induce) the forces of nature, and then from these forces to demonstrate (deduce) other phenomena."

3. (1) Do not use more hypotheses than necessary.
(2) To the same effects, assign the same causes.
(3) Properties of nearby bodies are assumed to be true of distant bodies also.
(4) Hypotheses (propositions) based on observations are accepted until refined by new observations.

4. Every body in the universe attracts every other body.

5. The same set of rules (laws) applies on the earth and in the heavens.

6. The forces exerted on the planets are always directed toward the single point where the sun is located.

7. the formula for centripetal acceleration

8. that the orbit was circular

9. No, he included the more general case of all conic sections.

10. 4×, 25×, 16×
11. $1/(60)^2 = 1/3{,}600 = 0.00028 = 0.028\%$
$9.8\ m/sec^2 \times 0.00028 = 0.0027\ m/sec^2$
12. That one law would be sufficient to account for both.
13. He thought it was magnetic and acted tangentially.
14. the physics of motion on earth and in the heavens under one universal law of gravitation
15. No. He thought it was sufficient to simply describe and apply it.
16. An all pervasive ether transmitted the force through larger distances. He did not wish to use a hypothesis that could not be tested.
17. phenomenological and thematic
18. discussion
19. (a) The forces are equal.
(b) The accelerations are inversely proportional to the masses.
20. (a) $2F$
(b) $3F$
(c) $6F$
21. (b) $F_{AB} = 4F_{CD}$
22. Otherwise, calculation of the force would require considering all parts of the bodies and their distances separately. Experimental evidence and mathematical analysis justify the use of "point masses" for rigid bodies. Only simple systems can really be studied. The conclusion that a simple law worked was very important.
23. the values of the constant in Kepler's third law $T^2/R^3 = k$ as applied to satellites of each of the two planets to be compared
24. the numerical value of G
25. F_{grav}, m_1, m_2, R
26. the period of the moon and the distance between the centers of the earth and the moon, or the ratio T^2/R^3
27. similar information about Saturn and at least one of its satellites
28. 1/1,000; inversely proportional to the masses
29. On the near side, the water is pulled away from the solid earth; on the far side, the solid earth is pulled away from the water. Since $F \propto 1/R^2$, the larger R is, the smaller the corresponding F.
30. all of them
31. As the moon orbits, its distance to the sun is continually changing, thus affecting the net force on the moon due to the sun and the earth. Also, the earth is not a perfect sphere.
32. Comets travel on very elongated ellipses.
33. no

Chapter 9

1. false
2. No. Don't confuse mass with volume or mass with weight.
3. Answer C
4. The total mass is 15 g.
5. No. Change speed to velocity and perform additions by vector techniques.
6. (a), (c), and (d) (Their momenta before collision are equal in magnitude and opposite in direction.)
7. (a) The total momentum does not change.
(b) The total initial momentum is equal to the total final momentum.
(c) Nothing can be said about the individual parts of the system.
8. Least momentum: a pitched baseball (small mass and fairly small speed)
Greatest momentum: a jet plane in flight (very large mass and high speed)
9. (a) about 4 cm/sec; faster ball delivers more momentum to girl.
(b) about 4 cm/sec; more massive ball delivers more momentum to girl.
(c) about 1 cm/sec; with same gain in momentum, more massive girl gains less speed.
(d) about 4 cm/sec; momentum change of ball is greater if its direction reverses.
(These answers assume the mass of the ball is much less than the mass of the girl.)
10. It can be applied to situations where only masses and speeds can be determined.
11.
$$\vec{F}\,\Delta t = \Delta\vec{p}$$
$$\vec{F} = \frac{\Delta\vec{p}}{\Delta t} = \frac{50}{15} = 3.3\ \text{N}$$
12. Conservation of mass: No substances are added or allowed to escape.
Conservation of momentum: No net force from outside the system acts upon any body considered to be part of the system.
13. None of these is an isolated system. In cases (a) and (b), the earth exerts a net force on the system. In case (c), the sun exerts a net force on the system.
14. You know that the total mass of the system will be 22 kg and the total momentum of the system will be 30 kg · m/sec up. You cannot tell what the individual masses or momenta will be.
15. Answer (c) (Perfectly elastic collisions can only occur between atoms or subatomic particles.)
16. Answer (d) (This assumes mass is always positive.)
17. Answer (c)
18. (a) It becomes stored as the object rises.
(b) It becomes "dissipated among the small parts" that form the earth and the object.

Chapter 10

1. Answer (b)
2. Answer (b)
3. Answer (c)
4. Answer (c); the increase in potential energy equals the work done on the spring.
5. Answer (e); you must do work on the objects to push them closer together.
6. Answer (e); kinetic energy increases as gravita-

tional potential energy decreases. Their sum remains the same (if air resistance is negligible).

7. Potential energy is greatest at extreme position where the speed of the string is zero. Kinetic energy is greatest at midpoint where the string is unstretched.
8. The less massive treble string will gain more speed although both gain the same amount of kinetic energy (equal to elastic potential energy given by guitarist).
9. Multiply the weight of the boulder (estimated from density and volume) by the approximate distance above ground level.
10. None. Centripetal force is directed inward along the radius, which is always perpendicular to the direction of motion for a circular orbit.
11. same, if initial and final positions are identical
12. same, if frictional forces are negligible; less if frictional forces between skis and snow are taken into account.
13. Answer (c)
14. Answer (c)
15. Answer (b)
16. Nearly all. A small amount was transformed into kinetic energy of the slowly descending weights; the water container would also have been warmed.
17. Answer (b)
18. Answer (d)
19. Answer (a)
20. Answer (b)
21. Answer (a)
22. Answer (e)
23. The statement means that the energy that the lion obtains from eating comes ultimately from sunlight. The lion eats animals, which eat plants, which grow by absorbed sunlight.
24. Answer (c)
25. Answer (a)
26. Answer (c)
27. Answer (c)
28. ΔE is the change in the total energy of the system; ΔW is the net work (the work done on the system minus the work done by the system); ΔH is the net heat exchange (heat added to the system minus heat lost by the system).
29. 1. heating (or cooling) it
    2. doing work on it (or allowing it to do work)
30. Answer (b)

Chapter 11

1. Answer (c)
2. true
3. false
4. a working model; a theoretical model; discussion
5. 10^{18} particles; 10^{-18} m in diameter; 10^2 m/sec; disordered velocities and directions distributed uniformly; these particles might be atoms, molecules, or dust particles.
6. Answer (b)
7. In gases, the molecules are far enough apart so thta the rather complicated intermolecular forces can safely be neglected.
8. Answer (b)
9. Answer (b)
10. Answer (d)
11. The atom had been described formerly in vague terms. His estimate put some real numbers into the theory and showed its usefulness in making predictions.
12. Answer (c)
13. When the piston is pushed in, work is done on the individual particles, thus increasing their kinetic energy. Since the temperature is proportional to this kinetic energy, the temperature must rise.
14. Low-density gases not near a phase change to a liquid or solid.
15. Answer (a)
16. An irreversible process is one in which order decreases and therefore the entropy increases.
17. Answers a, b, and c are correct.
18. The stove could warm up even more as the water molecules passed their kinetic energy to it. The second law of thermodynamics affirms that heat will not flow from a cool body to a hot body by itself. The second law (a statistical law) does not apply to individual molecules.
19. (a) unbroken egg
    (b) a glass of ice and warm water
20. (a) true
    (b) false
    (c) false
21. Answer (b)

Chapter 12

1. transverse, longitudinal, and torsional
2. longitudinal; fluids can be compressed but they are not stiff enough to be bent or twisted.
3. transverse
4. No. The movement of the bump in the rug depends on the movement of the mouse; it does not go on by itself.
5. energy (Particles of the medium are *not* transferred along the direction of the wave motion.)
6. the stiffness and the density
7. less stiff: slower propagation
   less mass: faster propagation
8. (1) wavelength, amplitude, polarization
   (2) frequency, period
9. the distance between any two successive points that have identical positions in the wave pattern
10. (1) 100 Hz
    (2) $T = \dfrac{1}{f} = \dfrac{1}{100 \text{ Hz}} = 0.01 \text{ sec}$
    (3) $\lambda = \dfrac{v}{f} = \dfrac{10 \text{ m/sec}}{100 \text{ Hz}} = 0.1 \text{ m}$
11. Answer (b)
12. $A_1 + A_2$
13. Yes. The resulting displacement would be $5 + (-6) = -1$ cm; superposition

14. cancellation
15. Antinodal lines are formed by a series of antinodal points. Antinodal points are places where waves arrive in phase and maximum reinforcement occurs. (The amplitude there is greatest.)
16. Answer (a)
17. when the difference in path lengths to the two sources is an odd number of half wavelengths ($\frac{1}{2}\lambda, \frac{3}{2}\lambda, \frac{5}{2}\lambda$, etc.)
18. (a) no motion at the nodes
    (b) oscillates with maximum amplitude
19. $\frac{\lambda}{2}$
20. $2L$, so that one-half wavelength just fits on the string
21. No. Only frequencies that are whole-number multiples of the fundamental frequency are possible.
22. All points on a wave front have the same phase; that is, they all correspond to crests or troughs (or any other set of similar parts of the wavelength pattern).
23. Every point on a wave front may be considered to behave as a point source for waves generated in the direction of the wave's propagation.
24. If the opening is less than one-half a wavelength wide, the difference in distance to a point P from the two edges of the opening cannot be equal to $\lambda/2$.
25. As the wavelength increases, the diffraction pattern becomes more spread out and the number of nodal lines decreases until the pattern resembles one-half of that produced by a point source oscillator.
26. yes to both (Final photograph shows diffraction without interference; interference occurs whenever waves pass each other.)
27. A ray is a line drawn perpendicular to a wave front and indicates the direction of propagation of the wave.
28. The angles are equal.
29. parabolic
30. The reflected wave fronts are parallel wave fronts.
31. (1) stays the same
    (2) becomes smaller
    (3) changes so that the wave fronts are more nearly parallel to the boundary (Or its direction of propagation becomes closer to the perpendicular between the media.)
32.

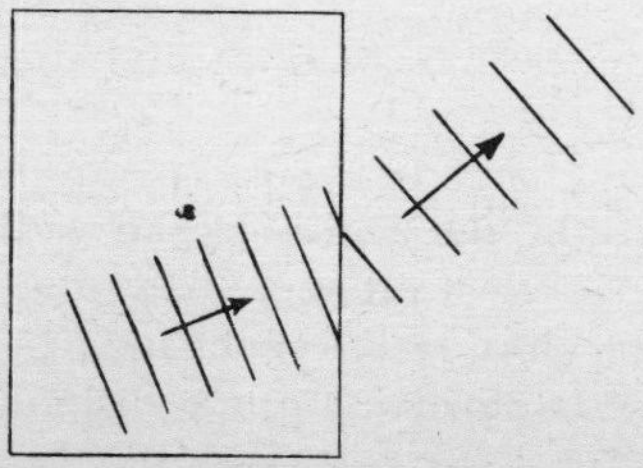

33. (1) $f\lambda = v$ relationship
    (2) reflection
    (3) refraction
    (4) diffraction
    (5) interference
34. Sound waves are longitudinal.

Chapter 13

1. No. Eventually diffraction begins to widen the beam. This property is called superposition.
2. Römer based his prediction on the extra time he had calculated it would require light to cross the orbit of the earth.
3. Römer showed that light has a finite speed.
4. Experiments carried out by Foucault and Fizeau showed that light has a *lower* speed in water than in air, whereas the particle model required that light have a *higher* speed in water.
5. When light enters a more dense medium, its wavelength and speed decrease, but its frequency remains unchanged.
6. Young's experiments showed that light could be made to form an interference pattern; such a pattern could be explained only by assuming a wave model for light.
7. It was diffraction that spread out the light beyond the two pinholes so that overlapping occurred and interference took place between the two beams.
8. Poisson applied Fresnel's wave equations to the shadow of a circular obstacle and found that there should be a bright spot in the center of the shadow.
9. Newton passed a beam of white light through a prism and found that the white light was somehow replaced by a diverging beam of colored light. Further experiments proved that the colors could be recombined to form white light.
10. Newton cut a hole in the screen on which the spectrum was projected and allowed a single color to pass through the hole and through a second prism; he found that the light was again refracted, but no further separation took place.
11. A shirt appears blue if it reflects mainly blue light and absorbs most of the other colors that make up white light.
12. The "nature philosophers" were apt to postulate unifying principles regardless of experimental evidence to the contrary, and were very unhappy with the idea that something they had regarded as unquestionably pure had many components.
13. The amount of scattering of light by tiny obstacles is greater for shorter wavelengths than for longer wavelengths
14. The "sky" is sunlight scattered by the atmosphere. Light of short wavelength (the blue end of the spectrum) is scattered most. On the moon, the sky looks dark because there is no atmosphere to scatter the light to the observer.

15. Hooke and Huygens had proposed that light waves are similar to sound waves. Newton objected to this view because the familiar straight-line propagation of light was so different from the behavior of sound. In addition, Newton realized that polarization phenomena could not be accounted for in terms of spherical pressure waves.

16. reflection, refraction, diffraction, interference, polarization, color, finite speed, and straight-line propagation. (This last would be associated with plane waves.)

17. No.

18. Light had been shown to have wave properties, and all other known wave motions required a physical medium to transmit them, so it was assumed that an "ether" must exist to transmit light waves.

19. Because light is a transverse wave and propagates at such a high speed, the ether must be a very *stiff solid*.

Chapter 14

1. Lodestone continuously attracts or repels only lodestone or iron. It has two poles (N and S) and orients itself in a north–south direction. The N and S poles cannot be isolated. The effects of lodestone are magnetic.

After being rubbed, amber will attract or repel many different types of material; the effect depends on what type of material is used to rub the amber. These effects are electrical.

2. He showed that the earth and the lodestone affect a magnetized needle in similar ways.

3. Amber attracts many substances; lodestone only a few. Amber needs to be rubbed to attract; lodestone always attracts. Amber attracts toward its center; lodestone attracts toward either of its poles.

4. 1. *Like charges* repel each other. A body that has a *net positive charge* repels any body that has a *net positive charge*. That is, two glass rods that have both been rubbed will tend to repel each other. A body that has a *net negative charge* repels any other body that has a *net negative charge*.
2. *Unlike charges* attract each other. A body that has a *net positive charge* attracts any body that has a *net negative charge* and vice versa.

5. A cork hung inside a charged silver can was not attracted to the sides of the can. (This implied that there was no net electric force on the cork: a result similar to that proved by Newton for gravitational force inside a hollow sphere.)

6. $F_{el} \propto 1/R^2$ and $F_{el} \propto q_A q_B$

7. F_{el} will be one-quarter as large.

8. No, the ampere is the unit of current.

9. $F = k\left(\frac{Q_1Q_2}{R}\right)$ (toward each other)

$k = 9 \times 10^9\ \text{N} \cdot \text{m}^2/\text{C}^2$

Therefore,

$F = 9 \times 10^9\ \text{N} \cdot \text{m}^2/\text{C}^2 \left(\frac{1\ \text{C} \times 1\ \text{C}}{1\ \text{m}^2}\right)$

$= 9 \times 10^9\ \text{N}$

10. Each point in a scalar field is given by a number only, whereas each point in a vector field is represented by a number and a direction. Examples of scalar fields: sound field near a horn, light intensity near a bulb, temperature near a heater. Examples of vector fields: gravitational field of earth, electric fields near charged bodies, magnetic fields near magnets.

11. To find the gravitational field at a point, place a known mass at the point, and measure both the direction and magnitude of the force on it. The direction of the force is the direction of the field; the ratio of the magnitude of force and the mass is the magnitude of the field.

To find the electric field, place a known positive charge at the point, and measure the direction and magnitude of the force on the charge. The direction of the force is the direction of the electric field. The ratio of the magnitude of the force and the charge is the magnitude of the field.

Note: To determine the force in either case, one could observe the acceleration of a known mass or determine what additional force must be introduced to balance the original force.

12. The corresponding forces would *also* be doubled and therefore the *ratios* of force to mass and force to charge would be unchanged.

13. The negative test body will experience a force *upward*.

14. $\vec{E} = \frac{\vec{F}}{q}$

$= \frac{1 \times 10^{-2}\ \text{N}}{3 \times 10^{-5}\ \text{C}}$

$= 3.3 \times 10^2\ \text{N/C up}$

15. $\vec{F} = q\vec{E}$

$= (5 \times 10^{-3}\ \text{C})(3.3 \times 10^2\ \text{N/C})$

$= 1.7\ \text{N up}$

16. The field concept avoids the necessity of using the Coulomb force law to find the force between every pair of charged particles when you want to find the force on a charge at a particular point.

17. If the droplets or spheres are charged *negatively*, they will experience an electric force in the direction opposite to the field direction.

18. Charge comes in basic units: the charge of the electron.

19. A negative charge (−) must also appear somewhere inside the same closed system. (For example, an electron separates from an atom, leaving the atom positively charged.)

20. It produced a steady current for a long period of time.

21. The voltage between two points is the work done in moving a charge from one point to the other divided by the magnitude of the charge.

22. No. The potential difference is independent of both the path taken and the magnitude of the charge moved.

23. An electron volt is a unit of energy.

24. If the voltage is doubled the current is also doubled.

25. It means that when a voltage is applied to the ends of the resistor and a current flows through it, the ratio of voltage to current will be 5×10^6.

26. Apply several voltages to its ends and measure the current produced in each case. Then find the ratios V/I for each case. If the ratios are the same, Ohm's law applies.

27. The electrical energy is changed into heat energy and possibly light energy. (If the current is *changing*, additional energy transformations occur; this topic will be discussed in Chapter 16.)

28. Doubling the current results in four times the heat production (assuming the resistance is constant).

29. The charges must be moving relative to the magnet. (They must in fact be moving *across* the field of the magnet.)

30. It was found to be a "sideways" force!

31. Forces act on a magnetized (but uncharged) compass needle placed near the current. The magnetic field at any point near a straight conductor lies in a plane perpendicular to the wire and is tangent to a circle that is in that plane and has its center at the wire. The general shape of the magnetic field is circular.

32. Ampère suspected that two currents should exert forces on each other.

33. (b), (c), (d)

34. (b), (c), (e)

35. The magnetic force is not in the direction of motion of the particle: It is directed off to the side, at an angle of 90° to the direction of motion. The magnetic force does *not* do any work on the particle, since the force is always perpendicular to the direction of motion.

36. Gravity always acts toward the center of the earth and is proportional to the mass. (It is independent of the velocity.)

The electric field acts in the direction of the field (or opposite to that direction for negative charges), is proportional to the charge on the object, and is independent of the velocity of the object.

The magnetic field acts perpendicularly to both the field direction and the direction of motion, is proportional to both the charge and the velocity, and depends on the direction in which the object is moving.

Chapter 15

1. The single magnetic pole is free to move, and it follows a circular line of magnetic force around the current-carrying wire.

2. Faraday is considered the discoverer of electromagnetic induction because he was the first to publish the discovery, and because he did a series of exhaustive experiments on it.

3. the production of a current by magnetism

4. The loop is horizontal for maximum current, vertical for minimum. The reason is that the coil is cutting lines of force most rapidly when horizontal, and least rapidly when vertical.

5. It reverses the connection of the generator to the outside circuit at every half-turn of the loop.

6. It comes from the mechanical device that is turning the coil in the magnetic field.

7. Use a battery to drive current through the coil.

8. Batteries were weak and expensive.

9. An unknown workman showed that the dynamo could run as a motor.

10. too glaring, too expensive, too inconvenient

11. an improved vacuum pump

12. A small current will have a large heating effect if the resistance is high enough.

13. Cities became larger, since easy transportation from one part to another was now possible; buildings became taller, since elevators could carry people to upper floors; the hours available for work in factories, stores, and offices became much longer.

14. There is less heating loss in the transmission wires.

15. A current is induced in the secondary coil only when there is a *changing* current in the primary coil.

Chapter 16

1. a magnetic field

2. the small displacement of charges that accompanies a changing electric field

3. The four principles are:

(1) An electric current in a conductor produces magnetic lines of force that circle the conductor.

(2) When a conductor moves across externally set up magnetic lines of force, a current is induced in the conductor.

(3) A changing electric field in space produces a magnetic field.

(4) A changing magnetic field in space produces an electric field.

4. It was practically the same as the speed of light determined by Fizeau; they differed by only a little more than 1%!

5. "Maxwell's synthesis" is his electromagnetic theory in which he showed the relationship between electricity, magnetism, and light.

6. that electromagnetic waves exist, that they travel at the speed of light, and that they all have the ordinary properties of light, such as reflection, refraction, ability to form standing waves, etc.

7. a loop of wire

8. They have very great wavelengths (from tens to thousands of meters)
9. The signals travel in nearly straight lines and would otherwise pass into space instead of following the earth's curvature.
10. the higher the frequency, the greater is their penetration of matter
11. A radar wavelength of 1 m is about 2×10^6 (2 million) times that of green light, which is about 5×10^{-7} m.
12. X rays are produced by the sudden deflection or stopping of electrons; gamma radiation is emitted by unstable nuclei of radioactive materials.
13. It was almost unthinkable that there could be waves without a medium to transmit them.
14. Albert Einstein's (in his theory of relativity)

Chapter 17

1. The combining capacity (or valence) of sulfur in H_2S is -2. In SO_3, sulfur has a combining capacity of $+6$.
2. In Al_2O_3, aluminum has a combining capacity of $+3$.
3. When elements combine, the atomic mass A (as compared to hydrogen, $A = 1$) is related to the reacting mass m (in grams) and the combining capacity v as follows:

   $A = mv$

   In CH_4, the combining capacity of carbon is $v = 4$. Therefore, the atomic mass of carbon is given by

   $A = mv$
   $= 3\text{ g} \times 4$
   $= 12$ (compared to one atom of hydrogen)

4.

Element	Atomic Mass (A)	Combining Capacity (v)	Mass (grams)
Hydrogen	1.008	1	1.008
Chlorine	35.45	1	35.45
Oxygen	16.00	2	8.00
Copper	63.54	2	31.77
Zinc	65.37	2	32.69
Aluminum	26.98	3	8.99

Chapter 18

1. They could be deflected by magnetic and electric fields.
2. The mass of an electron is about 1,800 times smaller than the mass of a hydrogen ion.
3. (1) Identical electrons were emitted by a variety of materials; and (2) the mass of an electron was much smaller than that of an atom.
4. All other values of charge he found were multiples of that lowest value.
5. Fewer electrons are emitted, but with the same average energy as before.
6. The average kinetic energy of the emitted electrons decreases until, below some frequency value, none are emitted at all.
7.

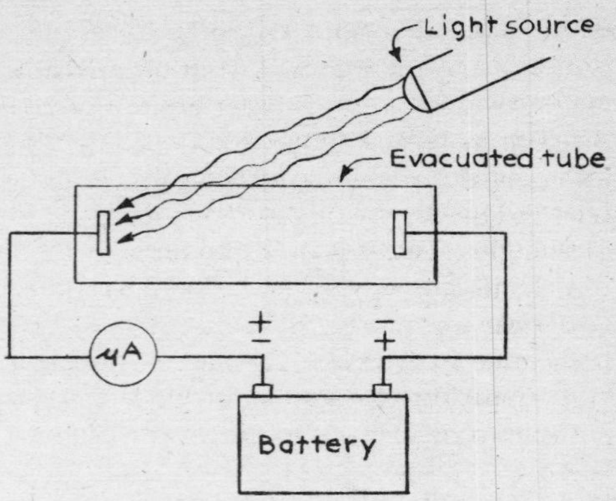

8. The energy of the quantum is proportional to the frequency of the wave, $E = hf$.
9. The electron loses some kinetic energy in escaping from the surface.
10. The maximum kinetic energy of emitted electrons is 2.0 eV.
11. When X rays passed through material, say air, they caused electrons to be ejected from molecules, and so produced + ions.
12. (1) not deflected by magnetic field; (2) show diffraction patterns when passing through crystals; (3) produce a pronounced photoelectric effect
13. (1) diffraction pattern formed by "slits" with atomic spacing (that is, crystals); (2) energy of quantum in photoelectric effect; (3) their great penetrating power
14. For atoms to be electrically neutral, they must contain enough positive charge to balance the negative charge of the electrons they contain; but electrons are thousands of times lighter than atoms.
15. There are at least two reasons: Firstly, the facts *never are* all in, so models cannot wait that long. Secondly, it is one of the main functions of a model to suggest what some of the facts (as yet undiscovered) might be.

Chapter 19

1. The source emits light of only certain frequencies, and is therefore probably an excited gas.
2. The source is probably made up of two parts: an inside part that produces a continuous spectrum; and an outer layer that absorbs only certain frequencies.
3. Light from very distant stars produces spectra that are identical with those produced by elements and compounds here on earth.
4. None. He predicted that they would exist because the mathematics was so neat.
5. careful measurement and tabulation of data on spectral lines, together with a liking for mathematical games
6. At this point in the development of the book, one cannot say what specifically accounts for the correctness of Balmer's formula. (The explanation requires atomic theory, which is yet to

come.) But the success of the formula does indicate that there must be something about the structure of the atom that makes it emit only discrete frequencies of light.

7. They have a positive electric charge and are repelled by the positive electric charge in atoms. The angle of scattering is usually small because the nuclei are so tiny that the α particle rarely gets near enough to be deflected much. However, once in a while there is a close approach, and then the forces of repulsion are great enough to deflect the α particle through a large angle.

8. Rutherford's model located the positively charged bulk of the atom in a tiny nucleus; in Thomson's model the positive bulk filled the entire atom.

9. It is the number (Z) of positive units of charge found in the nucleus, or the number of electrons around the nucleus.

10. three positive units of charge

11. Atoms of a gas emit light of only certain frequencies, which implies that each atom's energy can change only by certain amounts.

12. None. He *assumed* that electron orbits could have only certain values of angular momentum, which implied only certain energy states.

13. All hydrogen atoms have the same size because in all unexcited atoms the electron is in the innermost allowable orbit.

14. The quantization of the orbits prevents them from having other arbitrary sizes.

15. Bohr *derived* his prediction from a physical model. Balmer only followed a mathematical analogy.

16. According to Bohr's model, an absorption line would result from a transition within the atom from a lower to a higher energy state (the energy being absorbed from the radiation passing through the material).

17. (a) 4.0 eV (b) 0.1 eV (c) 2.1 eV

18. The electron arrangements in noble gases are very stable. When an additional nuclear charge and an additional electron are added, the added electron is bound very weakly to the atom.

19. Period I contains the elements with electrons in the K shell only. Since only two electrons can exist in the K shell, Period I will contain only the two elements with one electron and two electrons, respectively. Period II elements have electrons in the K (full) and L shells. The L shell can accommodate eight electrons, so those elements with only one through eight electrons in the L shell will be in Period II; and so forth.

20. It predicted some results that disagreed with experiment; and it predicted others that could not be tested in any known way. It did, however, give a satisfactory explanation of the observed frequency of the hydrogen spectral lines, and it provided a first physical picture of the quantum states of atoms.

Chapter 20

1. It increases, without limit.

2. It increases, approaching ever nearer to a limiting value, the speed of light.

3. Photon momentum is directly proportional to the frequency of the associated wave.

4. The Compton effect is the scattering of light (or X ray) photons from electrons in such a way that the photons transfer a part of their energy and momentum to the electrons, and thus emerge as lower-frequency radiation. It demonstrated that photons resemble material particles in possessing momentum as well as energy; both energy and momentum are conserved in collisions involving photons and electrons.

5. by analogy with the same relation for photons

6. The regular spacing of atoms in crystals is about the same as the wavelength of low-energy electrons.

7. Bohr invented his postulate just for the purpose. Schrödinger's equation was derived from the wave nature of electrons and explained many phenomena other than hydrogen spectra.

8. It is almost entirely mathematical; no physical picture or models can be made of it.

9. It can. But less energetic photons have longer associated wavelengths, so that the location of the particle becomes less precise.

10. It can. But the more energetic photons will disturb the particle more and make measurement of velocity less precise.

11. They are regions where there is a high probability of quanta arriving.

12. As with all probability laws, the *average* behavior of a large collection of particles can be predicted with great precision.

Chapter 21

1. It was phosphorescent. Becquerel wrapped a photographic plate in thick black paper to keep light out. Then he placed a small piece of the uranium compound on top of the black paper and allowed sunlight to fall on it. Upon developing the plate he found the silhouette of the mineral sample recorded on the plate. When he tried putting metallic objects between the sample and the plate, he found their outlines recorded even when a layer of glass was also introduced to eliminate possible chemical action.

2. No treatment was needed: the emission was spontaneous.

3. They were puzzling because they needed nothing to start them, and there was nothing that could stop them. They were similar to X rays in that both were very penetrating radiations, and both could ionize.

4. It is not, although slight differences might be observed because of the other element *absorbing* some of the radiation.

5. The radioactivity was much greater than expected for the amount of uranium in the ore.
6. separating it from barium, which is almost identical chemically
7. From most to least penetrating: γ, β, α. Penetrating power is inversely related to ionizing power because rays that are easily stopped (have low penetrating power) are so because they are expending their energy ionizing many atoms of the stopping material (high ionizing power), and vice versa.
8. Beta particles were found to have the same q/m ratio as electrons.
9. Alpha rays were deflected much less than β rays by a magnetic field.
10. Its emission spectrum, when caused to glow by an electric discharge, was the same as helium's.
11. It occurs when only a single pure element is present, and is not affected by chemical combinations of that element.
12. An example would be the decay of radon into polonium with the emission of an α particle (Rn $\rightarrow$ Po + He). It was contrary to the ideas of indivisibility of atoms held by nineteenth century chemists.
13. (1) Many of the substances in a series have similar chemical properties.
    (2) There are only small percentage differences in atomic mass.
    (3) Many of the substances decayed very rapidly into something else; all three kinds of rays are given off by the mixture.
14. At the start, the emission will be relatively slow and will consist entirely of α particles. Later, the emission will be greater and will contain, besides alpha particles, β and γ rays.
15. The law of radioactive decay is a statistical law; it says nothing about how long it will take any given atom to decay. To specify a "life time" would be to predict when the last atom would decay. Scientists do not know any way of doing that.
16. 1/16 of it.
17. We do not know. The statistical half-life laws do not apply to small numbers of atoms, and no other laws make predictions about individual atoms, or even about small numbers of atoms.

Chapter 22

1. They were *chemically* the same as previously known elements.
2. The atomic mass equals 12 amu. It occupies position 6 in the list of elements.
3. decreases 4 units; stays essentially the same
4. Decreases by 2 + charges; increases by 1 + charge
5. The rules are:
   (1) In α decay, the mass number decreases by 4, and the atomic number decreases by 2.
   (2) In β decay, the mass number remains the same, and the atomic number increases by 1.
   (3) In γ decay, both the mass number and the atomic number remain the same.

   In the Rutherford–Bohr model of the atom, the entire positive charge and almost the entire mass are contained in the nucleus. Since α, β, and γ rays are ejected from the nucleus, they will carry away from it both mass and charge. The α particle carries 2 positive charges and 4 amu; hence rule (1). The β particle carries 1 negative charge and negligible mass; hence rule (2). The γ ray has no mass and is uncharged; hence rule (3).
6. by subtracting α particle masses from the mass of the parent of the decay series
7. It must have a "velocity selector" that will allow only ions of a single speed to enter the magnetic field. This can be done with crossed electric and magnetic fields.
8. (1) faint second line in mass spectrum of pure neon
   (2) different atomic masses of samples of neon separated by diffusion
   (3) more intense second line in mass spectrum of one of the samples separated by diffusion
9. More massive atoms have a lower average speed and so diffuse more slowly than the less massive ones.
10. ${}_{78}Pt^{194}$; platinum
11. (A − 4). The rule is: Emission of an α particle results in a decrease in A of 4 units.
12. (Z + 1). The rule is: Emission of a negative β particle results in an increase in Z of 1 unit.
13. an isotope of hydrogen with twice the atomic mass of ordinary hydrogen
14. Heavy water is the compound D_2O. In other words, it is made with heavy hydrogen (deuterium) rather than ordinary hydrogen.
15. The third isotope has a *very* low abundance.
16. ${}_6C^{12}$ is the current standard. It was chosen mainly because it readily forms many compounds and so is available for measuring other masses by mass spectrograph techniques, which are much more accurate than chemical methods.

Chapter 23

1. Several atomic masses (which were not recognized as the average of several isotopes) were not close to whole multiples of the atomic mass of hydrogen.
2. 12 protons and 6 electrons
3. Yes, roughly. ${}_2He^4$ would contain 4 protons and 2 electrons inside the nucleus. (It does *not* work out, however, when very careful mass measurements are made.)
4. The number of tracks observed in a cloud chamber did not include any that would correspond to the original α particle breaking up into fragments.
5. For γ rays, the way it knocked protons out of paraffin would be a violation of the principles of energy and momentum conservation.

6. A neutron has no charge, and so is not deflected by magnetic or electric fields; nor does it leave a track in a cloud chamber.
7. The laws of conservation of momentum and kinetic energy were applied to neutron – proton and neutron – nitrogen head-on collisions. This yielded four equations in the four variables: m_n, v_n, v_n' (proton collision), and v_n' (nitrogen collision). The latter three were eliminated, and m_n found.
8. 7 protons and 7 neutrons
9. a nucleus of 2 protons and 2 neutrons, surrounded by 2 electrons
10. A neutron in the nucleus changes into a proton and a β particle, which immediately escapes.
11. Without the extra particle, there was no way to explain the disappearance of energy in β decay.
12. The repulsive electric force exerted by the large charge of the heavy nucleus on an α particle prevents it from reaching the nucleus.
13. Protons have only a single charge.
14. Some devices for producing projectiles are: Van de Graaff generators, linear accelerators, cyclotrons, synchrotrons. Devices that detect nuclear reactions are: cloud chambers, spark chambers, photographic emulsions, and bubble chambers.
15. They have no electric charge and so are not repelled by nuclei.
16. $_{14}Si^{28}$
17. $_6C^{13}$; 7 protons, 6 neutrons before; 6 protons, 7 neutrons after

Chapter 24

1. No. In some nuclear reactions energy is absorbed.
2. It can go off as γ rays or as the kinetic energy of the product particles.
3. The binding energy of the deuteron nucleus is the energy that would be required to break up the nucleus into its constituent particles: a proton and a neutron.
4. A nuclide with a high *average* binding energy is more stable.
5. No. Light nuclei are lower on the curve than heavy nuclei.
6. capture of a neutron by a uranium nucleus, followed by the β decay of the new nucleus
7. neutrons
8. a substance that slows down neutrons
9. It slows down neutrons well (because of the abundance of H atoms), but it also absorbs many (to form "heavy" water).
10. by control rods made of a material that absorbs neutrons; the farther in the rods, the slower the reaction.
11. The positively charged nuclei repel each other; high speeds are necessary for the nuclei to come near enough in collisions to fuse.
12. Since at very high temperatures the gas is ionized, a properly shaped magnetic field could deflect the charged particles away from the walls.
13. decreasing
14. The protons in a nucleus repel each other with intense electric forces.
15. The average binding energy curve suggests that each particle in the nucleus is bound only by its immediate neighbors.
16. An excited nucleus becomes distorted in shape; electric repulsion between bulges then forces them apart.
17. In the case of U^{238}, the excitation energy due to neutron capture alone is less than the activation energy required for fission. For U^{235}, the excitation energy is greater than the activation energy.
18. They correspond to completed shells (or sets of energy states) of protons and neutrons in the nucleus.
19. Neither. They each have different strengths and weaknesses.

Brief Answers to Study Guide Questions

Chapter 1

1. information
2. speed $= v = \Delta d/\Delta t$
   Uniform speed occurs when the ratio distance/time is constant.
   average speed $= v_{av}$ = total distance/total time
   slope $= \Delta y/\Delta x$
   Instantaneous speed (the speed at a given moment) is shown by the slope of the speed line at that moment.
   average acceleration $= a_{av}$ = change in speed/time interval for the change
3. average speed $v_{av} = 20$ m/sec
4. (a) 6 cm/sec (b) 24 km (c) 0.25 min (d) 3 cm/sec, 24 cm (e) 64 km/hr (f) 64 km/hr; 192 km (g) 5.5 sec (h) 8.8 m
5. She falls 228 m in 19 sec. After a further 25 sec, she has fallen a total of 528 m.
6. (a) 1.7 m/sec (b) 3.0 m/sec
7. The rabbit wins by 368 sec.
8. 3.6×10^3 km
9. discussion
10. (a) 9.5×10^{15} m (b) 2.7×10^8 sec or 8.5 yr (c)–(g) discussion
11. For the blue bicycle, $v_{av} = 16.7$ m/sec.

12. a_{av} for 5 sec is 6m/sec. After an additional 5 sec (a total of 10 sec), v = 60 m/sec.
13. (a) (1) 5 m/sec, (2) 5 m/sec, (3) 5 m/sec, (4) 5 m/sec (b) The straight line for distance versus time shows that the speeds were equal at all times. (c) The instantaneous speed is 5 m/sec. The slope is the same at all times.
14. (a) fastest in section CD; slowest in section BC (b) v_{AB} = 44 m/sec; v_{BC} = 3.6 m/sec; v_{CD} = 136 m/sec; v_{AD} = 37.9 m/sec (c) Instantaneous speed at point f = 136 m/sec.
15. (a) v_{inst} = 0.5 m/sec (at 10-sec mark); v_{inst} = 1.5 m/sec (at 25-sec mark) (b) a_{av} = 0.06 m/sec^2
16. 25.6 m
17. graph
18. discussion
19. Between 1 and 4.5 sec, 1.3 m/sec (b) 0.13 m/sec (c) 0.75 m/sec (d) 1.0 m/sec (e) 0.4 m (approx.)
20. (a) DE was covered fastest; BC was covered slowest. (b) The line representing EF (a resting interval) should be parallel to the horizontal axis. (c) v_{av} for 8 weeks is 75 km/week. (d) The instantaneous speeds at points P and Q are: v_P = 60 km/week, v_Q = 200 km/week.
21. (a) graph (b) graph
22. 795,454 cm/sec
23. graphs; d versus t: 0, 9, 22, 39.5, 60.5, 86 cm (approx.) at intervals of 0.2 sec; v versus t: 45, 65, 87.5, 105, 127 cm/sec (approx.) at intervals of 0.2 sec
24. discussion
25. discussion

Chapter 2

1. information
2. discussion
3. discussion
4. discussion
5. discussion
6. (a) Distance is 36 m, average speed is 6 m/sec, and distance for uniform acceleration is 36 m. (b) The equations used assume uniform acceleration.
7. derivation
8. (a), (b), (c)
9. discussion
10. discussion
11. derivation
12. 17 years; $10,000
13. (a) 57 m/sec^2 (b) 710 m (c) −190 m/sec^2
14. (a) true (b) true (based on measurements of six lower positions) (c) true (d) true (e) true
15. derivation
16. (a)

Position	d	v
A	+	+
B	+	+
C	+	0
D	+	−
E	−	−

(b) derivation (c) discussion
17. discussion
18. (a) −5.0 m (b) −10 m/sec (c) −15 m
19. (a) 120 m/sec (b) 15 m (c) 2 sec (d) 20 m (e) −20 m/sec
20. (a) 20 m/sec (b) −20 m/sec (c) 4 sec (d) 80 m (e) 0 (f) −40 m/sec
21. (a) −2 m/sec^2 (b) 2 m/sec (c) 2 m/sec (d) 4 m (e) −2 m/sec (f) 4 sec
22. discussion
23. (a) 4.3 welfs/surg^2 (b) 9.8 m/sec^2
24. derivation
25. derivation
26. derivation
27. discussion
28. discussion
29. discussion
30. (a) a = 2.5 m/sec^2 (b) d = 397 m (if a_g = 10 m/sec^2, d = 405 m) (c) The block slides for 4.5 sec, and its final velocity is 9 m/sec. (d) The distance is 330 m north and the final velocity is 58 m/sec north. (e) The final speed is 4 m/sec. (f) The final speed is 8 m/sec. (g) The acceleration is 15 m/sec^2. (h) The initial speed of the ball thrown upward is 31 m/sec.
31. (a) The average speed v_{AB} = 2.5 m/sec and v_{CD} = 7.5 m/sec. (b) The average speed v_{BC} = 2.5 m/sec. The average acceleration a = 2.5 m/sec^2. (c) discussion
32. (a) graph A: constant velocity; graph B: acceleration; graph C: acceleration; graph D: negative acceleration (deceleration) (b) graph A: backward; graph B: forward; graph C: backward; graph D: backward

Chapter 3

1. information
2. discussion
3. (a) Mechanics deals with forces. Dynamics deals with forces producing motions. Kinematics deals with motion with no concern for the forces acting. (b) (1) d, scalar (2) $\vec{f}$, vector (3) v, scalar (4) f, scalar (5) m, scalar (6) $\vec{d}$, vector (7) t, scalar (8) $\vec{a}$, vector (9) v, vector
4. (a) 3 blocks east from the starting point (b) 15 blocks (c) Part (a) deals with vectors; part (b) deals with scalars.
5. (a) construction (b) 2.4 units west
6. proof
7. Air resistance equals the earth's gravitational attraction of 750 N.
8. discussion
9. (a) In equilibrium, the net force on a body is zero. (b) In equilibrium, a body may be (1) at rest or (2) in uniform motion.
10. discussion
11. discussion
12. discussion
13. discussion
14. (a) Acceleration is (1) proportional to the force acting, (2) in the direction of the force, and (3) inversely proportional to the mass. (b) The acceleration is 15 m/sec^2 east.
15. 2.8 × 10^{-4} hr/sec
16. discussion
17. discussion

18. discussion

19. (c) 24 N out (d) 14.8 N left (e) 0.86 N north (f) 9.0 kg (g) 0.30 kg (h) 0.20 kg (i) 3 m/sec^2 east (j) 2.5 m/sec^2 left (k) 2.50 m/sec^2 down

20. (a) 2.0×10^2 m/sec^2; 7.8×10^2 m/sec (b) discussion (c) 2.4×10^2 m/sec^2

21. discussion

22. 2.0 kg

23. discussion

24. (a) 1 kg, 9.81 N in Paris, 9.80 N in Washington (b) individual calculations

25. (a) Since the pound is a unit of force (weight) and the kilogram is a unit of mass, they cannot be directly converted. Weight is a measure of the earth's gravitational attraction at its surface and therefore comparisons can only be made on earth. (b) Student answers will vary. (c) Student answers will vary. (For each 1 kg of mass lifted, 9.8 N of force are required.)

26. discussion

27. (a) -5×10^{-23} m/sec^2 (b) 10 m/sec (c) 10×10^{-23} m/sec

28. The acceleration of the girl is 2 m/sec^2. The force on the boy is 80 N, and his acceleration is 1.14 m/sec^2.

29. discussion

30. discussion

31. (a) diagram (b) 1.7×10^{-24} m/sec^2 (c) $(6 \times 10^{24})/1$ (d) diagram

32. (a) 862 N, 750 N, 638 N (b) The same as in (a) for a scale calibrated in newtons. (c) discussion

33. hints for solving motion problems

34. (a) The object will move 150 m to the right. (b) The speed will be 40 m/sec. (c) The net force is 16 N. (d) The mass is 5 kg. (e) 16 m/sec^2 (f) 5 sec (g) zero

Chapter 4

1. information

2. 3.8 m/sec^2, 5.1 sec, mass decreases

3. discussion

4. (a)

	2	5	10 sec
Horizontal distance ($v_x = 4$ m/sec)	8	20	40 m
Vertical distance ($v_y = 3$ m/sec)	6	15	30 m
(b) Total distance	10	25	50 m
(c) Total velocity	5	5	5 m/sec

5. derivation

6. 1.3 m, at an angle of 67° below the horizontal; 5.1 m/sec, 79° below the horizontal

7. (a) $t_x = 2.5$ sec (b) $d_y = 30.6$ m (c) v_x 12.5 m/sec

8. $t = 3.1$ sec; time of fall does not depend on horizontal velocity.

9. (a) $t = 4.1$ sec (b) The time of fall is independent of the horizontal velocity. (c) $v_y = 40$ m/sec (d) v_x remains at 8 m/sec.

10. discussion

11. discussion

12. discussion

13. discussion

14. 6.0×10^{-2} min, 3.0×10^{-2} min, 2.2×10^{-2} min, 1.3×10^{-2} min

15. discussion

16. discussion

17. table completion

18. (a) 2.2×10^{-10} m/sec^2 (b) 4×10^{20} N (c) approximately1/100

19. approximately 10^3 N

20. discussion

21. (a) $v = 6.2$ m/sec (b) $a_c = 19.2$ m/sec^2 (c) $F_c = 38.4$ N

22. $v = 3.2$ m/sec

23. $T = 3.4$ sec; because no force is considered, no mass is involved.

24. $a = 1{,}970$ m/sec^2

25. (a) Syncom 2 (b) Lunik 3 (c) Luna 4 (d) does not change

26. $v = 7{,}690$ m/sec; the orbital speed does not depend on mass.

27. $F = 683$ N

28. The accelerating force is 114 N. $T = 931$ min

29. 5.1×10^3 sec or 85 min, 7.9×10^3 m/sec

30. discussion

31. 7.1×10^3 sec or 120 min

32. (a) 3.6×10^2 sec (b) 36 km (c) discussion

33. $\Delta t = (m/F)(v_0 - v)$

34. discussion

35. essay

Chapter 5

1. information

2. discussion

3. (a) 674 sec (b) 0.0021%

4. table

5. discussion

6. discussion

7. discussion

8. 102°, 78°, 78° 102° starting with the upper right quadrant

9. (a) 15° (b) geometric proof and calculation; about 12,100 km

10. a, b, c, d, e, f

11. discussion

12. discussion

Chapter 6

1. information

2. diagram construction

3. discussion

4. 11 times; derivation

5. Copernicus calculated the distances of the planetary orbits from the sun and the periods of planetary motion around the sun. The Copernican system was more simple and harmonious than that of Ptolemy. In addition, the orbits began to seem like the paths of real planets, rather than mathematical combinations of circles used to compute positions.

6. The Copernican system led to a change in the order of importance of the earth and sun. The sun became dominant while the earth became "just another planet." These philosophical results were more important than the geometrical change.

7. discussion
8. discussion
9. 2.8×10^5 AU
10. discussion
11. discussion
12. discussion
13. discussion
14. discussion

Chapter 7

1. information
2. about 1/8 of a degree; about 1/4 of the moon's diameter
3. discussion
4. 4%
5. discussion
6. $a + c$
7. If the two foci (tacks) are placed at the same point, you will draw a circle. As the foci are separated, the ellipse becomes longer and thinner.
8. $e = c/a = 5/9 = 0.556$
9. (a) same as sketch in question 6 (b) $c = \text{A} - \text{P}$, $a = \text{A} + \text{P}$, $\text{A} = (a + c)/2$, $\text{P} = (a - c)/2$ (c) $R_{av} = a/2 = (\text{A} + \text{P})/2$
10. The second focus is empty.
11. discussion
12. (a) 0.43 (b) P = 2.5 cm; A = 7.5 cm (c) A = 45 cm, c = 40 cm, a = 50 cm
13. 0.209
14. 0.594/1
15. (a) 17.9 AU (b) 35.3 AU (c) 0.54 AU (d) 66/1
16. T = 249 years
17. k = 1.0 for all three planets
18. discussion
19. $T^2/R^3 = 8.9 \times 10^{-14}$ sec²/m³
20. $T_2 = 8\,T_1$
21. $R = 4.03 \times 10^8$ m
22. 8.9×10^{-14} sec²/m³ for each satellite
23. (a) student sketches (b) student calculations (c) Yes, Kepler's law of periods applies.
24. discussion
25. discussion
26. Kepler expected the theory to lead to predictions agreeing closely with new observations. Kepler sought algebraic patterns rather than geometric patterns. Kepler sought physical causes for observed motions.
27. Empirical laws are generalizations based on observations. They are reliable bases for theoretical speculations.

Chapter 8

1. information
2. (a) a straight line at uniform speed (b) caused the planets to deviate from a straight line (c) directed toward a center (the sun) (d) varies inversely with the square of the distance ($1/R^2$)
3. (a) It did not fall "down," but "toward the center of the earth." (b) $1/60^2$ of the earth's attraction on the apple: $2.7 = 10^{-3}$ m/sec² (c) $a_c = 2.71 \times 10^{-3}$ m/sec². The two values almost agree.
4. The Newtonian question is really "What holds the moon down?"
5. Every object in the universe attracts every other object with a gravitational force $F = G(Mm/R^2)$, where F is the force between objects, M and m are the masses of the objects, R is the distance between the centers of the objects, and $G = 6.67 \times 10^{-11}$ N·m²/kg². Available evidence shows no change in G with time or position.
6. yes, to about 1% agreement
7. discussion
8. discussion
9. derivation
10. $T^2 = \left(\frac{4\pi^2}{G}\right)\frac{R^3}{m}$
11. $F = 6.67 \times 10^{-8}$ N; $a_{1,000} = 6.67 \times 10^{-11}$ m/sec²; $a_{100} = 6.67 \times 10^{-10}$ m/sec²; these accelerations are very small; they might be measurable with a torsion pendulum or a modern optical device.
12. Unexplained irregularities in the motion of Uranus led to the prediction that one or more outer planets existed. Calculations led to predicted positions that were rather accurate for Neptune and approximate for Pluto. The belief that such planets existed led to the systematic search for them.
13. (a) 1.05×10^3 days²/AU³ (b) discussion (c) discussion
14. 42,600 km
15. 5.98×10^{24} kg
16. 6.04×10^{24} kg
17. The moon's tidal force on the water on the distant side of the earth is less than on the solid earth. As a result, the earth is pulled away from the water.
18. The inertial motion is along a straight line, $x = vt$. The accelerated motion is toward the earth's center, $y = \frac{1}{2}at^2$.
19. (a) 5.52×10^3 kg/m³ (b) discussion
20. 7.30×10^{22} kg
21. (a) 5.99×10^3 sec, or 1.66 h (b) 3.55 km/sec (c) collisions
22. table
23. about 170 times as great
24. 17.7 AU; 0.60 AU; 34.8 AU
25. derivations
26. discussion
27. discussion; no
28. discussion
29. It is useful today.
30. discussion

Chapter 9

1. information
2. discussion

3. (a) yes (b) the solar system
4. discussion
5. no
6. discussion
7. (a) 220.2 g (b) The mass of the solids before and after the reactions are equal.
8. (a) The mass is 60 g on the earth and on the moon. (b) Mass is an attribute of material. Weight describes a gravitational attraction. (c) Nothing is reported about masses.
9. (a) no change in total momentum
(b) Disk A 40 kg·m/sec west
Disk B 150 kg·m/sec north
Disk C 20 kg·m/sec east
Disk D 20 kg·m/sec east
(c) The final velocity v_2 = 6 m/sec north. (d) Momentum is a vector quantity, and opposing vectors can cancel, as in part (b) where A cancels (C + D).
10. (a) all except v_A' (which $=v_B'$)
(b) $v_A' = \dfrac{m_A v_A}{m_A + m_B}$
(c) 0.8 m/sec
11. dictionary comment
12. 3.3×10^{-6} kg
13. discussion
14. discussion
15. yes
16. (a) Δt = 4 sec (b) Final velocity = 20 m/sec by both solutions. (c) Some problems are easier to solve with the momentum formula, but it is not more basic.
17. discussion
18. 1.2×10^3 kg·m/sec; 4×10^2 N; 30 m
19. (a) about 100 m/sec (b) about 4.6 kg·m/sec (c) less than 0.003 sec (d) at least 1.5×10^3 N
20. yes
21. derivation
22. (a) $\Delta t = \dfrac{m(v_0 - v)}{F}$
(b) $m(v_0 - v)$
(c) $\dfrac{m(v_0 - v)}{v_e}$
23. derivation
24. 10 m/sec
25. 10.5×10^8 kg·m/sec
26. discussion
27. discussion
28. discussion
29. discussion
30. (a) 0.8 × mass of ball (b) −0.8 × mass of ball (c) 1.6 × mass of ball (d) depends on system considered
31. discussion
32. (a) The total kinetic energy of an isolated system involving only elastic collisions is constant. (b) The total kinetic energy does not change. (c) KE = 85 J (d) The total kinetic energy does not change.
33. table
34. (a) The total mass is 18 g. The total momentum is 18 g·cm/sec east. The total kinetic energy is 1,227 J.

(b)

	Mass	Momentum	Kinetic Energy
1. open, elastic	18 g	unknown	unknown
2. open, inelastic	18 g	unknown	unknown
3. closed, elastic	18 g	18 g·cm/sec east	1,227 J
4. closed, inelastic	18 g	18 g·cm/sec east	1,227 J

35. derivation
36. Both speeds equal $v/2$, but in opposite directions.

Chapter 10

1. information
2. discussion
3. (a) $v_1 - u$ and $v_2 - u$ (b) no (c) no (d) no (e) yes (f) iii (g) discussion
4. 5×10^{-15} J, 2×10^{14} electrons
5. d = 10 m
6. (a) 67.5 J (b) 4.5×10^9 J (c) 3.75×10^3 J (d) 2.7×10^{33} J
7. (a) 2 m/sec²; 30 sec; 60 m/sec (b) 60 m/sec
8. (a) −90 J (b) 90 J (c) 18×10^2 N
9. 2.3×10^2 J
10. (a) 2.2×10^{-3} J (b) 5.4×10^{-2} J
11. (a) 0.2 m (b) 7×10^9 J
12. discussion
13. (a) PE = mgh, where h is stretch (b) Δ(PE) = $-\Delta$(KE) (c) The energy is changing between KE, gravitational PE, and elastic PE.
14. (a) 1.1×10^{12} sec (about 3×10^4 years) (b) 1.6×10^{-25} m (c) The PE relates the rock to the earth. Potential energy is always relative to a frame of reference.
15. discussion
16. discussion
17. derivation
18. discussion
19. sketch
20. proof
21. (a) 96×10^8 J (b) 8.8×10^2 m (c) 48×10^5 N (d) discussion (e) discussion
22. discussion
23. (a) d = 102 m (b) 93 W (approx.)
24. Engine A: 50% efficient, 5 W power; Engine B: 40% efficient, 8 W power
25. discussion
26. discussion
27. discussion
28. (a) 3.5% (b) 0.75 metric ton/sec
29. (a) 15.5 (b) 7.3 (c) 5.8
30. no
31. c, a, e, b, d, f
32. no heat loss; reversible; 100% efficient
33. To be 100% efficient, the cold side of a heat engine must be at absolute zero.
34. (a) >1000 (b) discussion
35. 1/8°C; no
36. (a) fossil fuel: 0.60; nuclear: 0.50 (b) fossil fuel: 0.67 MW; nuclear: 1.00 MW

37. 1/4 kg
38. 21.5 days
39. discussion
40. discussion
41. discussion
42. discussion
43. $H(T_1 - T_2)T_1T_2$
44. (a) discussion (b) greater in lower orbit (c) less (d) less (e) discussion
45. (a) discussion (b) i: all three; ii: all three; iii: ΔH; iv: ΔH; v: all three; vi: ΔH
46. discussion
47. derivation
48. ice: 12 kJ/°K; water: −12 kJ/°K; no change in the universe

Chapter 11

1. information
2. discussion
3. discussion
4. A distribution is a statistical description.
5. discussion
6. no
7. discussion
8. (a) 10^{-9} m (b) 10^{-9} m
9. (a) 10^{21} (b) 10^{18}
10. *Zero* meters
11. 10.5 km
12. (a) $P = 66$ N/m^2 (b) $T = 100$°C
13. shoes: about 1/7 atm; skis: about 1/60 atm; skates: about 3 atm
14. derivation
15. $P = kDT$, therefore, $P \propto D$, $P \propto T$, and $D \propto 1/T$ when the other property is constant. The ideal gas law does not apply to very dense gases, or to gases when they liquify.
16. discussion
17. discussion
18. discussion
19. derivation
20. no change
21. pressure, mass, volume, temperature
22. discussion
23. discussion
24. discussion
25. discussion
26. discussion
27. derivation
28. discussion
29. Temperature will rise.
30. no
31. discussion
32. discussion
33. discussion
34. The melting of ice is an irreversible process because the ordered arrangement of molecules in the ice crystals is lost and entropy increases.
35. discussion
36. discussion
37. discussion
38. discussion
39. discussion
40. derivation
41. derivation

Chapter 12

1. information
2. discussion
3. discussion
4. construction
5. construction
6. discussion
7. discussion
8. construction; F (beats) $= F_2 - F_1$
9. discussion
10. discussion
11. construction
12. derivation
13. no; discussion
14. (a) $v = 343$ cm/sec (b) 5.86 m (c) $T = 3.9 \times 10^{-3}$ sec everywhere
15. (a) 3/4 L (b) 2/3 L (c) 1/2 L
16. $f = 80$ Hz
17. (a) $\lambda = 4L$

(b) $\lambda = \dfrac{4L}{2n+1}$

(c) $\lambda = 2L,\ \lambda = \dfrac{4L}{n+1}$ (n = 0, 1, 2, 3, etc.)

18. discussion
19. maximum
20. 100 and 1,000 Hz; yes
21. discussion
22. discussion
23. $\lambda = 6.25 \times 10^{-5}$ cm
24. construction
25. straight line
26. $\dfrac{R}{2}$
27. construction
28. $\dfrac{1}{4k}$
29. discussion
30. Speed is independent of frequency. Speed in a medium depends on stiffness and density of the medium.
31. Two straight-line waves inclined toward each other.
32. (a) $\theta_A = \angle BAD$ (b) $\theta_B = \angle CDA$ (c) $\lambda_A = BD$ (d) $\lambda_B = AC$ (e) derivation (f) derivation
33. $\lambda_D = 0.035$ m; $\lambda_s = 0.025$ m
34. $v = 2.5$ cm/sec
35. discussion
36. (a) 1.27×10^{-11} W (b) 8×10^{12} mosquitoes (c) subway train
37. $2d = vt$
38. 1,000 Hz: { air 0.3375 m; sea water 1.44 m; steel 4.8 m }
One tenth of each of these values for 10,000 Hz; discussion.
39. 3×10^5Hz; 2.5×10^7 Hz

Chapter 13

1. information
2. 7.5 cm
3. discussion
4. discussion
5. (a) 4.4×10^9 m (b) 3.0×10^8 m/sec (c) positive deviations; conjunction cycle
6. 1.8×10^{11} m
7. (a) 9.5×10^{15} m (b) 4,300 years (c) 30 times as great
8. derivation
9. discussion
10. 0.9 m; no; no
11. discussion
12. diagrams
13. (a) diagram (b) discussion
14. proof
15. (a) $(m + 1/2)$ when $m = 0, 1, 2$ (b) greater (c) increased separation of fringes (d) increased separation of fringes (e) fainter but more extensive
16. for violet, $f = 7.5 \times 10^{14}$/sec; for red, $f = 4.2 \times 10^{14}$/sec
17. $d = 1.6 \times 10^{-4}$ m
18. $d = 1 \times 10^{-3}$ m
19. discussion
20. discussion
21. discussion
22. 6×10^{14} Hz; 10^9 times AM frequencies; 10^7 times FM frequencies
23. vertical
24. discussion

Chapter 14

1. information
2. (a) tripled (b) halved (c) no change
3. 95 km
4. discussion
5. yes; discussion; sketches
6. (a) 1.6 N/kg (b) 4.2×10^{18} N/kg (c) directly proportional to $\sqrt{r}$
7. discussion
8. vector diagrams
9. (a) $F = k(Q_1Q_2)/R^2$
   When $Q_2 = 3\,q_e$, $F = 2.76 \times 10^{-11}$ N east
   $Q_2 = 6\,q_e$, $F = 5.5 \times 10^{-11}$ N east
   $Q_2 = 10\,q_e$, $F = 9.2 \times 10^{-11}$ N east
   $Q_2 = 34\,q_e$, $F = 3.1 \times 10^{-10}$ N east
   (b) 5.76×10^7 N/C. The same values for the force F are found as given in (a) for the various charges Q_2. (c) The field concept specifies the field at any point that will interact with a charge at that point, as in (b).
10. (a) 10^6 C (b) 10^{-9} C/m²
11. sketch (normal to surfaces)
12. help
13. 6.25×10^{18} electrons
14. 3.4×10^{42}
15. (a) $1/2\,mv^2 = 1/2kq^2/R$ (b) 1.2×10^{-18} J (c) 1.5×10^6 m/sec
16. Metals are conductors.
17. 30 V
18. same or zero
19. derivation
20. 3×10^6 V/m
21. 10^7 V/m
22. (a) 12 V (b) zero (c) 12 V
23. (a) 1.6×10^{-17} J (b) 5.7×10^6 m/sec
24. (a) $I = E/R = 12/3 = 4$ A; if the voltage is doubled, the current I is doubled to 8 A. (b) Voltage describes a potential *difference* between two points. Current describes a flow of charge through a conductor.
25. $R = 25\ \Omega$; when $I = 2$ A, $= E$ 50 V. If Ohm's law does not apply, and you do not know another relation between voltage, current, and resistance, you cannot relate current to voltage.
26. $P = 150$ W; $R = 16.7\ \Omega$
27. Power before the cut is 5,825 W, current is 48.5 A. Power after the 5% cut is 5,534 W, current is 48.5 A. You are not cheated, since you pay for power in watts.
28. The power limit of the circuit breaker is 1,200 W. You can add four lamps drawing 150 W each.
29. (a) 4 A (b) 5 Ω (c) 15 V
30. (a) 10^7 V (b) 5×10^8 J
31. discussion
32. 20 W
33. (a) 8 W (b) 20 W (c) 45 W
34. magnetic field vertical at surface
35. (a) north (b) 1 A, north
36. (a) An ampere is defined as the amount of current in each of two long, straight parallel wires, set 1 m apart, that causes a force of exactly 2×10^{-7} N to act on each meter of each wire. The unit of force between the wires is N/m·A². (b) $F = 4.8 \times 10^{-5}$ N
37. (a) derivation (b) v, B, and R
38. derivation
39. west
40. discussion

Chapter 15

1. information
2. discussion
3. yes
4. all except (d)
5. sketch
6. (a) exercise (b) upward (d) downward
7. Lenz's law
8. outside magnet
9. opposite
10. discussion
11. (a) the series circuit (b) the series circuit (c) the parallel circuit (d) In a parellel circuit, added resistors decrease the total effective resistance and the total current and power increase.

12. In series, each resistor carrries 5 A. In series, the total resistance is 8 Ω and the total current is 1.5 A. For each resistor, the current is 1.5 A at 6 V. In parallel, the total resistance is 2 Ω and the current is 6 A. For each resistor, the voltage is 6 V and the current is 4 A.
13. In series, the total resistance is 15 Ω and the current is 3.3 A in each resistor. The voltage across the 5 Ω resistor is 16.7 V at 3.3 A. The voltage across the 10 Ω resistor is 33.3 V and the current is 3.3 A. In parallel, the total voltage is 50 V and the total current is 15 A. The voltage is 50 V across both resistors. The current through the 5 Ω resistor is 10 A and through the 10 Ω resistor it is 5 A.
14. (a) 1 A (b) 10 Ω (c) burn out
15. (a) 1/12 A (b) 1,440 Ω, the same
16. (a) 1 A (b) 1/5 W (c) 1/2 A; 1/20 W (d) 0.97 A; 0.19 W; 5.6 W, 0.50 A, 0.05 W, 6 W
17. 5 A
18. derivation
19. The constantly changing magnetic field of the primary coils induces a constantly changing current in the transformer core and the coils of the secondary circuit by electromagnetic induction.
20. low voltage coil
21. discussion
22. discussion
23. report
24. discussion
25. The efficiency of electric power plants is limited by the second law of thermodynamics. Modern power plants can achieve about 38–40% efficiency for fossil fuel plants and 30% for nuclear plants (the maximum possible efficiency is 60% for fossil fuel plants and 50% for nuclear plants).

Chapter 16

1. information
2. symmetry
3. no
4. accelerating charge, mutual induction
5. (a) height (b) pressure (c) field strength
6. deflector orientation
7. light properties
8. discussion
9. discussion
10. 5×10^6 m; 600 m and 193 m; 11 m
11. 10 m to 100 m
12. discussion
13. discussion
14. (a) TV or FM: 10^8 Hz (frequency), 1 m (wavelength); red light: 10^{14} Hz, 10^{-6} m; infrared: 10^{13} Hz, 10^{-5} m; electric wires: 10^2 Hz, 10^6 m (b) TV or FM: little diffraction; red light: sharp shadow; infrared: sharp shadow; electric wires: great diffraction
15. discussion
16. ionospheric reflection of shorter wavelength radiation
17. 42,400 km
18. phase difference between direct and reflected waves
19. 2.6 sec
20. absorption
21. evolution
22. ultraviolet and infrared
23. discussion
24. unnecessary
25. discussion
26. discussion
27. discussion
28. essay
29. essay

Chapter 17

1. information
2. 80.3% zinc; 19.7% oxygen
3. 47.9% zinc
4. 13.9 times mass of H atom; same
5. 986 g nitrogen; 214 g hydrogen
6. (a) 14.1 (b) 28.2 (c) 7.0
7. Na;1 Al;3 P;5 Ca;2 Sn;4
8. graph; discussion
9. 8.0 g; 0.895 g
10. (a) 0.05 g Zn (b) 0.30 g Zn (c) 1.2 g Zn
11. (a) 0.88 g Cl (b) 3.14 g I (c) discussion (d) discussion
12. 35.45 g
13. discussion
14. discussion

Chapter 18

1. information
2. (a) 2.0×10^7 m/sec (b) 1.8×10^{11} C/kg
3. proof
4. discussion
5. discussion
6. 2000 Å; ultraviolet
7. 4×10^{-19} J; 4×10^{-18} J
8. 2.6×10^{-19}; 1.6 eV
9. 4.9×10^{14}/sec
10. (a) 6×10^{14}/sec (b) 4×10^{-19} J (c) 2.5×10^{20} photons (d) 2.5 photons/sec (e) 0.4 sec (f) 2.5×10^{-10} photon (g) 6.25×10^{17} electrons/sec; 0.1 A
11. 1.3×10^{17} photons
12. (a) 6.0×10^{23} electrons (b) 84×10^{21} copper atoms/cm³ (c) 1.2×10^{-23} cm³ (d) 2.3×10^{-3}
13. (a) $2x = n\lambda$ (b) $2x$ = any odd number of half wavelengths (c) $\cos\theta = 2d/\lambda$ for first order
14. 1.2×10^{19}/sec
15. discussion
16. 1.2×10^5 V; 1.9×10^{-14} J; 1.2×10^5 eV
17. glossary
18. discussion

Chapter 19

1. information
2. discussion
3. Five are listed in the *Text*, but theoretically an infinite number. Four lines in visible region.

4. $n \times 8$; $\lambda = 3880$ Å
$n \times 10$; $\lambda = 3790$ Å
$n \times 12$; $\lambda = 3740$ Å
5. (a) yes (b) $n_i \times \infty$ (c) Lyman series 910 Å; Balmer series 3650 Å; Paschen series 8200 Å (d) 21.8×10^{-19} J, 13.6 eV
6. discussion
7. discussion
8. 2.6×10^{-14} m
9. (a) discussion (b) $10^{-4}/1$
10. 3.5 m
11. derivation
12. discussion
13. list
14. diagram
15–21. discussion
22. essay

Chapter 20

1. information
2. 0.14 c or 4.2×10^7 m/sec
3. 3.7×10^{-14} N
4. $p = m_0 v$ and $KE = m_0 v^2/2$
5. (a) changes are too small (b) 1.1×10^{-12} kg
6. (a) 2.7×10^{33} J (b) 3.0×10^{16} kg (c) 5×10^{-7}% (d) rest mass
7. (a) 1.2×10^{-22} kg·m/sec (b) 1.1×10^{-22} kg·m/sec (c) 2.4×10^{-22} kg·m/sec (d) 1.1×10^{-22} kg·m/sec
8. $p = 1.7 \times 10^{-27}$ kg·m/sec; $v = 1.9 \times 10^3$ m/sec
9. discussion
10. diagram
11. 6.6×10^6 m/sec
12. 3.3×10^{-33} m
13. λ becomes larger
14. (a) 3.3×10^{-25} m/sec (b) 5.0×10^{-8} m/sec (c) 3.3×10^{-6} m/sec (d) 3.3×10^6 m/sec
15. discussion
16. 3×10^{-31} m
17–24. discussion

Chapter 21

1. information
2. discussion
3. (a) 1.2×10^{-13} J (b) 0.75 MeV
4. (a) 5.7×10^{-2} m (b) 210 m (c) $R = 3{,}700\ R_e$
5. Charges are positive; field is into the page.
6. (a) 1.8×10^4 N/C (b) 1.8×10^3 V (c) undeflected
7. (a) γ (b) α (c) α (d) γ (e) γ (f) α or γ (g) β (h) α (i) α (j) β
8. discussion
9. (a) one-half (b) three-quarters (c) discussion
10. 10%
11. (a) graph (b) proof (c) 5.0×10^{20} atoms
12. (a) 5.7×10^{-13} J/disintegration (b) 45 W
13. 3.70×10^5 disintegration/sec
14. 2The number remains constant.
15. (a) about 4 days (b) discussion
16–19. activity

Chapter 22

1. information
2. discussion
3. discussion
4. (a) discussion (b) discussion
5. (a) $R = \frac{mv}{Bq}$, Rvm (b) 5.4 cm (c) 5.640 m (d) 0.0048 m
6. equations
7. chart
8. diagram
9. diagram
10. 4,000 years; 23,000 years
11. (a) 12.011 amu (b) 6.941 amu (c) 207.2 amu
12. 4.0015 amu
13. (a) $\times$ 1/4 (b) about $\times$ 1/2 (c) about 2.25×10^9 years (d) yes

Chapter 23

1. information
2. discussion
3. 235 protons; 143 electrons
4. equations
5. equations
6. (a) γ (b) Al^{28} (c) Mg^{24} (d) Mg^{25}
7. (a) discussion (b) in Unit 3 under conservation laws.
8. 1.10 amu, 5.2%
9. table
10. (a) 78 (b) 79 (c) 80 (d) 80
11. (a) $_{11}Na^{24}$ (b) $_{11}Na^{24}$ (c) $_{11}Na^{24}$ (d) $_{11}Na^{24}$
12–15. discussion
16. Less by 0.02758 amu
17. activity

Chapter 24

1. information
2. 4.95 MeV
3. 7.07 MeV/nucleon
4. opposite directions, each 8.65 MeV
5. absorbed, 1.19 MeV
6. 0.56 MeV
7. 8.61 meV
8. neutron capture, β-decay, β-decay
9. Ba^{141} is 1180 MeV; Kr^{92} is 800 MeV; U^{235} is 1790 MeV. discussion
10. 208 MeV
11. diagram
12–14. discussion
15. 26.7 MeV
16. (a) 4.33×10^9 kg/sec (b) 5.23×10^{23} horsepower
17. equations
18. 1.59 MeV released
19. U^{233} is fissionable
20. Pu^{241} is fissionable
21. discussion
22. essay
23. activity

INDEX

Acceleration, caused by gravity (film loop), 41–42
centripetal, 34
in free fall, 18–21
of gravity (experiment), 21–24
measurement of (experiment), 22–24, 25–28
(film loop), 47
see also Motion; Speed
Accelerometers (activity), 35–38
automobile, 37–38
calibration of, 36–37
damped-pendulum, 38
liquid-surface, 35–37
Advance of perihelion, 101
Air, standing wave in (film loop), 170–171
Air track, in bullet speed experiment, 122
in collision experiments, 103, 120–121
Aircraft takeoff, and conservation of energy (film loop), 167–168
Alpha (α) particles, range of (experiment), 240–242
Rutherford's scattering experiment with, 230–231, 234–235
Altitude, in astronomy, 9
Amplifying circuit, 194
Amplitude modulation, 198
Analemma, plotting of (activity), 88
Analysis of a hurdle race (film loop), 47–48
Andromeda, Great Nebula in, 66
Angle(s) of incidence and refraction, 176
Angular measurements, making of (activity), 83–84
Antacid tablet, in conservation of mass activity, 147
Antinodal lines, 141
Armature, construction of, 207
Ascending node, 72
Ascension, right, 86
Astrolabe, 9–10, 55
Astronomical unit (AU), 88
Astronomy, naked-eye (experiments), 7–12, 50–54
observations (experiment), 10–12
Atmospheric pressure (activity), 149–150
Atom(s), black box (activity), 231
hydrogen, energy levels of, 225
ionization of (activity), 229–230
modeling with magnets (activity), 230–231
Rutherford nuclear model of, 230–231
Rutherford scattering experiment (film loop), 234–235
Thomson model of (film loop), 233–234
Automobile accelerometer, 37–38
Autoradiography (experiment), 250
Azimuth, 7–8

Balance(s), 214
construction and use of (experiment), 181–183
Ballistic cart projectiles (activity), 39–40
Ballistic pendulum, in bullet speed experiment, 160
Band saw blade, standing waves on (activity), 231
Battery, construction of (activity), 204
eleven-cent (activity), 204
Beaker and hammer (activity), 35
Beam projector, construction of, 175–176
Beta (β) particles, range of (experiment), 240–242
Beta (β) radiation, measuring energy of (experiment), 250–252
Beta (β) ray spectrometer, 250–252
Bicycle generator (activity), 209
Big Dipper, 9
Billiard ball, dynamics of (film loop), 163–164
Biographical Encyclopedia of Science and Technology (Asimov), 89
Black box atoms (activity), 231
Bullet, measuring speed of (experiment), 122–124
(film loop), 160–162

Calibration, of accelerometer, 36–37
of oscillator, 28
Calibration curve, 37
Calorie, defined, 129
Calorimeter, 128
Calorimetry (experiment), 128–132
Camera(s), Polaroid Land, 4–5
see also Photography
Cannon, recoil velocity of, 162
Cannon ball, path of, 32
"Can Time Go Backward?" (Gardner), 168
Capacitor, 244
Car, weighing with tire pressure gauge (activity), 150
Carbon–14 atoms, 249
Carrier wave, 195
Cartesian diver (activity), 149–150
Catenary curve, 125
Cathode rays, 192
deflection of (activity), 227
in a Crookes tube (activity), 227
Celestial Calendar and Handbook, 11
Celestial sphere model (activity), 86–87
Celsius (centigrade) scale, 127
Centigrade scale, 127
Central forces (film loop), 98–99
Centripetal acceleration, 32–33
Centripetal force (experiment), 32–33
on a turntable (experiment), 33–34
Chadwick, and neutron, 253
Chain, in least-energy experiment, 124–125
Charge(s), electric
measurement of (experiment), 217–219
speed of, 192
see also Electric charge
Charge-to-mass ratio, for electron (activity), 227
equation for, 215
(experiment), 214–217
Checker snapping (activity), 35
Chemical change, and electric currents, 212–214, 233
Chemical reactions, tracers in, 250
Chladni figures, 173–179
Chladni plates, 153
Cloud chamber, 237
Cluster of objects, explosion of (film loop), 159–160
scattering of (film loop), 158–159
Collage, physics (activity), 209
Collector, 220
Colliding carts (film loop), 44–45
Colliding disk magnets (experiment), 111–112
Colliding freight cars (film loop), 162–163
Colliding pucks (experiment), 111
Collision probability, for gas of marbles (experiment), 133–134
Collisions, elastic, 103
inelastic, 106; (film loop), 157, 158
molecular (experiment), 132–135
with object of unknown mass (film loop), 253
one-dimensional (experiment), 102–109; (film loop), 156–157
partially elastic (film loop), 162–163
perfectly inelastic, 106
two-dimensional (experiment), 110–118; (film loop), 157–158
Collision squares, mean free path between (experiment), 134–135
Color(s), aspects of (activity), 201–202
Land effect and (activity), 202
refraction of, 176–177
see also Light; Spectrum(a)

Comet(s), 91
Halley's, orbit of (experiment), 79–82
orbits of (activity), 91; (experiment), 75–79
Comet Cunningham, 75
Communication, and waves (experiment), 195–199
Compass, floating, 209
Computer, programming orbit by, 97–98
"Computer Music," *Scientific American*, 210
Computers and the Human Mind (Fink), 210
Conduction, 132
Conic sections models (activity), 90–91
Conservation of energy (experiment), 118–122
airplane takeoff and (film loop), 167–168
pole vault and (film loop), 166–167
Conservation of mass (activity), 147
Conservation of momentum, 157–158
(activity), 147–148
(experiment), 110–112, 113–117
Constant pressure gas thermometer, 128
Constant speed, 26
Cooling, measuring rate of, 131
Coordinate system, 68
Copernicus, trial of (activity), 93
Copper atom, in electrolysis experiment, 212–213
Coulomb's law, and electric forces (experiment), 181–183
Crime of Galileo, The (de Santillana), 90
Crookes tube, cathode rays in (activity), 227
X rays from (activity), 228
Curie, 249
Current(s), electric, *see* Electric current(s)
Current balance, adjusting (experiment), 184–185

Dalton's puzzle (activity), 226
Damped-pendulum accelerometer, 38
Data, recording of, 19
variations in (experiment), 13–14
Davy, Humphry, and electrochemical reactions, 212
sodium production by electrolysis and, 233
de Broglie waves, and turntable oscillator patterns (activity), 231
Declination, lines of, 86
Descartes, René, 149
Dice, in random event experiment, 237, 243–244
Diffraction, ultrasound (experiment), 145
wave (experiment), 140
Diffraction grating, 223
Diffraction patterns, photographing of (activity), 200–201
Diode, characteristics of (experiment), 193–194
Direct fall, measuring gravity by (experiment), 21–22
Disk magnets, collisions of, 111–112
Dispersion of colors in refraction, 176
Distribution table, 238
Diver in a bottle (activity), 149–150
Doctor Faustus (Marlowe), 89
Drum, standing waves on, 152
vibrations of (film loop), 173
Dynamics of billiard ball (film loop), 163–164
Dynamics carts, in collision experiment, 102–103, 118–119

Earth, circumference of, 54–56
interaction with electric current, 189
orbit of (experiment), 61–63
size of (experiment), 54–56
Earth–sun distance (activity), 91
Ecliptic, pole of, 86
Einstein, Albert
photoelectric equation of, 222
writings by or about (activity), 227
Albert Einstein: Creator and Rebel (Hoffmann and Dukas), 227
Albert Einstein: Philosopher-Scientist (Schilpp, ed.), 227
Einstein (Bernstein), 227
Einstein, His Life and Times (Frank), 227
Ideas and Opinions (Einstein), 227
Elastic collisions, 103
Electric charge, measurement of (experiment), 217–219
Electric circuit, in half-life experiment, 244–245
Electric compass, construction of (activity), 203–204
Electric current(s), and chemical change, 212–214, 233
force between, 185–186
forces on (experiment), 183–187
generation of (activity), 207
interaction with earth, 189
and magnet, force between, 187–188
Electric field, deflection, 191–192
detection of (activity), 204
Electric force, measurement of, 185–186
(experiment), 179–181
Coulomb's law and (experiment), 181–183
Electric lamp, lighting with a match (activity), 227–228
Electrochemical reactions, 212–214
(activity), 226
(film loop), 233
Electrolysis (experiment), 212–214
production of sodium by (film loop), 233
of water (activity), 226
Electromagnetic waves, standing (film loop), 211
Electron, charge-to-mass ratio (experiment), 214–217
measuring charge-to-mass ratio for (activity), 227
velocity of, 193
Electron beam, deflection of (experiment), 214–217
deflection by electric field, 191–192
deflection by magnetic field, 192
focusing of (experiment), 192–193
reflecting of (experiment), 193–194
Electron beam tubes (experiment), 190–194
building and operating of, 191
Electron charge, speed of, 192
Electrons and Waves (Pierce), 210
Elementary charge, measurement of (experiment), 217–219
Eleven-cent battery, 204
Elizabethan World Picture, The (Tillyard), 89
Elongation, of Mercury, 73
Encke's comet, 91
Energy, of alpha (α) particles, 241–242
of beta (β) radiation (experiment), 250–252
conservation of (experiment), 118–122; (film loop), 166–168
conversion of (activity), 148–149
gravitational potential, 124, 164–165
heat, 128–132; *see also* Calorie
ionization, 248
kinetic (film loop), 160, 165–166; in pendulum swing activity, 124
least (experiment), 124–126
measurement of (film loop), 164
mechanical, conversion of heat to (activity), 148–149
potential, 124–126
Ephemeris, for Mars, 72
Epicycles, photographing of, 85–86
and retrograde motion (activity), 84–86
Exchange of momentum devices (activity), 147–148
Explosion of a cluster of objects (film loop), 159–160

Faraday, and electrochemical reactions, 212
Faraday disk dynamo (activity), 206–207
Force, between magnet and current (experiment), 187–190
central (film loop), 98–99
centripetal, 32–33
on currents (experiment), 183–187
electric (experiments), 179–183
inverse-square, 99
magnetic, 192
nuclear, 235
on a pendulum (activity), 92–93
variation in wire length and, 196
FORTRAN, 97
Foundations of Modern Physical Science (Holton and Roller), 204
Frames of reference (activity), 89–90
fixed and moving, 44–45
Franck–Hertz effect, 229
Free fall, 42–43
acceleration in, 18–21
from aircraft, 46–47
from mast of ship, 45–46
Frequencies, measuring unknown (activity), 42
of waves, 139
Friction, on rotating disk, 34

Galilean relativity (film loops), 45–47
Galileo (activity), 90
inclined plane experiment, 18–21
Galileo (Brecht), 90
"Galileo: Antagonist," *Physics Teacher*, 90
"Galileo Galilei: An Outline of His Life," *Physics Teacher*, 90
Galileo Quadricentennial Supplement, *Sky and Telescope*, 90
Galileo and the Scientific Revolution (Fermi), 90
Gas(es), behavior of (experiment), 137–138
pressure of, 137–138
standing waves in (film loop), 170–171
temperature of, 138
volume and pressure (experiment), 137–138
volume and temperature (experiment), 138
Gas thermometer, 128
Gauss, 205
Gay Lussac's law, relation between temperature and volume, 137
Geiger counter, 238
Generator, demonstration of (activity), 209
bicycle (activity), 209
Generator jump rope (activity), 207
Geocentric model of universe, 94
Gilbert's versorium (activity), 203–204
Gliders, collisions of, 103, 120–121
Graph, drawing of, 18
Gravitational potential energy (film loop), 164–165
Gravity, acceleration caused by (film loop), 42–43
measuring acceleration of (experiment), 21–24
Great Nebula, in Andromeda and Orion, 66

Half-life (experiments), 243–248
Halley's comet, plotting orbit of (experiment), 79–82
Handkerchief diffraction grating (activity), 200
Heat, conversion to mechanical energy (activity), 148–149
exchange and transfer of (experiment), 132
latent, of melting ice (experiment), 132
mechanical equivalent of (activity), 149–150
Heat capacity, measurement of, 130–131
Heat energy, conversion of (activity), 148–149
measuring of, 128–132
Height of Piton, a mountain on the moon (experiment), 57–60
Heliocentric model of universe, 94–95
Heliocentric system, 63
Histogram, 238–239
Horizontal motion, measurement of, 47
Horsepower, student (activity), 148
Hurdle race, analysis of, I and II (film loops), 47–48
Hyades, observations of, 66
Hydrogen atom, calculating energy levels for, 225

Ice, calorimetry (experiment), 131–132
latent heat of melting, 132
Ice lens, construction of (activity), 203
Icosahedral dice, 237
Images, real, 63–64
Incidence, angle of, 176
Inclination, of Mars' orbit, 70–72
Inclined air track, in energy conservation experiment, 120–121
Inclined-plane experiment, 18–21
Inelastic collisions, 106
one-dimensional (film loop), 157
two-dimensional (film loop), 158
Inertia, and gravitation, 28
Infrared photography (activity), 201
Instantaneous speed, 42
Interference, ultrasound (experiment), 146
wave, 139; (experiment), 140, 144
Interference pattern (experiment), 141–142
of light (experiment), 178
Inverse-square force, 99
Ionization, measurement of (activity), 229–230
Ionization energy, 229, 248
Ionosphere, 198
Irregular areas, measurement of (activity), 91
Isolated north magnetic pole (activity), 206
Iterated blows, 98–99
Iteration of orbits, 76

Julian Day, 53
Jupiter, mass of, 97
observations of, 65–66
positions of, 60–61
satellite orbit (film loop), 95–97

Kepler's laws (film loop), 99–100
satellite orbit and, 90
Kinetic energy (film loop), 165–166
calculation of (film loop), 160
see also Energy

Laboratory notebook report, 2–3
Land effect, 202
Land two-color demonstrations (activity), 202
Lapis Polaris Magnes (activity), 209–210
Latent heat, measurement of (experiment), 130–131
Latent Image (Newhall), 201
Latex squares, electron micrograph of, 217
Lead nitrate, in conservation of mass experiment, 147
Least energy (experiment), 124–125
Lenses, in telescope (experiment), 63–67
Light, angles of incidence and refraction, 176
diffraction of, 222
dispersion into spectra, 222–223
effect on metal surface (experiment), 219–222
interference patterns of (experiment), 178
photoelectric effect of (experiment), 219–222
polarized, 202–203
rainbow effect (activity), 201–202
scattered (activity), 201
wave and particle models of, 13–16
wavelength of (experiment), 177–179
see also Color; Wave(s)
Light beam, refraction of (experiment), 175–176

Line of nodes, 80
Liquids, mixing hot and cold (experiment), 129
Liquid-surface accelerometer, 35–37
Literature, Elizabethan world view in (activity), 89
Longitudinal wave pulse, 139
Lunar eclipses (table), 12

Mach, Ernst, and inertia, 28
Magnesium flashbulb, in conservation of mass activity, 147
Magnet(s), interactions of, 119–120
  modeling atoms with (activity), 230–231
Magnetic field, deflection of electron beam by, 192; (experiment), 214–217
  measuring intensity of (activity), 204–205
Magnetic pole (activity), 206
Marbles, collision probability for a gas of (experiment), 133–134
  inferring size of, 133–134
Mars, inclination of orbit (experiment), 70–72
  orbit of (experiment), 67–70
  positions of, 60–61
Mass, conservation of (activity), 147
  inertial, 28
  measuring of (experiment), 27–28
  neutron, 253
  weight and (experiment), 28
Mathew Brady (Horan), 201
Mean free path, between collision squares (experiment), 134–135
Measurement(s), of acceleration (experiment), 25–28
  of acceleration in free fall, 18–21
  of acceleration of gravity, 21–24
  angular (activity), 83–84
  of elementary charge (experiment), 217–219
  of energy (film loop), 164
  of energy of beta (β) radiation (experiment), 250–252
  of irregular areas (activity), 91
  of magnetic field intensity (activity), 204–205
  of mass and weight (experiment), 27–28
  of momentum, 156–157
  precision in, 16–18
  of speed of sound (experiment), 154–155
  of uniform motion (experiment), 14–18
  of unknown frequencies (activity), 42
  of wavelength (experiment), 178–179
Mechanical energy, conversion of heat to (activity), 148–149
Mechanical wave machines (experiment), 155
Melting, 130
Mercury, elongations of, 73
  orbit of (experiment), 72–75
Metal plate, vibrations of (film loop), 173–174
Meters, construction of (activity), 207–208
Method of beats, 195
Microwaves, interference of reflected, 197–198
  properties of (experiment), 196–197
  reflected, 197–198
  signals and (experiment), 198–199
Microwave transmission systems (activity), 210
Millikan, oil drop experiment, 217
Modeling atoms with magnets (activity), 230–231
Model of the orbit of Halley's comet (experiment), 79–82
"Moiré Patterns" (Oster and Nishijima), *Scientific American*, 154
Moiré wave patterns (experiment), 153–154
Molecular collisions, Monte Carlo experiment on, 132–135
Momentum, conservation of, 110–112, 157–158; (activity), 147–148
  measurement of, 156–157
Momentum devices, exchange of (activity), 147–148
Monte Carlo method in molecular collisions experiment, 132–135
Moon, crater names of (activity), 89
  distance to (experiment), 57
  height of mountain on (experiment), 57–60
  observations of, 66; (experiment), 10–11, 51
  phases of, 51
  surface of, 58–59
"Moon Illusion, The," *Scientific American*, 84
Motion, on inclined plane (experiment), 18–20, 20–21
  Newton's second law of (experiment), 25–28
  perpetual (activity), 150–151
  relative (film loop), 44–45
  retrograde (experiment), 60–61
  in a rotating reference frame (activity), 40–41
  uniform (experiment), 14–18
  see also Acceleration; Speed; Velocity
Motor, construction of (activity), 208
  demonstration of (activity), 209
Motor–generator demonstration (activity), 208–209
Music, and speech (experiment), 154
  wave patterns of (experiment), 154

Nails, in measurement of kinetic energy experiment, 165–166
Naked-eye astronomy (experiment), 7–12, 50–54
Net count rate, 245
Neutron, calculating mass of, 253
New Handbook of the Heavens, 65
Newton's second law (experiment), 25–28
Nodal lines, 141
Nodes, 141, 172
North magnetic pole (activity), 206
North star (Polaris), 9
Nuclear force, 235

Objective lens, 64
Occultation, 96
One-dimensional collisions (experiment), 102–109
  (film loop), 156–157
  stroboscopic photographs of, 104–110
Orbital eccentricity, calculation of, 74–75
Orbits, comet (activity), 91
  computer program of, 97–98
  earth's, 79
  five elements of, 72
  of Halley's comet (experiment), 79–82
  of Jupiter satellite (film loop), 95–97
  of Mars (experiment), 67–70
  of Mercury (experiment), 72–75
  parabolic (activity), 91–92
  pendulum, 92–93
  of planets, 53–54, 65–75
  satellite (activity), 90
  stepwise approximation to (experiment), 75–79
  unusual (film loop), 100–101
Orion, Great Nebula in, 66
Oscillator, calibration of, 28
Out of My Later Years (Einstein), 227

Parabola, in least energy experiment, 124–126
  waterdrop, photograph of (activity), 39
Parabolic orbit, drawing of (activity), 91–92
Parallax, 57, 186
Partially elastic collisions (film loop), 162–163
Particle and wave models of light, 13–16
Pendulum, ballistic, 122–123
  forces on (activity), 92–93
  measuring gravity by (experiment), 22–23

Pendulum accelerometer (activity), 38
Pendulum swing, energy analysis of (experiment), 124–126
Penny and coathanger (activity), 42
Perfectly inelastic collision, 106
Perihelion, advance of, 101
Periodic wave, 139
Perpetual Motion and Modern Research for Cheap Power (Smedile), 205
Perpetual motion machines (activity), 150–151, 205–206
"Perpetual Motion Machines," *Scientific American*, 205
Perturbation, 100–101
Photoelectric effect (activity), 227–228
(experiment), 219–222
Photoelectric equation, 222
Photoelectric tube, 219–220
Photographic activities, 201
Photography, history of (activity), 201
infrared, 201
measuring gravity by, 23
Polaroid, 4–5
Schlieren, 201
of spectrum (experiment), 223–224
stroboscopic, measuring gravity by (experiment), 24
of waterdrop parabola, 23–24
Physics collage (activity), 209
"Physics and Music," *Scientific American*, 153, 154
Physics for Entertainment (Perelman), 203
Physics of Television (Fink and Lutyens), 210
"Physics of Violins, The," *Scientific American*, 154
"Physics of Woodwinds, The," *Scientific American*, 154
Picket fence analogy, and polarized light, 203
Piton, height of (experiment), 57–60
Planck's constant, 222
Planet(s), locating and graphing (experiment), 53–54
observations of (experiment), 11
Planetary longitudes (table), 53
Planetary notes (table), 12
Pleides, observations of, 66
Poisson's spot (activity), 201
Polaris (North Star), 9
Polarized light (activity), 202–203
Polarized Light (Shurcliff and Ballard), 202
Polaroid Land camera, use of, 4–5
Pole vault (film loop), 166–167
Polonium, disintegration of, 247–248
Postage stamps honoring scientists (activity), 228–229
Potential energy, 124–126
Power, output of (activity), 148
Pressure, atomospheric (activity), 149–150
volume of gas and (experiment), 137–138
Principia (Newton), 91
Program orbit, I and II (film loops), 97–98
Projectile(s), ballistic cart (activity), 39–40
fired vertically (film loop), 47
speed of, 122–123
Projectile motion demonstration (activity), 38
Projectile trajectories, photographing of (activity), 39–40
Pucks, in collision experiment, 111
Pulls and jerks (activity), 35
Pulses, 139
see also Wave(s)

Quantum Electronics (Pierce), 210

Radiation, heat exchange by, 132
electromagnetic and microwave, 197–198
Radio transmitter, generation of electromagnetic waves by, 211
Radioactive decay, *see* Half-life
Radioactive isotopes, half-life of, 245–246
as tracers, 248–250
Radioactive tracers (experiment), 248–250
Rainbow effect (activity), 201–202
"Raisin pudding" atom model, 230
Random event(s) (experiment), 236–240
Random event disks, 236–237, 243
Random two-digit numbers (table), 136
Reading suggestions (activity), 210
Real images, 63–64
Recoil (film loop), 162
Record-keeping, 4–5
Rectification, of diode, 193–194
Reference, in astronomy, 7–10
Reflection, sound (experiment), 143
ultrasound (experiment), 145
wave (experiment), 140, 152–153
Refraction, angle of, 176
of colors, 176–177
of a light beam (experiment), 175–177
Refraction, sound (experiment), 143
wave, 139, 140
Regularity, and time (experiment), 12
Relatively principle, 46
Resonant circuits (experiment), 196
Retrograde motion (experiment), 60–61
epicycles and (activity), 84–86
geocentric model of (film loop), 94
heliocentric model of (film loop), 94–95
Right ascension, 86
Ripple tank, waves in (experiment), 140
"Role of Music in Galileo's Experiments, The," *Scientific American*, 20
Rotating disk, friction on, 34
Rotating reference frame, moving object in (activity), 40–41
Rubber hose, vibrations of (film loop), 172–173
Rubber tubing and welding rod wave machine, 155
Rutherford nuclear model, 230–231
scattering experiment and (film loop), 234–235

Satellite orbits, demonstrating of (activity), 90
of Jupiter, 95–96
"Satellite Orbit Simulator," *Scientific American*, 90
Saturn, observations of, 65
Scale model of the solar system (activity), 88
Scattered light (activity), 201
Scattering of a cluster of objects (film loop), 158–159
Schlieren photography (activity), 201
Science of Moiré Patterns, The (Oster), 154
"Science of Sounds, The," *Bell Telephone Laboratories*, 154
Science from Your Airplane Window (Wood), 202
Scientific American, activities from, 227
Scientific method, 4–5
Scientists on stamps (activity), 228–229
Seventeenth Century Background (Willey), 89
Shape of the earth's orbit (experiment), 61–63
Sidereal day (activity), 87
Signals, and microwaves (experiment), 198–199
Similarities in Wave Behavior (Shive), 155
Single-electrode plating (activity), 226–227
Sinusoidal curves, 168–170
Size of the earth (experiment), 54–56
Sodium production by electrolysis (film loop), 233
Solar eclipses (table), 12
Solar system, scale model of (activity), 88
Sound (experiment), 142–144
calculating speed of, 144
Sound waves, diffraction of (experiment), 143

reflection of, 143
refraction of (experiment), 143
speed of (experiment), 144, 154–155
transmission of, 143
see also Ultrasound; Wave(s)
Specific heat capacity, 131
Spectra, creating and analyzing, 222–225
Spectroscopy (experiment), 222–225
Speech, and music (experiment), 154
Speech wave patterns (experiment), 154
Speed, of bullet (experiment), 122–124; (film loop), 160–162
constant, 26
electron charge, 192
instantaneous, 42
of sound (experiment), 144, 154–155
of a stream of water (activity), 38–39
ultrasound (experiment), 146
Stamps, scientists on (activity), 228–229
Standard deviation, 237
Standard error, 237
Standing wave(s), on a band saw blade (activity), 231
on a drum and violin (experiment), 152, 153
electromagnetic (film loop), 211
in a gas (film loop), 170–171
on a string (film loop), 170
in a wire ring, 232
Stars, chart of, 8
observations of (experiment), 11
Steel balls, in collision experiment, 104–109, 157–159
Stepwise approximation of an orbit (experiment), 75–79
Stonehenge (activity), 88–89
Stonehenge Decoded (Hawkins and White), 89
"Stonehenge Physics," *Physics Today*, 89
Stopping voltage, 220
String, standing waves on (film loop), 170
Stroboscopic photography, measuring gravity by, 24
of one-dimensional collision, 104–110
of two-dimensional collision, 113–118, 121–122
Sun, earth's orbit around, 63
observations of (experiment), 10, 50–51
Sundial, building of (activity), 88
Sundials (Mayall and Mayall), 88
Sunspots, observation of, 66–67
Superposition (film loop), 168–170

Tagged atoms, 248–249
Telescope, aiming and focusing of, 64–65
making of (experiment), 63–67
observations with, 64–67
Temperature, of gas (experiment), 138
thermometers and (experiment), 126–128
Temperature scale, defined, 126–127
Terminator, 58
Thermometers, comparison of, 127–128
constant pressure gas, 128
making of, 126–127
temperature and (experiment), 126–128
Thin film interference (activity), 200
Thomson, J. J., and cathode rays, 192
model of the atom and (film loop), 233–234
"raisin pudding" atomic model and, 230
Thorium decay series, 246–247
Threshold frequency, 222
Time, and regularity (experiment), 12
reversibility of (film loop), 168
Tire pressure gauge, weighing a car with (activity), 150
Total internal reflection, 177
Tracers, in chemical reactions, 250
radioactive (experiment), 248–250
Trajectories, curves of (experiment), 28–30
prediction of (experiment), 30–32
Transistor amplifier (activity), 206
Transit, 96
Transmission, of sound (experiment), 143
ultrasound (experiment), 145
Transverse wave, 139
Trial of Copernicus (activity), 93
Triode, characteristics (experiment), 193–194
Turntable, centripetal force on (activity), 33–34
measuring gravity by (experiment), 24
Turntable oscillator patterns, and de Broglie waves (activity), 231
Turntable oscillators in wave-communication experiment, 195
Twentieth-century version of Galileo's experiment (experiment), 20–21
Two-dimensional collisions, I and II (experiment), 110–118; (film loop), 157–158
stroboscopic photographs of, 113–118, 121–122
Two New Sciences (Galileo), 45

Ultrasound (experiment), 144–146
speed of (experiment), 146
Uniform motion, measuring of (experiment), 14–18
Unusual orbits (film loop), 100–101

Vacuum tubes, characteristics of, 193–194
Vector addition (film loop), 43–44
Velocity, of a boat (film loop), 43–44
recoil, 162
see also Speed
Velocity time graph, 163
Venus, and earth–sun distance (activity), 91
observations of, 65
Vernal equinox, 67, 86
Versorium, Gilbert's (activity), 203–204
Vertical motion, measurement of, 46
Vibrations, of a drum (film loop), 173
of a metal plate (film loop), 173–174
of a rubber hose (film loop), 172–173
of a wire, 232; (film loop), 171–172
see also Wave patterns
Violin, standing waves on, 152, 153
wave patterns of, 170
Volta, and electrochemical reactions, 212
Voltaic pile, construction of (activity), 204
Volume, and pressure of gas (experiment), 137–138
temperature of a gas and (experiment), 138

Water, electrolysis of (activity), 226
interference pattern in, 141–142
speed of stream of (activity), 38–39
Waterdrop(s), measuring gravity by, 23–24
Waterdrop parabola, photographing of (activity), 39
Water wave(s), reflection of (experiment), 152–153
Wave(s), and communication (experiment), 195–199
de Broglie, 231
diffraction (experiment), 140
electromagnetic, 211
frequency, 139
interference (experiment), 140, 144
interference pattern (experiment), 141–142
longitudinal, 139
microwaves, interference of, 197–198
periodic, 139
properties (experiment), 139

reflection (experiment), 140, 197–198
refraction (experiment), 140
sinusoidal, 168–170
sound, *see* Sound waves
standing, 141, 143–144, 152; (film loop), 211
transverse, 139
water, 152–153
see also Light; Sound; Standing waves
Wave frequencies, 139
Wavelength, 139
of light, 177–179
measuring of (experiment), 141, 145, 178–179
of sound (experiment), 143–144
Wave machines (experiment), 155
Waves and Messages (Pierce), 210
Wave and particle models of light, 13–16
Wave patterns (experiment), 152, 153
moiré (experiment), 153–154
music and speech (experiment), 154
violin, 170
Weight, and mass, 28
measuring of (experiment), 27–28
Wire, vibrations of, 232; (film loop), 171–172
Wire ring, standing waves in (activity), 232
Work, output of (activity), 148

X rays, from a Crookes tube (activity), 228

Young's experiment: the wavelength of light (experiment), 177–179